T4-AJU-317

BASIC ELECTRONIC TEST PROCEDURES

2ND EDITION

No. 1927
$23.95

BASIC ELECTRONIC TEST PROCEDURES

2ND EDITION

IRVING M. GOTTLIEB

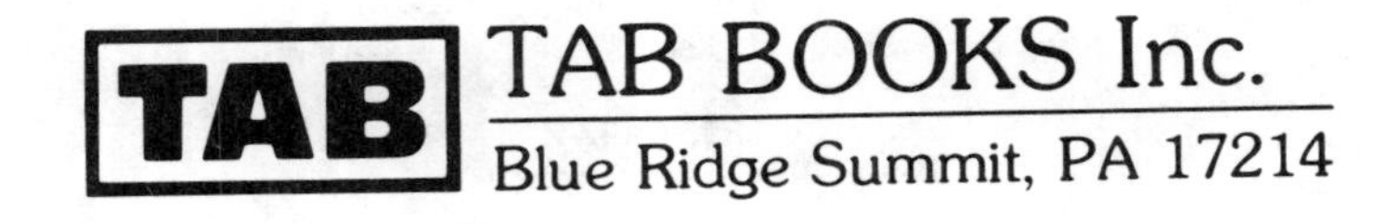
TAB BOOKS Inc.
Blue Ridge Summit, PA 17214

FIRST EDITION

FIRST PRINTING

Printed in the United States of America

Library of Congress Cataloging in Publication Data

Gottlieb, Irving M.
Basic electronic test procedures.

Includex index.
1. Electronic measurements. 2. Electronic instruments. I. Title.
TK7878.G67 1985b 621.381′043 85-10048

ISBN 0-8306-0927-X
ISBN 0-8306-1927-5 (pbk.)

Contents

Preface

It is common knowledge that test and measuring instrumentation of great complexity, sophistication, and expense is available for the evaluation of electronic circuits and systems. Even if we could afford the cost and were willing to exert the effort of using such elegant instruments, though, it is doubtful that the overall results would necessarily be more meaningful than those attainable from inexpensive meters and simple test procedures. Such a statement pertains to the vast majority of tests and measurements routinely made by engineers, technicians, servicemen, and hobbyists; it also has much validity for many of the determinations made in scientific and engineering laboratories. It is truly a fact of technical life that "reasonable accuracy" suffices for the majority of electronic and electrical tests, measurements, and evaluations.

It is, unfortunately, an all-too-common notion that reasonable accuracy is obtained in all cases by merely connecting the measuring instrument to the circuit undergoing evaluation and noting the reading. Such near involuntary behavior is, indeed, the intended objective of the instrument maker, and it is a noble one. But it remains necessary to know the nature of the measuring instrument, of the circuit undergoing test, and of the relationships between the two. Depending upon the extent of such knowledge, even the most trivial measurement procedure can yield information ranging from the grossly deceptive to the surprisingly useful. Meaningful interpretations do not automatically ensue from mechanical test rituals.

Many useful tests and measurements are covered in this book, and the emphasis is always on the deployment of commonly available instruments, rather than laboratory types. Test procedures and measurement techniques are reinforced by the appropriate *basic principles. Examples* of test and measurement setups are given to make concepts more practical. Each chapter is followed by 10 thought-provoking and relevant *questions and answers.* These should prove useful for students and are intended to enhance the rewards of self-study.

Finally, note that no particular favoritism is accorded to any vintage of instruments in this book; the basic idea is to stress the classic nature of basic test and measurement techniques. Thus, you can generally obtain meaningful information from both the older VOM's and the more modern digital multimeters. Indeed, the simpler meters may possess certain advantages over their counterparts. In most cases, it is the awareness of *what* is taking place and *why* that looms paramount in obtaining useful results. It is, accordingly, the declared objective of this book to help develop such awareness in the testing and measuring of electronic and electrical quantities.

Introduction

Because this book is intended for a wide variety of electronics practitioners, it would be most foolhardy to prescribe a standard method of optimizing its use. Although serious students would do well to read it from beginning to end, those who have already acquired a respectable amount of technical proficiency could hardly be expected to read this book as one would read a novel. Undoubtably, many will use it in the manner of a *reference* or *handbook*, consulting the chapters or pages that bear relevancy to the technical problem at hand. For this purpose, the extensive index should facilitate progress in finding the sought guidance.

For the most part, one can expect to achieve measurement accuracies within the 1 to 10 percent range, i.e., ± 10%. Digital multimeters, when properly utilized, permit fractional-percentage measurement accuracies. At the other extreme, some tests are useful, even though they are "ballpark" in nature and many even yield "go or no-go" information. In all cases, however, the underlying motif is to enable you to know what you are doing and why. Such knowledge leads to the ability to *interpret* test and measurement information—an ability not to be found in the most complex and costly instrumentation. For, even microprocessor and computer-controlled instruments can only make decisions based upon their programming. Such decisive ability may serve the limited purposes of a repetitive task on an assembly line, but it cannot favorably compare with the human intuition of a knowledgeable person employing simple instruments under various conditions.

Also, many computer-oriented workers find they lack substantial background in basic electronic and electrical technology. For such specialists, the experiments and questions and answers found in each chapter will help attain a practical "feel" of circuit parameters. Best of all, neither "old-salts" nor neophytes need feel restricted in their electronic endeavors because the instrumentation they possess is more mundane than sophisticated.

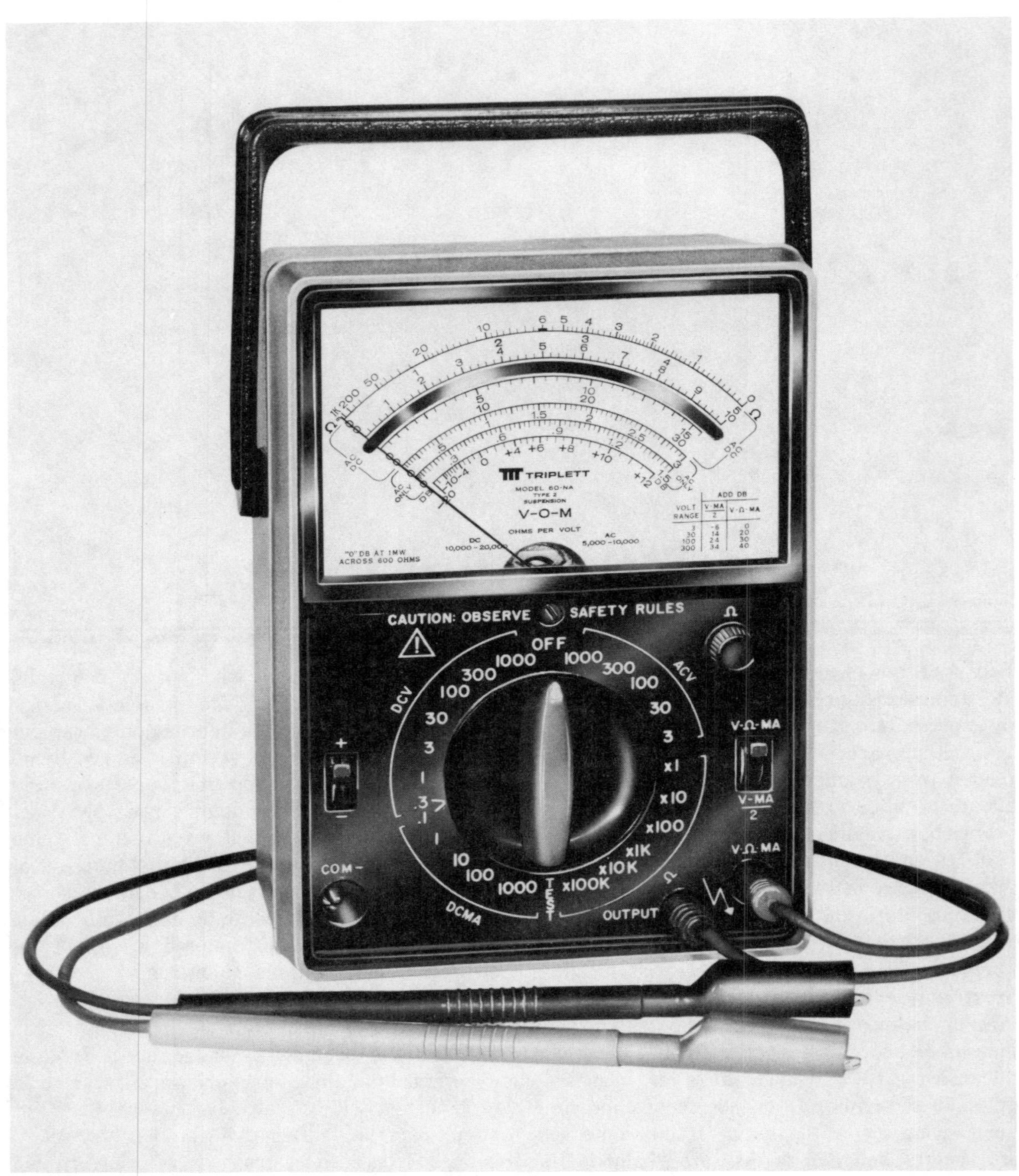

High-quality VOMs of this type have long been indispensable for both laboratory and field use. Even though digital meters have justifiably made serious inroads, it is not yet clear whether this venerable workhorse of instrumentation and testing is destined for obsolescence. Indeed, it appears to be well entrenched as a most useful tool for the everyday tasks of electronic technology. Courtesy of the Dynascan Corporation.

Chapter 1

Dc Resistance

It is assumed that the theory of operation of simple electric measuring instruments is understood. Such knowledge is, indeed, the proper starting point for the intelligent use of meters. However, it is necessary to acquire an awareness of matters beyond mere principles if we are to successfully use the meter for real tests. Simply knowing why the meter movement responds is essential, but not sufficient. Rather, we must deal with a wider domain, one involving certain practical aspects of the meter itself, the effects of the way in which it is used, and the nature of the device undergoing test. To begin a discussion of these matters it is logical to first investigate the most commonly encountered instrument for measuring dc resistance, the volt-ohmmeter, or as it is generally known, the VOM. This instrument is, in reality, a multimeter. Generally, it can be used to measure dc voltage and current, ac voltage, and dc resistance. When set for dc resistance measurements, its function is that of an ohmmeter. It is with this function of the VOM that we shall begin our investigation of simple test procedures.

THE SCALE

One of the most important features of a meter is its scale, for it is here that we read the quantity being measured. It happens that the ohmmeter scale of the VOM is not a simple one, nor does it lend itself to quick and easy interpretation of a reading. Careless regard for the nature of this scale often leads to sloppy test results. The scale of a series-type ohmmeter is shown in Fig. 1-1A. Not only is this scale extremely nonlinear, but its end points are zero and infinity! The attempt to cover this tremendous range with a scale of reasonable length results in the left-hand portion of the scale being cramped, and even more so as infinity is approached. You can immediately see that, in the interest of reasonable accuracy, it is desirable to make readings which are not too close to the left extreme of the scale. Somewhat more emphatically, it can be stated that one of the objectives in making resistance measurements is to avoid using the meter in such a way that pointer indications fall near the infinity extreme of the scale.

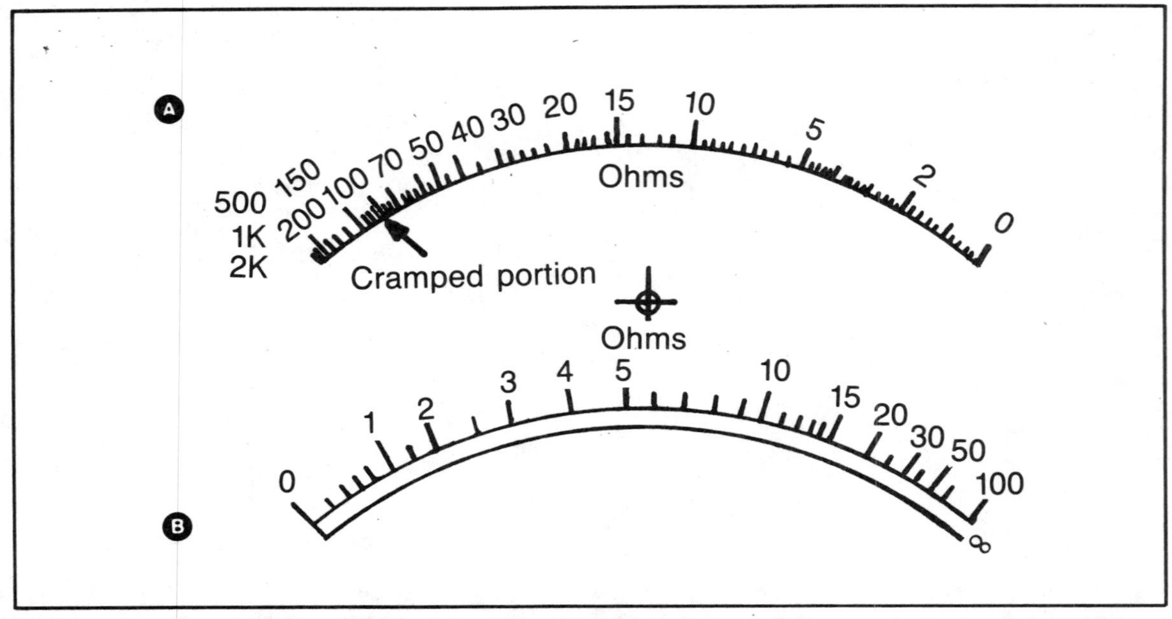

Fig. 1-1. Typical ohmmeter scales. (A) Series ohmmeter; (B) shunt ohmmeter (used in most electronic meters and in some VOMs intended for relatively low resistance measurements).

The scale of a typical shunt-type ohmmeter is illustrated in Fig. 1-1B. It is essentially a mirror image of the scale in the series-type ohmmeter. Other than reversed "right" and "left" characteristics, what has been said and what shall be said pertaining to the scale of the series-type instrument also applies to the scale of the shunt-type meter. Thus, the cramped portion of the scale shown in Fig. 1-1B is at its **right** end, etc. The attempt to avoid use of the cramped portion of the ohmmeter scale is a logical precaution and one we might even develop as a matter of habit. Somewhat less obvious, but of no lesser importance, is another consideration leading to further care in our use of the ohmmeter scale.

Another precaution in the use of the scale-pointer system: The factory-rated accuracy of a VOM is based on a full-scale dc voltage indication. The accuracy actually is related to a voltage range over which the voltage is uncertain. You might say that the meter lacks sufficient resolution to pinpoint values within the scale interval defined by the manufacturer's specified accuracy. But, what bearing does the accuracy of a **voltage** indication have on the indications read from the **ohmmeter** scale?

Referring to Fig. 1-2, we see the same ohmmeter scale as shown in Fig. 1-1A. In addition, however, a dc voltage scale has been included. Suppose that the factory-rated accuracy of the VOM is specified as plus or minus 3 percent of the full-scale **voltage** indication. This means that voltages read from the 250-volt scale are always characterized by a variation of plus or minus 7 1/2 volts. Thus, the total possible variation of 15 volts is depicted by the so-called "arcs of error" shown for two regions of the scale in Fig. 1-2. Concerning the dc voltage readings, it behooves us to make measurements with the meter pointer as close to full-scale deflection as is practical. In this way, the possible percentage error of the reading is minimized, tending to approach the best capability of the instrument more closely as full-scale deflection is achieved. Now, let us see how the arc of error affects accuracy on the ohmmeter scale.

Because of the nonlinearity of the ohmmeter scale, the arc of error is capable of causing a relatively great error near the zero end of this scale. Conversely, the very same arc, when applied near the central region of the ohmmeter scale, involves a much less severe inaccuracy. **We now have a reason for avoiding the region near the zero end of the ohmmeter scale.** There are other reasons to regard this portion of the scale as undesirable for making readings. For example, the effect of an imprecise zero adjustment is most pronounced near zero ohms. The same is true when the internal batteries become weak.

Is any portion of the ohmmeter scale trustworthy? It has been shown that the zero end of the ohmmeter scale should, whenever feasible, be avoided as well as the infinity end. The natural question arising from these restrictions is how much of the scale should be avoided, and how much of it can be used to take readings? It is not practical to provide an answer in terms of numerical scale markings, or in terms of arc segments, or even in deflection percentages. The general idea in the use of the ohmmeter is to cause the reading to occur as **close to mid-scale** as possible by selecting an appropriate range. This, indeed, should be regarded as the **primary function** of the ohms-range switch. And the same thing applies to shunt-type ohmmeters, intermediate or potentiometric types, and to the ohmmeter function of electronic meters.

OHMMETER SENSITIVITY

An interesting feature of the ohmmeter scale is that the mid-scale indication is, or is very close to, the internal resistance of the ohmmeter. As such it can serve as a figure-of-merit for specifying the sensitivity of the ohmmeter. If two meters have as their lowest ranges an R × 1 selection, a meter with a mid-scale indication of 10 is preferable to one with a mid-scale indication of 30 if we evaluate in terms of resolution for **low**-resistance readings. If the

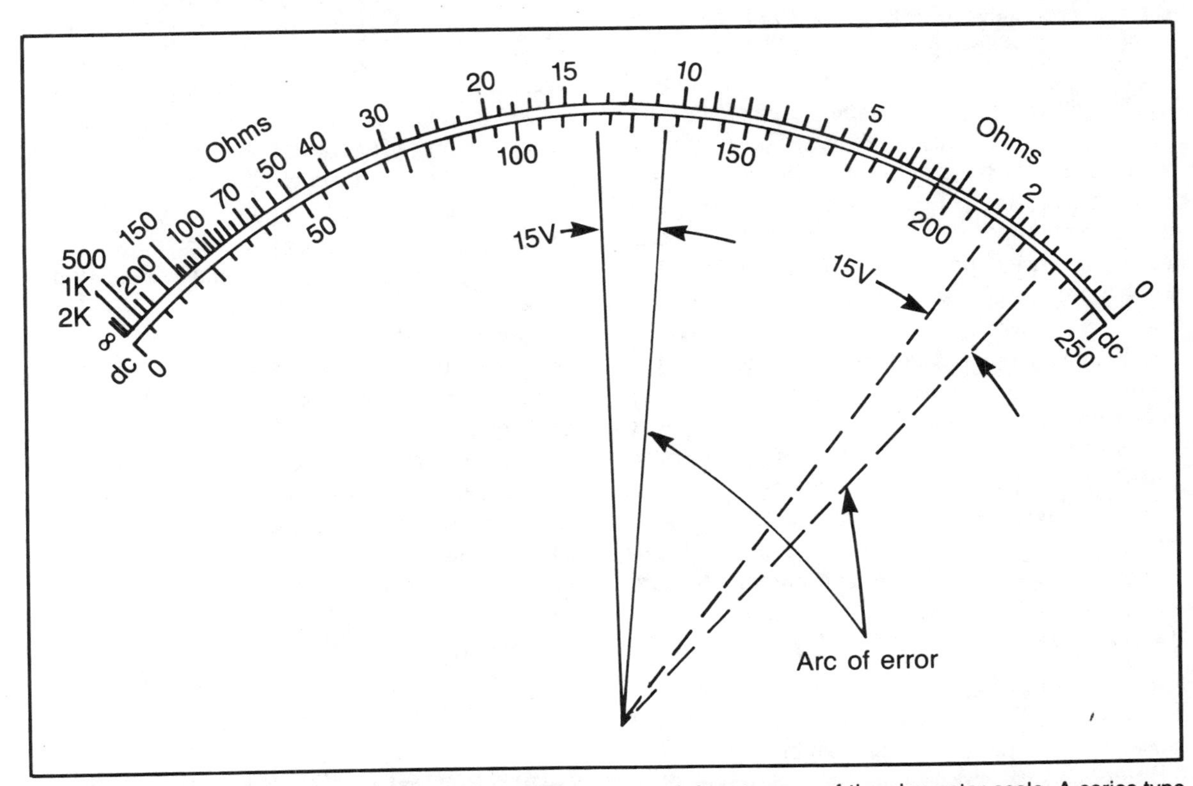

Fig. 1-2. Relationship between the arc of error for dc voltages and the accuracy of the ohmmeter scale. A series-type scale is shown. The same reasoning also applies to a shunt-type scale.

same two meters also have R × 100,000 scales as their highest ohms ranges, we have to look at the situation a little differently. The meter with the mid-scale indication of 30 will give the best resolution for **high**-resistance readings. So, it is seen that the sensitivity of ohmmeters is not the same as the sensitivity for current or voltage meters. All that really is accomplished with different ohmmeter "sensitivities" is to spread one end of the scale at the expense of increased cramping of the other end.

PARALLAX

The term parallax describes optical errors caused by angles other than 90 degrees between the line of sight and the plane of the meter scale. The meter scale should be read from a head-on viewing position. A parallax error will tend to be small if slight angular deviations from the head-on viewing position do not noticeably affect the apparent position of the pointer in relation to the closest scale markings. The best safeguard against parallax errors is the use of a meter having a **mirrored scale.** The mirror reflects the image of the pointer when the user looks at the meter from an angle. When the reflection of the pointer is not visible when viewed with one eye, the meter is read. As could be anticipated, the use of the mirrored scale pays an extra dividend in that the probability is greatly increased that different people will make meter readings in essentially the same way.

Formerly one of the features of laboratory instruments, mirrored scales are increasingly found on relatively inexpensive meters. It is well to cultivate the habit of making the proper use of the mirror; rather than being an incidental or decorative feature of such equipped instruments, the mirror provides the final refinement in communication between the indicating device and human observer.

Meter Position

The positioning of the meter can obviously help to avoid parallax errors. But the matter of scale observation is not the only consideration involved in the physical positioning of the meter. The manufacturer may specify the full-scale accuracy for **either** horizontal or vertical positioning, but not necessarily for both. With some instruments, it will be found that the mechanical zero-adjust does not accommodate both horizontal and vertical positions. Also, bearing friction tends to be a function of physical positioning. In order to make proper use of the meter during test procedures, the manufacturer's instructions concerning physical positioning should be followed. The more recently developed *suspension* or *taut-band* movements tend to be more tolerant with respect to the positioning of the instrument.

MAGNETIC FIELDS

Basically, the meter movement is actuated by the interaction of magnetic forces. Therefore, the accuracy of the meter can be affected by external magnetic fields. Not only do such fields spoil the accuracy of readings taken close together, but the magnetic disturbance tends also to degrade the long-term accuracy of the meter by reducing the strength of its internal magnet. Both ac and dc magnetic fields can distort readings. In setting up for a test procedure, the meter should never be situated near heavy-current-carrying apparatus such as motors, heaters, iron core components, or similar electrical equipment. Additionally, masses of iron or other ferromagnetic material should be avoided.

Meter shielding techniques incorporated by manufacturers are effective in varying degrees, but should not be relied on for complete protection from the effects of strong magnetic fields. The most common meter movement utilizes a permanent magnet, the poles of which supply the magnetic flux for the pivoted current-carrying coil. More recent constructions make use of a free-to-rotate soft iron solenoid enclosed by a ferromagnetic shell. Of the two types, the older construction with its horseshoe magnetic is more susceptible to external fields and masses of iron. It is also much more likely to influence other meters which are physically adjacent to it.

ELECTROSTATIC FIELDS

Under some atmospheric conditions, the scale-viewing window, as well as other parts of a test instrument, may accumulate an electric charge. The effect is overwhelming and test results can be rendered totally useless. The forces exerted upon the pointer of the meter can be very strong. Sometimes only a segment of the deflection is primarily affected and the existence of the condition is not immediately evident. Certain plastics tend to be more susceptible to charge retention than glass. There are available certain commercial preparations which discourage charge accumulation. By cleaning the window with one of these products, a conductive residue is deposited which lasts for a reasonable time. Sometimes, you can temporarily clear up the trouble by blowing your breath on the window. Passing a lighted match over the general area fronting the window often dissipates the accumulated charge by ionizing the adjacent air. Usually, a charge accumulation is caused by certain transient conditions and can be dissipated. At the other extreme, plastic windows subjected to sustained atmospheric electrification can accumulate a semi-permanent polarization which prevents normal functioning of the meter movement. Under such circumstances, you may be forced to await a change in weather conditions so that a charge tends to be dissipated from, rather than deposited on, objects.

VIBRATION AND MECHANICAL SHOCK

The useful life of a test meter is very much determined by the vibration and mechanical shock it is subjected to. Such disturbances represent the principal cause of demagnetization of the meter magnet. Also adversely affected are bearing friction and mechanical alignment of the movement. Vibration and shock are especially damaging because the effects tend to be cumulative as the abuse is continued. Even an inexpensive meter should be handled as a precision instrument and should not be moved about or transported in a reckless fashion. When the VOM is not in use, it is wise to leave the function selector switch on the highest dc current range. This provides the greatest electromechanical damping of the movement and inhibits any tendency for the pointer to respond with wide swings while the meter is being moved. (At the highest current range, the lowest-resistance internal shunt is used.) If the meter is to be transported, even greater damping can be provided by placing a short piece of heavy wire across the current terminals. Some meters actually have a short-circuit position on the function switch for this protective purpose.

More modern meters often incorporate a suspension or taut-band movement. This construction dispenses with pivots and bearings. It tends to be less sensitive to shock and vibration. It remains wise, however, to guard the instrument from exposure to mechanical forces as much as possible, for no measuring instrument can be immune to physical abuse.

TEMPERATURE

The effects of temperature changes are twofold. If the meter is subjected to above ambient temperatures, the internal magnet is likely to become demagnetized. Never situate the meter in the conduction, convection, or radiation field of a heater source. A second condition leading to degradation of meter accuracy is related to thermomechanical stresses which always accompany temperature changes. The meter works best and retains its accuracy longer in a room or laboratory where both short-time and long-time temperature fluctuations are minimal. Transporting a meter from one place to another through an area of high or low temperature can be damaging to the meter. Not only do extreme temperature changes play havoc with the magnetic and mechanical aspects of the meter, but the electrical stability of the associated components is also invariably affected. For example, multiplier resistors may change in value to the extent that they never do return to normal even when the meter is back in a normal environment.

HUMIDITY AND DUST

Under no conditions should a test meter be exposed to unusual conditions of humidity or dust. High levels of humidity or of atmospheric dust are damaging and total recovery cannot be expected when such abnormalities no longer exist. In practice, it is necessary to sometimes move the meter from one locale to another. But this should be done with caution. For example, the meter should be well protected from rain and wrapped or packaged to protect it from road dust.

ELECTRICAL OVERLOAD

In cases where a meter is subjected to an electrical overload which does not burn out the movement or bend the pointer, it is sometimes said that no damage will occur if the duration of overload is kept to a minimum by operator skill. Nothing can be farther from the truth. Every time the meter is pinned, either in the forward or reverse direction, the instrument suffers mechanical, magnetic, and electrical damage. And while set in the ohmmeter function, the inadvertent connection of the test leads to a source of electrical energy such as a charged capacitor, a battery, or an active circuit will result in some permanent damage. No matter how quickly the reflexes respond, there will be some damage to the meter. Nor should it be supposed that connecting the test prods to an ac source won't hurt the meter simply because the pointer is not forcibly slammed against one of its stops. The mechanical vibration imparted to the movement by the ac is damaging to the bearings and to their mechanical alignment. Also, there is a demagnetizing effect on the poles of the meter magnet. It is very important that an ohmmeter be used only for testing passive circuits.

THE INTERNAL BATTERY

It must not be supposed that the electrical zero-adjust of the ohmmeter can be used to overcome the effects of a dying battery. Normal batteries have small variations in their open-circuit voltage and, primarily, their internal resistance, as a result of age, amount of use, temperature, and other factors. A zero adjustment of the pointer made with bad batteries can produce deceptive readings. Although zeroing may be indeed accomplished, the accuracy of the meter will be impaired. This is bad enough, but where batteries need replacement, the zero adjustment will not hold once it is made. Obviously, any tests made when the zero-adjust drifts are useless. Once you become familiar with a meter, you can tell when excessive rotation of the zero-adjust is needed to position the pointer. If the zero adjustment does not hold steady for a few seconds as the meter prods are held together, the internal battery has probably exhausted its useful life. It also should be kept in mind that the ohmmeter generally employs more than one set of batteries, which are switched with the range selector. When replacing these batteries, you must be mindful of polarity. The battery clips and connectors must be free of corrosion and otherwise electrically and physically clean. Only quality brands units should be used and these should be fresh. Do not substitute batteries of different types than those recommended by the manufacturer of the instrument.

TEST LEADS

Although the general idea is to set the range switch so that readings are near mid-scale, there are situations where use of the scale extremities cannot be avoided. A case in point is the measurement of low resistance on the R × 1 scale. When making such measurements, the resistance of the test leads cannot be overlooked. When in good condition, the resistance of the test leads is generally in the vicinity of 0.1 ohm and the meter calibration properly absorbs this resistance so that the reading of the unknown resistance is correct. However, frayed leads or leads with corroded terminations can cause gross errors in low-resistance readings. It should not be supposed that the initial zeroing adjustment cancels out this error, because such is not the case. The presence of excessive resistance in the test leads causes an abnormally low reading. Therefore, when measuring components with resistances on the order of several ohms, it is possible that serious errors will exist because of excessive lead resistance. A practical test procedure for determining the condition of the test leads follows.

Test Procedure for Determining the Condition of VOM Test Leads

Objective of Test: To ascertain whether the test leads of a VOM have reasonable resistance.

Test Equipment Required: VOM with manufacturer's original test leads.

Test Procedure: Remove the test leads from the meter and substitute a short length of heavy copper wire. About three inches of No. 16 wire bridged across the ohmmeter terminals of the meter is a suitable jumper for this purpose. Zero the meter on the $\times 1$ scale. Next, remove the jumper and insert the test leads. With the prods of the test leads shorted together, observe the meter reading. If the indication is much higher than 0.1 ohm, the test leads have developed an abnormally high resistance.

Comments on Test Results: Although high-resistance test leads affect the accuracy of all the ohmmeter scales, the greatest departure occurs on the $R \times 1$ scale (or even more so on an $R \times 0.1$ scale if present). A too low indication will result from high-resistance test leads despite the fact that the meter has been initially zeroed. The described test procedure should be carried out periodically, since the effects of broken strands or corrosion films are not generally obvious in the physical appearance of the leads.

Test Procedure for Measuring Dc Resistance with a Simple VOM

Objective of Test: To demonstrate the test techniques covered in the general discussion.

Test Equipment Required:
A VOM with a series-type ohmmeter.
The following 10 percent 1/2-watt composition resistors: 10, 33, 100, 330, 1000, 3300, 10,000, 33,000, 100,000, 330,000, 1,000,000, and 3,300,000 ohms.

Test Procedure: Arrange the resistors in the general sequence of ascending values as listed above. Wherever feasible, the range switch should be set so that the reading falls closest to mid-scale. In order that this can be so, the meter must be zeroed prior to testing a resistor. Zeroing is accomplished by shorting the test prods together and carefully adjusting the zero control of the meter. Admittedly, this is tedious. Sometimes, a given zero adjustment will hold for more than one test procedure, but usually, only if it is unnecessary to change the range switch. In this test procedure, the range switch gets much attention and there will be little escape from the need to constantly zero the meter.

Generally, the first test does not result in the most accurate reading. Rather, the scale position of the pointer informs us how to change the range switch to cause the reading to occur closer to the mid-scale position. Only after we are satisfied that the best has been done in this respect do we make a value reading. And when making a reading, care should be exercised in multiplying the indicated value by the ratio indicated by the range switch. In this test procedure we have the advantage of experimenting with color-coded resistors. In general test work, this advantage is either lacking or cannot be accepted as being reliable.

This test procedure is easier with alligator clips rather than pointed prods. It is important, however, to refrain from touching or holding the resistor or the exposed metal of the clips or prods. Otherwise, the resistance of the skin will be measured in parallel with the resistor and the reading will be low. Even if this shunting effect is avoided, it is not good practice to expose the object undergoing test to body heat.

Although this test procedure is simple and straightforward, do not disregard what has been stated in the general discussion with regard to parallax, physical positioning of the meter, and mechanical shock and vibration.

Comments on Test Results: The resistors selected are fairly representative of the resistance range encountered in a wide variety of electronic testing. This is not to say that very much lower and higher values are not also found. However, you may already have noticed that the limitation of the simple VOM has been approached. Depending upon the meter used for this test procedure, either the 10-ohm or the 3.3-megohm readings may not have appeared at a mid-scale position. Just where the borderline values occur depends on the individual meter. In general, the shunt type is able to accommodate lower resistance values than is the series type such as we have been considering. The scale of the shunt-type VOM is similar to but reversed from that of the more common series type. This is, with the shunt-type meter, the zero end of the scale is at the left and the infinity end of the scale is at the right. Other than this, the test procedure for evaluating resistance is essentially the same for the two types. From the above, you might suspect that the series type is best for high resistance values and this is true. Depending again upon the individual meter, you can sometimes extend the basic test procedure to accommodate tens of megohms and higher.

OTHER TYPES OF MULTIMETERS

Although we are focusing our attention on the VOM with a series-type ohmmeter, the shunt-type instrument must be considered, too. Its essential distinguishing feature is the ohms scale with zero at the left end and infinity at the right end. When using this type of ohmmeter, the meter is adjusted to set the pointer to the infinity mark and this is accomplished by keeping the prods from making contact, rather than shorting them as in the use of the series ohmmeter. The shunt ohmmeter is more capable of measuring very low values of resistance. A third type, the potentiometric ohmmeter, behaves essentially the same as a series type, but offers some advantage in a more favorable distribution of indications near the high end of the scale.

The ohmmeter function of an electronic multimeter may have either a scale similar to a series ohmmeter or a reversed scale. Depending upon which type of scale it has, the right-hand adjustment of the pointer is made the same as for series or shunt-type ohmmeters. Usually, the scale of the electronic meter will simulate that of the shunt ohmmeter, so that the prods are initially kept apart and the pointer is positioned at infinity. Additionally, electronic meters generally have an additional left-end scale adjustment. This adjustment establishes the zero end of scale deflection. This zero adjustment is made by shorting the test prods. In the majority of electronic multimeters both the infinity and the zero adjustment must be made before the ohmmeter mode of operation can be properly used. Because of interaction between the zero and infinity adjustments, it may be necessary to repeat both adjustments several times.

Electronic multimeters tend to subject the circuit or device under test to lower current than is the case with simple VOMs. More modern types also apply lower voltages to the external circuit, but each instrument must be evaluated on the basis of its individual characteristics. Electronic multimeters tend to be more vulnerable to disturbance in rf fields than is the case with simple VOMs. Therefore, it is wise to avoid placing the meter close to an active rf source such as a transmitter or oscillator. If the rf enters through the ac line, a filter may have to be inserted in the line. Such filters are commercially available. Effective filters can sometimes be constructed by tightly winding several tens of turns of, say, No. 18 solid hookup wire on a 3/8-or 1/2-inch form. Two of these should be made and one inserted into each line lead. 0.1 μF 600-volt Mylar capacitors should then be connected to the ends of the rf chokes as shown in Fig. 1-3. Actually, the circuit is a balanced low-pass filter, comprising a single pi section. The line filter should be contained in a metal box and the box should be connected to the ground terminal or to the metal case of the electronic multimeter. It may also be found that the line cord itself may have to be run through metallic shielding braid. In such a case, the braid should be grounded to the instrument.

It is not infrequently found that the rf is picked up because the test leads act as an antenna. This calls for either shielded leads or very short leads. The effectiveness of filter and shielding techniques

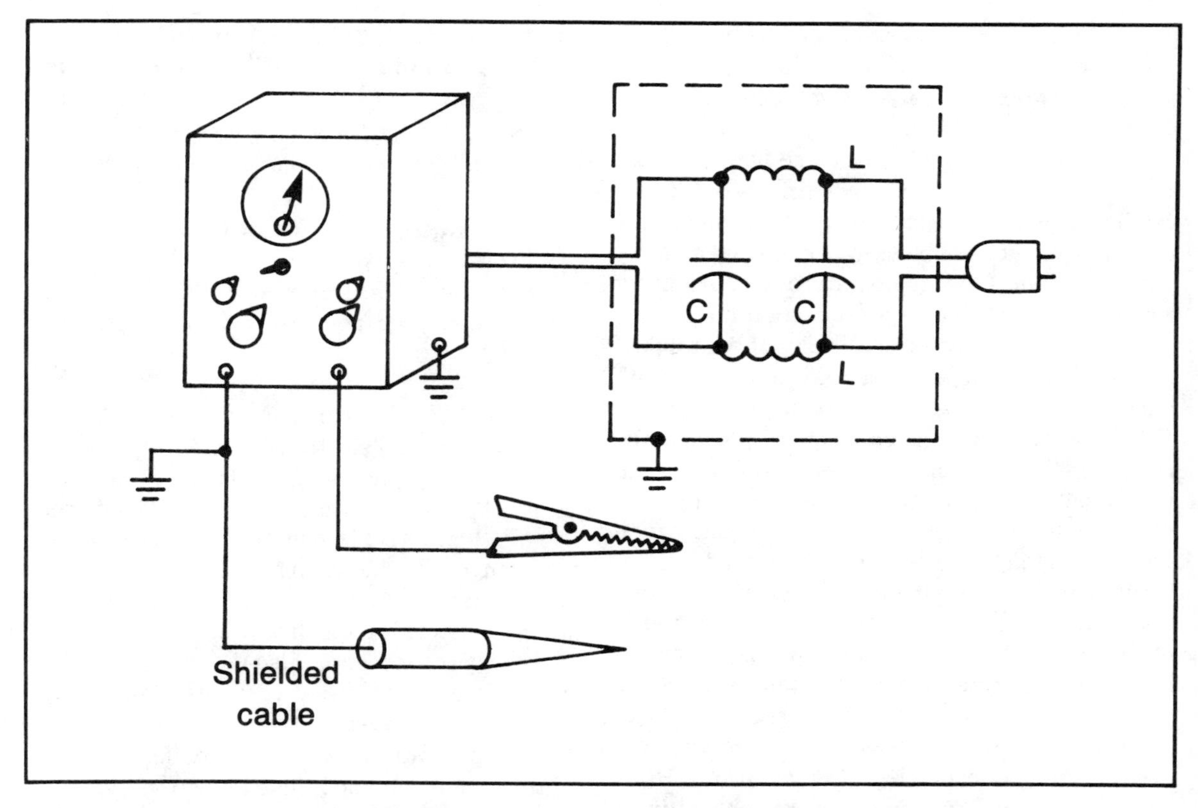

Fig. 1-3. Filter designed to prevent rf pickup when using a line-operated electronic meter.

is quite variable and some experimentation usually is necessary.

It is often found that the way in which the instrument is grounded is quite important. Generally speaking, grounding must be short and should derive its ground connection from a conduit or from a cold-water pipe. Grounding of ac-operated electronic multimeters is also desirable in order to keep the "hot" test prod from assuming an ac potential. In this regard, it may be necessary to experiment with the ac plug in the receptacle both ways. Better type instruments have line plugs with a third grounding prong. An oscilloscope can be used to check these matters, but make sure that the oscilloscope chassis and the chassis of the electronic multimeter both connect to the ground side of the ac line.

The use of modern battery-operated electronic multimeters is strongly recommended. These instruments are less susceptible to interference from the power line or to rf fields than the older vacuum-tube voltmeter types. Also, they are much more stable in their adjustment and alignment. There is considerably less likelihood of making erroneous tests because drift characteristics are orders of magnitude less than in the VTVM. Various combinations of FETs, bipolar transistors, and integrated circuits are used, but all of these instruments are capable of the short- and long-term stability typical of solid-state equipment.

Digital Meter

The digital ohmmeter can also be classified as an electronic meter. However, it is a unique test instrument when compared with conventional analog types. Its main feature is the absence of the parallax error and errors associated with the interpretation of a pointer and scale reading. Less obvious is the greater accuracy and higher resolution which can be designed into the operation of digital VOMs. The overall result of these factors is that more refined tests can be made than is commonly feasible with conventional meters. For example, the digital ohmmeter can be used to obtain meaningful data regarding the temperature coefficient of resistors. When used as a production line instrument, it ensures greater uniformity of the product than is likely when performance is monitored with analog instruments. A typical middle-quality digital VOM provides ohmmeter ranges in which the resolution is plus or minus 0.1 percent of the highest indication available on that range. The accuracy is plus or minus 0.2 percent of the highest indication on that range. Therefore, resistance readings are accurate within plus or minus 0.3 percent of the full-scale readout. This greatly exceeds the capability of analog instruments. Because there is no bearing friction or other mechanical hindrances, readings are repeatable to a very high degree.

Digital ohmmeters are active devices with voltage and current available at the test prods. However, the current availability tends to be relatively low. The specifications of a typical digital VOM show the current availability on all resistance ranges to be less than 1.5 milliamperes. Although these instruments often have built-in filters to remove incoming transients, the speed of response, especially of the ohmmeter function, exceeds that of analog instruments wherein meter damping imposes an objectionable time delay for some types of rapid test techniques.

Surely, the digital VOM must possess some disadvantages, and one is cost. In the practical world of electronics, it turns out that much can be done quite satisfactorily with less costly electronic analog instruments and even with the simple nonelectronic VOM. The digital VOM tends to be sensitive to transients and some care is often needed to obtain reliable readouts when there are SCRs, motors, defective fluorescent lamps, and other spike-generating apparatus in the test area.

Procedure for Determination of the Active Characteristics of the Simple VOM

Objective of Test: To see the VOM from the viewpoint of the circuit undergoing test.

Test Equipment Required:
VOM
Electronic meter (additional VOM if the electronic meter does not have a current-measuring mode).

Test Procedure: The current and voltage functions of the electronic meter are used to measure the output voltage and the current supply ability of the VOM in its ohmmeter mode of operation (see Fig. 1-4). The ohmmeter is zeroed on each of its ranges. The dc voltage available at the prods is then measured and recorded. Similarly, the dc current injected into the milliammeter of the measuring instrument is also recorded. In making such voltage and current readings, always begin with the high ranges on the electronic meter and work down until a satisfactory deflection is obtained. Under no conditions allow the electronic meter to be in its ohmmeter mode. Notice that the voltage and current measurements are necessarily made separately. The voltage readings approximate the open-circuit voltage of the VOM. The current readings can approximate the short-circuit current capability of the ohmmeter under test if the readings are made on a high-current range of the measuring VOM. We are less interested in ultimate accuracy here than to demonstrate the potential of the ohmmeter for inflicting damage on passive circuit components. Also, record the polarity of the voltage available from the prods. This data is part of the meter's characteristics and should not become separated from it.

Comments on Test Results: This test procedure immediately reveals the VOM for what it is. Far from being a passive instrument for making tests on other passive objects, it is now seen to be an active device with sufficient reserve to alter, modify, or destroy the object undergoing test! Unawareness of this basic fact can result in punctured dielectrics, destroyed pn junctions, burned-out thermocouples and precision shunts, and at best, erroneous test results from polarity effects and from induced heating in small parts. The dynamo-like nature of the ohmmeter is not a trivial thing and no testing should ever be undertaken without due consideration of the possible effects of the voltage and current applied to the tested circuit. It is not uncommon for the unwary technician to "discover" burned-out meter movements, fuses, semiconductors, etc. in equipment being tested with an ohmmeter. It is sad to say that some of the "defects" perhaps did not exist prior to the test procedure!

Procedure for Determining Resistance by the Voltmeter-Ammeter Method

Objective of Test: To demonstrate the measurement of resistance by the voltage and current present in a circuit.

Test Equipment Required:
dc milliammeter with 0-1 ma range.
dc voltmeter with 0-2.5 volt range.
Single-pole-single-throw switch.
8.2K 5 percent 1-watt composition resistor.
2.2K 5 percent 1-watt composition resistor.
9.0-volt transistor radio battery.

Test Procedure: Connect the components as shown in Fig. 1-5. If not obvious from looking at the switch, the open and closed positions should be identified with an ohmmeter before the switch is connected into the circuit. With the switch in its open position, carefully check the polarity of the meters and of the battery. (The small terminal of the battery is positive.) Close the switch and record the readings of the milliammeter and voltmeter. Having done this, open the switch and compute R_x from the Ohm's Law relationship, R_x equals E divided by I, where E is the voltage developed across the resistance R_x, and I is the current through R_x.

Comments on Test Results: Although this

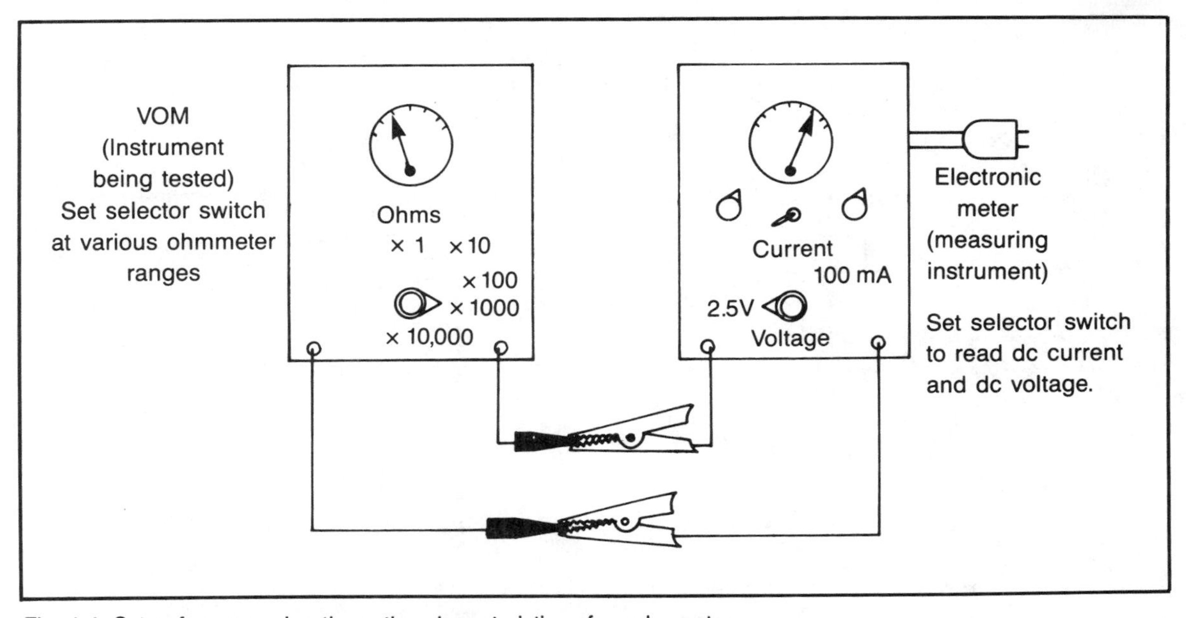

Fig. 1-4. Setup for measuring the active characteristics of an ohmmeter.

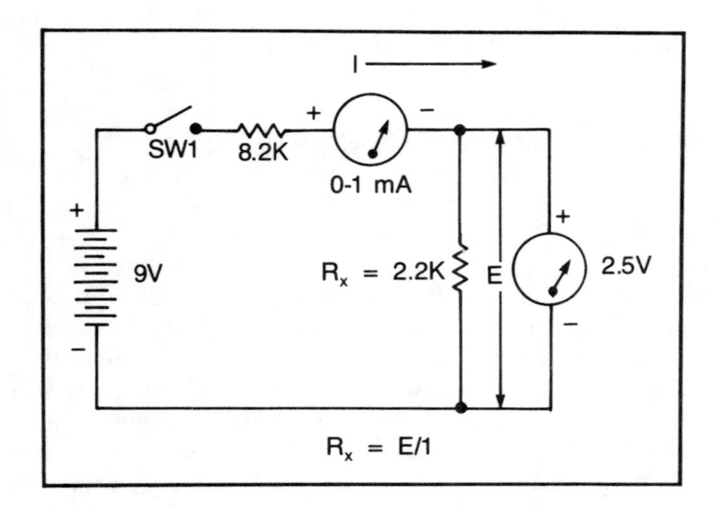

Fig. 1-5. Typical circuit for determining R_x from the Ohm's Law relationship.

is a simple method to determine an unknown resistance, the validity of the results depends on the sensitivity of the voltmeter. For example, suppose a 1000-ohms-per-volt voltmeter had been used on its 2.5-volt range. Such a meter would look like a 2,500-ohm resistance connected across R_x. The equivalent circuit would then be that shown in Fig. 1-6. You can see that instead of measuring the voltage drop across the nominal 2.2K R_x, we are now measuring the voltage developed across an equivalent resistance comprised of the parallel combination of R_x and the internal resistance of the voltmeter. This turns out to be 1.17K and will lead to completely erroneous results. If we were to use

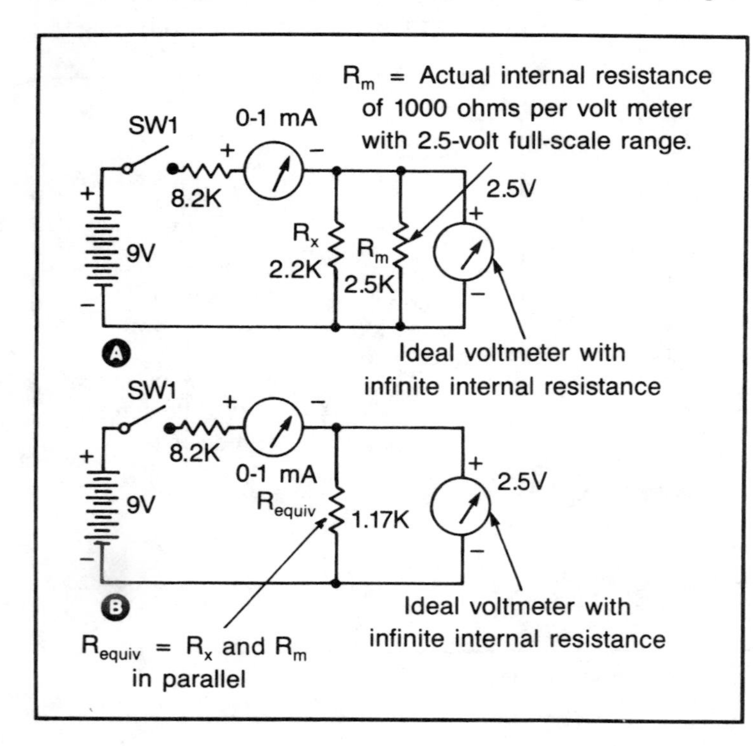

Fig. 1-6. Equivalent circuits showing the loading effect of a 100 ohms-per-volt voltmeter. Circuit A represents an ideal voltmeter with the R_m of the actual voltmeter. B is a simplified version of the same circuit.

a 5000-ohms-per-volt meter, the loading effect would still be drastic and the test results would be unacceptable. It turns out that for this particular circuit, we could just get by with a sensitivity of 20,000 ohms per volt. The conclusion is that the internal resistance of the voltmeter must be much higher than the unknown resistance to which it is connected. Indeed, the voltmeter resistance should be at least 20 and preferably greater than 50 times the resistance of the element represented by R_x in this test procedure.

This being the case, it is possible to use common dc voltmeters if we ascertain that the circuit loading effect is negligible. If the current meter indication changes much beyond a just perceptible amount when the voltmeter is connected, we can be certain that the loading effect of the voltmeter is excessive and a voltmeter with a higher internal resistance should be used. The best way around this problem is to use an electronic voltmeter. Most usually have an internal resistance of 10 or 11 megohms on all dc ranges. This is sufficiently high so that the loading is negligible over a wide range of resistance measurements in a circuit such as that in Fig. 1-5.

It should be realized that the internal resistance of the current meter does not affect the accuracy of this kind of resistance measurement. Although its presence does reduce the current circulating in the circuit, it still indicates the actual current flowing through R_x, which is vital to the accuracy of the procedure. Although it is possible to account for the loading disturbance caused by a voltmeter of inadequate sensitivity, the basic simplicity of the test procedure would thereby be lost.

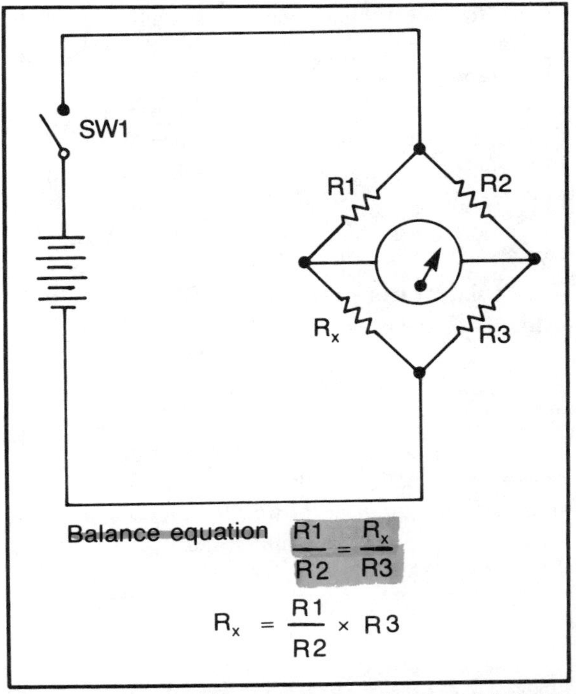

Fig. 1-7. Basic Wheatstone bridge circuit and balance equations.

WHEATSTONE BRIDGE

The Wheatstone bridge is a simple circuit comprising four resistances, a source of dc voltage, and null-sensing instrument, usually a galvanometer or microammeter. The basic Wheatstone bridge circuit is shown in Fig. 1-7. Notice that one of the four resistances is the unknown resistance. Wheatstone bridges are available commercially, and the laboratory grades tend toward accuracies of plus or minus 5 percent over a typical range of a fraction of an ohm to a megohm. In some models the range extends from 0.001 ohm to 10 megohms, and special models measure considerably higher values of resistance.

One of the reasons for the inherent accuracy of the Wheatstone bridge is the fact that the readout meter does not have to yield a numerical indication of the current flowing through it; instead, it is used to indicate the lack of current or a null. Since the direction of meter deflection is opposite on either side of the null, the null indication is a function of the meter's sensitivity rather than being dependent upon its linearity or the accuracy of its scale. Because of the mode of operation, null-indicating meters have the zero at center scale. If the meter is mechanically aligned, the null indication will occur at the zero mark. However, the null is actually established by adjusting the resistance arms of the bridge so that no motion of the pointer is noticeable when connecting and disconnecting the dc voltage source to the bridge. Such a null is further checked by making one bridge arm very slightly higher and very slightly lower in resistance. Such a procedure confirms the null by causing the meter pointer to deflect in one direction, then the other. It is obvious that a null adjustment can be expected to be a much more accurate procedure than the reading of a meter indication in the conventional way. When the null has been established, the unknown resistance is calculated from the resistance arm settings. These are read from calibrated dials or step switches.

From Fig. 1-7, it is seen that the basic equation for achieving a null (bridge balance) is R1 divided by R2 equals R_x divided by R3. With practical bridges the test procedure and computation are simplified by ganging R1 and R2 together in such a way that their **ratio** is indicated by a calibrated dial. This dial thereby becomes a ratio switch. Since the selectable ratios are integral numbers or powers of ten, a quick multiplication leads to the measured value of the unknown resistance. Resistance arm R3 is variable and its resistance is indicated by a calibrated dial. The consequence of this arrangement is that the ratio switch is used to "ball-park" the balance adjustment. Then R3, known as the "rheostat arm," is carefully adjusted to establish the null. The reading of R3 is then multiplied by R1/R2, as indicated by the setting of the ratio switch.

Actually, the rheostat arm, R3, consists of several tapped resistances in series. This is shown in the typical commercial version of the Wheatstone bridge in Fig. 1-8. In Fig. 1-8, the R3 arm comprises RA, RB, RC, and RD. These four resistance sections provide tapped increments of 0.1 ohm, 1.0

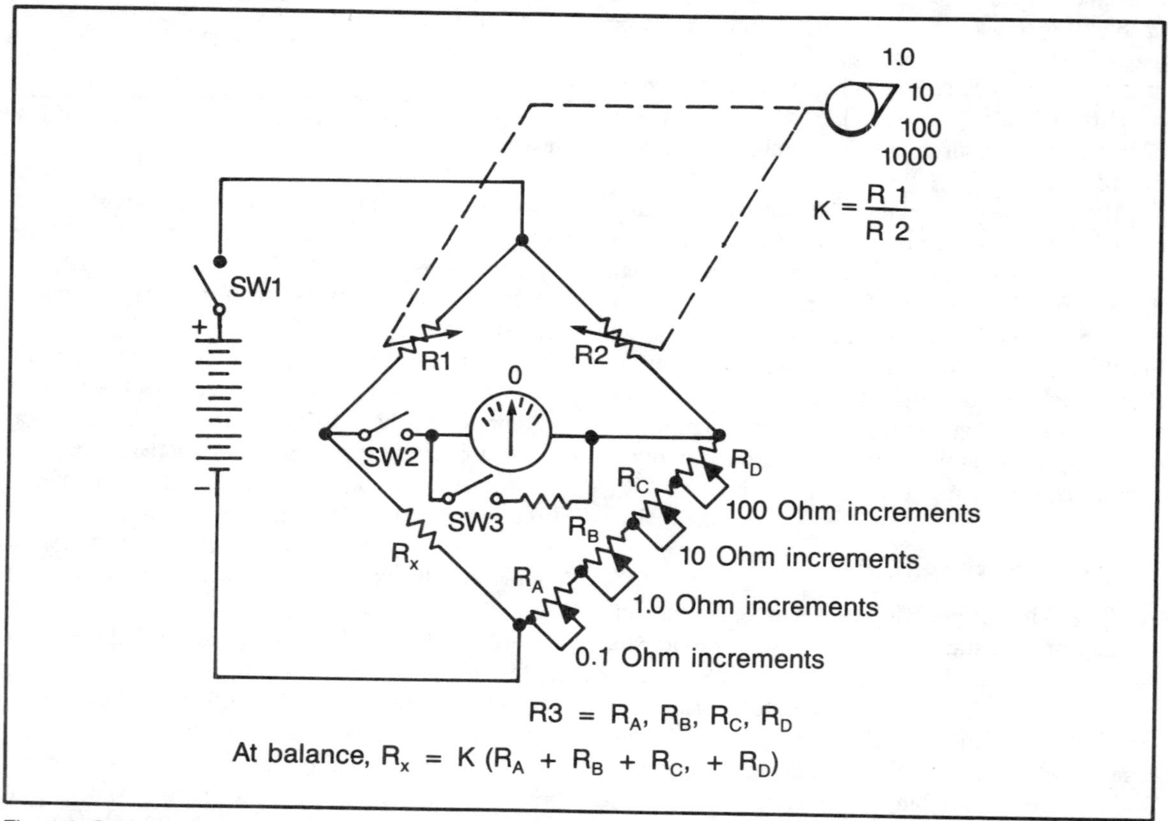

Fig. 1-8. Schematic of a typical commercial Wheatstone bridge.

ohm, 10 ohms, and 100 ohms, respectively. With such an arrangement, the net value of R3 is the arithmetic sum of the 4-section cascade. Obviously, the use of such a cascade for the rheostat arm of the bridge makes possible greater accuracy in the measurement of the unknown resistance.

Another feature generally found in commercial bridges is a provision for desensitizing the null indicator, such as the switch-controlled shunt shown in Fig. 1-8. Instead of a shunt, a series resistance is often used with the sensitivity-control switch connected across the series resistance. In any event, the basic idea is to provide a means to spare the meter from violent deflection during the initial stages of the balancing procedure. Although not shown in Fig. 1-8, commercial bridges usually provide more than one dc voltage source. The voltage sources are switched by the range selector, thus enabling the null sensitivity to be maintained when measuring high values of R_x. Other things being equal, null sensitivity is greater with higher voltages. However, when measuring lower resistances, there is the danger that the resistance elements will be heated by the high circulating current. Changing the voltage source with the range avoids this undesirable situation.

For practical work, it will be found that the bridge is quite flexible. Specifically, an unknown resistance can be nulled at more than one setting of the ratio switch. Theoretically, the maximum sensitivity exists when the resistance of the four arms and that of the null detector are equal. In actual use, a reasonable approach to maximum sensitivity generally results from setting the ratio switch as near to a unity ratio as possible and, secondly, from having a condition of equality between R1, R2, and R3. This second condition is not always possible to achieve because most bridges provide control of the ratio R1/R2, but not necessarily of the individual values of R1 and R2. This second condition may be worthy of consideration when setting up a bridge circuit from resistance boxes and variable resistances, however. In most test procedures, satisfactory results come from establishing the null with the ratio switch as near to the unity ratio as possible.

Some bridges are ac-excited by a low audio-frequency sine-wave oscillator and headphones are used to establish the null. With a pure 1000-Hz sine wave, good results are readily obtained because both the ear and the headphones are quite sensitive in the vicinity of this frequency. The value of a resistance so measured will be, for all practical purposes, the same as the actual dc resistance. However, when measuring high resistances, say over 50K, it becomes increasingly important that the harmonic content of the ac source be kept very low. Otherwise, differential phase shift from stray reactance will tend to make the null less sharp. This is because the harmonics will still be heard when the fundamental is nulled out. If it is found that this condition exists, its effect can be somewhat overcome by practice. The idea is to concentrate on the fundamental tone only.

Procedure for Setting Up a Simple Wheatstone Bridge

Objective of Test: To use readily available equipment and components to set up a Wheatstone bridge suitable for a wide range of resistance measurements.

Test Equipment Required:

One small 6.3-volt filament transformer.

10,000-ohm 10-turn linear potentiometer.

Resistance box with three tapped resistance elements having adjustable steps of 10 ohms, 100 ohms, and 1000 ohms.

An alternative resistance arm can be made up of a switching arrangement which selects either a 10-ohm, a 100-ohm, a 1000-ohm, a 10,000 ohm, or a 100,000-ohm 1 percent resistor.

An inexpensive oscilloscope.

A 10-turn counting-type dial for the linear potentiometer.

Single-pole single-throw switch.

Test Procedure: Connect the apparatus as shown in Fig. 1-9. The linear potentiometer should be mounted on a panel with its counting dial. The rheostat arm should initially have a 10K, 1 percent

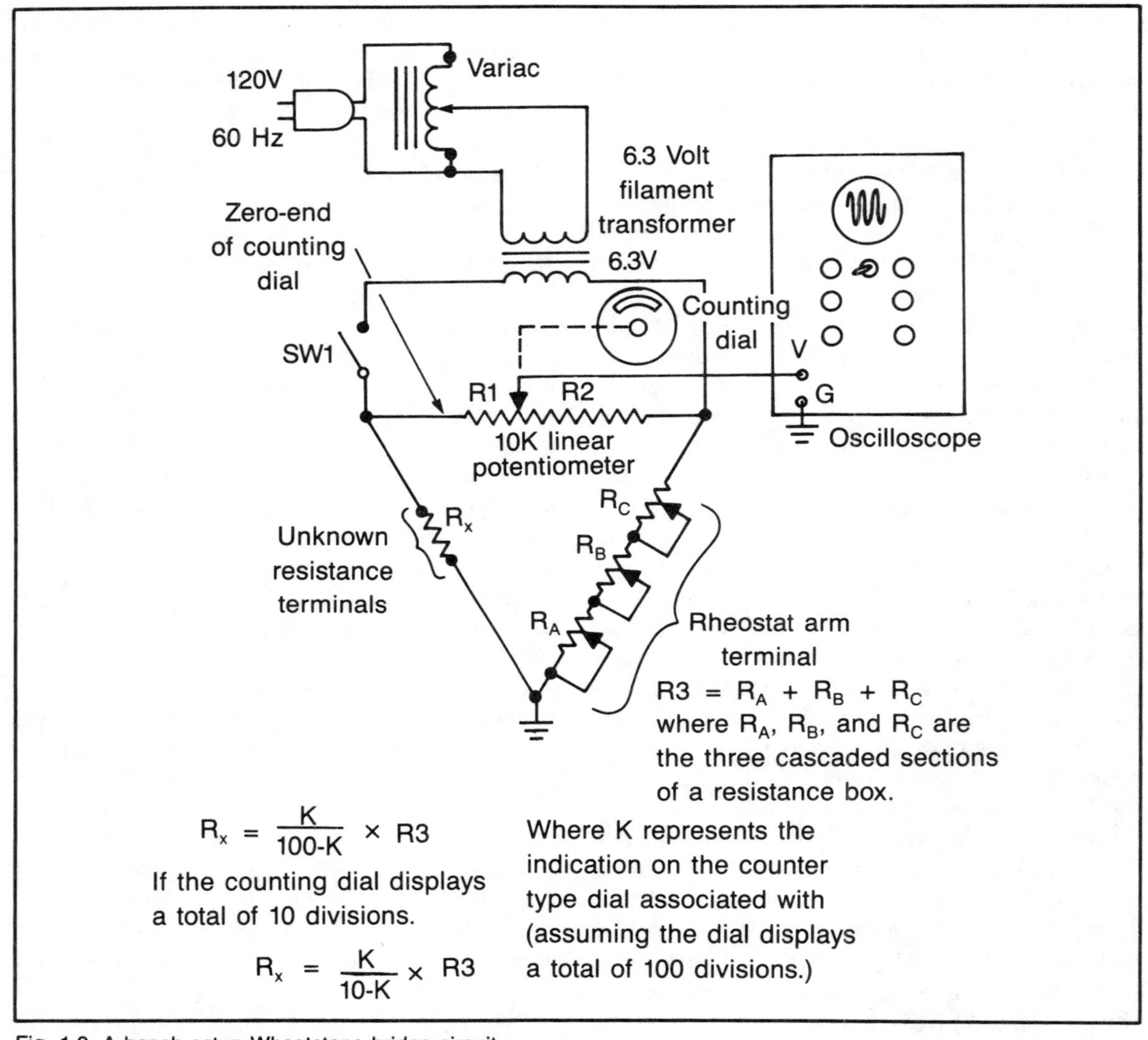

Fig. 1-9. A bench setup Wheatstone bridge circuit.

resistor connected to its terminals and an identical resistor should be connected to the unknown resistance terminals. The initial position of the ratio dial should be about 3 or 4 on its counting dial. Close switch SW1 and adjust the oscilloscope for a display of several cycles of the 60-Hz sine wave. Carefully adjust the ratio dial for a null of the pattern on the oscilloscope screen. When doing this, the vertical gain of the oscilloscope should be increased as the null is approached. It will be observed that there is a 180-degree change of phase on either side of the null. When the best possible null is attained, the counting dial should be set to its exact mid-position (the number 50 for most dials) for the linear pot adjustment corresponding to the null condition. Once this has been done, the bridge is calibrated. The number 50 will denote a ration of R1/R2 equals unity. When the dial indicates 60, the ratio R1/R2 is 60:40 or 3:2. For the number 70, R1/R2 equals 70:30, and so on. Similarly, when the dial indicates 40, the ratio R1/R2 is 40:60 or 2:3. With the dial on 30, R1/R2 is 30:70 and so on. (If this sequence does not exist, the end connections to the linear potentiometer should be transposed.)

After this calibration has been made, the 10,000-ohm precision resistors should be removed. Connect the decade box or its suggested alternate to the rheostat arm terminals. Connect an unknown resistance to the unknown resistance terminals. The bridge is then balanced preferably with the linear potentiometer as close to its mid-position as practical. The unknown resistance will be (R1/R2) × R3 for the condition of a nulled display on the oscilloscope screen.

Comments on Test Results: Notice that ac rather than dc excitation of the bridge is used. However, because of the relatively low frequency (60 Hz), the bridge performs essentially the same as it does with dc insofar as the measurement of resistors is concerned. The use of ac excitation allows the use of an ordinary oscilloscope as the null detector. Also, there is no need for a zero-center microammeter. (Of course, such meters are available commercially. They are not, however, often found to be part of the equipment of the average laboratory or service facility.) The exact value of the linear potentiometer is not important. Indeed, a value other than the prescribed 10,000 ohms could be used.

This bridge is intended for making tests on unknown resistors. Its use for measuring the resistance of inductive windings is not recommended in general for coils with inductances greater than several millihenrys. If dc excitation is desired, a 9-volt battery can be used as the source and the null detector can be a 50-0-50 microammeter. The transformer then would not be needed. Meter protection for off-null adjustment can be provided by a high resistance in series with the microammeter. A pushbutton switch connected across the high resistance is closed when the null is being approached. Conversely, a dc oscilloscope can be used as a null detector. With dc excitation, the resistance of inductive windings can be readily measured.

Of course, there are other ways of setting up a Wheatstone bridge circuit in t[illegible] straightforward procedure is to us[illegible] boxes for the arms, R1, R2, an[illegible] in the basic bridge circuit in Fi[illegible] point is that bridge resistance measurement tech[illegible] ques are highly accurate if the ratio of two arms and the absolute value of the third arm are accurately known. The low-resistance limit is not reached until you approach the resistance of the actual interconnecting leads. The measurement of resistance can be extended to such high ranges that you begin to encounter practical difficulties from leakage currents and time constants.

In general, the Wheatstone bridge, although an active circuit, can be designed to be less likely to inflict damage on delicate components than an ohmmeter. This is accomplished by the use of a low excitation voltage and a very sensitive null detector. The null detector is made sensitive by means of electronic amplification. Oscilloscopes and electronic meters with a zero-center provision are good examples of sensitive null detectors.

Procedure for Determining the Resistance of a Sensitive Meter

Objective of Test: To measure the internal resistance of a sensitive meter without exposing it to the dangerous currents available at the prods of an ohmmeter.

Test Equipment Required:

1.5-volt C or D flashlight cell.
A high resistance potentiometer
Microammeter or milliammeter to be measured
A low resistance potentiometer

Test Procedure: The circuit is shown in Fig. 1-10. To determine the maximum resistance required in the series potentiometer, divide the voltage of the flashlight cell by the full-scale reading of the meter; then increase the value obtained by 10 to 100 percent. For example, if we have a 10-microampere meter, 1.5 divided by 10 × 106 gives us a value of 150,000 ohms. But for the sake safety, we actually select a 200,000-ohm potentiometer. In general, the next available resistance above the calculated value will be adequate.

The shunt potentiometer value is not so easy to predict, but its maximum value will always be much lower than that of the series potentiometer. A 5,000- or 10,000-ohm potentiometer might be selected for the first try. The more sensitive the meter, the greater will be the required maximum resistance of the shunt potentiometer. Although results can be obtained with a higher than necessary resistance, you get better resolution when the maximum resistance of the shunt potentiometer is not excessive. In any event, proceed as follows:

With switch SW1 in its open position, make certain that the series potentiometer is at its maximum resistance position before connecting the flashlight cell. Be absolutely certain that no mistake is made, because if this potentiometer is in its zero position, or close to it, the meter will be damaged or destroyed. An ohmmeter should be used to check this situation, since a visual inspection of the potentiometer shaft or wiper can easily be transposed in the mind to its very opposite electrical condition. Connect the flashlight cell and very carefully adjust the series potentiometer to produce an exact full-scale deflection of the meter. Next, close switch SW1 and adjust the shunt potentiometer until the meter deflection is exactly half scale. Having done this, disconnect the flashlight cell. Remove the shunt potentiometer without disturbing its adjustment. Use an ohmmeter to measure the resistance of the shunt potentiometer, being careful to perform the measurement on the same two terminals which were previously connected across the sensitive meter. The reading obtained from the ohmmeter measurement is essentially equal to the internal resistance of the sensitive meter.

For most practical purposes, the accuracy of the above procedure will be satisfactory. Accuracies on the order of plus or minus one percent are possible. A very slight error does exist because the procedure is based upon the assumption that a constant current is delivered to the sensitive meter and that the current supplied by the flashlight cell does not change when the shunt potentiometer is connected across the sensitive meter.

In the event that a precision bridge or a digital ohmmeter is to be used to measure the resistance of the shunt potentiometer, a refinement can be added to the test procedure which tends to make the inherent error approach zero. Referring to Fig. 1-11, a second current meter is used. The full-scale indication of this meter should at least be equal to that of the meter undergoing test, but can be somewhat higher. The test procedure remains similar, except that now the initial value of current supplied to the sensitive meter (while switch SW1 is open) is kept constant by readjusting the series potentiometer. Thus, we maintain the same current flow out of the flashlight cell during the full-scale and the half-scale deflection settings of the meter.

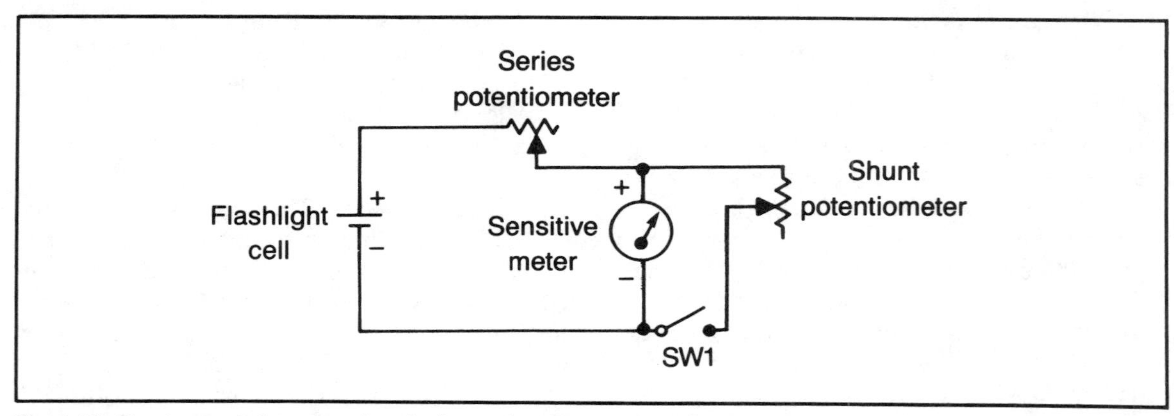

Fig. 1-10. Simple circuit for measuring the internal resistance of a sensitive meter.

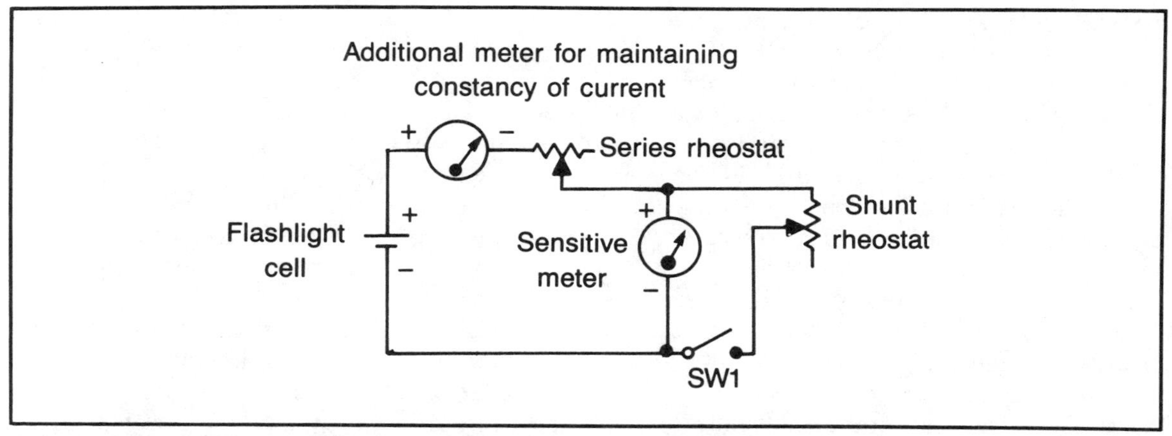

Fig. 1-11. Refined circuit for measuring the internal resistance of a sensitive meter.

The sole function of the additional current meter has been to indicate the existence of this constant-current condition. Its accuracy is not important because we are interested in the condition of constancy rather than in the absolute value of the current.

Procedure for Determining Internal Resistance of a Battery

Objective of Test: To measure a battery parameter which reliably indicates its condition.

Equipment Required for Test:

VOM

Rheostat with a maximum resistance depending on the type of battery to be checked. For size D flashlight cells, a maximum resistance of one to three ohms is appropriate. For size C flashlight cells, a 5-ohm rheostat is suitable. For 9-volt transistor batteries, use a 200-ohm rheostat.

Test Procedure: Connect the cell or battery in the circuit shown in Fig. 1-12. Make certain that the rheostat is initially at its *maximum* resistance. With switch SW1 open, read the open-circuit voltage indicated on the voltmeter. Close switch SW1 and adjust the rheostat to produce a reading of one half the open-circuit voltage. This can only be approximated because of the tendency of the battery voltage to continue to decrease while supplying current to the rheostat load. However, if the battery is good, the half-voltage operation will be substantially steady for a few seconds.

There will be a significant difference between fresh and depleted batteries in the amount of resistance needed to approach half-voltage operation, despite the fact that the open-circuit voltages might have been the same, or nearly so. A fresh battery will accommodate a relatively low-resistance load, whereas the terminal voltage of a depleted battery will drop with a much higher load resistance. A weak battery will also be less likely to display a reasonably steady voltage under load. The evaluation requires judgment and is essentially related to the type of battery under consideration. The evaluation is best made by comparison to a cell or battery known to be in good condition. The actual internal resistance is equal to the rheostat resistance needed to produce the half-voltage operation.

Comments on Test Results: Although the test might appear to be imprecise, it nevertheless is a reliable way to determine the condition of a cell or battery. As a cell or battery ages, its ability to provide current to an external circuit diminishes. But this is not due to a decreased open-circuit terminal voltage. Therefore, a simple voltmeter test is not a suitable indication of the electrochemical condition, except perhaps for cells or batteries which have approached total exhaustion. Conversely, a measurement of internal resistance under load is a very good indication of a battery's ability to perform properly. For example, when a depleted cell with a high internal resistance is used in an ohmmeter, the accuracy of low resistance indications is substantially degraded. In such cases, the internal resistance of the cell produces the same effect as test leads with inordinately high resistance. The zero-adjust control may still zero the pointer, but cannot overcome the effect of the high internal resistance of the cell when the ohmmeter is used to measure resistance.

The internal resistance measurement also provides a direct indication of the ability of the cell or battery to provide current. The mere consideration of the open-terminal voltage is misleading in this respect. An automobile storage battery behaves similarly; it may show an open-circuit terminal voltage of 12.5 or 13.0 volts, but fail completely to operate the starter motor. When it is discovered that the internal resistance of a cell or battery is substantially higher than that of a fresh unit, any circuit requiring current may not operate properly, or at all.

Transistor radios and other electronic circuitry are often very sensitive to the internal resistance of the cells or battery from which they are powered. This is not initially due to an inadequate current supply, although the decrease of current availability does limit maximum performance levels. More significant is the fact that the internal resistance in the power source acts as a mutual impedance which couples together various stages of the apparatus. For example, a battery with high internal resistance generally causes "motorboating" and other feedback problems in a transistor radio. Thus, the end of the useful life of the battery is approached long before it lacks the current capacity to operate the radio at satisfactory sound levels. This can be readily proved by connecting a 100-μF capacitor across the terminals of a weak battery. The capacitor bypasses the internal resistance of the battery and prolongs its useful life until the internal resistance greatly limits the available current.

In order to measure the rheostat resistance for large cells, such as size D units, a Wheatstone bridge may be needed for the fractional-ohm values involved. On the other hand, since we are "ballparking" rather than making a precise measurement, the physical position of the rheostat wiper arm will give us the desired information.

The principle underlying this test procedure is illustrated in Fig. 1-13. Here, the internal resistance of the battery is represented by R1, a physical resistance in series with the battery. The voltage dropped across this resistance is indicated by a dc voltmeter as V1. Recall that we initially measured the no-load terminal voltage of the battery, V2. We assume that switch SW1 is open and the dc voltmeter imposes no load; that is, it consumes no

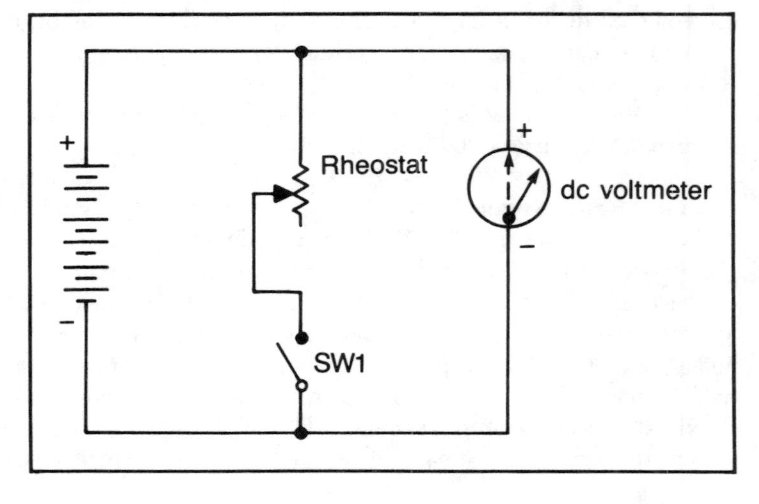

Fig. 1-12. Circuit for measuring the internal resistance of a cell.

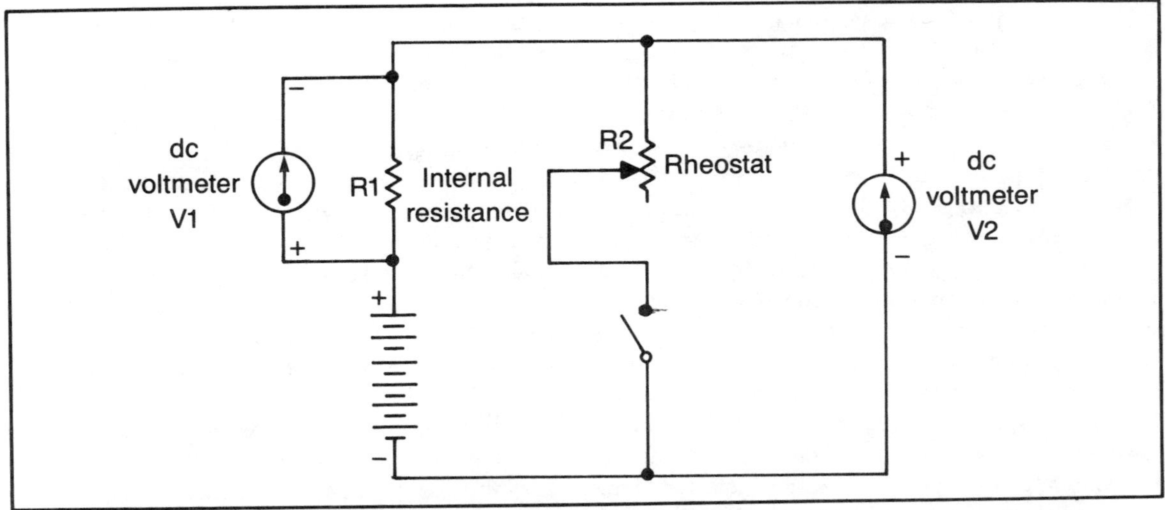

Fig. 1-13. Circuit showing the logic of the procedure used to measure the internal resistance of a cell or battery.

current. That being the case, there is no voltage drop across R1 and V2 is the true voltage developed at the electrodes of the battery. Now, we close switch SW1 and adjust the rheostat load to produce a half open-circuit voltage reading of V2 divided by 2. Assuming that the developed voltage across the electrodes of the battery has not been affected, where has the remaining voltage, V2 divided by 2, gone? As would be suspected, this voltage now appears across R1 and would be so indicated were it physically possible to connect a voltmeter to indicate V1. But if V1 and V2 divided by 2 are now equal, both being one half of V2, the open-circuit battery voltage, then it must follow that R1 is equal to R2. This must be so because the same current flows through R1 and R2. Therefore, we can correctly state that, even though we cannot physically measure R1 or V1, R2 simulates the internal resistance, R1, when the test is conducted as described.

CAPACITOR LEAKAGE TESTS

One of the most elusive yet most important servicing techniques is the ability to measure excessive dielectric leakage in capacitors. High leakage, that is, abnormally low dielectric resistance, upsets the biasing, polarity of applied voltages, and alignment of circuits because it is generally assumed during design that capacitors have perfect dielectric insulation and no resistive path exists across the terminals. Most vulnerable to the effects of leakage in a capacitor are high-impedance circuits such as the grid circuit of a vacuum tube and the gate circuit of an FET. However, dielectric leakage can simulate relatively low resistive values and thereby disturb lower impedance circuits such as the biasing of bipolar transistors.

It is not easy to say how much leakage is allowable. Much depends on how the capacitor is used. Interstage coupling capacitors can cause trouble when the leakage can be detected only on the highest ohms scale of the VOM, or even an electronic multimeter. On the other hand, capacitors used in certain bypass functions can be quite leaky without disturbing the circuit in which they are connected. It is a good practice, however, to replace any capacitor with a substantially lower leakage resistance than is normal for that type and capacity. When you detect such a condition it almost surely signifies a developing defect which is very likely to further deteriorate with time. This is especially true when the circuit voltage applied to the capacitor is much higher than the voltage across the prods of the ohmmeter.

Another difficulty involved in any attempt to

define unacceptable leakage is that such a number would necessarily have to be expressed as so many units per microfarad. A measurable leakage value for a 10,000-pF capacitor could be acceptable, but the same leakage in a 500-pF capacitor could cause trouble. It turns out that the normal leakage resistance of capacitors up to 0.1 μF which use such dielectrics as ceramic, Mylar, polystyrene, polycarbonate and other plastic materials, is too high to be measured on common ohmmeters. On the other hand, paper capacitors can readily show leakages of several hundred megohms. These capacitors should be suspect when an ohmmeter indicates leakage corresponding to, say, 100 megohms or less, regardless of the capacity. Oil dielectric capacitors have very high internal resistance when new, but often deteriorate with age. If the dielectric resistance of an oil capacitor is not much higher than several-hundred megohms under an ohmmeter test, such a capacitor should be considered failure prone, especially if, as usual, the capacitor is operating in a circuit where 500 volts or more is present.

Mica capacitors have traditionally had a reputation as near-perfect components and it is not surprising to find that some technicians feel that these capacitors can't become leaky. However, mica capacitors are not immune to various forms of electrical and chemical deterioration. Often moisture plays a part in the development of leakage paths. Sometimes an ohmmeter test of a mica capacitor will show that it has developed a relatively low resistive path across the terminals. The action of radio frequencies in transmitters is one fairly common cause of such leakage. The net effect is similar whether the leakage path exists in the the mica dielectric itself or in the encapsulating material.

Capacitors using metallized dielectrics will often be found to be quite leaky. Whether this should be considered grounds for discarding the capacitor depends upon the way it is used in the circuit. Such capacitors tend to be self-healing; that is, electrical rupture of the dielectric usually clears itself. However, even though an internal flashover in one of these capacitors will generally fail to develop a short circuit, the after-result is often a readily detectable resistive path. Such flashovers also tend to reduce the capacity, but this is usually negligible.

Air-dielectric capacitors should show infinite resistance on the highest range ohmmeter test. Sometimes sufficient dust or foreign material will collect between the plates to produce a measurable leakage path. In transmitters the simultaneous presence of radio frequency voltages and moisture will show up as a measurable leakage path. When such a discovery is made, some kind of remedial action must be taken, for such resistance paths tend to become lower with time.

Numerous other dielectrics are encountered in capacitors. Among these are porcelain, glass, and various plastic materials. Although the minimum allowable resistive path must necessarily remain a matter of judgment based on experience and on comparison with a duplicate unit known to be good, the testing of capacitors is not a waste of time. In a considerable portion of capacitor failures, the leakage will be low enough to be read on the lower scales of the ohmmeter. Frequently, a virtually zero-resistance short will be found. With a little practice, an ohmmeter is quite effective in locating defective capacitors.

Electrolytic capacitors are inherently leaky compared to the types just discussed. The leakage depends on age, temperature, and applied voltage. The problem is deciding how much leakage to tolerate. This depends to a considerable extent on the circuit in which the electrolytic capacitor operates. For most electrolytic capacitors between 1 and 100 μF, a discernible leakage on the R × 10,000 ohms scale of the ohmmeter tends to be reasonable. A more than negligible leakage on the R × 1000 ohms scale may be excessive leakage. That is, the pointer should ultimately come to rest at or very near the infinity mark on the scale (assuming the ohmmeter has a mid-scale indication of 30). As with other capacitors, and perhaps even more so, an ohmmeter test of an electrolytic will frequently identify units with shorts or near-shorts so that there is no gray area to ponder over. So-called "solid-state" tantalum capacitors usually show a lower leakage (higher resistance) than conventional liquid or "dry" electrolytics.

The test procedure for capacitors is significally different from that followed when testing resistances, largely due to the energy-storage capability of capacitors. For one of several reasons, a capacitor may store an appreciable charge that is capable of damaging or destroying the ohmmeter. The charge may have been placed in the capacitor when it was last energized in a circuit, but charges are developed in other ways, too. Some capacitors develop a charge from physical deformation as a result of being jostled about. And it is well known that capacitors can accumulate a charge from the atmosphere. Many dielectric materials exhibit dielectric absorption or dielectric hysteresis characteristics. When capacitors using such dielectrics are shorted—even for an appreciable length of time—they do not dissipate all of their energy. Surprisingly, some time later it may be found that the capacitor has again developed appreciable voltage across its terminals. Temperature changes, chemical effects, and the effects of radiation fields may also play a part in the storage of a charge in a capacitor. In any event, the ohmmeter prods must never be placed across the terminals of a capacitor without first making sure that any residual charge stored in the dielectric is dissipated. This is accomplished by shorting the terminals of the capacitor not merely once but several times. Although exceptions to this rule could be applied to very small capacitors, it is well to allow for no possible exceptions. Even a tiny capacitor can store a high level of energy if previously exposed to relatively high voltage. It never should be assumed that a capacitor is "dead" because of the lapse of time since it was knowingly subjected to a voltage. The retention of a charge is often surprisingly long, days and weeks not being uncommon.

Other than the above discussed procedure, the ohmmeter is zeroed as in resistor testing. Initial testing should start with a high-range ohmmeter scale. When the ohmmeter prods are connected to the capacitor, the capacitor begins a charge cycle due to the voltage applied from the ohmmeter battery. The pointer will tend to initially swing towards zero, because at the first instant the capacitor behaves as a short circuit. As charge pours into the capacitor, its apparent impedance increases and the ohmmeter pointer moves towards the infinity end of the scale. How fast this movement occurs depends on the size of the capacitor and the internal resistance of the meter. Sometimes you have to wait a number of seconds to a number of minutes until the capacitor becomes charged to the voltage applied to it. The capacitor is fully charged when the pointer stops moving. Before this, no useful leakage reading can be taken. The pointer will ultimately stop at the infinity mark if leakage is negligible, or at a finite resistance if there is detectable leakage in the capacitor. For capacitors up to several hundred pF, the charging cycle takes place almost instantly. At the opposite extreme, capacitors of hundreds of microfarads can take minutes to complete the cycle. This is due not only to the time constant of such large capacitors, but since these are invariably electrolytic types, some time may have to be allowed for the electrochemical forming of the dielectric film.

Except in the case of special types, electrolytic capacitors are polarized. An ohmmeter test is valid only if the positive test lead is connected to the positive terminal of the capacitor. If the test is attempted with reverse polarity, the swing of the needle will be faster than it is for correct polarization and the resistance indication will be abnormally low. Such reverse-polarity testing is likely to falsely suggest the need for replacement. Even worse, certain electrolytic capacitors with low voltage ratings can be destroyed by the application of a reverse-polarity voltage.

Although the ohmmeter test is largely employed to find leaky and shorted capacitors, it can also be used to detect open units, providing the capacity is not too small. Fortunately, the open-circuit defect is found predominantly in the larger value capacitors. If the charge cycle is not noticed when the prods are connected to the capacitor terminal, an open lead inside of the capacitor is indicated. With most ohmmeters you can reliably check capacitors as small as 0.01 μ this way. Just how small the capacitor can be depends upon the sensitivity, the voltage, and the damping of the meter. In any case, the capacitor should always be

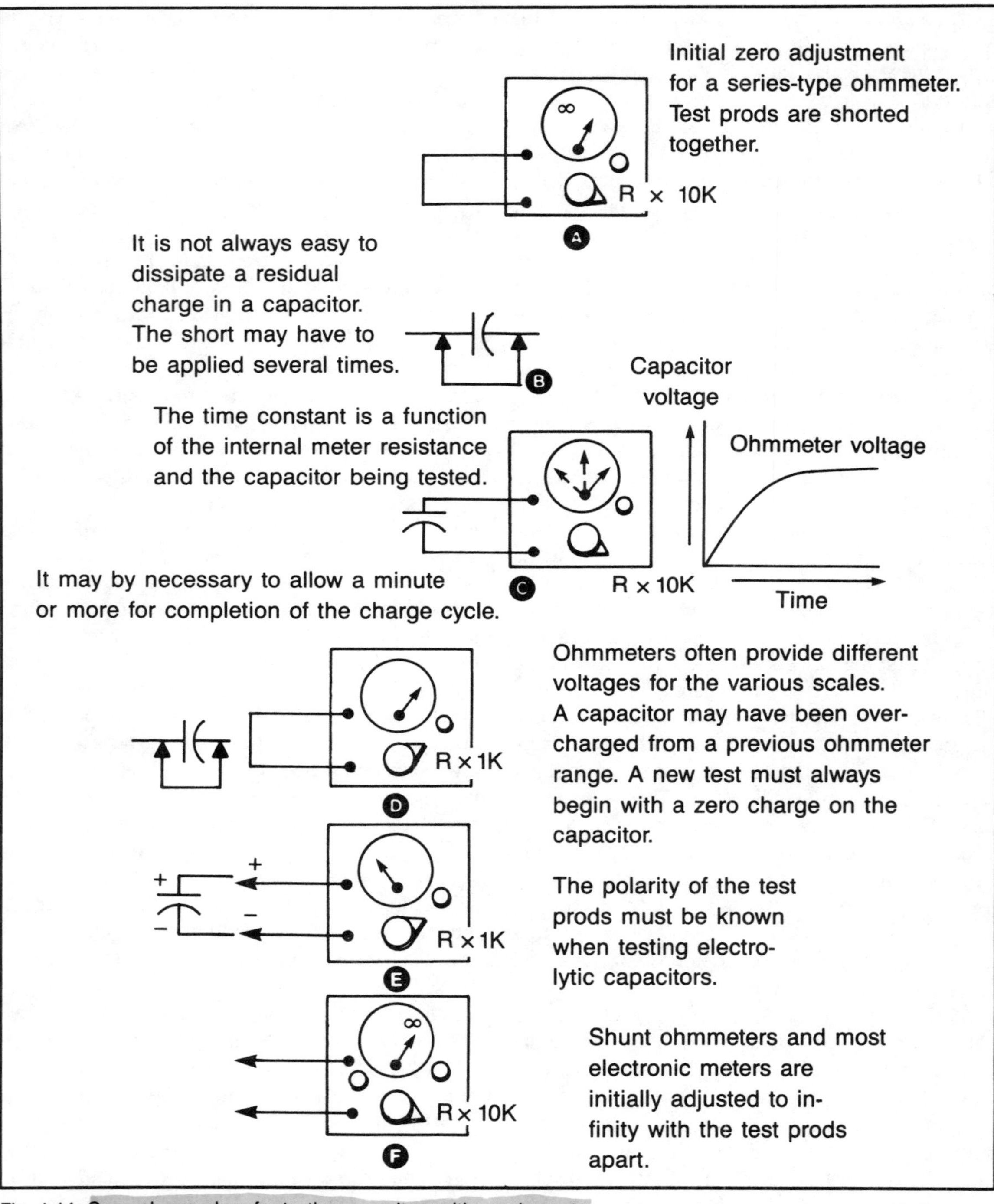

Initial zero adjustment
for a series-type ohmmeter.
Test prods are shorted
together.
∞
R × 10K
A
It is not always easy to
dissipate a residual
charge in a capacitor.
The short may have to
be applied several times.
B
Capacitor
voltage
The time constant is a function
of the internal meter resistance
and the capacitor being tested.
Ohmmeter voltage
It may by necessary to allow a minute
or more for completion of the charge cycle.
C
R × 10K
Time
Ohmmeters often provide different
voltages for the various scales.
A capacitor may have been over-
charged from a previous ohmmeter
range. A new test must always
begin with a zero charge on the
capacitor.
R × 1K
D
+
+
−
−
R × 1K
E
The polarity of the test
prods must be known
when testing electro-
lytic capacitors.
∞
R × 10K
F
Shunt ohmmeters and most
electronic meters are
initially adjusted to in-
finity with the test prods
apart.

Fig. 1-14. General procedure for testing capacitors with an ohmmeter.

shorted several times prior to making this test. The highest ohmmeter range offers the greatest sensitivity and permits this test to be performed on the smallest capacitors. The open capacitor will show negligible leakage, but in such a case the leakage evaluation is meaningless.

A final precaution in the testing of capacitors stems from the emphasis placed on dissipating a stored charge prior to making the test. Suppose, as has been suggested, the highest ohmmeter range is initially used. Subsequently, it may be desired to repeat the test on the next lower range. Such an occasion is likely to arise in the testing of electrolytic capacitors because of the wide range of leakages found in these units. Because ohmmeters often use different internal battery voltages on different ranges, a capacitor being tested may have been charged to an excessively high voltage on the higher scales and this charge will be dumped into the meter when it is switched to a lower scale. This can be damaging to the meter and this risk should be avoided by first removing the capacitor and shorting its terminals several times before testing on a lower ohmmeter range. A summary of the test procedures discussed for capacitors is shown in Fig. 1-14.

The ohmmeter tests for capacitors provide reliable evaluations in the majority of cases. This is particularly so after you learn to temper the readings with judgment developed from experience. Occasionally, the ohmmeter test will not reveal defects that are present during actual circuit operation. Sometimes dielectric breakdown occurs intermittently, or even continuously, under the higher operating voltage applied to the capacitor in the circuit. The relatively low voltage from the ohmmeter may result in a "clear" indication. Similarly, the leakage of electrolytic capacitors is sometimes unacceptably high during actual operation, but does not appear to be excessive during the ohmmeter test. Other defects which are not revealed by the ohmmeter include corona, lead and foil inductance, microphonic behavior, unsuitable temperature coefficient, and voltage-dependent capacitance.

AUTOMOTIVE COMPONENTS

An ohmmeter will also reveal defects in automobile distributor caps. These sometimes develop leakage paths which are not readily discernible from a visual inspection. On the other hand, the ohmmeter can be used to confirm the nature of a visible apparent defect on the insulation surface. Such leakage paths often result in engine malperformance which is difficult to analyze. A suspected distributor cap should be removed from the automobile and placed on the test bench. The ohmmeter should be set for its highest range and properly zeroed. Then the prods should be systematically employed to test between all adjacent spark-plug terminals and between all spark-plug terminals with respect to the center (coil) terminai. See Fig. 1-15. Any leakage indication should be further investigated. Sometimes a film of carbon dust on the interior surface of the cap causes such leakage. Any film should be removed with a clean, dry rag. Actually, no dust, dirt, or moisture should be tolerated on any portion of the cap. Not too infrequently, a leakage path will be indicated by an ohmmeter, but there will be no visual evidence of it. Such a leakage path is caused by "bubble" or internal "blister" within the insulation material and is a valuable discovery, for otherwise much time can be consumed in attempting to correct a misfiring engine.

Ohmmeter tests of the spark plugs often prove worthwhile, too. The plugs are not removed from the engine, but the cables are disconnected. With the ohmmeter zeroed at its highest range, test between the center terminal of the plugs and the engine block or other ground connection that is clean and will permit good contact. A resistance or leakage indication on the meter can be caused by such things as chemical deposits on the porcelain support of the center electrode, a crack in the porcelain insulator, fouled points from oil or carbon, or from water leaking into the compression chamber as a result of a blown head gasket. This test is not all-inclusive in the evaluation of the plugs. Burned electrodes or incorrect gaps have to be discovered by visual inspection. However, plugs which

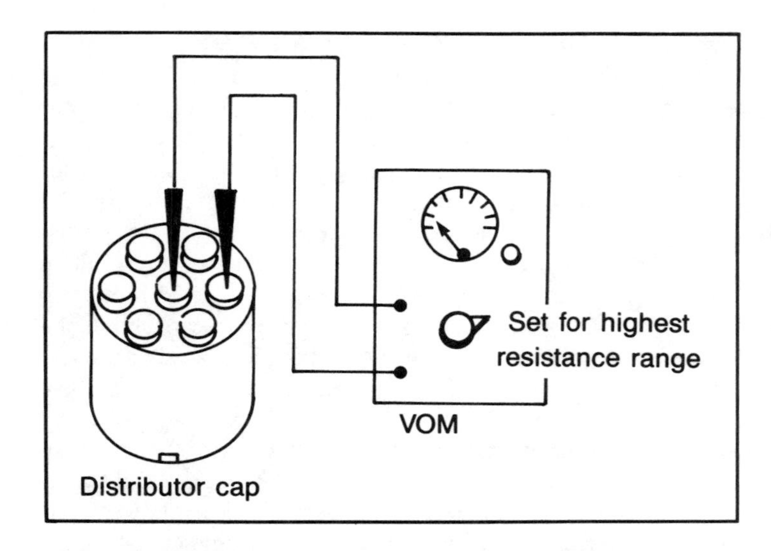

Fig. 1-15. Insulation leakage test procedure for an automobile distributor cap. Although an R × 100,000 or R × 1 meg range is desirable for this test, bad leakage paths may be detected with a VOM in which the highest ohms range is R × 10,000.

show low resistivity most certainly should be removed and investigated. It is often wise to remove excessive chemical deposits and this is best done by sandblasting with special equipment at service stations. The indication of a resistance path across the plug is very likely to adversely affect ignition performance.

The ohmmeter can also be used to investigate the condition of interference-suppression cable. When erratic readings are obtained from flexing such cable, it is a definite indication that it should be replaced.

Yet another elusive cause of ignition trouble is the capacitor ("condenser" in automotive circles) across the ignition breaker points in the distributor. The capacitor need not be physically removed from the distributor, but its center lead should be disconnected. Then, a leakage test can be made with respect to body or engine metal. Be sure to dissipate any charge in the capacitor by shorting it several times. Any leakage in the capacitor means that it should be replaced. However, this test may fail to uncover other defects such as the development of a high series resistance. Be alert, also, for the possibility of intermittent lead connections. This shows up as an erratic charging cycle and can sometimes be discovered by flexing the lead. Another worthwhile test is to check between the metal capacitor case and the distributor plate upon which the capacitor is mounted. With the ohmmeter at R × 1 or at the lowest range, any reading which is thought to differ from that obtained with the ohmmeter prods shorted together is cause for concern. A fraction of an ohm can prevent the capacitor from properly performing its function. To eliminate such resistance, clean the metallic surfaces and make sure that the capacitor is firmly gripped in its clamp. When making the aforementioned tests, make certain that the ohmmeter prods never inadvertently contact the "hot" side of the electrical system. The ignition switch must be in its "off" position during such tests.

INDUCTORS AND TRANSFORMERS

Ohmmeter tests are very effective for determining the condition of inductors and transformers. Although there are some defects which are not always readily detectable with the ohmmeter, most abnormalities are revealed by simple resistance and continuity test procedures. A burned-out winding, one which no longer provides a conductive path between its terminals, is indicated by an infinity or unreasonably high resistance reading. A high-resistance reading may be due to some conductance through carbonized insulation material. The normal dc resistance of inductor and transformer windings employed in electronics can usually be read on the R × 1, R × 10, or the R × 100 ranges. An R × 0.1 ohmmeter range is often desirable, but not absolutely necessary. The ohmmeter is much used to simply establish the existence of, or the lack of, conductivity across the terminals of a winding. Though simple, the tests should be made with judgment based on simple logic. For example, filament windings do not have much dc resistance and even a small 6.3-volt filament transformer can show a resistance of a fraction of one ohm across the filament winding. Even if we cannot read such a low resistance accurately, the near zero reading on the R × 1 range usually (but not necessarily) indicates the winding is good.

Similarly, line-voltage windings often have resistances of several ohms to several tens of ohms, and sometimes somewhat higher. The resistance of the line-voltage winding (the primary) can usually be read on either the R × 1 or the R × 10 ohmmeter ranges. This pertains to the most often encountered transformers used in general electronics work. Of course, judgment must be exercised when dealing with very large or tiny iron-core components. If the windings on a transformer are not open, the tentative assumption should be "so far, so good." Audio-type transformers differ from power and filament transformers in that the resistance of the windings tends to be higher. This is due to the extremely fine wire which is often used.

If a reasonable resistance can be measured across the terminals of a winding, the isolation between windings should be tested. In general, a near-infinity indication on the highest ohmmeter range should exist. A lower resistance reading indicates breakdown in the insulation between windings. Such a condition can affect the performance of the transformer and can surely be expected to deteriorate further with the passage of time. A similar test should always be made between the windings and iron core. This is a frequently encountered breakdown path and can be responsible for the leakage measured between the windings themselves.

If the windings of a transformer or inductor show continuity and perfect isolation between windings and from windings to core, you've learned some important facts about the electrical state of the component. We cannot yet assign a clean bill of health, however, because of the possibility of shorted turns. A single shorted turn in a transformer can make it useless. If it is a power or filament transformer, it will run hot and will not develop normal voltages or currents. An audio transformer will not operate properly in its circuit because a shorted turn reduces the effective inductance of the windings. The same applies to simple inductors and chokes. But can we detect shorted turns with the ohmmeter? If a duplicate component known to be good is available, a comparative check with the ohmmeter can reveal the abnormally low resistance in a shorted winding. But the detection of a single shorted turn is not generally possible. Fortunately, it often happens that a whole layer or a substantial portion of a layer of a winding is shorted. Under this condition, an ohmmeter, used to compare the transformer under test with an identical one known to be good, can reliably indicate the abnormally low resistance in the shorted winding.

It must be emphasized that all of the described tests should be made only under the condition that no winding of the transformer is connected to an active source of voltage. Moreover, the tests cannot be assumed to be reliable unless all windings are completely disconnected from associated circuitry. Otherwise, the presence of diodes, capacitors, resistors, etc., can mask the results of the tests. Indeed, it will very often be found that it is not the transformer that is defective, but rather that it is operating in an overloaded state due to a faulty component in the associated circuitry.

It is also important to remember that when connecting or disconnecting the ohmmeter probes to one winding, high voltages can be developed in other windings. Therefore, hands should not be carelessly in contact with any winding terminals during testing. Otherwise, unpleasant shocks can result. Do not connect a charged capacitor, a battery, or any other voltage source to any winding while the ohmmeter is connected to another winding. The voltage thereby induced by transformer action in the winding undergoing test can damage the ohmmeter.

Dc resistance tests on other inductors, such as rf coils and rf chokes, are carried out in essentially the same way as the tests on the larger iron-core components. However, the emphasis usually will be on continuity. For example, burned-out or otherwise open windings on i-f transformers are occasionally found in radios and TV sets. Figure 1-16 summarizes the test procedures for transformers and inductors.

ALTERNATE METHODS FOR DETERMINING RESISTANCE

Figure 1-17 shows three methods of determining resistance which are similar to, but simpler than, some already discussed. However, these are likely to include various effects that can lead to

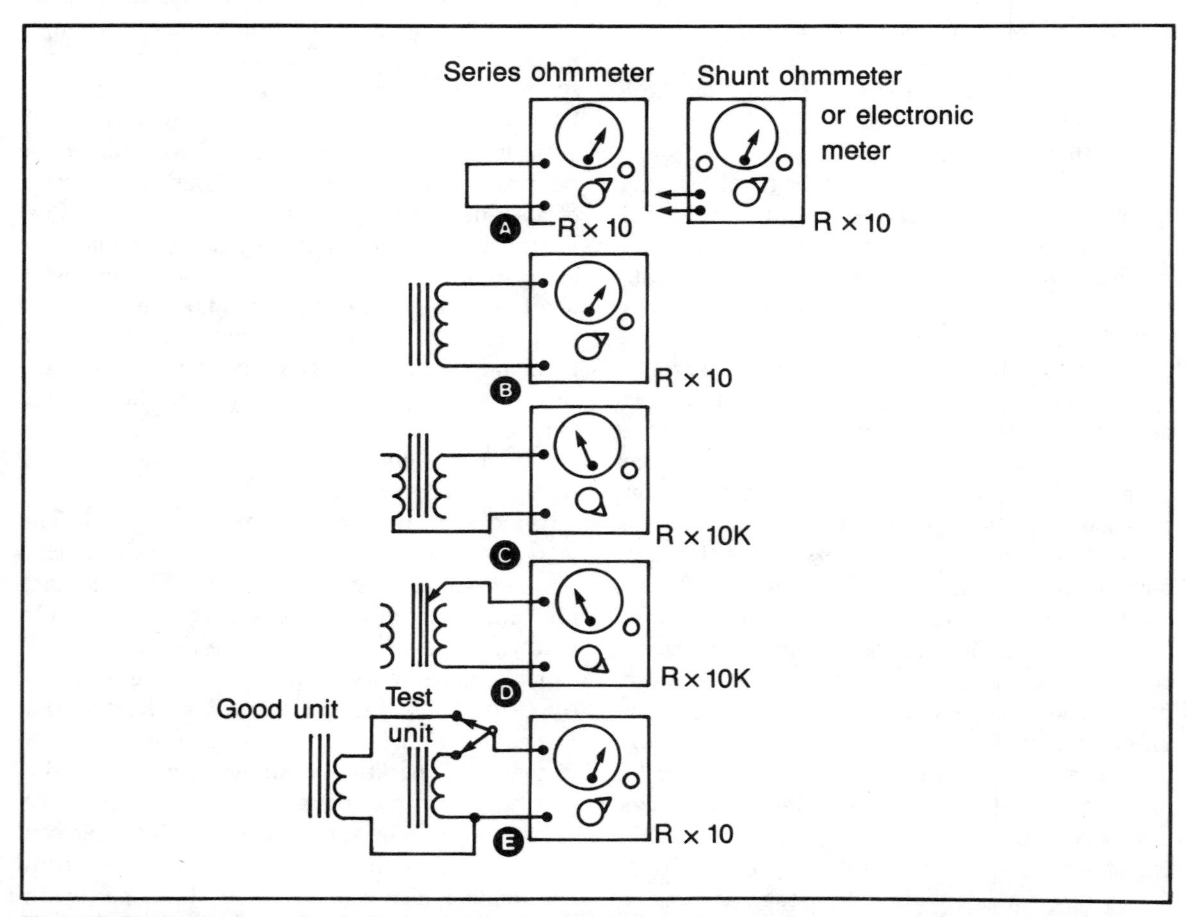

Fig. 1-16. General procedure for testing transformers and inductors with an ohmmeter. (A) Make appropriate initial adjustments of meter. (B) Make a continuity test on each winding. The R × 1, R × 10, or R × 100 ranges will most often be found suitable. Record the actual resistance indications and determine whether they appear to be reasonable. (C) Test the isolation between windings. This is done on the highest ohmmeter range and is essentially an insulation measurement. A near-infinity indication should be obtained. (D) Make a similar test between the iron core and all windings. Again, negligible evidence of current leakage should be given. The reading should be infinity, or nearly so. (E) Shorted turns in a winding can often be detected by comparing the test winding with an identical winding on a unit known to be good. Shorted turns decrease the dc resistance of the defective winding.

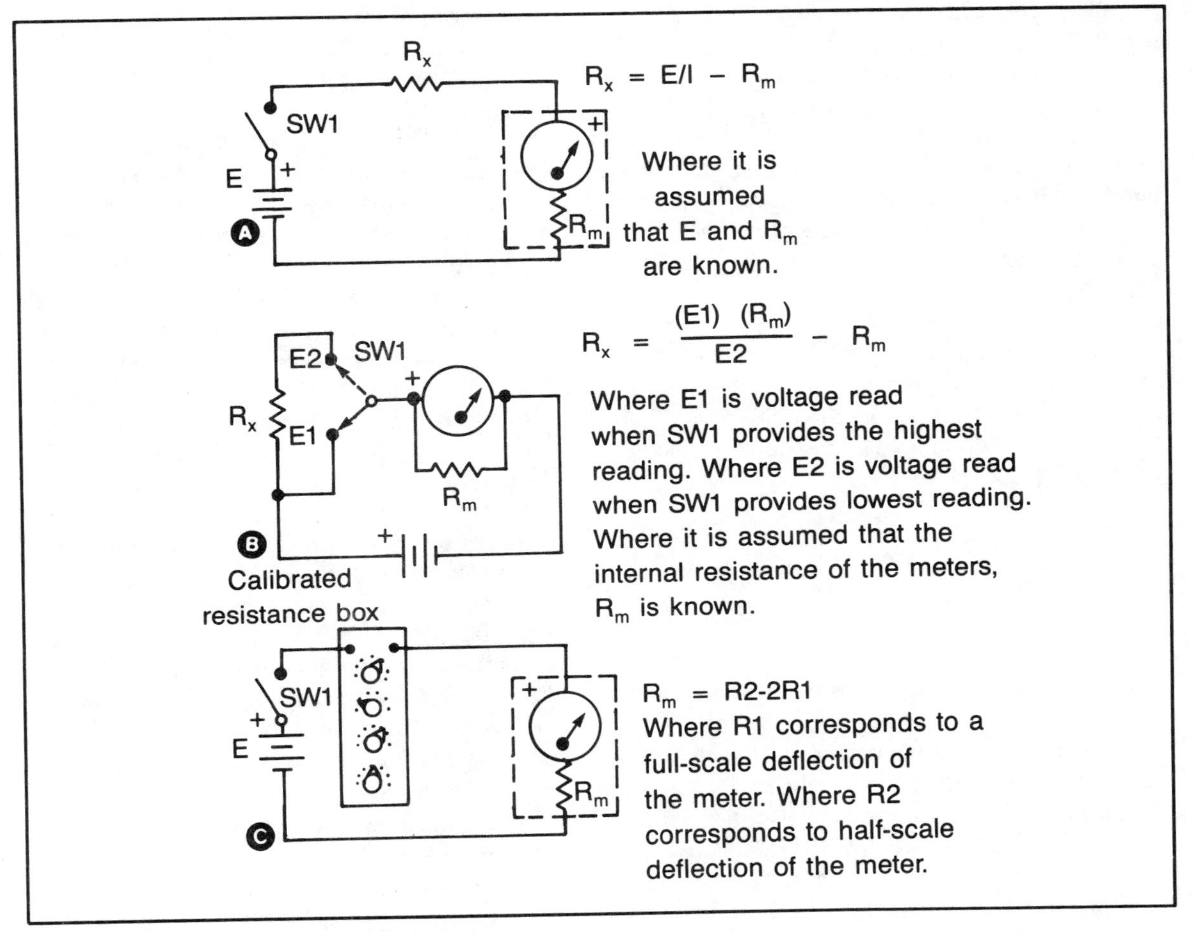

R_x
SW1
E
A
$R_x = E/I - R_m$
R_m
Where it is assumed that E and R_m are known.
E2
SW1
R_x
E1
R_m
B
Calibrated resistance box
$R_x = \frac{(E1)\ (R_m)}{E2} - R_m$
Where E1 is voltage read when SW1 provides the highest reading. Where E2 is voltage read when SW1 provides lowest reading. Where it is assumed that the internal resistance of the meters, R_m is known.
SW1
E
C
R_m
$R_m = R2\text{-}2R1$
Where R1 corresponds to a full-scale deflection of the meter. Where R2 corresponds to half-scale deflection of the meter.

Fig. 1-17. Alternate ways to measure resistance.

some inaccuracies. But the very simplicity of these methods is their chief virtue. They are worthy of study as examples of ways of getting information with a bare minimum of equipment. In practical electronics work, it is not uncommon to encounter a shortage of instruments and components. Under such circumstances the accomplishment of a test procedure can be particularly rewarding.

In Figure 1-17A, a single current meter and a battery are used to provide the needed information from which the unknown resistance, R_x, can be calculated. We must know the battery voltage and the internal resistance, R_m, of the meter. If a mercury cell, or a battery composed of mercury cells, is used, voltage E is known quite accurately, providing excessive current drain is not imposed on the battery. Mercury batteries tend to maintain a near-constant terminal voltage over a long period of time. This, then, is a good method to use if R_x is relatively high and the current meter is a microammeter or low-scale milliammeter. Of course, with batteries of large current capacity and higher range current meters, such resistance measurements can be extended to quite low values of R_x. Of importance is the fact that the current drain be relatively light in terms of the battery capacity. The current meter shows its internal resistance, R_m. R_m must be known to perform this test procedure. The unknown resistance, R_x is obtained by subtracting R_m from the quotient E/I.

With the setup depicted in Fig. 1-17B, the value of an unknown resistance, R_x, can be determined when only a battery and a voltmeter are available. The voltmeter's internal resistance, R_m, is shown in shunt with an ideal meter (one with infinite resistance). Such a condition has been employed in earlier discussions and may appear not to square with the physical facts of voltmeter construction. Nonetheless, the electrical equivalent is valid. R_m is calculated by multiplying the sensitivity of the meter (i.e., the ohms-per-volt rating) by the full-scale indication of the range being used. For example, if the meter is a 20,000-ohms-per-volt unit and its range selector is on 10 volts, R_m is 20,000 × 10 or 200,000 ohms. The unknown resistance, R_x, is given by:

$$\frac{E1 \times R_m}{E2} - R_m$$

Here, E1 is the highest of the two voltage indications and E2 is the lowest.

In Figure 1-17C the internal resistance, R_m, of a current meter is determined from two settings of the calibrated resistance box. Notice that the battery voltage does not enter into the calculation. The important thing here is to be extremely cautious that the meter is not damaged. Initially, the resistance box should be set at its maximum resistance. Proceed slowly and get the "feel" of the technique. First, the resistance box is slowly decreased in value until an exact full-scale deflection is indicated on the current meter. In bringing this about, give thought to the incremental changes occurring in the resistance box. The resistance should be varied downward in small increments to avoid the danger of pinning the meter. When the full-scale deflection if obtained, record the amount of resistance in the resistance box. This resistance is R1. Next, add resistance to the box until an exact half-scale deflection is indicated on the current meter. The box resistance corresponding to the half-scale reading is recorded as R2. The internal resistance of the meter, R_m, is given by R2—2R1.

Resistance Measurement Using Constant-Current Supplies

A novel way to measure resistance is to use a constant-current source to supply the current, I, through an unknown resistance. The voltage, E, is measured across the resistance and the resistance is then equal to E divided by I. On the surface, this appears to be just an old-hat Ohm's Law application. However, two aspects of the technique make it worthy of special attention. First, the ohms scale is linear when the voltmeter is calibrated to indicate resistance. In a simple case, if the constant current is one milliampere, then the value of the unknown resistance is the voltage reading times one thousand. Such a setup is shown in Fig. 1-18A. Secondly, modern laboratory instruments now often include regulated power supplies with various degrees of sophistication. Most, moreover, can be adjusted for a constant-current as well as constant-voltage output. More costly types provide accurate programming of the value of constant current. The programming can be done with a resistance or a voltage. Also, the availability of FETs and constant-current diodes also tends to make this method of measuring resistance much more practical than it was some years ago when the concept was feasible but the implementation was difficult and hardly straightforward.

The arrangement shown in Fig. 1-18A makes use of a regulated power supply which can be operated in the constant-current mode. It should be realized that current-regulated sources are characterized by the so-called compliance voltage, which is the highest output voltage that can be provided when the output current remains constant. A viewpoint perhaps more related to the application under discussion is that the load resistance (R_x) can range from zero to a certain maximum resistance. This tends to restrict the use of this method to the measurement of resistances up to, say, 50,000 ohms. It is possible to design such a system to test considerably higher resistances, but the voltage and current required would, in most cases, involve impractical levels. In order to easily measure resistances less than several hundred ohms, a voltmeter with a fractional volt full-scale range is desirable, whereas resistances in the tens of ohms can be measured with a millivoltmeter. For all resistance ranges that can be accommodated, the linear scale calibration will prevail. Notice that despite practical limitations of range, this technique covers the resistance ranges which are generally found in bipolar transistor circuits. Also, the voltage impressed across the unknown resistance tends to be greater than that from an ohmmeter, but the current injection is much less than from most ohmmeters.

In Fig. 1-18B the constant-current supply is an FET. When connected in the configuration shown, the FET functions as a constant-current diode. The variable resistance in the source circuit provides an adjustment of the constant-current output. Although an n-channel FET is shown, a p-channel unit can be used if the battery and meter polarities are reversed. The basic idea is to use an FET with as high a transconductance as possible because this parameter governs the constancy of the output current. Also, a high transconductance will allow a wider range of current values to be selected by adjusting the source resistance. In any case, the higher the source resistance, the poorer the current regulation will be. Therefore, it is best to use as little source resistance as is consistent with the desired current output. It may be necessary to select FETs in order to obtain optimum performance.

Shown in Fig. 1-18C is a two-FET version of the circuit in Fig. 1-18B. This arrangement is capable of much closer current regulation and can

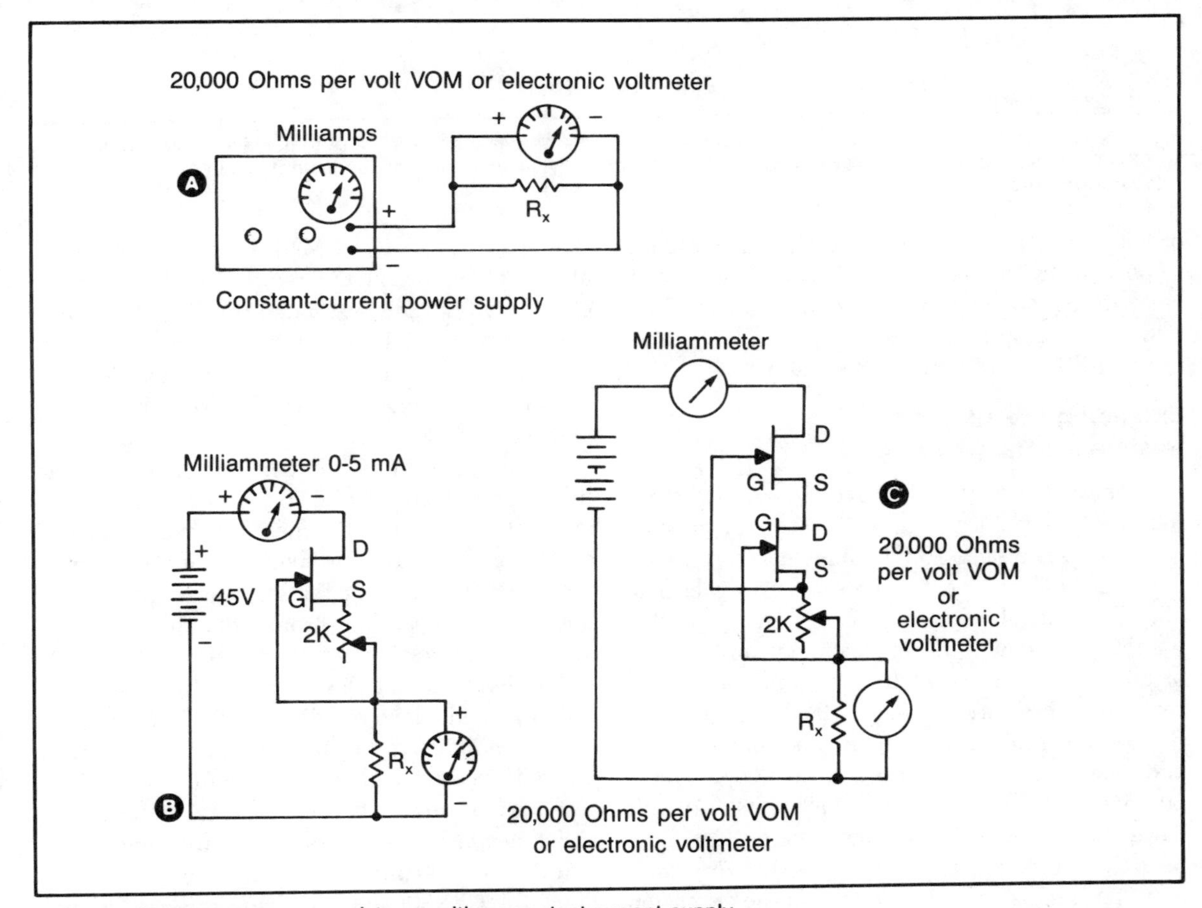

Fig. 1-18. Setups to measure resistance with a constant-current supply.

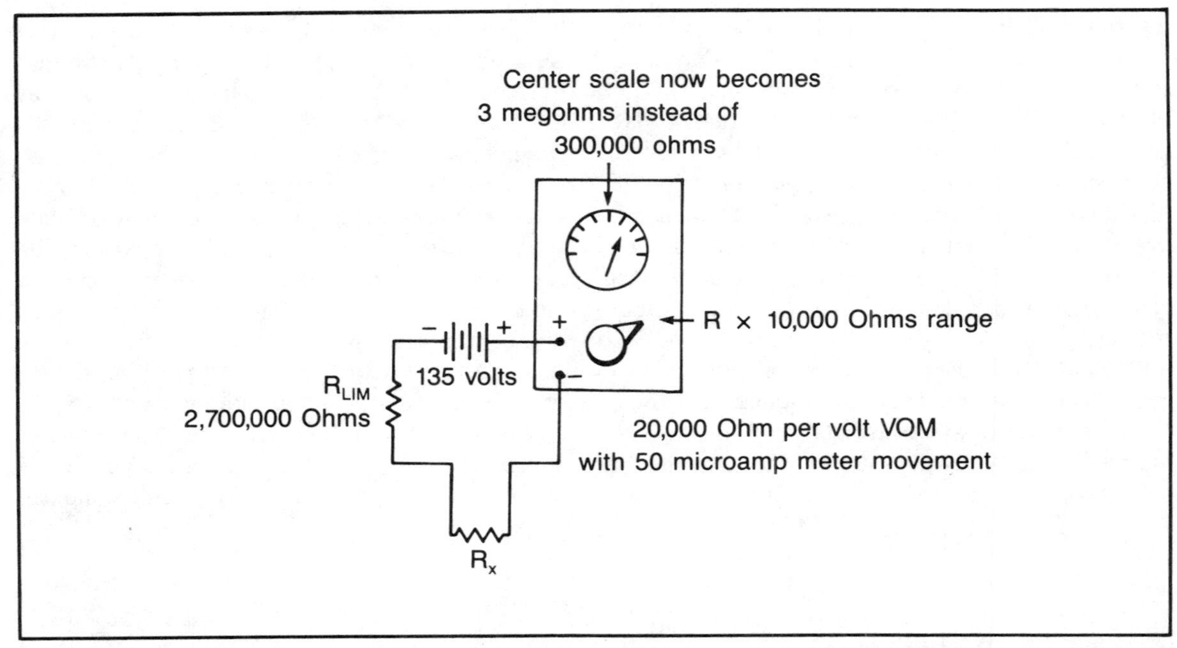

Fig. 1-19. Method for extending the high-resistance range in the ohmmeter of a VOM. A tenfold increase in range to R × 100,000 ohms is attained. However, this setup is not generally suitable for semiconductor tests because of the high voltage at the prods.

provide a greater selection of output currents than is possible with the single FET circuit. It also depends less on the transconductance of the FETs. However, for optimum results, it is still desirable to use FETs with a high transconductance.

Extending the High Resistance Range in VOMs

Many VOMs are limited in their high-resistance measuring capability by an R × 10,000 range. With the usual scale on a 20,000 ohms-per-volt instrument, the measurement of a resistance higher than about 10 megohms is not readily accomplished without rapidly diminishing accuracy, and with 40 or 50 megohms the measurement approaches the realm of guesswork. Figure 1-19 shows a method for extending this range tenfold. The added battery and resistance values apply to a VOM with a 50-microampere meter, and which uses a 15-volt battery for its R × 10,000 ohm range. (It so happens that meters of this description are very common, but the basic logic involved in the extension can be applied to all VOMs.) Although it might be all too obvious that the resistance-measurement capability can be extended by the technique depicted in Fig. 1-19, there is more than first meets the eye if you are to avoid the necessity of recalibrating the ohms scale. The reasoning is developed as follows:

In order for a 50-microampere meter to read full scale with an internal 15-volt battery, the total internal resistance of the meter circuit must be 15 divided by 50×10^{-6} or 300,000 ohms. (The zero-resistance indication on the scale corresponds to 50 microamperes passing through the meter.) We do not concern ourselves with how this 300,000 ohms is distributed among the meter coil, the limiting resistance, and the zero-adjust variable resistance. Rather, our objective is to **duplicate the current condition** producing full-scale deflection with the test prods shorted, but with the total meter circuit resistance being ten times greater than under normal R × 10,000 ohms operation. If we succeed in doing this, the 300,000 ohm half-scale reading will become 3 megohms. Advantage is taken of the fact that the half-scale reading of an ohmmeter represents the internal resistance of the meter circuit.

A little contemplation of these matters will show that two modifications must be implemented to achieve this objective. First, an external limiting resistance, R_{LIM}, of 3 megohms minus 300,000 ohms must be added in series with one lead. Secondly, in order to reestablish the former current condition, the total battery voltage acting in the meter circuit must now be ten times the former battery voltage, or 10 × 15 equals 150 volts. But since 15 volts is internally contained, we only have to add 135 volts of a properly polarized battery supply to the external circuit. This is exactly what has been done in Fig. 1-19. The polarities appear to be "bucking," but in most series-type meters, the positive-pin-jack corresponds to the **negative** pole of the internal battery. Therefore, the connections depicted produce the desired series-aiding arrangement of internal and external dc sources.

From what has been said, it is obvious that we have converted the R × 10,000 ohms range to an R × 100,000 ohms range without the need for making internal changes in the VOM, or for recalibration of the ohms scale! To implement this technique you should be sure that there is no possibility that the external 135-volt battery can accidentally become connected across the VOM terminals. External limiting resistance, R_{LIM} should be a 2,700,000-ohm, 1 percent, metallic-oxide type. A quarter- or half-watt rating will suffice to keep temperature effects negligible.

Using the Electronic Meter for Measuring Very High resistances

Figure 1-20 shows a useful test procedure for measuring unknown resistances in the gigohm ranges. With some meters, this can be extended into the tens of gigohms. The usefulness of this meth-

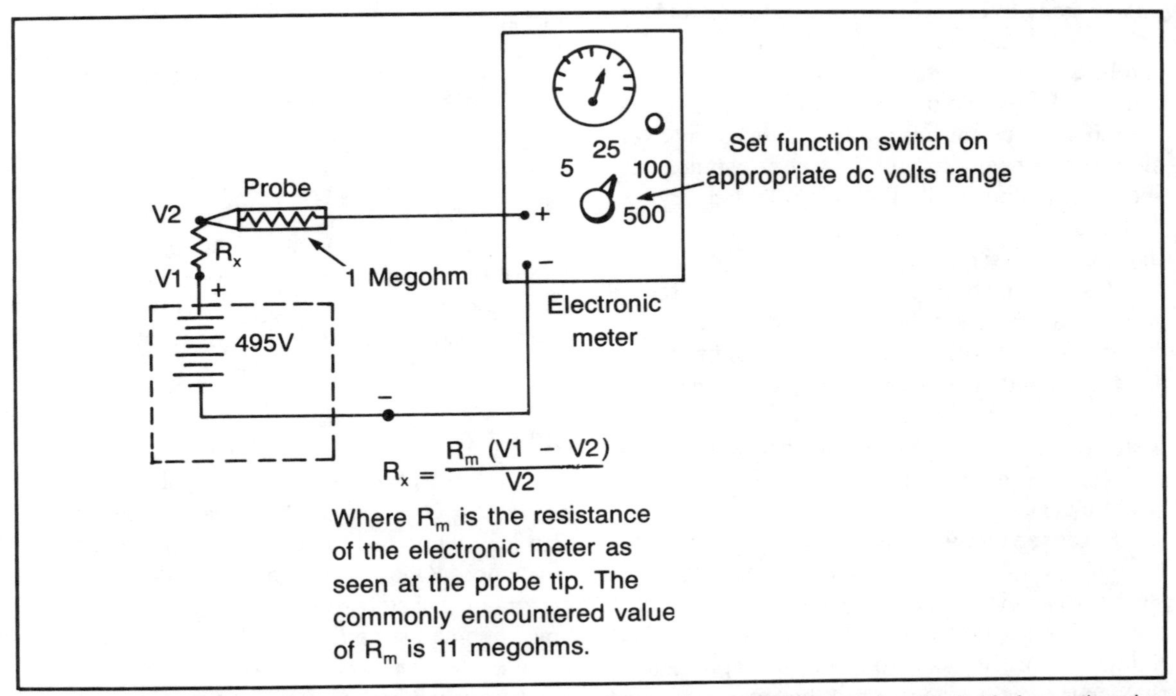

Fig. 1-20. Method for measuring very high resistances with an electronic meter. This setup is not suitable for semiconductor measurements.

od stems from the fact that the ohmmeter scale of most meters is not reliable for reading or estimating resistances beyond several hundred, or at most one gigohm. In order to get the most usefulness of the electronic meter when deployed in this fashion, the external power supply should develop about 500 volts. Voltages of this magnitude can be lethal, so the performance of this test should not be undertaken by anyone not versed in the techniques and safety precautions which apply to working with high voltages. Because of the danger involved, this test procedure is more appropriate to the activities of the technician and engineer in the R & D laboratory than to the serviceman. This is particularly true since such extremely high resistance measurements are not usually required in service and maintenance work. On the other hand, this voltage is comparable to those encountered in TV sets, transmitters, and other electronic equipment. Therefore, the setup in Fig. 1-20 does not present any unreasonable risk to those who habitually practice electrical safety. A 500-volt source can be obtained from five small 90-volt batteries and one small 45-volt battery all connected in series; alternately, eleven small 45-volt batteries can be series-connected. Read voltage V1 as quickly as possible.

To make use of this method of measuring very high resistances, the unknown resistance, R_x, should be supported by Plexiglas, Teflon, or other high-resistivity insulating material. All surfaces, those of the resistor itself and the supporting material, should be thoroughly clean and dry. It is only necessary to know the two voltages, V1 and V2, to compute the resistance of R_x. In the interest of safety, the 500-volt battery can be assembled from the tiniest units commonly available. The V1 measurement can be dispensed with, since it is being assumed that the terminal voltage is, for practical purposes, 500.

Because voltage V2 is measured on one of the low-voltage ranges, no inaccuracy need generally exist with a pointer indication which is a small fraction of full-scale deflection. Of course, as the highest limitations of the test procedure are approached, the accuracy must necessarily become somewhat uncertain. Electronic meters with fractional-volt ranges allow reasonably accurate measurements to be made throughout the tens of kilohm range because a substantial deflection is always possible for the V2 measurement.

As an example, suppose an 11-megohm electronic meter is available. Such a meter provides ten megohms internal resistance on all dc voltage scales, but because of the 1-megohm isolating resistor in the probe, the effective meter resistance, R_m, becomes 11 megohms. Suppose that the external dc supply is 495 volts, comprised of 11 small 45-volt batteries connected in series. This is close enough to 500 volts and should not introduce a significant error for this type of measurement. When the probe is in the V2 position, the voltage is found to be 2.0 volts on the 2.5-volt range of the meter. What is the value of R_x? Substituting in the equation,

$$R_x = \frac{R_m\ (V1 - V2)}{V2}$$

we have:

$$R_x = \frac{11 \times 10^6\ (500 - 2.0)}{2.0}$$

$$R_x = 11 \times 10^6\ (249) = 2739 \text{ megohms}$$

Despite the assumptions and approximations involved, the probability of reasonable accuracy is much greater than would ordinarily be obtained from an estimated reading in the tightly crowded end portion of the ohms scale.

QUESTIONS

1-1. Series-type and shunt-type ohmmeters are known to give substantially the same resistance indications when used to measure a wide range of resistances. One day, it is discovered that the two instruments are far from agreeing when the resistance of small lamp filaments are tested. Explain.

1-2. It has been determined that many useful ohmmeter tests can be made on a circuit board without removing resistors. The board also contains many silicon diodes. What characteristic should be specified for the ohmmeter in order to maximize the number of valid tests that can be made on the resistors?

1-3. An ohmmeter is being used to test a 1K resistor on a circuit board. A very low resistance is measured, but it is then observed that a silicon diode is connected across the resistor. Assuming the diode is good, what might be done other than disconnecting a resistor or diode load?

1-4. The voice coil of a dynamic speaker is tested for continuity on the R × 1 scale of an ohmmeter. Besides continuity, what other indication might be noted?

1-5. An ohmmeter is used to service a nonoperational unit. It is discovered that a meter movement, a tunnel diode, a thermocouple, and a tiny fuse are all defective. These components are replaced, but normal operation still does not return. Finally, it is found that a wire connection has an internal break that is not visible. Replacement of this lead results in normal operation. Explain the probable meaning of the defective components.

1-6. An oil impregnated capacitor is to be checked for leakage. A test lead is used to first short its terminals. This results in a luminous spark accompanied by a loud noise. The ohmmeter is than connected across the terminals of the capacitor. The pointer is pinned so violently that it is bent. Explain what has happened.

1-7. An apprentice technician has been doing well on an assignment where shipments of incoming capacitors must be tested for leakage with the ohmmeter function of a VOM. The capacitors are Mylar and ceramic types. It is found that a small percentage of these must be rejected because of excessive leakage. Later it is decided to change to solid-state tantalum capacitors. The apprentice, upon testing a number of these, discovers a high percentage of units with unacceptable leakage. In most of these rejects, the leakage resistance is a tenth or less of the average value of the acceptable units. What quick conclusion might be drawn from his report?

1-8. A Wheatstone bridge has been used to measure a number of 10K, 10 percent composition resistors. The objective is to separate them into three resistance groups: high, nominal and low resistance. The bridge appears to operate properly, but a routine maintenance check shows that the battery is weak. It is replaced by a fresh battery. Will it be necessary to repeat the measurements of the resistors?

1-9. The voltmeter-ammeter method is used to test a quantity of metal-film resistors which are suspected of containing a small percentage of incorrectly marked units. The resistance range is from several thousand ohms to ten megohms. A 45-volt B battery is used as a current source. The various milliampere and microampere ranges of a VOM are used for current measurement. Voltage is indicated by an electronic meter. The test procedure appears to be satisfactory for measurements approaching a megohm, but beyond this some kind of trouble is encountered. Indeed, no resistor above a computed value of about 5 1/2 megohms can be found. What is happening?

1-10. Series- and shunt-type ohmmeters make use of Ohm's Law to indicate the value of an unknown resistance. The constant-current method of determining resistance also is based upon Ohm's Law. Why do ohmmeters have a highly nonlinear scale as contrasted to the linear scale of the meter employed in the constant-current method?

ANSWERS

1-1. The different readings need not be related to the fact that one instrument is a series-type ohmmeter and the other is a shunt-type. More basically, it will be found that different model or brand meters provide different voltages and currents at the test prods. This is of no consequence in the testing of ordinary resistors. However, the resistance of a lamp filament is very dependent upon its temperature, which in turn is a function of the current passing through it. A tungsten filament increases in resistance as more current is allowed to flow through it. Meters with differing current capabilities will, therefore, give different resistance indications. However, each resistance is a "truthful" value in that it corresponds to the amount of current supplied to the lamp filament.

1-2. For this purpose, you need an ohmmeter with as many resistance ranges as possible, providing no more than approximately 0.3 volt at the test prods. The presence of nondefective silicon diodes will not affect resistance readings because virtually no conductance will exist in the diodes when the voltage is well below their threshold of forward conduction, which is in the vicinity of 0.6 volt. Such ohmmeters are generally electronic meters and are designed to provide this specific advantage. On the other hand, such an instrument suffers a disadvantage when you want to test the forward conduction of silicon diodes and transistors.

1-3. Reverse the test prods. This reverses the voltage applied across the diode, thereby inversely biasing it. Under this condition, you should be able to make a substantially accurate measurement of the 1K resistor.

1-4. The clicking or noise heard when the ohmmeter prods are connected and disconnected from the voice-coil terminals indicates that the voice coil moves.

1-5. The mentioned components are all of a delicate nature and are subject to damage or burnout from current injected by the ohmmeter prods. Most likely, these components were destroyed during testing before the broken lead was discovered.

1-6. Because of dielectric absorption, it is often difficult to dissipate the stored charge in capacitors. One shorting is not sufficient. The shorting technique should be repeated several times with a short, heavy section of copper wire. The short should not be momentary, but sustained for perhaps a half minute at a time. If the capacitor does not give up its charge, as attested by continued sparking, the technique of shorting may have to be extended in both number and duration. Capacitors differ greatly in this characteristic. But a single application of the shorting wire should never be considered safe.

1-7. The probability is that the apprentice ignored the fact that solid-state tantalum capacitors are polarized types and can be properly tested for leakage only if the test prod polarities coincide with the terminal polarities of the capacitors. In his previous experience with Mylar and ceramic capacitors, polarity could be ignored.

1-8. It is indeed wise to replace the weak battery. However, the null indication of the bridge is not voltage dependent, but is entirely a function of the resistances in the four arms of the bridge. A very exhausted battery would ultimately cause the null to be less sharp and accuracy would tend to be degraded because of the loss of balancing resolution. However, in the case described, it would be unnecessary to repeat the measurements. This is particularly true because the measurement objective obviously does not require laboratory precision.

1-9. The use of the electronic type voltmeter is strongly recommended for this method of resistance measurement. Such a voltmeter has a high internal resistance, usually about 11 megohms. Therefore, it imposes a negligible degree of circuit loading when measuring the voltage drop across resistors having values of a small fraction of 11 megohms. However, even the 11 megohms of the electronic meter is not high enough when we wish to extend the tests to the megohm range. For example, an attempt to measure a 10-megohm resistor produces totally erroneous results. The circuit "sees" the 10 megohm resistor and the 11 megohm resistance of the electronic meter as a parallel resistance equivalent of 5.24 megohms.

1-10. Ohm's Law is used in the ohmmeters in the following way: I equals E divided by R where I is the current flowing through the meter (and shunts), E is a source of essentially constant voltage from a battery or cell, and R is the unknown resistance. Thus, meter current is inversely proportional to the value of the unknown resistance. The inverse relationship yields a nonlinear scale of resistance values. For sake of illustration, let E equal 1.0 volt. Now, suppose we have resistors arranged in the sequence, 10 ohms, 20 ohms, 30 ohms, 40 ohms, 50 ohms, and 60 ohms. Notice that the spacing between any adjacent pair of these values is always the same that is, ten ohms. By Ohm's Law, the corresponding currents are, respectively, 100,50,33.3,25,20,and 16.6 milliamperes. It is readily seen that the spacing between the currents is non-uniform. This results in a nonlinear scale.

In the case of the constant-current method, the meter actually indicates the voltage developed across the unknown resistor. Thus, we have E equals I × R. For the sake of illustration, let I be constant at 100 milliamperes. Then, for the same sequence of resistors, the corresponding voltages are 1.0,2.0,3.0,4.0,5.0, and 6.0 volts. It is readily seen that the spacing between the voltages is uniform. This results in a linear scale. In this case, meter voltage is directly proportional to the voltage drop across the unknown resistor. The direct proportion relationship yields a linear scale of resistance values.

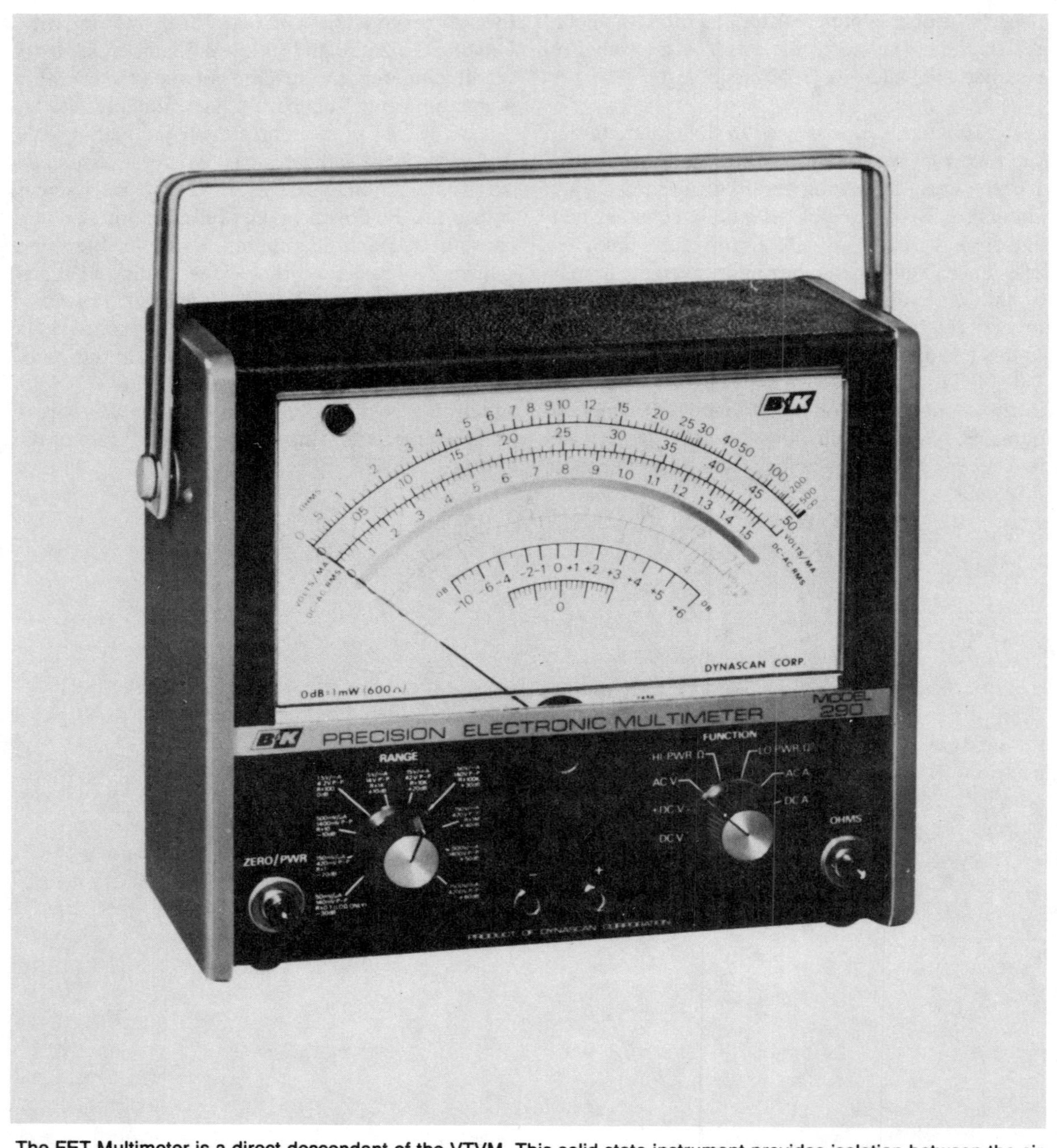

The FET Multimeter is a direct descendant of the VTVM. This solid-state instrument provides isolation between the circuitry being tested and the meter providing the analog readout. Thus, measurements are relatively undisturbed by the instrument itself. Courtesy of the Dynascan Corporation.

Chapter 2
Voltage and Current

Voltage and current measurements are two of the main activities used in designing, servicing, and evaluating electronic circuits. Other circuit parameters, such as resistance, impedance, transconductance, etc., are introduced for the basic purpose of causing specific voltages and currents to exist. Consider, for example, the biasing network in the base circuit of a silicon transistor. To operate the transistor as a Class A amplifier, the base-biasing resistors must apply a voltage in the vicinity of 0.6 volt to the base-emitter junction. If the transistor circuit is not amplifying a signal the first thing we check is whether or not the correct voltage is applied to the base-emitter junction and whether or not a reasonable bias current flows into the base. If the base-emitter voltage is too low, the transistor cannot function as an active element. If an excessively high voltage is applied to the base-emitter junction, the transistor will be at or near saturation, and can at best provide highly distorted amplification. Closely allied with the base-emitter voltage and current is the collector voltage and current. Unless these four quantities are right, various troubles will affect the operation of the transistor as a linear amplifier of small signals. The most direct way of evaluating the basic operating conditions in a transistor stage is to measure these voltages and currents.

The simplest voltage and current measurements—those which involve direct current only—are not necessarily trivial procedures. To begin with, the insertion of a current-indicating meter may give a true indication of the current flowing through it, but the circuit current without the meter may be appreciably greater. This is because of the inherent resistance of the microammeter, milliammeter, or ammeter. The meter resistance decreases the flow of current just as would an equivalent resistor. The effect of such a circuit disturbance may be anything from negligible to the most important factor in current flow. It all depends upon the relationship of the meter resistance to the circuit resistance. This point is illustrated in the circuit shown in Fig. 2-1. A milliammeter with an in-

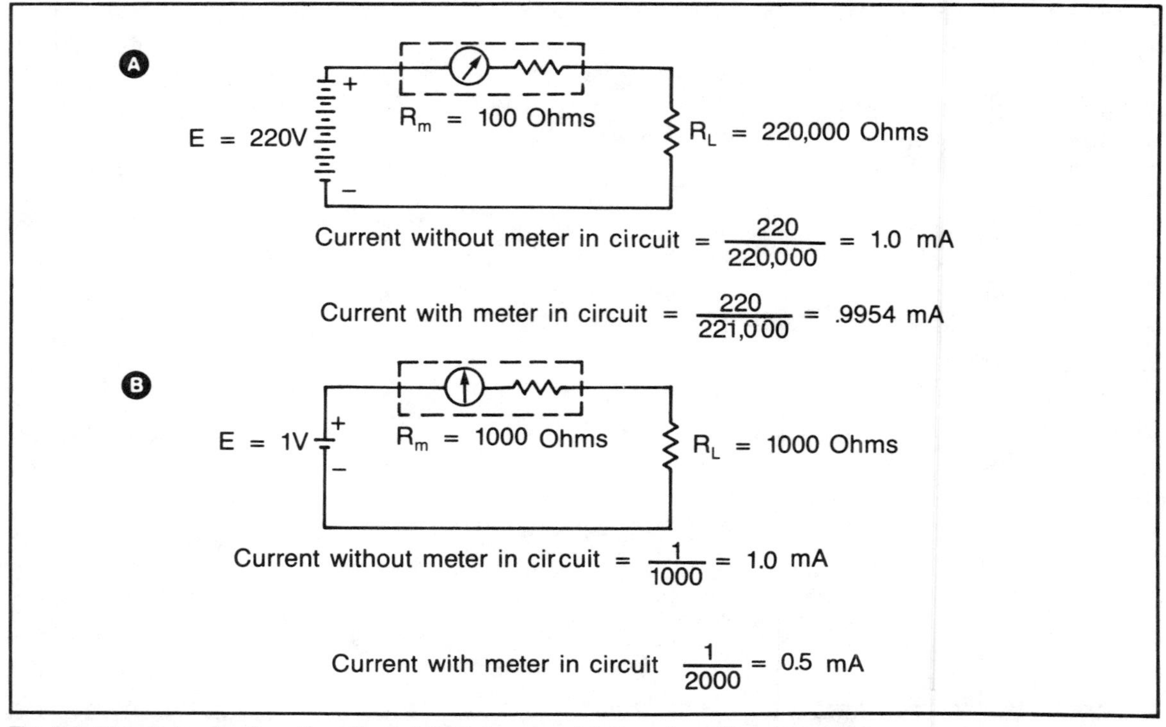

Fig. 2-1. These circuits show the effect of the internal resistance of a milliammeter while measuring current.

ternal resistance, R_m of 1000 ohms is used to measure the current flowing through the load resistance, R_l in the two circuits, A and B. In circuit A, the disturbance of the milliammeter is negligible, but in circuit B the insertion of the meter must reduce the current through R_l to one half the current that would flow without the meter in the circuit. In circuit B, the meter reading is accurate—the current through load resistance R_l is indeed one-half milliampere. However, we have a situation we did not bargain for, namely that the insertion of the meter has changed the circuit conditions. In terms of our initial objective, to measure the circuit current with a one-volt source and a 1000-ohm load resistance, the indication on the milliammeter can lead to an unacceptably erroneous conclusion.

Since real current meters always have an internal resistance, we can only go so far in reducing the type of error just demonstrated. A more costly instrument may have a much lower resistance than an inexpensive one and will have other features which enable more accurate measurements to be made. However, there will always be circuit conditions in which the lowest resistance that can be designed into a current meter will significantly disturb the circuit under test. In order to avoid the effect of such circuit disturbances you should become familiar with a simple concept and an easy-to-apply formula for correcting the indication of the current meter in circuits where its disturbance is appreciable. Naturally, it is always desirable to insert the current meter in the circuit and note the reading without anymore ado. Such a simple procedure may or may not be justified. The concept and formula we are about to investigate can quickly guide us in deciding whether or not a correction should be applied to the meter indication.

Referring to Fig. 2-2A, we see an apparently more complex circuit than those in Fig. 2-1. The additional resistances, R_a and R_b, have been added to make the circuit more like those you'll find in practical situations. However, the method of compensating for the error caused by the current meter is not complicated. For example, we need not deal with the voltage, E, at all. It will be assumed that after the reading, I_m is recorded from the current meter, the battery or power source is replaced by a short circuit. The important point now is that we have read I_m and know the meter's internal resistance, R_m. Our objective is to set up a simple relationship from which we can compute the current that would have been indicated if the meter had been an ideal one; that is, with zero internal

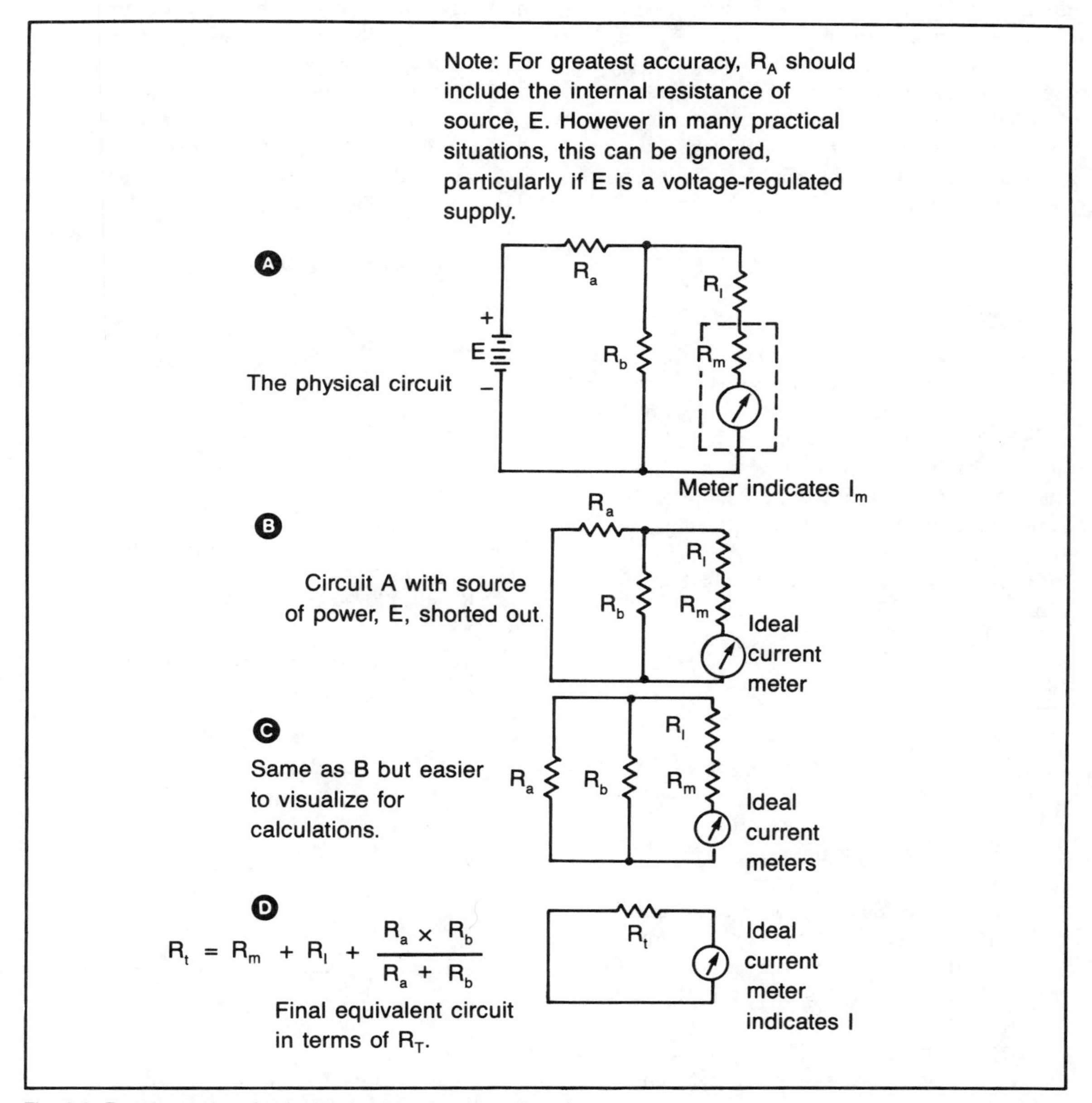

Fig. 2-2. Representative circuit showing how to correct I_m to I.

resistance. (Such an ideal meter would not disturb the circuit and would always indicate the true current flow in the portion of the circuit where it was inserted.)

The first step in this procedure is to redraw the circuit as in Fig. 2-2B. Notice that all we have really done is short out the dc power source. Now, the basic idea is to put ourselves in the place of an ideal meter, such as is represented by the circle symbol alone, and ask the question, "What total resistance do I see?" The answer for this circuit is R_m and R_l in series, and these, in turn, in series with the parallel combination R_a and R_b. Therefore, the value, R_t, of this network of resistance is:

$$R_t = R_m + R_l + \frac{R_a \times R_b}{R_a + R_b}$$

In Fig. 2-2C, the circuit is drawn in such a manner as to make the situation more obvious. Finally, when we have worked out the value of R_t, the circuit is simplified to that in Fig. 2-2C. The absence of the dc power source is of no consequence. Our objective has been only to derive a single equivalent resistance, R_t, which would be seen by an ideal current meter. We are now ready to make use of a formula which corrects the current reading, I_m. This formula is:

$$I = I_m \frac{(R_t)}{(R_t - R_m)}$$

I is the true, undisturbed current flowing through load resistance R_l.

I_m is the current through load resistance R_l, indicated by the current meter.

R_t is the equivalent resistance of the entire circuit as seen by an ideal current meter (one with zero resistance). R_t is calculated with the dc power source shorted out. Among the resistances seen by the ideal current meter is the internal resistance, R_m, of the actual current meter.

Let us summarize what has been done: In the circuit of Fig. 2-2A, the current through load resistance R_l was measured with a current meter. Because of the internal resistance R_m of the current meter, it is known that some error is introduced into the circuit when this measurement is made. The current indicated by the meter was I_m. I_m must be corrected to give I, the current which would have been indicated by an ideal current meter. Having read I_m, the rest of the procedure involves manipulation of the circuit on paper and the application of simple formulas. No further work need be done with the physical circuit.

Let's "plug" some practical numbers into the circuit in Fig. 2-2A. Let R_a equal 5000 ohms and R_b equal 2000 ohms. The meter resistance, R_m, is 500 ohms and load resistance R_l is 1000 ohms. When the circuit was set up, the current meter indicated 2.0 milliamps (ma). Although a certain voltage, E, was required to produce the indicated current, we no longer need concern ourselves with this voltage. Solving for R_t:

$$R_t = R_m + R_l + \frac{R_a \times R_b}{R_a + R_b}$$

$$R_t = 500 + 1000 + \frac{5000 \times 2000}{5000 + 2000}$$

$$R_t = 1500 + 1428 = 2928 \text{ ohms}$$

Now we have enough data to calculate the true current, I, in milliamperes:

$$I = I_m \left(\frac{R_t}{R_t - R_m}\right)$$

$$I = 2.0 \left(\frac{2928}{2928 - 500}\right)$$

$$I = 2.0 \left(\frac{2928}{2428}\right) = 2.41 \text{ milliamperes}$$

We see that we have corrected an appreciable error. A second look at the arithmetic shows that a negligible error would have existed if the meter resistance had been, say, 50 ohms. On the other hand, the error would have been negligible with the 500-ohm meter if the load resistance, R_l, had been much higher, say, 30,000 ohms or more. Still another quick observation indicates a reduction in error as R_a and R_b are increased in value. (A method for safely measuring the internal resistance of sensitive meters is described in Chapter 1.)

The technique just described for correcting errors introduced by a current meter resistance is directly related to a commonly used procedure for determining current in a circuit. Suppose that it is desired to monitor the current flowing through R_l in Fig. 2-3. If an appropriate current meter is not available, the current can be calculated by measuring the voltage drop across a small resistance, R_s, inserted in series with R_l as shown. R_s has the same effect on the circuit as did R_m in the previous discussion dealing with current meter resistance. Therefore, the same concept and the same formulas apply to ascertain just how low R_s should be. It is only natural to suppose that the lower R_s is, the better. However, if R_s is decreased too much beyond whatever value results in negligible disturbance, it may not prove practical to measure the voltage drop across it.

The voltmeter should have an internal resistance of 50 and preferably 100 or more times that of R_s.Otherwise, the voltmeter reading must be corrected before applying Ohm's Law to calculate current from the value of R_s and the voltage developed across it. The nature of this technique is such that is is not difficult to obtain this relationship if enough current is available to flow through R_s. In certain circuits where both R_s and the current through it are inordinately small, the voltage drop across R_s will be too small to be read on ordinary dc voltmeters. In such cases, a dc oscilloscope, an electronic meter, or some form of dc amplification must be provided. In any event, R_s should be evaluated by the method described for meter resistance. The trick is to have R_s high enough to produce a measurable voltage drop, but low enough to have a negligible limiting effect on the current through R_l.

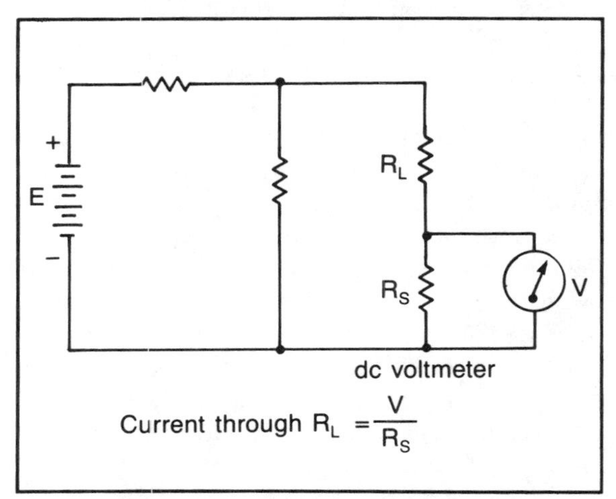

Fig. 2-3. A voltmeter and a small resistance can be used to measure current.

THE IMPERFECT VOLTMETER

What has been demonstrated for the current meter also applies to a voltmeter. It would be most naive to suppose that connecting a voltmeter across a source of voltage will necessarily indicate the voltage that actually exists when the meter is not connected. It is true that it is a common practice to unthinkingly apply the test prods from a voltmeter to a circuit and blindly assume that an accurate or even usable measurement has been made. Admittedly, you can achieve a fairly good batting average in certain types of work with such a simple procedure. But this is a case of pure luck, where the relationship of the voltmeter characteristics to those of the circuit are sometimes favorable. We must, however, ultimately pay for an unawareness of the actual possibilities, for highly erroneous indications are also possible.The ability to know when to expect error and how to correct for is, indeed, important.

The dc voltmeter is less than perfect because of its natural resistance. Contrary to the ideal current meter, we want the ideal voltmeter to have *infinite* resistance. Because practical voltmeters have finite resistances, they necessarily consume current from the voltage source being measured. Such a loading effect tends to lower the voltage across this source. The indicated voltage on the meter is true

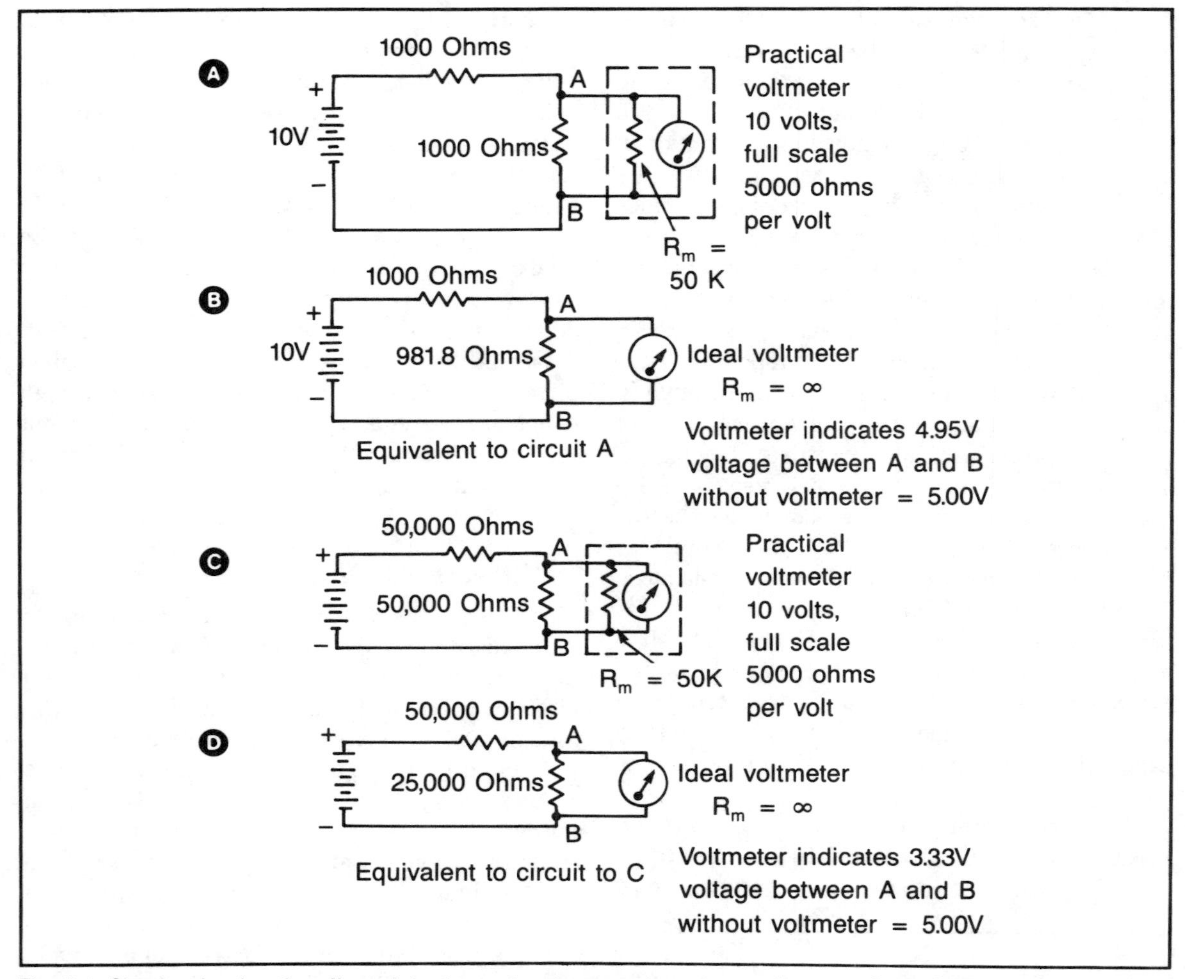

Fig. 2-4. Circuits showing the effect of the internal resistance of a voltmeter in measuring voltage.

in the sense that it is the voltage that now exists in the portion of the circuit being monitored. But the presence of the meter has altered the circuit by adding its internal resistance to it. The question is simply whether the circuit disturbance has been negligible or appreciable.

In Fig. 2-4, the same 5000 ohms-per-volt meter is used to measure voltage in two circuits. In circuits A and C, the potential across points A and B is 5 volts without the voltmeter connected. This is obvious, because in both circuits a 10.0-volt source feeds a potentiometer arrangement which divides the applied voltage exactly in half. The voltmeter is shown with its internal resistance, R_m, in parallel with its terminals. This is a valid representation, for it is certainly the situation that exists in the circuit when we connect the voltmeter. (The fact that this comes about via a multiplier resistor connected in series with the moving coil, which also has resistance, is not of consequence when looking at the terminals of the voltmeter. At the terminals, R_m may be said to be in parallel with an ideal moving-coil instrument having infinite resistance.)

Circuits A and C, together with their equivalent

circuits, show that the voltage indicated in circuit A is substantially correct, whereas the voltage indicated in circuit C is seriously in error. Realizing the reason for this drives home the fact that you can easily imagine situations where the errors in the indicated voltage readings would be very high. On the other hand, it is clear that a voltmeter with a *higher* internal resistance would have produced less error. For example, suppose we had used a commonly available 20,000 ohms-per-voltmeter with a full-scale indication of 10 volts. R_m in such a meter is 200,000 ohms and this paralleled with the 50,000-ohm resistor between points A and B in circuit C produces an equivalent resistance of 40,000 ohms. This meter will then indicate the voltage:

$$10 \frac{(40,000)}{(40,000 + 50,000)} = 4.44 \text{ volts.}$$

In order to more closely approach the correct voltage of 5.00 volts, a meter with still greater sensitivity is needed; that is, a meter with an even higher internal resistance. In a pinch, it often proves wise to make the measurement on a higher than otherwise necessary scale. For example, if the 5000 ohms-per-voltmeter initially used in circuit C had been switched to its 50-volt scale, R_m would have been increased to 250,000 ohms with obviously beneficial results. (Of course, a meter is not used advantageously at the near-zero end of its scale and we cannot put too much reliance on this technique without introducing another source of inaccuracy.

It is obvious that a high internal resistance in a voltmeter is a virtue and we lessen the probability of obtaining erroneous indications if the voltmeter resistance *greatly* exceeds the equivalent circuit resistance. For this reason, electronic voltmeters are very desirable. Usually, they provide a fixed resistance of 11 megohms on all scales. Extremely sensitive moving-coil instruments are also available. For example, one of the claims of a maker of a 200,000 ohms-per-volt moving-coil meter is that the resistance exceeds that of most electronic meters when measuring dc voltages above 55 volts.

Correction Procedure

Our objective is to set up a simple relationship from which we can compute the voltage that *would have been indicated if* the voltmeter had been an ideal one; that is, with infinite resistance. Such an ideal meter would not disturb the circuit and would always indicate the true voltage across two circuit points.

The first step in the procedure is to redraw the circuit in Fig. 2-5A as shown in Fig. 2-5B. Notice that all we have really done is short out the voltage source. Now, the basic idea is to determine the resistance seen by an ideal voltmeter, such as is represented by the circle symbol alone. The resistance for this circuit is R_a, R_l, and R_m in parallel. Therefore, the value, R_t, of this network of resistance is:

$$R_t = \frac{1}{\frac{1}{R_a} + \frac{1}{R_l} + \frac{1}{R_m}}$$

which can also be written,

$$R_t = \frac{R_a \times R_l \times R_m}{R_l \times R_m + R_a \times R_m + R_a \times R_l}$$

In Fig. 2-5C, the circuit is drawn in such a manner as to make the situation more obvious. Finally, when we have worked out the value of R_t, the circuit configuration simplifies to that in Fig. 2-5D. The absence of the voltage source is of no consequence. Our objective has been only to derive a single equivalent resistance, R_t, which would be seen by an ideal voltmeter. We are now ready to make use of a formula which corrects the voltage reading V_m, which was initially made. This formula is:

$$V = V_m \frac{(R_m)}{(R_m - R_t)}$$

V is the true, undisturbed voltage across resistance R_l.

V_m is the voltage across R_l indicated by the voltmeter.

R_t is the equivalent resistance of the entire circuit as seen by an ideal voltmeter (one with infinite resistance). R_t is calculated with the battery voltage source shorted out. Among the resistances seen by the ideal voltmeter is the internal resistance, R_m, of the actual voltmeter.

Notice that the basic approach is similar to that used to correct the indication of the imperfect current meter in the preceding discussion.

Let's substitute some numbers into the circuit in Fig. 2-5A. Let R_a be 100,000 ohms and R_l be 100,000 ohms. The voltmeter sensitivity is 20,000 ohms per volt and its full-scale voltage is 10 volts. Therefore, R_m is 20,000 × 10 or 200,000 ohms. When the circuit was set up, the voltmeter indicated 8.0 volts. Although a certain battery voltage, E, was required to produce the indicated voltage, we no longer need concern ourselves with voltage E.

Solving for R_t:

$$R_t = \frac{R_a R_l R_m}{R_l R_m + R_a R_m + R_a R_l}$$

$$R_t = \frac{1 \times 10^5 \times 1 \times 10^5 \times 2 \times 10^5}{1 \times 10^5 \times 2 \times 10^5 + 1 \times 10^5 \times 2 \times 10^5 + 1 \times 10^5 \times 1 \times 10^5}$$

$$R_t = \frac{2 \times 10^{15}}{5 \times 10^{10}} = 0.4 \times 10^5 = 40 \times 10^3 = 40{,}000 \text{ ohms}$$

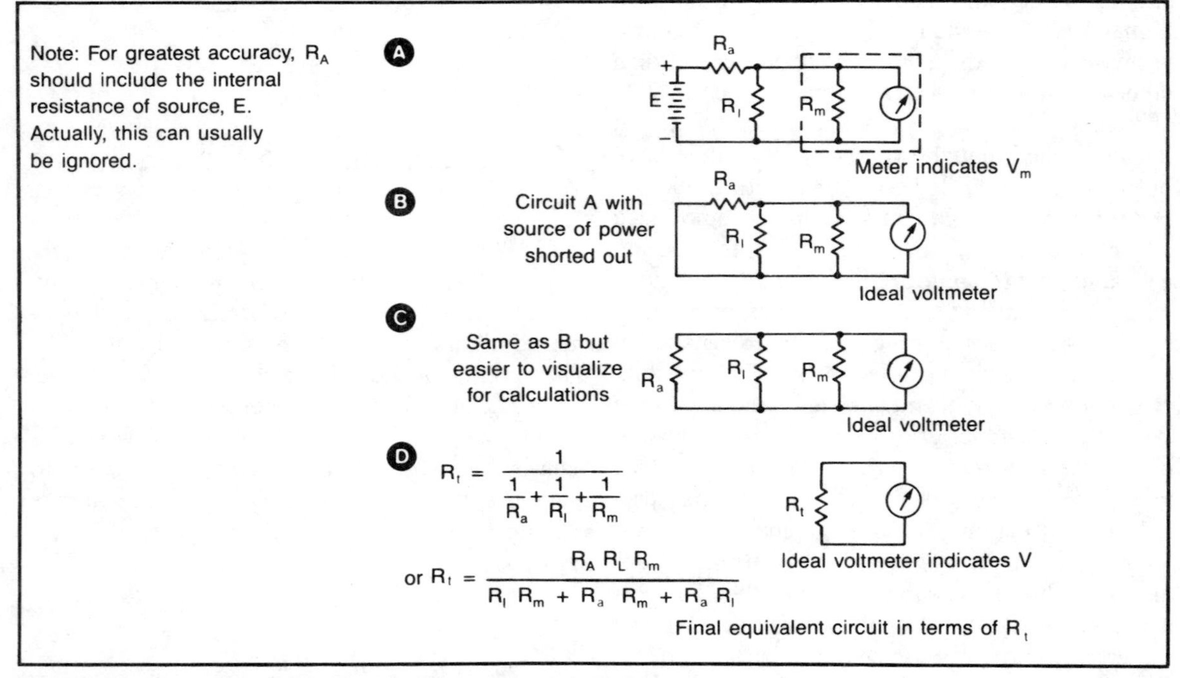

Fig. 2-5. Representative circuit showing how to correct V_m to V.

Now we have sufficient data to solve for the true voltage, V.

$$V = V_m \frac{(R_m)}{(R_m - R_t)}$$

$$V = 8.0 \frac{(200{,}000)}{(200{,}000 - 40{,}000)}$$

$$V = 8.0 \frac{(200{,}000)}{(160{,}000)} = 8.0 \frac{(5)}{(4)} = 10.0 \text{ volts.}$$

We see that we have corrected an appreciable error. A second look at the arithmetic shows that a negligible error would have existed had we used an electronic meter with an internal resistance of 11 megohms. On the other hand, the error would have been negligible with the 20,000 ohms per volt meter if R_a or R_l has been in the vicinity of, say, 5000 ohms. Notice that in this circuit, the meter resistance could actually be lower than R_l providing that the resistance of R_a was very low (a small fraction of R_m). Under this condition there would also be negligible error.

Procedure to Demonstrate the Circuit Disturbance Caused by a Current Meter

Objective of Test: To show both negligible and serious circuit disturbances with the same current meter used under two different conditions.

Test Equipment Required:
0-10 milliampere dc meter
20,000 ohms-per-volt VOM
1/2-watt, 10 percent composition resistors: 39 ohms, 120 ohms, two 2700 ohms.
Two single-pole, single throw switches.
1.5-volt battery and a 45-volt battery.

Test Procedure: First, if you can't tell by looking, use an ohmmeter to identify the on and off positions of the switches. Place the switches in their off positions. Wire the circuit shown in Fig. 2-6A. Place switch SW1 in its on position. Make sure that an up-scale indication is displayed by the milliam-

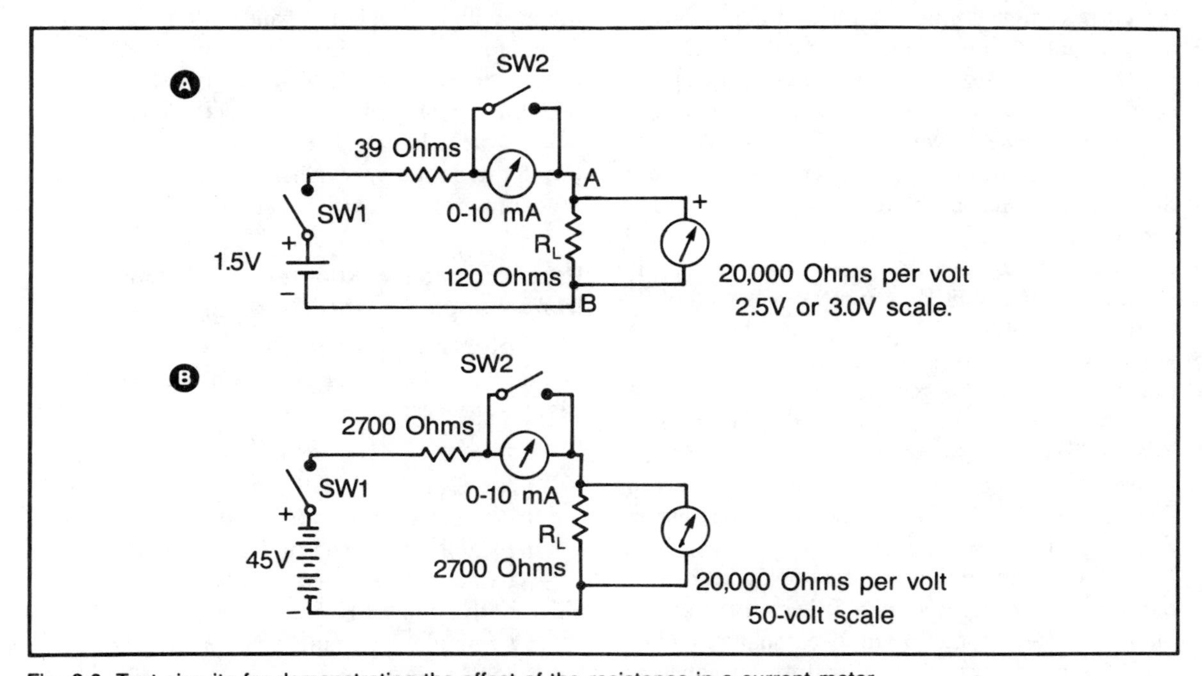

Fig. 2-6. Test circuits for demonstrating the effect of the resistance in a current meter.

meter, but its reading need not be recorded. Record the reading of the voltmeter. Next, place switch SW2 in its on position, thereby shorting out the milliammeter. The reading on the milliammeter should now become essentially zero. Record the new reading from the voltmeter. Terminate the experiment by opening switch SW1. The fact that the voltmeter substantially changed its indication clearly demonstrates that the insertion of the milliammeter, as would be done for the purpose of measuring the current in this circuit, results in a large disturbance in the circuit.

Next, wire the circuit shown in Fig. 2-6B. Follow the same procedure as with circuit A.This time you will see that the voltmeter indicates substantially the same reading for both positions of switch SW2. This clearly shows that the milliammeter maybe inserted in this circuit for a current measurement with a negligible disturbance of the circuit conditions.

Comments on Test Results: This test procedure is not used to derive exact data, but it is quite evident that care has to be exercised in making current measurements. The voltage indicated across points A and B depends on the current flowing through resistor R_1. If this voltage changes, it suggests that the current through R_1 has changed. If the milliammeter were an ideal type with zero resistance, its presence in the circuit would not produce the voltage change across R_1. The significance of this test procedure is that although circuits A and B have the same configuration, the parameters of circuit B permit the insertion of the 0-10 mA meter with negligible disturbance, whereas a gross upset is produced by the same meter in an attempt to measure current in circuit A.

The circuit disturbance observed for circuit A is actually due to two causes. First, the internal resistance of the current meter reduces the current flow through the 39-ohm and the 120-ohm resistors. Secondly, the insertion of the meter resistance alters the voltage dividing ratio of R_1 with respect to the total circuit resistance. It is as if the adjustment on a potentiometer had been changed. (This could be demonstrated by substituting a constant-current power supply in place of the 1.5-volt cell. Such a supply would maintain the same current in R_1 whether the milliammeter was in the circuit or was shorted out. Nonetheless, we would observe a change in the voltmeter as the result of actuating switch SW2.

This test procedure demonstrates somewhat indirectly that the resistance of the current meter may have to be reckoned with. We did not attempt to obtain the true current, that is, the current in the circuit with the meter shorted out. This can be done and is explained in the general discussion for making current measurements.

In circuit B, the same upsetting tendencies exist as in circuit A, but to a negligible degree. The behavior of circuit B shows that a milliammeter can be inserted for the purpose of direct measurement of current. The current indicated on the meter is the true circuit current and no corrective computations are necessary. A test procedure requiring correcting calculations is not always necessary. With a knowledge of the resistance of the current meter and of the general nature of the circuit, you develop the ability to sense when the effect of the current meter will be negligible and when the current reading must be modified to account for the effect of meter resistance. The meter resistance of the 0-10 mA movement was not specified because circuit A was deliberately designed to be unsatisfactory for even the most sensitive (lowest resistance) milliammeter ordinarily available.

Procedure to Demonstrate the Circuit Disturbance Caused by a Voltmeter

Objective of Test: To show both a serious and a negligible circuit disturbance caused by the same voltmeter used under two different conditions.

Test Equipment Required:

0-100 microampere dc meter

0-10 milliampere dc meter

1/2-watt, 10 percent composition resistors: 68K, 220K, two 1.2K.

Two single-pole, single-throw switches

9-volt battery.

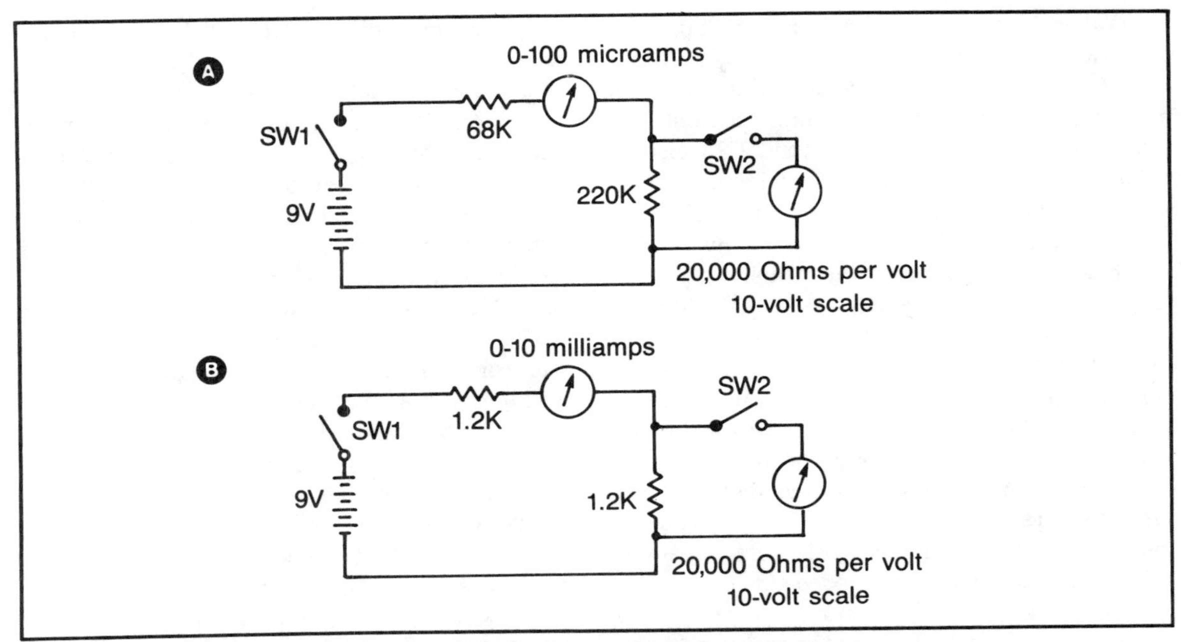

Fig. 2-7. Test circuits for demonstrating the effects of the resistance in a voltmeter.

Test Procedure: First, if it is not obvious from inspection, use an ohmmeter to identify the on and off positions of switches. Place the switches in their off positions. Connect the components as shown in Fig. 2-7A. Be extremely careful that the microammeter is not accidentally misconnected. Such a sensitive meter is very easily damaged. When the circuit is completed, place SW1 in its on position. Record the reading on the microammeter. Next, close switch SW2. An upscale indication should be obtained on the voltmeter, but this need not be recorded. Record the new reading on the microammeter. The fact that the circuit current was considerably changed by the voltage measurement clearly shows that the voltmeter produces a large disturbance in the circuit. From this, we deduce that the voltage reading cannot be accurate.

Next, wire the circuit shown in Fig. 2-7B. Follow the same procedure as with circuit A.This time you will see that connecting the voltmeter to the circuit results in a negligible change in circuit current. From this, we see that the voltage measurement does not disturb the circuit and the reading on the voltmeter is valid.

Comments on Test Results: This test procedure represents a quick way to ascertain whether or not a voltmeter reading can be accepted as such, or must be corrected because of a circuit disturbance. (The method of computing such a correction, when needed, is described in the general discussion.) It is interesting to note that a commonly available 20,000 ohms-per-volt meter was used for the voltage tests. It is often assumed that meters with this sensitivity rating allow voltage measurements to be made with almost reckless abandon. Yet the circuit disturbance produced by this meter in the high-resistance circuit in Fig. 2-7A is quite like conditions found in practical electronics. In vacuum tube and FET circuits, even higher resistance levels are common and very erroneous voltage readings are possible. On the other hand, the low-impedance circuit in Fig. 2-7B shows that the 20,000 ohms-per-volt meter can be used for making substantially accurate voltage measurements. Therefore, a large portion of the voltage measurements in bipolar transistor circuitry can be reliably made with the 20,000 ohms-per-volt meter.

When an appreciable circuit disturbance accompanies a voltage test, as in circuit A, our initial response must be to disbelieve the voltmeter reading. The correct voltage can be computed as explained in the general discussion. In circuits which are not simple, or which may not be accessible, such calculations may not be straightforward or practical. Often, the better approach is to use a more sensitive voltmeter. This usually implies an electronic meter. Contrasted to the 200,000 ohms internal resistance of the voltmeter used in this test procedure, the electronic meter generally provides 11 megohms of internal resistance on all voltage scales. The electronic meter tends to make voltage measurements more of a routine matter, for the likelihood of introducing any circuit disturbance is materially reduced. You should be ever mindful, however, that even the electronic meter is capable of circuit disturbance and erroneous readings if the circuit being tested contains resistance levels of several megohms or higher. On the other hand, some electronic meters have internal resistances very much higher than the commonly encountered 11-megohm types.

THE DC OSCILLOSCOPE AS A VOLTMETER

An oscilloscope with a dc amplifying capability can serve as a very useful voltmeter in various test situations. Not only is the input resistance of such scopes generally comparable to electronic meters, but the vertical amplifier permits measurements of a small fraction of a volt, frequently in the millivolt range. When using a dc oscilloscope to measure many voltages in complex circuits, you are relieved of the apprehension of damaging the instrument when probing test points. No matter what the gain or range setting of the oscilloscope, voltages up to several hundred and of either polarity can be contacted by the test probe without fear of damage. Nor are ac voltages harmful. This feature alone can make it worthwhile to measure dc voltages with the oscilloscope. Circuit loading tends to be slight as the input impedance generally ranges from 2.7 megohms to 27 megohms, depending on the model and on the probe used.

Another real advantage of the dc oscilloscope is the overall circuit information you get while measuring dc voltage levels. Intermittents, transients, and noise riding on the dc level will show up if they exist. In a large portion of electronic circuits ac voltages are superimposed on a dc level. Consider, for example, a transistor audio-amplifier stage. If the collector is monitored with a dc scope, you see the no-signal quiescent voltage; that is, the collector operating point. If a signal is applied at the input of the amplifier, the amplified ac signal is shown superimposed on the dc collector voltage. As the input signal is increased in amplitude, you can see the amplified collector signal as it approaches its peak-to-peak limit, because at either saturation or cutoff, the peaks begin to flatten. In the distortion region, the scope displays the signal in such a way that the nature of the trouble can be observed and the cause of it determined. Even if both ac and dc meters were connected in the collector circuit, it would require some plain guesswork to understand the way in which the amplifier was operating. Not only does the dc oscilloscope reveal both dc and ac parameters, but it clearly shows the interaction between the two. For example, the overdriven condition mentioned can shift the dc bias level at the base. With the probe on the base, a glance indicates the situation. Also, the scope can often reveal the presence of such hard-to-identify conditions as parasitic oscillation and stray signal pickup.

It is well to realize that not just any dc oscilloscope is suitable as a dc voltmeter. Many of the older models, as well as some of the inexpensive varieties, do not have stable dc amplifiers. This causes the sweep line, i.e., the "zero-setting," to drift. This drift can be quite severe, and it not only degrades the accuracy of measurements, but can lead to altogether erroneous test results. It should not be left unsaid that the response of the scope should be linear on all ranges.If you have the opportunity to select from a number of brands and models, attention should be also given to the bandwidth. For making exclusively dc voltage measurements, it is true that the bandwidth could be very narrow, even a few hundred or thousand hertz. However, in order to be able to see the additional characteristics of circuit operation, a wide bandwidth is desirable. In some laboratory scopes, the bandwidth is a function of vertical gain. If millivolt dc levels are being measured, the bandwidth is narrow. It becomes progressively wider as the range switch is turned up to volts, tens of volts, and hundreds of volts.

When evaluating a dc scope, short the probe to ground and set the vertical gain control and range switch for maximum sensitivity. Adjust the sweep line to the center position after the instrument has had time to warm up. This might be on the order of 20 minutes or so with a vacuum-tube type; five minutes should suffice for a scope using solid-state circuitry. If the sweep line is constantly in a state of vertical drift, such an oscilloscope had best not be employed as a dc voltmeter, at least not at the sensitivity setting where the instability prevails. A good quality solid-state scope with dc capability can be a joy to use. The ability to display voltages of either polarity together with the immunity to damage from overload, make certain test programs speedy and decisive.

When using a dc scope as a voltmeter, the sweep control is set at AUTOMATIC and the sync control is set at INTERNAL. This is because we want a steady horizontal line across the screen. (Older instruments, as well as the inexpensive types, are always in automatic sweep operation, since such scopes do not provide the triggered sweep option.) The sweep speed or sweep frequency control adjustment is not critical. The frequency or repetition rate selected by this control should not be so slow that the development of the horizontal sweep can be seen or there is noticeable flicker in the intensity of the line. On the other hand, it should not be so fast that the desired brightness cannot be attained. Between these two limits, there are a number of available sweep speeds that can be used satisfactorily. (Make certain that the scope is not in its triggered sweep mode or the horizontal line will not be present. This admonition applies to certain laboratory instruments.) The horizontal line is focused, and is vertically centered on the screen with the vertical position control. Except for calibration, the oscilloscope is now ready to function as a dc voltmeter. A positive voltage at the probe moves the baseline upward from its center position. A negative voltage at the probe moves the baseline downward from its center position (Fig.

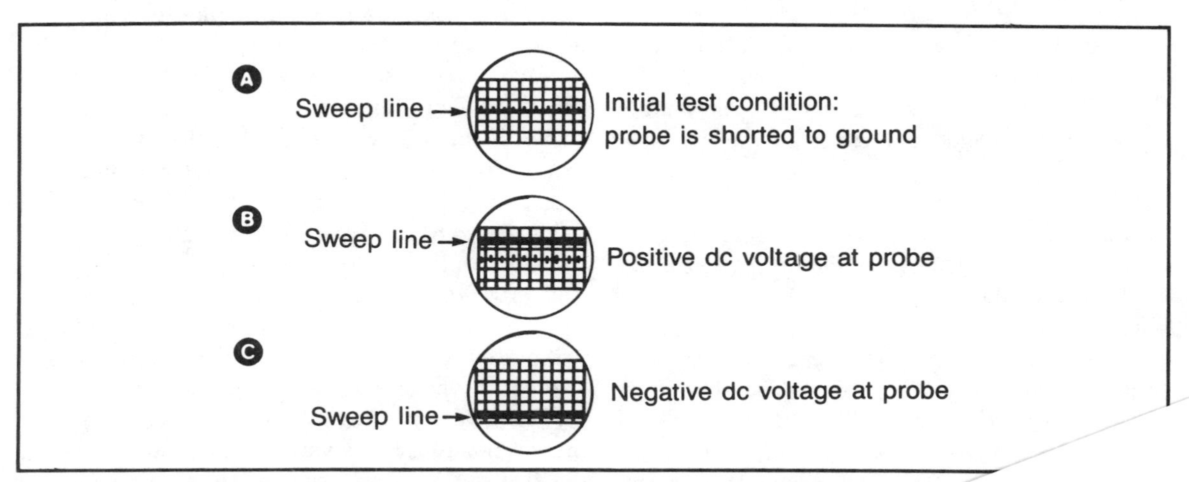

Fig. 2-8. The dc oscilloscope can be used as a voltmeter. The sweep line is initially set on the the graticule. (A) The probe or test prod is shorted to ground. A positive dc voltage deflects the a negative dc voltage deflects it downward (C).

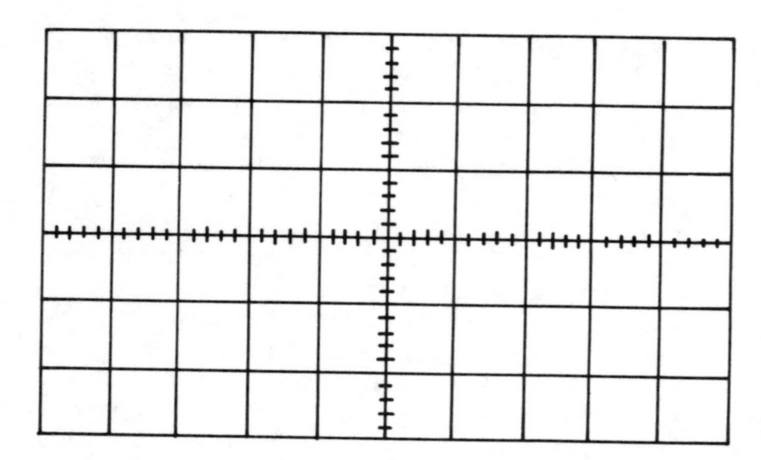

Fig. 2-9. Drawing of a typical oscilloscope graticule. Major divisions may represent centimeters or inches. Each major division is divided into five parts.

2-8). It is only necessary to know how many volts per division the screen graticule represents (Fig. 2-9).

In order to measure voltages precisely from the deflection on the screen, some calibration technique is usually necessary. We must know the vertical sensitivity of the oscilloscope in volts per graticule division. When this is known, the value of the dc level, indicated by the movement of the sweep line from the center position on the screen, can be read on the screen graticule. However, the vertical sensitivity changes with the setting of the vertical gain potentiometer. Therefore, some calibration technique is necessary. Generally, a known reference voltage must be used to establish the volts per division sensitivity at the setting of the vertical gain control for which a measurement has been made. Voltage calibrators are available for this purpose. Such devices provide a range of dc voltages which can be read from dials. Calibrations are often made for specific settings of the vertical gain potentiometer and recorded in chart form. Then it is not necessary to go through the calibration procedure every time a test is made. If the precalibration involves a sufficient number of settings of the vertical gain control, it is easy to find one such setting which will permit a reasonable display of the dc voltage level during the test procedure. When making such readings from the screen of the oscilloscope, do not neglect to multiply reading by factors required by the probe and the vertical gain attenuator on the panel of the scope.

The voltage calibrator may deliver either sustained dc levels or square waves. When square waves are provided, the indicated calibrations are peak-to-peak amplitudes. If the sweep frequency of the oscilloscope is made much higher than the repetition rate of the square-wave calibrating signal, the display on the oscilloscope will consist of two parallel horizontal lines. This type of display is often found to be particularly convenient for calibration (Fig. 2-10).

Many oscilloscopes have a self-contained square-wave calibrator. Generally, this is a 1-Hz multivibrator associated with a multiposition attenuator. The attenuator is panel-mounted and provides a number of amplitude-calibrated square waves from a connector which is also mounted on the panel. Usually, a potentiometer is included which provides a continuous, but a noncalibrated variation of the square-wave amplitude between attenuator steps. Oscilloscopes which are designed primarily for TV servicing often provide a single calibrated voltage reference. This usually is a 1.0-volt p-p square wave at the power line frequency. Here, you must rely on the accuracy of the input attenuator for the vertical amplifier. However, for TV service work, such calibration appears to be satisfactory.

There are also oscilloscopes in which the vertical gain control is factory calibrated. When a

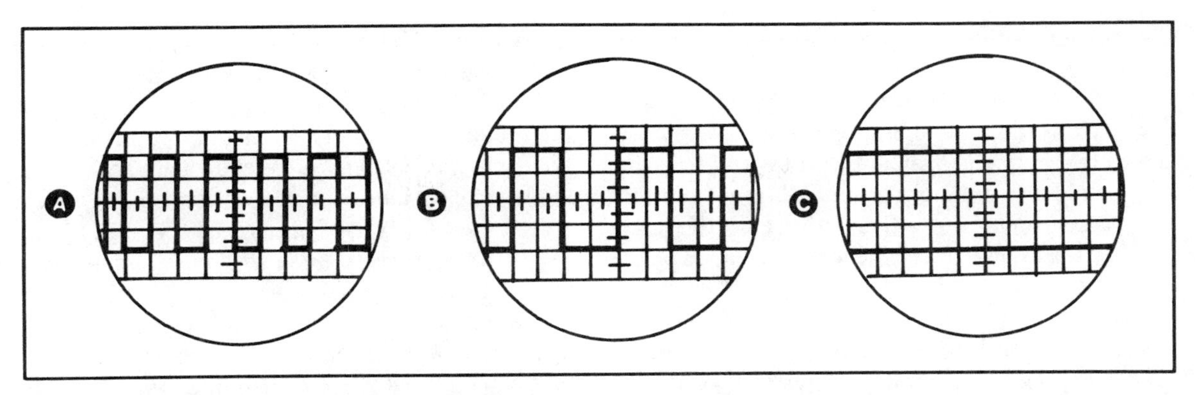

Fig. 2-10. The oscilloscope can be calibrated with a square wave. In A, the sweep frequency is approximately one-sixth of the repetition-rate of the square wave. In B, the sweep frequency has been made higher. In C, the sweep frequency has been made higher yet, yielding the convenient pattern of two parallel lines.

signal, whether it is an ac wave or the deflection of the base line, is displayed, its amplitude is quickly determined by the sensitivity which is indicated by the vertical gain control. Such a provision simplifies the test procedure and allows rapid evaluation of a display. There are some who would question the accuracy of this technique, but it certainly can be adequate for general test procedures in electronics work. This is particularly true if the oscilloscope is a quality, solid-state instrument. Although all oscilloscopes use step attenuators calibrated for vertical gain, there has been a reluctance to calibrate the associated vertical gain potentiometer. Now, more scopes are being provided with this feature. If a good quality linear potentiometer and a dial with mechanical accuracy are used, the scheme is entirely suitable.

Some sophisticated laboratory oscilloscopes provide a digital readout of the input signal. Such instruments are very expensive and are not commonly available. However, their existence is worthy of mention, for it has frequently happened that instrument features which appear to be economically out of reach for general use today become commonplace at a modest cost tomorrow. The triggered sweep is an example.

The calibration of an oscilloscope can be accomplished by a simple "brute-force" method if an audio oscillator and an ac VTVM or electronic meter are available. In Fig. 2-11, you can see that

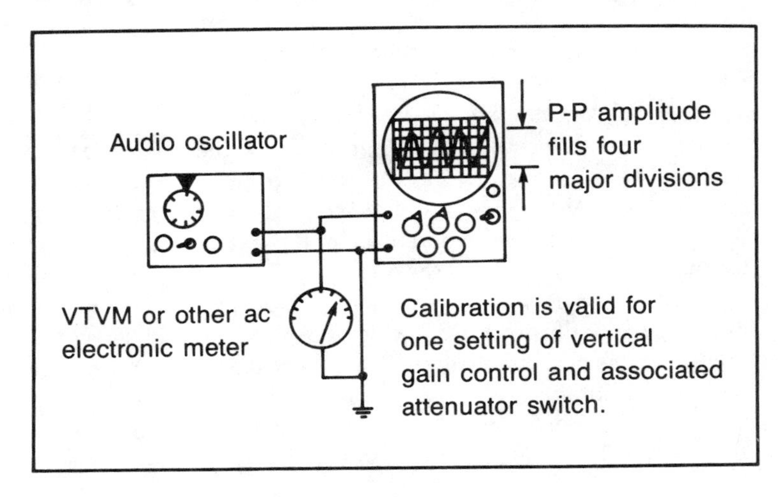

Fig. 2-11. Calibration setup for establishing the volts per major division sensitivity of the scope.

the voltage at the input of the scope is measured. If the audio oscillator generates a good sine wave and the electronic meter is a good quality or laboratory type, the calibration should be accurate. An audio frequency range of a hundred to a few thousand hertz is generally suitable. The output of the audio oscillator is adjusted to cause the peak-to-peak amplitude of the sine wave displayed on the oscilloscope to occupy an integral number of major divisions. It is easiest to deal with a display consisting of several to perhaps ten sine wave cycles. In performing this calibration do not change the setting of the vertical gain potentiometer after the display of the dc signal has been completed. Record the reading on the ac voltmeter and the number of major divisions filled by the sine wave. This is all there is to such a spot calibration, but some care must be exercised in the interpretation of the results.

In order to make proper use of our calibration data, it should be realized that the reading on the ac voltmeter is in RMS values and the reading we take from the oscilloscope display is a peak-to-peak indication. Therefore, we must first convert the ac voltmeter reading to its peak-to-peak equivalent. This is accomplished by multiplying the indicated RMS value by 2.28. Thus, if the amplitude of the audio-oscillator signal is measured as ten volts on the ac meter, the peak-to-peak equivalent of ten volts RMS is 22.8 volts. If, the display of the oscillator signal occupies four major divisions as illustrated in Fig. 2-11, the sensitivity of the oscilloscope is determined by dividing the peak-to-peak amplitude indirectly obtained from the ac voltmeter by the number of major divisions the oscillator signal occupies on the graticule of the oscilloscope. In Fig. 2-11 the peak-to-peak amplitude of the sine wave occupies exactly four major graticule divisions. Therefore, for this example, the vertical sensitivity of the oscilloscope is 22.8 divided by 4 or 5.70 volts per division. This is valid for the specific adjustments of the vertical gain control and the vertical attenuator which prevailed during the calibration. Provided these are not disturbed, the scope can now be used to measure dc voltage. The departure of the sweep line from the center position on the screen is simply evaluated on the basis of 5.70 volts per major vertical division. Resolution is improved by also making use of the minor divisions. These divide the major divisions into five equal parts, as shown in Fig. 2-9. Obviously, an easier-to-use calibration would be for a sensitivity of 5.0 volts per major division.

If the scope is to be useful over a wide range of dc voltages, the above described calibration procedure has to be carried out for a number of adjustments of the vertical gain control and the vertical attenuator. The calibrations should be charted or graphed so that they may be used when needed.

Often a probe is used during dc measurements. Since the test setup in Fig. 2-11 does not take into account the attenuation of the probe, the scope reading must be multiplied by the attenuation factor of the probe. This is easy because the probe invariably has an attenuation factor of ten. It makes the scope one-tenth as sensitive as it would be without the probe. Therefore, we multiply the volts per division value by ten. For the case cited above, the use of the probe changes the calibration from 5.70 volts per major division to 57.0 volts per major division. (50.0 volts per major division would be an easier sensitivity calibration to use in this case.) The reason the probe is not used in Fig. 2-11 is that most audio oscillators do not deliver enough output for much of the calibration that would be involved.

USING THE DC OSCILLOSCOPE AND THE ELECTRONIC METER TO MEASURE CURRENT

It has been demonstrated already that the internal resistance of current meters often causes appreciable circuit disturbances. Despite the fact that a current meter may quite accurately indicate the current flowing through it, this may not necessarily be even a good approximation to the circuit current without the meter. Although much has been accomplished by manufacturers to make available voltmeters with a very high resistance, relatively little has been done to make the very low resistance current meter a common measuring instrument. Dc ammeters, milliammeters, and microammeters with great sensitivity, which is to say, very low resistance, can be obtained, but only at high cost. In the practical situations encountered in electronics, this sometimes poses a problem.

It is not always realized that available instruments may be used to measure dc current with a minimum of circuit disturbance. Particularly if you have a dc oscilloscope or a dc electronic meter, and especially if these instruments have scales or sensitivities calibrated in fractional or millivolt levels. The technique is similar for the scope and electronic meter. Essentially, it consists of monitoring the voltage drop across a low resistance. By a low resistance we mean a value much lower than those found in readily available current meters for the same current range. In Fig. 2-12, we see both the dc oscilloscope and the dc electronic meter connected to function as current meters in a circuit. The heart of the technique is resistance R. The more accurately its value is known, the closer is the possible accuracy. The voltage indications of the measuring instruments are converted to current by the relationship I equals E divided by R where E is the indicated voltage. For example, if R is 10 ohms and the electronic meter is set for full-scale range of 100 millivolts, this converts the meter into a 0-10 milliampere current meter:

$$\frac{100 \times 10^{-3}}{10} = 10 \times 10^{-3} \text{ ampere, or } 10 \text{ milliamperes}$$

Similarly, if 0-10 millivolts can be read on the measuring instruments, a 0-10 milliampere current meter calls for a 1.0-ohm value for R. Notice that on the 0-1 volt scale, the 1.0-ohm resistor produces a current meter within the 0-1.0 ampere range.

Also notice in Fig. 2-12 that the circuit and the instruments are connected to a common ground. The fact that a ground symbol is shown at the negative side of the battery circuit does not have much significance. The use of the battery here represents any dc source of power. Probably the most commonly encountered source will be a dc power supply operated from the ac line. This being the case, you must be sure that in making ground connections that there is no inadvertent short circuit. For example, the grounding situation shown in Fig. 2-12 would result in a short circuit if resistance R and the instruments connected across it were in the positive line of the circuit, because the dc power source would be shorted to ground. However, if one terminal of the instrument were not grounded as shown, the current could be measured in the positive line of the circuit. This would not be desirable, though, because the chassis of the instruments would then be "hot" with respect to the circuit ground. Some inexpensive instruments do not have an isolation transformer between the circuitry and the power line. The use of such instruments without checking the ground situation can lead to short circuits, the application of power-line voltage to the circuit undergoing test, and to dangerous electric shock. Battery-powered instruments allow greater flexibility and eliminate the possibility of short circuits or the presence of ac line voltage. However, it is still highly desirable

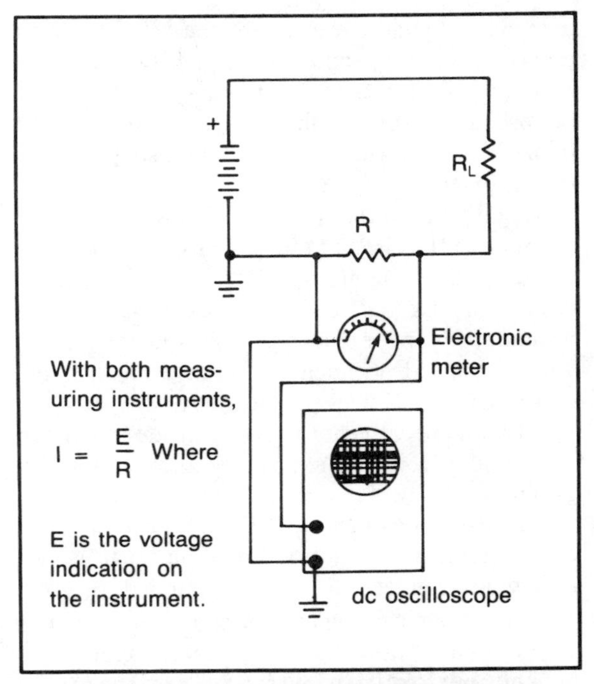

Fig. 2-12. Circuit demonstrating the use of a dc electronic meter or dc oscilloscope as a current meter.

to avoid making the instrument chassis "hot" with respect to circuit ground. Whenever possible, the resistance R and its associated instrument should be inserted in the circuit as depicted in Fig. 2-12 in order to keep the chassis of the instruments at circuit ground.

It might seen that in a situation where neither terminal of the circuit voltage source is grounded, the above discussion is not relevant. If, indeed, the circuit and its voltage source are isolated from ground, the danger of a short circuit is not a factor when R and an instrument are connected in one side of the circuit. However, the mere fact that the positive and negative terminals of the dc source are ungrounded does not exclude the possibility of a ground at the center tap of the dc source. And if we make tests on the dc source and find no evidence of any grounding, we must also be sure that the circuit which connects to it is ungrounded.

The above cautions are cited not to make the use of this current-measuring technique appear formidable, but to drive home the fact that the grounding situation must be kept in mind. In actual practice, with instruments having isolation transformers and polarized ac line plugs, and with a little common sense, the advantages of this scheme are not difficult to obtain. Good results are attainable with either the dc scope or the electronic meter.

Procedure Using a Dc Oscilloscope as a Null Detector

Objective of Test: To show some of the advantages of achieving bridge balance with a dc oscilloscope rather than a zero-center meter.

Test Equipment Required: A good quality oscilloscope with dc capability.

a 10-turn, 10K linear potentiometer

Two 1/2-watt, 10 percent 10K composition resistors

22 1/2-volt battery

Single-pole, single-throw pushbutton switch

Test Procedure: Connect the apparatus as shown in Fig. 2-13. Before connecting the scope to the bridge,allow it to warm up until it becomes stable as evidenced by a minimal tendency for the sweep line to drift. The oscilloscope is adjusted as when using it as a dc voltmeter. Initially, the sensitivity should be reduced by means of the vertical attenuator and/or the vertical gain control. With most scopes, it will be necessary to make a number of zeroing adjustments of the sweep line, particularly as the bridge balance approaches the best attainable null. This is done by closing SW2 while SW1 is open. As a better condition of bridge

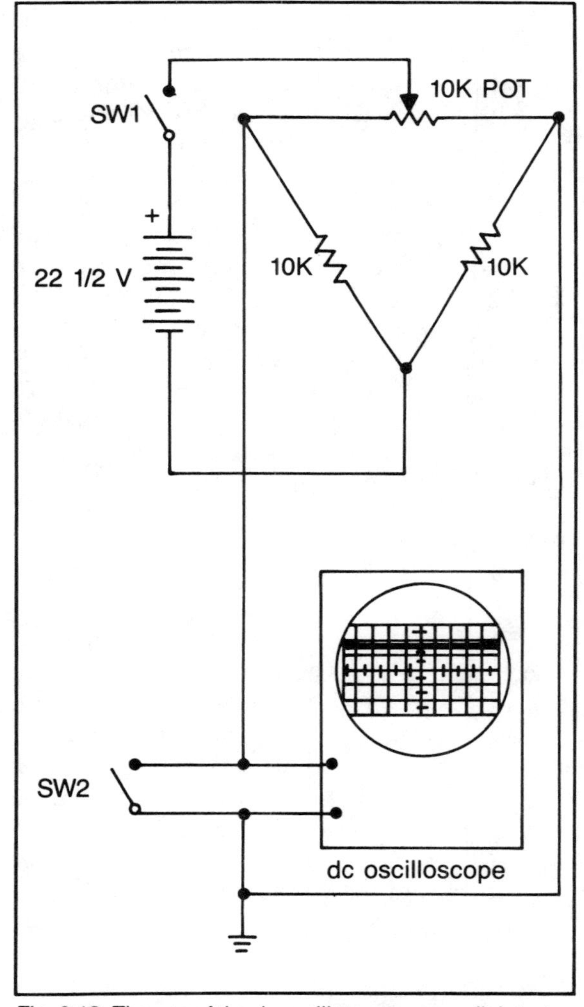

Fig. 2-13. The use of the dc oscilloscope as a null detector.

balance is approached by adjustment of the 10K potentiometer, the oscilloscope is adjusted for ever greater sensitivity. It is not the purpose of this test procedure to make actual measurements of unknown resistances, but rather to demonstrate the use of the dc oscilloscope as a null detector. In this connection, observe whether a deep null will hold its zero on the oscilloscope for several minutes or more.

Comments on Test Results: It is obvious that a stable dc oscilloscope makes an excellent null detector. Its immunity to overload damage from a gross bridge imbalance condition is a decided advantage over sensitive meters. When dealing with bridge arms which are on the order of tens and hundreds of kilohms, it may be found that sufficient power line pickup as well as other noise voltages tend to broaden the otherwise sharply focused sweep line. This may call for shielding techniques. Also, you can try a capacitance across the scope probe. (Too large a capacitor can cause an objectionable lag in the response of the scope.) Often, the effect of noise voltages can be made less objectionable at the higher sweep speeds. For this particular application, a wide bandwidth is not needed.

The application of the dc oscilloscope to other balancing techniques may involve difficulties because of the ground situation. For example, a differential amplifier stage could be investigated for balance at the plates, collectors, or drains of the paired active elements only if it can be definitely ascertained that there is no conflict of grounds. Whether or not a problem exists depends on the power supply used with the stage. Differential input, dual-beam, or dual-trace oscilloscopes are often useful for achieving balance of dc voltage levels because the grounds of such instruments do not have to connect to circuitry points that are "hot" with respect to a ground system common to both circuitry and scope. Battery-operated oscilloscopes are also immune to ground problems.

If the drift characteristic of the dc oscilloscope is slow and unidirectional, rather than fast and erratic, its potential as a sensitive null detector may not be appreciably degraded. This is because the eye is primarily sensitive to the vertical movement of the sweep line as switch SW1 is actuated. Thus, a no-motion indication of null may be more reliable as an indication of bridge balance than trying to ascertain that the sweep line is not displaced from its absolute center position.

THE MEASUREMENT OF AC VOLTAGES AND CURRENTS

Much of what has been discussed relative to the measurement of dc voltages and currents pertains also to ac measurements when appropriate ac instrumentation is used. For example, it is possible that circuit disturbances introduced by imperfect ac voltmeters and ammeters will have an effect similar to that prevailing during dc measurements. With sine waves ac in resistive circuits, the situation is identical. With nonsinusoidal voltages or currents, or where there is inductance and capacitance in the circuit, further complications are introduced because of phase and harmonic considerations. However, the basic fact remains that if voltmeters consume negligible current and current meters produce a negligible voltage drop, the effect on the circuit will not be significant.

The major concern involved in the measurement of ac voltages and currents has to do with three vital factors. We must know the waveshape of the ac, the quantity to which the meter responds, and the kind of calibration used for the scale. Otherwise, erroneous measurements can be obtained. For example, a peak-responding FET voltmeter generally has an RMS scale, since this is the most basically meaningful way to designate ac values. Such a meter will read correctly on sine waves, but if we naively use it in a circuit in which there are square waves or triangular waves, or for that matter any waveshape other than sinusoidal, the reading will not be correct. A similar situation is encountered with rectifier-type ac meters such as are associated with VOMs. These meters are also calibrated to read RMS voltage. However, they are responsive to the average value of the wave. The calibration is valid for sine waves. If, however, the rectifier meter is used with square or triangular waveshapes, the indicated RMS reading will be in error.

This also applies to many digital electronic meters (DMMs). As time goes on, an increasing number of DMMs with "true" RMS capability are being used. These instruments indicate RMS regardless of waveform and even if there is a dc level or a dc component. The barrier to these very desirable meters has been expense, but new signal processing techniques, as well as market forces, have reduced prices significantly. Such meters, once rarely seen outside of the engineering laboratory, are clearly becoming popular for everyday test and measurement needs.

Actually, a volume could be written concerning the errors you'll run into with various types of ac instruments in the measurement of nonsinusoidal ac waves. You cannot be too cautious in this important area of testing. The following material deals with commonly encountered test meters and representative waveforms.

Table 2-1 shows the waveform characteristics of the important ac instruments. Strictly speaking, the peak and the peak-to-peak responding types sense voltage, whereas current is the actuating force for such analog types as the d'Arsonval and taut-band meter in VOMs. (An FET analog instrument, on the other hand, can be said to be responsive to voltage.) Whether current or voltage is actually sensed is not of great importance. Other things being equal, it is easier for the instrument designer to achieve high input-impedance with voltage sensing. However, modern bipolar op-amps require such low actuating current, that, they too, can readily yield the required high input impedance. More important, we have to pay attention to the waveform response of ac measuring instruments.

Widely differing waveforms which have the same RMS value have a very basic common characteristic. Namely the ability to generate the same amount of heat energy in a given resistance. Extending the significance of this standardization further, the RMS value of a wave is exactly the value of a sustained dc level which generates an equal amount of heat energy in the same resistance.

Some electronic instruments have blocking capacitors in their input circuits. Such instruments cannot indicate true RMS values when the measured waveform has a dc component, or is superimposed upon a dc level. In order to function

Table 2-1. Waveform Characteristics of ac Measuring Instruments.

TYPE	OSCILLOSCOPE METER (mostly analog types)	ELECTRONIC-METER (anal.)	ELECTRONIC METER (anal.) (ALSO, SOME EARLY DIGITAL TYPES)	IRON VANE THERMOCOUPLE DYNANOMETER SOME ANALOG AND DIGITAL ELECTRONIC METERS	RECTIFIER METER (Includes VOMs, some analog and digital electronic meters.)	RECTIFIER METER (Includes VOMS, some analog and digital electronic meters.)
RESPONSE	PEAK TO PEAK	PEAK	PEAK	RMS*	AVERAGE	AVERAGE
SCALE OR READOUT	PEAK TO PEAK	PEAK	RMS	RMS*	RMS	AVERAGE
CALIBRATION WAVEFORM	SINE, TRIANGLE, OR SQUARE	SINE, TRIANGLE, OR SQUARE	SINE ONLY	SINE with possibility of other waveforms	SINE ONLY	SINE ONLY

* Iron vane, dynamometer, and thermocouple instruments respond to I^2_{RMS}. Therefore their RMS scales are non-linear. Analog electronic meters may have non-linear RMS scales for the same reason but in some models, the RMS scale is electronically linearized. (The concept of linearity does not apply to the digital readout of DMMS.)

as a "true" RMS meter, there must be a precision dc amplifier and no dc blocking capacitor. Those analog ac instruments which respond the same to all, or to most waveforms, have not been among the most commonly-encountered types. There are various reasons why this has been so.

Some of these instruments have quite low resistances when used as a voltmeter, high resistances when used as an ammeter, or are extremely sluggish, or are delicate with a low overload capability, or are quite limited in frequency response, etc. Cost is a factor in some instances. In any event, the ac measuring instruments most generally found in electronics work are the rectifier meter and the "ordinary" ac electronic meter. The rectifier meter is none other than the ac voltmeter found in most VOMs.

Both of these ac instruments are capable of reasonable accuracy and they both provide the advantage of a linear, or a near-linear scale. But both can give you very misleading information if used to measure waveforms other than sine waves. It is of utmost importance that you know what to expect in such cases, for nonsinusoidal waveforms are very common in electronics work.

In order to know how to correctly interpret readings made on rectifier meters and ac electronic meters when dealing with other than sine waves, it is necessary to resort to both Table 2-1 and Table 2-2. A few words about Table 2-2 should clear up certain kinds of confusion which tend to occur when waveform values are discussed. In Fig. 2-14 we see a sine wave and the important values by which waves are measured. Your first thought about these values could well be that the peak, RMS, and average values are all zero because in the course of a full cycle there are identical excursions above and below the base line. We could, indeed, advance valid arguments to substantiate this. However, the classic approach to definitions of these amplitude levels assumes that the negative half cycle is folded up or turned over to become a positive half cycle. At first there may seem to be some quite arbitrary gamesmanship going on, but there is a logical reason for this. (Engineering texts speak of dealing with a rectified wave. Even this description can create confusion because half-wave rectification does not fold over one of the half cycles. Rather, it deletes half of the original sine wave, making for an entirely different situation. However, if we condition ourselves to think always of the effect of *full-wave rectification*, we will be on the right track.)

Now, let us see why the wave is modified before designating its amplitude levels. Our objective is to see the wave from the viewpoint of an ac meter such as an electrodynamometer. This is a venerable instrument which was popular when the principles of alternating current circuits were first being defined. This instrument develops torque in the same direction over the entire period of the sine wave (as opposed to a dc meter which develops zero net torque over the full cycle). When we use the electrodynamometer, it is as if the wave were unidirectional insofar as the effect it has on the pointer of the instrument. Another way of looking at this situation is to realize that the negative portion of the cycle is every bit as effective in developing heat in a resistance as is the positive portion. This being the case, it would be self-defeating to reason from the shape of the sine wave that the various amplitudes have a net value of zero. This technique of dealing with the full-wave rectified version of the sine wave applies to other ac waveshapes as well. However, as we shall see, the RMS and the average values then are different from their counterparts in the sine wave. Another aspect of standardized convention for dealing with ac waves is that the RMS and the average values are always understood to be defined in terms of the peak value. Thus, in the sine wave, the average value is 0.636 times the peak amplitude. Similarly, the RMS value of the sine wave is 0.707 times the peak amplitude. Let us see what all this leads to when we attempt to measure ac voltages or currents.

Figure 2-15 shows a simple circuit which we want to test to determine whether or not it is basically OK. Let us suppose that any one of several instruments might be available and that the test will consist of monitoring the voltage across resistance R. To depict this situation, Fig. 2-15 shows the following instruments connected across R: an ac

Sine Wave	Waveform	Average Value	RMS or Effective Value	Peak-to-Peak Value
Sine Wave	Peak	0.636 x Peak Value	0.707 x Peak Value	2.00 x Peak Value
Square Wave	Peak	1.000	1.000	2.00
Triangular Wave	Peak	0.500	0.576	2.00
Half Sine Wave	Peak	0.318	0.500	Not defined
Unipolar Square Wave	Peak	0.500	0.709	Not defined
Rectangular Wave	Peak, t, T	$\frac{2t}{T}$	$\sqrt{\frac{2t}{T}}$	2.00
Unipolar Rectangular Wave	Peak, t, T	$\frac{t}{T}$	$\sqrt{\frac{t}{T}}$	Not defined

Table 2-2. Ac Wave Values.

oscilloscope, a dc oscilloscope, an electrodynamometer voltmeter, a VOM set to read dc, a VOM set to read ac, and a peak responding electronic voltmeter set to read ac. For simplicity, assume that the diode, CR1, is ideal in that the forward voltage drop across it is zero and its reverse resistance is infinite. Also, this diode produces no waveform distortion other than that of rectification. What should appear across R and what voltages are actually indicated by the various instruments?

The action of the rectifier diode produces the half-wave positive-going pulse shown in Fig. 2-15. The waveform seen on the ac oscilloscope could be misleading if the peak voltage is measured as shown. Instead of 100 volts, this measurement would be 78.2 volts because the baseline of a unidirectional wave is displaced from its average position after passing through the capacitors of an ac scope. We can escape this trap by measuring the peak as the distance from the flat part of the wave to a rounded tip of any cycle after the first one. This is not good practice because in more complex waves various erroneous assumptions can be made. The evaluation of the wave with a dc oscilloscope does produce accurate results and conveys all the information needed. A half sine wave with a peak

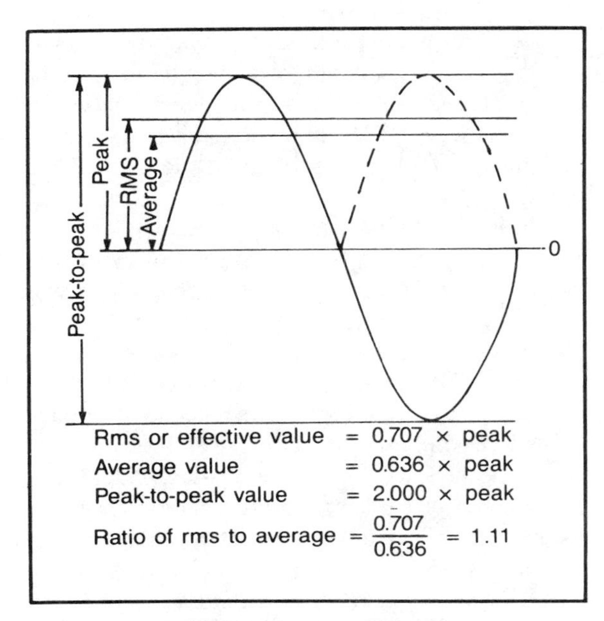

Fig. 2-14. Amplitude levels in the sine wave.

amplitude of 100 volts tells all.

Let us now investigate the situation with the various meters. Starting with the meter we would be least likely to have, notice that the electrodynamometer indicates 50 volts RMS. From Table 2-1, we see that this is an RMS-respónding and RMS-indicating instrument. There are no problems thus far; the existence of 50 volts RMS across R is expected from Table 2-2, where the half sine wave is tabulated as having an RMS value of 0.500 times the peak value.

The VOM set to read dc indicates 31.8 volts. This is a correct indication, but it may not be immediately seen how it comes about. Dc instruments are not listed in Table 2-2, but the common D'Arsonval meter movement is an average-responding and average-indicating instrument. Table 2-2 does show that the RMS value of the half sine wave is 0.500 times its peak. Thus, the RMS equivalent of the 31.8-volt reading on the dc VOM is 0.500 × 100, or 50.0 volts. This confirms the dynamometer RMS indication. For this waveform the ratio 0.500/0.318 equals 1.57 is useful. The indication on the dc meter multiplied by 1.57 gives the RMS voltages. Thus, 31.8 × 1.57 equals 50 volts RMS.

The ac VOM produces an indication of 35 volts on its RMS scale. Strange? Recall that all ac meters with RMS scales were originally calibrated from a sine wave. This, then, is true for the ac VOM.

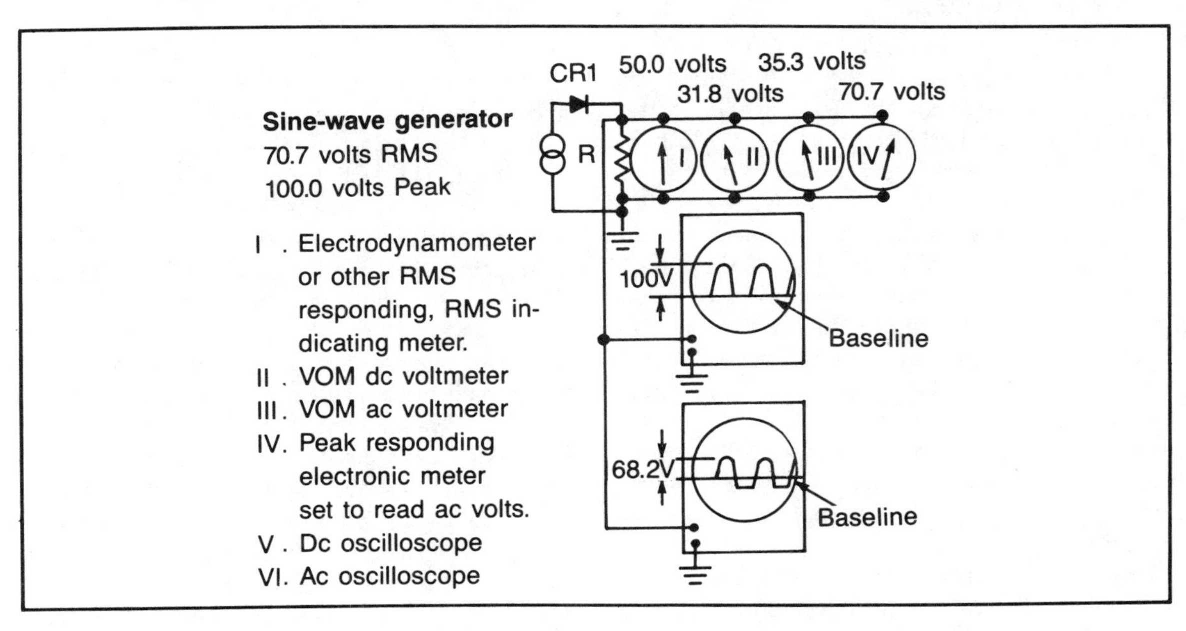

Fig. 2-15. Ac voltage indications on various instruments when used to measure a half sine wave.

When it was calibrated, it responded to 0.636 of the sine wave peak amplitude. But the calibration was 0.707 of the sine wave peak amplitude. These numbers are, respectively, the average and RMS values of the sine wave. The calibration was higher than the quantity actuating the meter by the ratio 0.707/0.636 or 1.11. The quantity actuating the meter was the average value of the sine wave.

In the present situation, the ac VOM is being actuated by an average value which, from Table 2-2, is 0.318 for the half sine wave. But the scale has been calibrated to read 1.11 times the average value. Therefore, the reading is 1.11 × 0.318 × 100 or 35.3 volts. The figures above pertain to ac VOMs with full-wave rectifiers. For an ac VOM with half-wave rectification, the 0.636 average value would be 0.318. However, when such a meter is factory calibrated,the value 2.22 is used in place of 1.11. Therefore, the reading obtained will be the same for both meters. However, with the half-wave rectification meter, you could get a reading of zero by reversing the leads in this case. (Also, the ohms-per-volt sensitivity of the half-wave rectifier type is one half that of the type employing full-wave rectification.)

The ac electronic meter is the type with no input capacitor ("series-diode type"). It responds to positive peaks and has no way of "knowing" that the half sine wave is not a full sine wave. It indicates 70.7 volts which is the RMS value of a full sine wave such as was originally used in calibrating the meter. In Table 2-2, we see that the RMS value of the full sine wave is 0.707 × peak, whereas for the half sine wave the RMS value is 0.500 times peak. Our reading is too high by the factor 0.707 divided by 0.500 or 1.41. Therefore, the corrected reading is 70.7 volts divided by 1.41 or 50.0 volts. This corrected RMS value confirms the electrodynamometer indication. Notice that with this type of ac electronic meter, a zero reading would be obtained in this case if the leads had been reversed. It is an interesting fact that with sine-wave measuring instruments (with the diode short-circuited) all indicating meters, except the dc VOM, would read the correct RMS value of 70.7 volts. The dc VOM would read zero. (Actually, dc meters should be protected from such situations because the ac produces damaging vibrational forces in the meter movement and tends to demagnetize the magnetic fields). Both scopes would report identical and accurate peak-to-peak indications of 200.0 volts.

It is not to be assumed that all electronic meters, other than the RMS responding types, are peak responding. The peak-responding meter was selected for discussion because the other instruments provided examples of response to RMS, average, and peak-to-peak values. Actually, many of the popular electronic meters respond to average values, resembling the VOM or rectifier meter in this respect. However, the average-responding electronic meter places a higher internal resistance (less loading) in the circuit than the rectifier instruments commonly associated with VOMs. Even extra sensitive VOMs, such as those with dc voltmeter sensitivities of 200,000 to 1,000,000 ohms per volts, still are limited to 20,000 ohms per volt or so when used as a rectifier type ac voltmeter.

DEALING WITH THE CHOPPED WAVES OF PHASE-CONTROL SYSTEMS

The SCR and the triac have revolutionized the control of power to a load. These semiconductor devices are extremely efficient compared to older methods which, in some form or another, dissipated relatively large amounts of power in the control element itself. The SCR and triac are phase control rather than resistive control devices. Indeed, their "on" resistances are exceedingly low, which is the main feature. At first, the single SCR was used in half-wave circuits in which the conduction angle of the applied sinusoidal voltage was controlled. This technique continues to have widespread use in low-power applications. However, full-wave control of power is advantageous in a number of respects. This is not always easy to accomplish with back-to-back SCRs, but once certain difficulties are overcome, it is easier to obtain smooth control over a wide power range. Furthermore, the overall efficiency and power factor of such a circuit exceed the performance of the simpler half-wave circuit. With the advent of the triac, such full-wave operation became more practical. Partial schematics of full-wave configurations are shown in Fig. 2-16.

The measurement of voltages and currents in these phase control circuits is a very risky task with ac VOMs and peak-responding electronic meters. The metering of harmonic-rich waves with these commonly available instruments can become rather complicated. With the information in Table 2-3, we can devise a simplified approach which avoids the use of these instruments. The use of the data in Table 2-3 enables us to determine the RMS value of voltage delivered to a load, R, for six different conduction angles. And from this information, the RMS value of load current and the average power delivered to the load are easily computed.

An oscilloscope can be used to monitor the peak value, E_m, of the voltage applied to the load at 180 degrees conduction. If half of the sine waveform is adjusted on the scope screen so that the 180 degrees of the wave encompasses six major horizontal divisions, each of these divisions will then represent 30-degree increments. This allows evaluation of the circuit for the six conduction angles listed in Table 2-3. However, once E_m is determined, it is only necessary to consult the Table to derive the RMS value of load voltage, the RMS value of load current, and the average power delivered to the load for the six conduction angles. The RMS value of load voltage is E_m multiplied by the factor in Column 3. The RMS value of load cur-

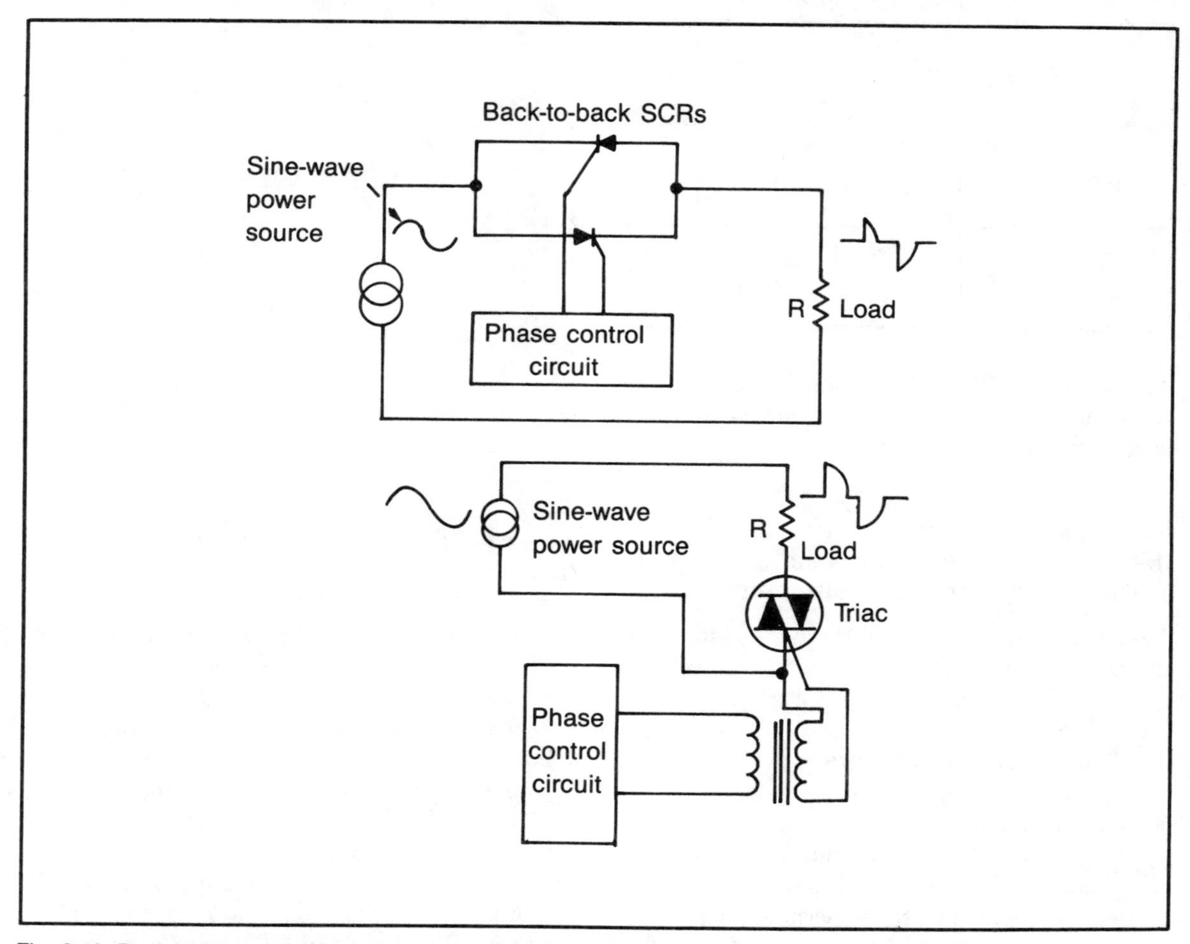

Fig. 2-16. Partial schematics of full-wave phase control systems.

1 CONDUCTION ANGLE	2 WAVEFORM	3 RMS VOLTAGE ACROSS LOAD
30°		0.1200 E_m
60°		0.3126 E_m
90°		0.5000 E_m
120°		0.6342 E_m
150°		0.6968 E_m
180°		0.7071 E_m

Table 2-3. Factors for Calculating the RMS Load Voltage for Six Conduction Angles in a Full-Wave Circuit.

rent is the RMS load voltage divided by load resistance R. The average power delivered to the load is either E_{RMS}^2 or I_{RMS}^2R.

Procedure for Determining the Non-Sinusoidal Response of Commonly Used Ac Instruments

Objective of Test: To test the response of the ac VOM and the peak-responding electronic meter to square and triangular waves.

Test Equipment Required:
ac VOM
Peak-responding electronic meter
RMS-responding electronic meter or electrodynamometer voltmeter. (If neither of these instruments is available, the experiment can still be performed by substituting an analysis for the true RMS indication these two instruments would provide.)
Audio function generator
ac or dc oscilloscope

Test Procedure: Connect the equipment as shown in Fig. 2-17. An RMS-responding electronic meter or the electrodynamometer voltmeter is desirable, but optional. The oscilloscope is used to maintain a constant peak-to-peak amplitude. (The wave amplitude sometimes changes when a function generator is switched from one waveform to another.) Once the experiment is under way, the vertical sensitivity of the scope should not be changed. If the peak-to-peak amplitude changes from one wave to another, the amplitude control on the function generator should be used to correct it. An amplitude of several volts peak-to-peak will be satisfactory. Higher amplitudes are acceptable if not accompanied by wave distortion.

First, set the function generator for a sine-wave output. A frequency of a few hundred hertz should be satisfactory. We do not want to get up into the higher audio frequencies where the frequency capability of the ac VOM might be exceeded. All meters should show substantially the same reading.

Next, switch the function generator to deliver square waves. Observe the peak-to-peak indication on the scope and adjust the amplitude control of the function generator if necessary. Record the indication of all of the meters. The indication on the peak-responding electronic meter should show no change. The indication on the RMS-responding electronic meter or the dynamometer voltmeter should now be 1.41 times its sine-wave indication, and the indication on the ac VOM should now 1.57 times its sine-wave reading.

Finally, switch the function generator for a triangular-wave output. Observe the peak-to-peak scope indication and make any necessary adjustments of the amplitude control on the function generator. Record the readings on all the meters. The indication on the peak-responding electronic meter should show no change. The indication on the RMS-responding electronic meter or the dynamometer voltmeter should now be 0.814 times its sine-wave indication. The indication on the ac VOM should now be 0.786 times its sine-wave readings.

Comments on Test Results: This test should correct some widely held notions concerning ac measurements. The fact that the peak-reading electronic meter maintained its indication throughout the test might suggest that it alone, provides the correct readings. Of course, this is far from the truth. Only the RMS-responding electronic meter and the electrodynamometer voltmeter produce indications which remain essentially accurate for the three commonly encountered waveshapes. On the other hand, this test should drive home the fact that all of the ac instruments respond properly to sine waves. The implication is that more than one type of instrument is satisfactory if you know that sine waves exist in the circuit to be tested. It is important that the sine waves be quite pure; that is, have negligible harmonic content. You cannot

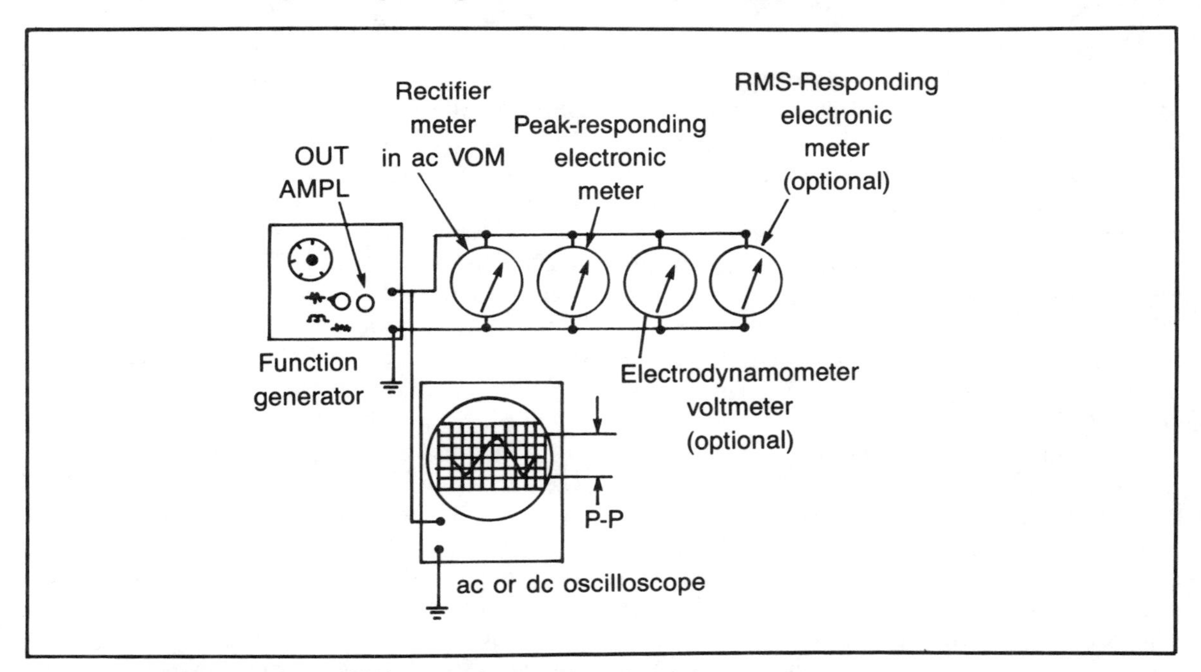

Fig. 2-17. Waveform test for ac VOM and peak-responding electronic meters.

immediately tell whether or not a sine wave is pure from a casual inspection of the waveshape on the oscilloscope. Many smooth appearing waves may look like pure sine waves, but can contain abundant harmonic energy. If so, the indications on the ac VOM and the peak-responding electronic meter will necessarily contain some error. Measuring and testing will have evolved to a higher state of the art when most instruments in use will possess true RMS capability. Although that day has not yet arrived, it is clearly on the way.

In general, it is desirable to use an oscilloscope to show the nature of the wave being indicated on an ac VOM or a peak-responding electronic meter. Also, the RMS value of a complex or irregularly shaped wave can be computed from a Polaroid picture of the scope pattern and this is sometimes done. However, the technique is not simple, and the computations can be quite tedious. In circuits where a knowledge of the peak value of an ac wave is suitable for the kind of circuit analysis involved, the oscilloscope provides this information. Also, the peak-responding electronic meter indicates the peak amplitude for any of the three waves discussed if its RMS indication is multiplied by 1.41.

Another implication of this test is that all of the meters can be used to indicate the RMS value of square and triangular waves. The values are read directly from the scales of the RMS-responding electronic meter and the electrodynamometer voltmeter. The scale indications of the peak-responding electronic meter and the ac VOM have to be multiplied by the appropriate conversion factors to yield the correct RMS values. For accuracy with the latter two meters, you must be certain that the square or triangular waves have the proper shape insofar as symmetry and duty cycle are concerned.

The danger of gross error is great when using the ac VOM and the peak-responding electronic meter on waves which are irregular, asymmetrical, or which are superimposed on a dc level. There are significant differences in different models of the same type instrument also. For example, some ac VOMs have an internal dc blocking capacitor while others do not. These meters also employ various full-wave and half-wave rectifier arrangements. Peak-responding electronic meters use both series-diode and shunt-diode input circuits. For the waves discussed in this test, these differences in meter circuitry are of no importance. For other waves, however, you must proceed with caution.

From all that has been said, the peak-responding electronic meter might appear to be a poorly adapted instrument for measuring ac voltage. However it has two features which have accounted for the popularity of this instrument. First, the sampling of the peak values of a wave results in minimal circuit loading. In other words, it behaves as a high-impedance voltmeter. Secondly, by putting the input diode and associated components in a probe, the frequency capability is extended to hundreds of megahertz.

THE DECIBEL

The use of decibel units of measurement must be of considerable importance; otherwise we would not find decibel scales on so many differing types of meters. See Fig. 2-18. Indeed, the proper use of decibel measurements leads to certain conveniences not readily provided by ordinary arithmetic measurement units. On the other hand, erroneous use of the decibel is far from uncommon. The original use of the decibel was to describe the ratio between two power levels, P1 and P2. Thus, P2:P1 is the numerical ratio between P2 and P1 and is the "natural" or arithmetic way of expressing this relationship. In the typical case, P1 might represent the input power to an amplifier and P2 would represent the output power. It is simple enough then to visualize the power gain of the amplifier as the ratio of power level P2 to power level P1.

In actual practice, it is not usually convenient to employ wattmeters to measure power levels. Another inconvenience exists when you want to determine the power gain or loss from a number of amplifiers and/or attenuators. This is particularly true when gains and losses are large numbers and are not nice rounded-off values or simple integers times powers of ten. For example, if we cascade two amplifiers, one with a power gain of 15.89 and

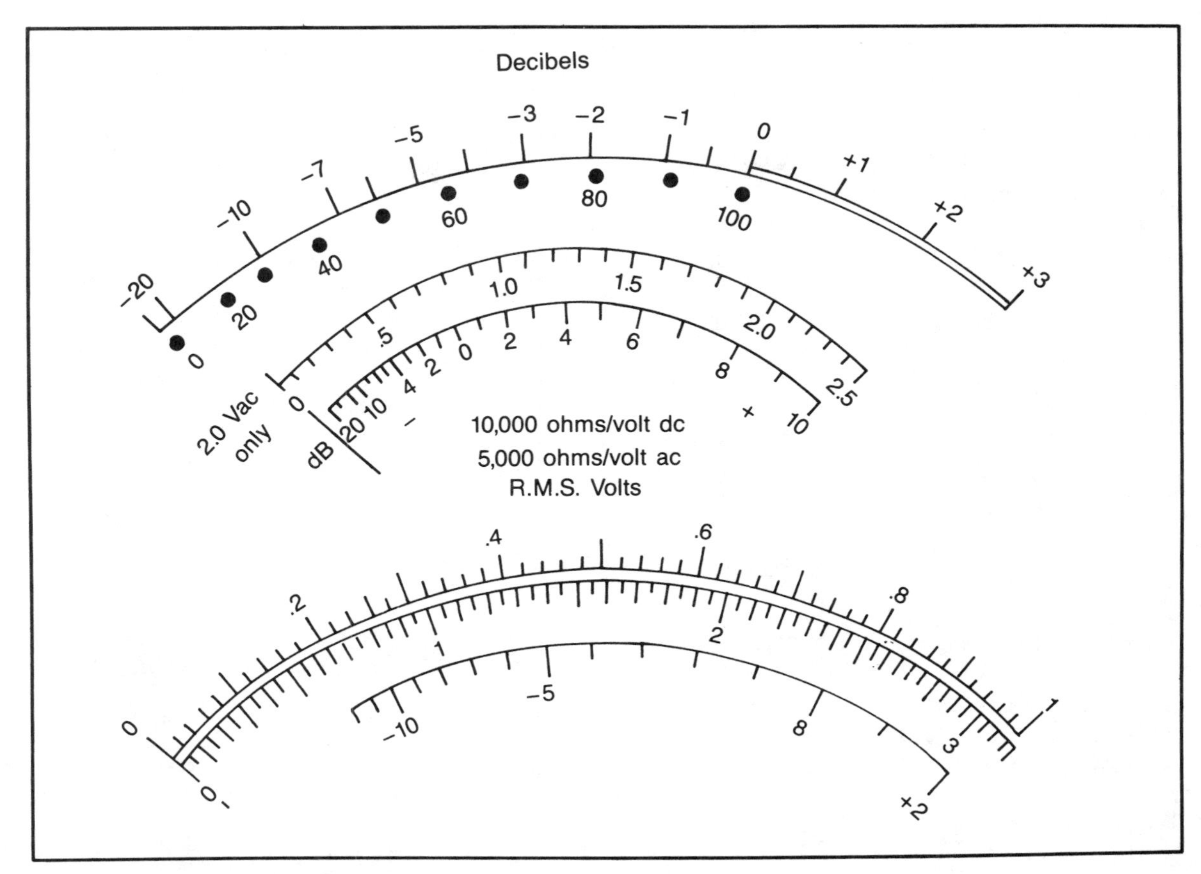

Fig. 2-18. Typical decibel scales.

the other with a power gain of 1474, we "intuitively" know that the overall gain is the product of the individual gains. (Even this is not necessarily true because there may be a reflection loss due to mismatch between the output of amplifier 1 and the input of amplifier 2. However, for the sake of simplicity, it is assumed here that the power gain of amplifier 1 is specified for the condition of feeding into the input impedance amplifier 2.) The point we wish to make here is that the evaluation of overall gain or loss can become an awkward procedure if the ordinary arithmetic approach is used.

For the purposes of review and emphasis, a discussion of some of the fundamental ideas about decibel notation is appropriate. First, consider the equation, x equals a^y. Both for simplicity and for our purposes here, assume that a is always the number ten. Then we have the mathematical statement that the number x is equal to ten raised to the exponent y. Thus, if y equals 2, this equation tells us that x equals 10^2 or 100. Similarly, if y is 3, x is 1000. If y is 6, x becomes 10^6, or one million. Notice the power of the exponent and the tremendous range that can be covered merely by changing the exponential power of ten! Actually, an even more useful system would result if we could represent the entire series of numerical values instead of merely integral powers of ten. Then we could cover the range between 100 and 1000, between 1000 and 10,000, etc. You might suppose that this can be accomplished by decimal or fractional values of the exponent. Thus, $10^{2.537}$ must correspond to some number between 100 and 1000 because the exponent is between 2 and 3. But now the system

Table 2-4. Four-Place Logarithms. N Also Represents Antilogarithms.

N	0	1	2	3	4	5	6	7	8	9	N	0	1	2	3	4	5	6	7	8	9
10	0000	0043	0086	0128	0170	0212	0253	0294	0334	0374	55	7404	7412	7419	7427	7435	7443	7451	7459	7466	7474
11	0414	0453	0492	0531	0569	0607	0645	0682	0719	0755	56	7482	7490	7497	7505	7513	7520	7528	7536	7543	7551
12	0792	0828	0864	0899	0934	0969	1004	1038	1072	1106	57	7559	7566	7574	7582	7589	7597	7604	7612	7619	7527
13	1139	1173	1206	1239	1271	1303	1335	1367	1399	1430	58	7634	7642	7649	7657	7664	7672	7679	7686	7694	7701
14	1461	1492	1523	1553	1584	1614	1644	1673	1703	1732	59	7709	7716	7723	7731	7738	7745	7752	7760	7767	7774
15	1701	1790	1818	1847	1875	1903	1931	1959	1987	2014	60	7782	7789	7796	7803	7810	7818	7825	7832	7839	7846
16	2041	2068	2095	2122	2148	2175	2201	2227	2253	2279	61	7853	7860	7868	7875	7882	7889	7896	7903	7910	7917
17	2304	2330	2355	2380	2405	2430	2455	2480	2504	2529	62	7924	7931	7938	7945	7952	7959	7966	7973	7980	7987
18	2553	2577	2601	2625	2648	2672	2695	2718	2742	2765	63	7993	8000	8007	8014	8021	8028	8035	8041	8048	8055
19	2788	2810	2833	2856	2878	2900	2923	2945	2967	2989	64	8062	8069	8075	8082	8089	8096	8102	8109	8116	8122
20	3010	3032	3054	3075	3096	3118	3139	3160	3181	3201	65	8129	8136	8142	8149	8156	8162	8169	8176	8182	8189
21	3222	3243	3263	3284	3304	3324	3345	3365	3385	3404	66	8195	8202	8209	8215	8222	8228	8235	8241	8248	8254
22	3424	3444	3464	3483	3502	3522	3541	3560	3579	3598	67	8261	8267	8274	8280	8287	8293	8299	8306	8312	8319
23	3617	3636	3655	3674	3692	3711	3729	3747	3766	3784	68	8325	8331	8338	8344	8351	8357	8363	8370	8376	8382
24	3802	3820	3838	3856	3874	3892	3909	3927	3945	3962	69	8388	8395	8401	8407	8414	8420	8426	8432	8439	8445
25	3979	3997	4014	4031	4048	4065	4082	4099	4116	4133	70	8451	8457	8463	8470	8476	8482	8488	8494	8500	8506
26	4150	4166	4183	4200	4216	4232	4249	4265	4281	4298	71	8513	8519	8525	8531	8537	8543	8549	8555	8561	8567
27	4314	4330	4346	4362	4378	4393	4409	4425	4440	4456	72	8573	8579	8585	8591	8597	8603	8609	8615	8621	8627
28	4472	4487	4502	4518	4533	4548	4564	4579	4594	4609	73	8633	8639	8645	8651	8657	8663	8669	8675	8681	8686
29	4624	4639	4654	4669	4683	4698	4713	4728	4742	4757	74	8692	8698	8704	8710	8716	8722	8727	8733	8739	8745
30	4771	4786	4800	4814	4829	4843	4857	4871	4886	4900	75	8751	8756	8762	8768	8774	8779	8785	8791	8797	8802
31	4914	4928	4942	4955	4969	4983	4997	5011	5024	5038	76	8808	8814	8820	8825	8831	8837	8842	8848	8854	8859
32	5051	5065	5079	5092	5105	5119	5132	5145	5159	5172	77	8865	8871	8876	8882	8887	8893	8899	8904	8910	8915
33	5185	5198	5211	5224	5237	5250	5263	5276	5289	5302	78	8921	8927	8932	8938	8943	8949	8954	8960	8965	8971
34	5315	5328	5340	5353	5366	5378	5391	5403	5416	5428	79	8976	8982	8987	8993	8998	9004	9009	9015	9020	9025

N	0	1	2	3	4	5	6	7	8	9
35	5441	5453	5465	5478	5490	5502	5514	5527	5539	5551
36	5563	5575	5587	5599	5611	5623	5635	5647	5658	5670
37	5682	5694	5705	5717	5729	5740	5752	5763	5775	5786
38	5798	5809	5821	5832	5843	5855	5866	5877	5888	5899
39	5911	5922	5933	5944	5955	5966	5977	5988	5999	6010
40	6021	6031	6042	6053	6064	6075	6085	6096	6107	6117
41	6128	6138	6149	6160	6170	6180	6191	6201	6212	6222
42	6232	6243	6253	6263	6274	6284	6294	6304	6314	6325
43	6335	6345	6355	6365	6375	6385	6395	6405	6415	6425
44	6435	6444	6454	6464	6474	6484	6493	6503	6513	6522
45	6532	6542	6551	6561	6571	6580	6590	6599	6609	6618
46	6628	6637	6646	6656	6665	6675	6684	6693	6702	6712
47	6721	6730	6739	6749	6758	6767	6776	6785	6794	6803
48	6812	6821	6830	6839	6848	6857	6866	6875	6884	6893
49	6902	6911	6920	6928	6937	6946	6955	6964	6972	6981
50	6990	6998	7007	7016	7024	7033	7042	7050	7059	7067
51	7076	7084	7093	7101	7110	7118	7126	7135	7143	7152
52	7160	7168	7177	7185	7193	7202	7210	7218	7226	7235
53	7243	7251	7259	7267	7275	7284	7292	7300	7308	7316
54	7324	7332	7340	7348	7356	7364	7372	7380	7388	7396
80	9031	9036	9042	9047	9053	9058	9063	9069	9074	9079
81	9085	9090	9096	9101	9106	9112	9117	9122	9128	9133
82	9138	9143	9149	9154	9159	9165	9170	9175	9180	9186
83	9191	9196	9201	9206	9212	9217	9222	9227	9232	9238
84	9243	9248	9253	9258	9263	9269	9274	9279	9284	9289
85	9294	9299	9304	9309	9315	9320	9325	9330	9335	9340
86	9345	9350	9355	9360	9365	9370	9375	9380	9385	9390
87	9395	9400	9405	9410	9415	9420	9425	9430	9435	9440
88	9445	9450	9455	9460	9465	9469	9474	9479	9484	9489
89	9494	9499	9504	9509	9513	9518	9523	9528	9533	9538
90	9542	9547	9552	9557	9562	9566	9571	9576	9581	9586
91	9590	9595	9600	9605	9609	9614	9619	9624	9628	9633
92	9638	9643	9647	9652	9657	9661	9666	9671	9675	9680
93	9685	9689	9694	9699	9703	9708	9713	9717	9722	9727
94	9731	9736	9741	9745	9750	9754	9759	9763	9768	9773
95	9777	9782	9786	9791	9795	9800	9805	9809	9814	9818
96	9823	9827	9832	9836	9841	9845	9850	9854	9859	9863
97	9868	9872	9877	9881	9886	9890	9894	9899	9903	9908
98	9912	9917	9921	9926	9930	9934	9939	9943	9948	9952
99	9956	9961	9965	9969	9974	9978	9983	9987	9991	9996

appears to lose its simplicity because we cannot mentally evaluate ten raised to nonintegral exponents.

Returning to the simple equation, x equals a^y, we write its equivalent logarithmic form, y equals $\log_{10}x$. (Log is the commonly used abbreviation of logarithm.) This equation explains the concept of the logarithm. Notice first that it is stated that y is equal to something involving the number x. From our previous discussion of the original equation, we know that y is an exponent. We also know that x is the number that we are representing by ten raised to exponent y. Now, the logarithmic equation is read as follows: y equals the log of x to the base ten—or better, y equals the log, to the base ten, of x. We are being informed that there is an exponent, y for any value of x such that 10^y equals x. If we refer to a table of logarithms to the base 10 and look for the number represented by x, a corresponding number will be found which represents y. Thus, logarithms are exponents. More specifically, logarithms to the base 10 are exponents to which ten is raised in order to produce the number x.

Pointers on Reading the Log Table

The logarithm is usually a 4- or 5-place number. Some tables show seven places or more in the interest of even more accurate computations. For many of the purposes of practical electronics, the use of three or four places (Table 2-4) certainly suffices. Accompanying the logarithms are the numbers (antilogarithms) that represent the actual value of ten raised to the logarithm. The number, or antilogarithm, is x in the basic equation, and equals a^y. Normally, we look up this number, then see what the corresponding logarithm, y, is. There is a logarithm for any number you can think of, or very nearly so. Yet the log table is not nearly so extensive as this fact would suggest. The reason is that the logarithm of any given number, say 3, enables us to determine the logarithm for all similar numbers, such as 0.3, 30, 300, 3000, etc. But the logarithm as listed is, in general, not complete. Something must be done to it to show where the decimal of the antilogarithm actually falls. That part of the logarithm which is listed in the table is called the **mantissa**. To this, we must add the characteristic. Think of it in this way: The mantissa is a decimal and it generally requires the characteristic in front of it. This is very easy to do:

For numbers from 1 to 10, the characteristic is zero. Here the listed logarithms are used, for the complete logarithm consists of the mantissa only. For example, the logarithm of the number 3 is 0.47712 directly from the table. For numbers from 10 to 100, the characteristic is 1. Therefore, the logarithm of the number 30 is 1.47712. For numbers from 100 to 1,000, the characteristic is 2. Therefore, the logarithm of the number 300 is 2.47712. For numbers from 1,000 to 10,000, the characteristic is 3. Therefore, the logarithm of the number 3000 is 3,47712. And so on.

Going the other way, the numbers from 0.1 to 1.0, the characteristic is negative 1, written 1. Thus, the logarithm of the number 0.3 is 1.47712. For numbers from 0.01 to 0.1, the characteristic is negative, 2 written 2. Thus, the logarithm of the number 0.03 is 2.47712.

The manipulation of logarithms with a negative characteristic is sometimes found to be a stumbling block, although there are definite rules which lead to correct results. However, it is also true that the logarithm of a decimal number or of a fraction is equal to the negative logarithm of the reciprocal of that number. Thus, log 0.3 equals -log 3.333. Now, looking up the log of 3.333 is easy and involves no negative characteristic. The implication of the minus sign in front of log 3.333 is simply that the power transaction is a loss, not a gain. Indeed, whenever we are dealing with the log of a number less than unity, a loss is involved. (Notice that the logarithm of 1.000 is zero.)

A few examples should clear up any questions. Consider y equals $\log_{10}100$. If we know how to read the log table, we can quickly determine that the log to the base ten of 100 is 2.0000. Therefore, 2.0000 is the exponent to which ten is raised to yield 100. That is, 10^2 equals 100. Again, suppose that y equals $\log_{10}$ 1,000,000. By the same process, we find that y equals 6. By substitution, 6 equals $\log_{10}$ 1,000,000. And this is also expressed, 10^6 equals

1,000,000. Looking up numbers like 100 or 1,000,000 is a cinch. But we now have a powerful means of dealing with any number, even those which must of necessity be represented by nonintegral exponential powers of ten.

Suppose, that we have the number 125. What exponential power of ten is equivalent to 125? Since 125 is now x, we find this number in the log table. It is 2.09691. (The .09691 is found in the table; 2 is the characteristic since 125 is between 100 and 1,000. We have just determined that $10^{2.09691}$ equals 125. The equivalent logarithmic expression is 2.09691 equals $\log^{10}$, 12^5.

In an algebra text we find the following equivalence: (a^m) (a^n) (a^o) equals $a^{m + n + o}$. This shows that when exponents are involved, a problem in multiplication can be converted to one of addition—a much easier process in many instances. Because logarithms are exponents, we can apply this principle. Let's do a multiplication problem with logarithms. How much is 131 × 17 × 2583? We write down the following data:

No.	Logarithm
131	2.11727
17	1.23045
2583	3.41211
	6.75983

So the product of 131 × 17 × 2583 is $10^{6.75983}$. To obtain x, the corresponding number, we find 6.75983 under logarithms and notice the number it represents as a power of ten. This process is termed finding the *antilogarithm*. In our case, we find it to be 5,752,060. Recapping, 131 × 17 × 2583 equals $10^{2.11727}$ + 1.23045 + 3.41211 equals $10^{6.75983}$ or 5,752,060, which is to say, 6.75983 equals $\log_{10}$ (5,752,060).

What may not be obvious from these computations and referrals to the log tables is the speed and ease with which arithmetically awkward operations can actually be carried out once you are familiar with the procedure. Decibels also obey the laws of exponents.

Basic Concept of the Decibel

The original use of the decibel was to provide an exponential method of representing the ratio of one power level to another. Thus, P2/P1 could represent the power gain of an amplifier if P1 is the input power required to produce output power P2. The formula which relates the number of decibels (dB) to power ratio is:

$$dB = 10 \log_{10} (P2/P1)$$

This looks familiar, because it resembles the logarithmic form of x equals a^y, which is y equals $\log_{10x}$. Indeed, if we write the formula for the **bel**, from which the decibel is derived, we have, bels equals $\log_{10}$ (P2/P1), which is identical to the logarithmic form of x equals a^y. Incidentally, the bel is an impractically large unit for the purposes of electronics, so we use the decibel, a tenth of a bel.

So far, the decibel is a *power ratio* expressed by an exponent. The overall power gain in dB of a complex system is the sum of the individual gains expressed in dB (provided that the gain of one amplifier is specified for the load condition represented by the input of a subsequent amplifier or circuitry). Attenuation, or *power loss*, is represented by negative gain, so losses are subtracted. Thus, a power gain of 20 dB followed by 3 dB of attenuation amounts to a net power gain of 17 dB. The overall result of many gains and losses in a complex system, such as radar, yield a net figure—a number of plus or minus decibels which is representative of system performance, but still on a power-ratio basis. For example, the radar return signal will be minus so many dB, which means that the power level of the echo (P2) is a tiny fraction of the outgoing power level (P1). Notice that nothing has thus far been said about voltage, current or resistance. These parameters do not appear in the decibel formula so they can assume any value.

Finding Power Ratio When Decibels are Given

In many situations, power gain or loss can be

stated in decibel form. For example, decibel meters connected to the input and output of a device or system indicate the power gain which is expressed as a numerical ratio. If the resistances or impedances associated with the two measurements are equal, the power gain (or loss) expressed in decibels is simply the difference between the two meter readings, that is dB equals P2/P1. If not, the corrective term, 10 $\log_{10}$ (R1-R2) must be added. So, in general, the conversion of the two meter readings to a decibel power ratio is, dB equals P2/P1 + $\log_{10}$ (R1/R2). When R1 equals R2, the conversion reverts to the simple procedure of taking the difference between the two meter readings. We should be careful to deal with the proper algebraic sign when performing these computations. For example, in the simpler situation of P2/P1, the result is negative when P1 is larger than P2. Gain expressed by negative decibels signifies a power loss rather than gain. If R1 is less than R2, the term $\log_{10}$ (R1/R2) works out to be negative and must therefore be subtracted from P2/P1.

The above emphasizes the need to deal with true decibel values and not, in the general case, raw dB readings from the meters. This being the case, we know that the number of decibels of gain or loss is described by the basic relationship, dB equals 10 $\log_{10}$ (P2/P1). But what is the power ratio P2/P1 when we already have its value expressed in decibels? An algebraic manipulation of this formula provides a solution in terms of the desired power ratio:

$$P2/P1 = \text{Antilogarithm} \frac{(\text{dB})}{(10)}$$

We first divide the number of decibels by ten. Then we look up the corresponding antilogarithm in the log table. This is the numerical value of the power ratio, P2/P1. As an example, suppose the true power gain of an amplifier is 23.5 dB. Dividing 23.5 by ten, we have 2.35. Here, 2 is the *characteristic* and the remaining decimal is the *mantissa*, which actually appears in the log table. The antilog of 0.35 is approximately 224. The characteristic informs us that the number must be between 10 and 100. Therefore, the power ratio, P2/P1 is 22.4.

Conversions of this kind are often made more quickly by resorting to a decibel table (Table 2-5) which shows the relationship between decibels and power and voltage ratios. Decibel tables are not as detailed as log tables, but decibel computations are often carried out in round numbers and are used in many applications to roughly specify system gains and losses rather than to strive for precision. We have purposely avoided the decibel table thus far because the application of the decibel to domains other than the ratio of two power levels bears some discussion.

Finding the Voltage Ratio When Decibels Are Given

Let's take another look at the basic decibel formula, dB equals 10 $\log_{10}$ P2/P1. From Ohm's Law, we know that P equals E^2 divided by R. If we let resistance R1 be associated with P1 and E1, we can write more specifically, P1 equals $E1^2$ divided by R1. With similar reasoning, P2 equals $E2^2$ divided by R2. If we then substitute these equivalents into the basic decibel formula, we obtain:

$$\text{dB} = 10 \log_{10} \frac{E2}{E1} \times \frac{R1}{R2}$$

This is more convenient to use in its complete logarithmic form, which is:

$$\text{dB} = 20 \log_{10} \frac{E2}{E1} + 10 \log_{10} \frac{R1}{R2}$$

When R1 and R2 are equal, this simplifies to:

$$\text{dB} = 20 \log_{10} \frac{E2}{E1}$$

This is an interesting development, for we now have expressions for power ratio in terms of voltages. Practically, it is usually much easier to measure voltages than powers. It should be clear-

Voltage Ratio (Equal Impedance)	Power Ratio	← − db + →	Voltage Ratio (Equal Impedance)	Power Ratio
1.000	1.000	0	1.000	1.000
0.989	0.977	0.1	1.012	1.023
0.977	0.955	0.2	1.023	1.047
0.966	0.933	0.3	1.035	1.072
0.955	0.912	0.4	1.047	1.096
0.944	0.891	0.5	1.059	1.122
0.933	0.871	0.6	1.072	1.148
0.923	0.851	0.7	1.084	1.175
0.912	0.832	0.8	1.096	1.202
0.902	0.813	0.9	1.109	1.230
0.891	0.794	1.0	1.122	1.259
0.841	0.708	1.5	1.189	1.413
0.794	0.631	2.0	1.259	1.585
0.750	0.562	2.5	1.334	1.778
0.708	0.501	3.0	1.413	1.995
0.668	0.447	3.5	1.496	2.239
0.631	0.398	4.0	1.585	2.512
0.596	0.355	4.5	1.679	2.818
0.562	0.316	5.0	1.778	3.162
0.531	0.282	5.5	1.884	3.548
0.501	0.251	6.0	1.995	3.981
0.473	0.224	6.5	2.113	4.467
0.447	0.200	7.0	2.239	5.012
0.422	0.178	7.5	2.371	5.623
0.398	0.159	8.0	2.512	6.310
0.376	0.141	8.5	2.661	7.079
0.355	0.126	9.0	2.818	7.943
0.335	0.112	9.5	2.985	8.913
0.316	0.100	10	3.162	10.00
0.282	0.0794	11	3.55	12.6
0.251	0.0631	12	3.98	15.9
0.224	0.0501	13	4.47	20.0
0.200	0.0398	14	5.01	25.1
0.178	0.0316	15	5.62	31.6
0.159	0.0251	16	6.31	39.8
0.141	0.0200	17	7.08	50.1
0.126	0.0159	18	7.94	63.1
0.112	0.0126	19	8.91	79.4
0.100	0.0100	20	10.00	100.0
3.16×10^{-2}	10^{-3}	30	3.16×10	10^{3}
10^{-2}	10^{-4}	40	10^{2}	10^{4}
3.16×10^{-3}	10^{-5}	50	3.16×10^{2}	10^{5}
10^{-3}	10^{-6}	60	10^{3}	10^{6}
3.16×10^{-4}	10^{-7}	70	3.16×10^{3}	10^{7}
10^{-4}	10^{-8}	80	10^{4}	10^{8}
3.16×10^{-5}	10^{-9}	90	3.16×10^{4}	10^{9}
10^{-5}	10^{-10}	100	10^{5}	10^{10}
3.16×10^{-6}	10^{-11}	110	3.16×10^{5}	10^{11}
10^{-6}	10^{-12}	120	10^{6}	10^{12}

Table 2-5. Decibel Table.

ly understood that, even though voltages are "plugged into" these equations, the number of dB thereby obtained means exactly what is meant in the basic decibel formula. We compute so many dB and this still corresponds to the power ratio, P2/P1, even though P2 and P1 were neither measured nor used in the equations involving E2 and E1. But remember that if resistance R1 and R2 are not equal, the corrective term, $10 \log_{10}$ R1/R2 must be added to the last formula. With this caution in mind, the inverse relationship can be worked out by algebraic manipulation so that we can find the voltage ratio corresponding to a given number of decibels. For this purpose, we have:

$$\frac{E2}{E1} = \text{antilog } \frac{dB}{20}$$

We first divide the number of decibels by 20. Then the antilog of this value yields the desired voltage ratio, E2/E1.

Table 2-5 shows both voltage and power ratios in terms of decibels. Notice that the voltage ratio column carries the caption equal impedances; that is, R1 equals R2. However, the table is also valid when R1 differs from R2 when the corrective term, $10 \log_{10}$ R1/R2 is applied. In the table, we see that 20 dB corresponds to a power ratio of 100 and a voltage ratio of 10. This means that 20 dB of power gain corresponds either to a 100-to-one power ratio, or to a ten-to-one voltage ratio. The power ratio basis of the decibel system must always be kept in mind. Voltage measurements and voltage ratios are practically convenient but indirect methods of arriving at the corresponding number of decibels, which is a power ratio. If power measurements were easy to make and the appropriate power meters were commonly available, the decibel system would be easier to use because we would never have to be concerned about equal resistances or impedances.

When Is a Decibel Not a Decibel?

It has been shown that for the condition, R1 equals R2, the decibel formula can be expressed in terms of a voltage ratio, thus:

$$dB = 20 \log_{10} \frac{(E2)}{(E1)}$$

When R1 and R2 are not equal, going through the step-by-step derivation, it is as follows:

$$dB = 20 \log_{10} \frac{(E2)}{(E1)} + 10 \log_{10} \frac{(R1)}{(R2)}$$

Although more complicated, this formula does enable us to obtain the correct decibel value corresponding to voltages which are not developed across equal-value resistances.

The use of decibels to represent voltage ratios with a disregard of resistance values has developed a considerable following. This author questions the wisdom of this practice, for it is all too obvious that much confusion and erroneous interpretations are the result. An example where you could encounter such decibel use might be in the evaluation of the "voltage gain" of a radio receiver. Seemingly, no harm is done by expressing in "uncorrected" decibels the voltage ratio of the audio output to the number of microvolts present at the antenna terminal. The relative number of dB quoted appears to be a good way to express the sensitivity of a receiver if you consider, the more signal, the better. Fundamentally, the receiver cannot be compared to a voltage stepup transformer, but rather to a power amplifier. It is perfectly all right to deal with the input and output voltages, but the correction implied by the $10 \log_{10}$ (R1 divided by R2) term should not be ignored if we are interested in engineering accuracy. Otherwise, a comparison of the sensitivity of two different receivers is necessarily open to question.Some advocate the use of the decibel to indicate a ratio of one quantity to another. Such use may be said to be no less accurate than the strict mathematical approach, but this appears debatable. And in the meantime, the intermingling of the two types of decibels inevitably causes confusion.

An interesting insight into the nature of the two decibel definitions can be gained from the evaluation of a stepup transformer as shown in Fig. 2-19. If the secondary of a transformer delivers ten times the voltage across its primary, the ac impedance of the primary will be one hundredth that of the secondary; this is R1/R2. If the second term of the above formula is ignored, the transformer apparently has a gain of 20 dB. This corresponds to the voltage gain of 10, which appears to check out. However, recall that the classic use of the decibel is as a unit designating the ratio of one power level, P2, to another, P1. The power gain of any ideal

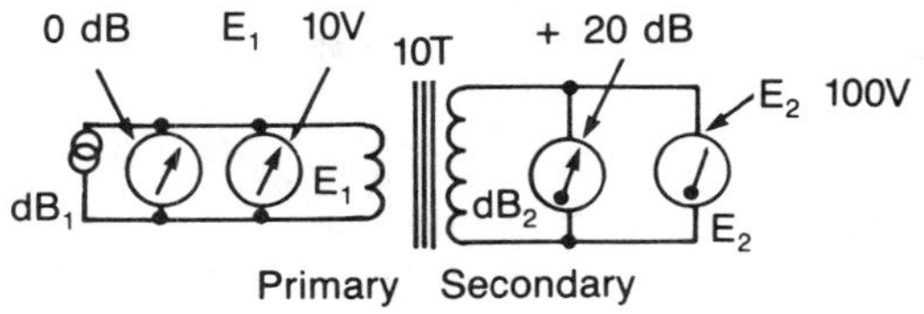

0 dB E$_1$ 10V 10T + 20 dB

E_2 100V

dB_1 E_1 dB_2 E_2

Primary Secondary

Contention 1: This transformer has a "voltage gain" of 20 dB because the voltage stepup ratio is ten to one. Thus, $dB = 20 \log_{10} (100/10)$

$dB = 20 \log_{10} (10)$

$dB = 20 (1) = 20$ which apparently proves contention 1.

Contention 2: If meters with dB scales are used, it is readily "proven" that the secondary developes 20 dB more voltage than is applied to the primary because we see this evidence directly on the dB scales.

Rebuttal: This system can conceivably serve useful purposes in dealing with ratios. However, some name other than *decibel* should be used to designate the unit. Otherwise, confusion and error will result when this system is used together with "classic" decibels! For example, a transformer of **any** ratio cannot exceed unity power gain or **zero** dB. **Both** the dB scale and the voltage scale of meters must be **corrected** when the two readings forming the "ratio" involve different resistances. In this transformer, the ac impedance of the primary is $1/10^2$ or 1/100 that of the secondary. These impedances can be considered to be resistive. The "corrected" formulas for decibel units are:

True dB from voltmeter readings $= 20 \log_{10} (E_2/E_1) + 10 \log_{10} (1/100)$

True dB from dB meter readings $= dB_2 - dB_1 + 10 \log_{10} (1/100)$

It will be seen that both formulas reduce to $20 - 20 = 0$. Because zero dB corresponds to unity power-gain, we see that our 10:1 transformer does not boost power level in the manner of an amplifier.

Fig. 2-19. Transformer "voltage gain" vs dB levels.

transformer is unity, regardless of the voltage ratio! Interestingly enough, when the second terms, $10 \log_{10} (R1/R2)$, of the formulas are used, the expressions for true dB levels reduce to the logarithms of zero. This, of course, corresponds to a power gain of one or unity. As previously stated, even though voltages are inserted into the dB formulas, the number of dBs computed should, by classic definition, designate the corresponding power ratio.

The author is aware that there will be those who see nothing wrong with stating that this transformer has a voltage gain, or a voltage ratio, of 20 decibels. If this philosophy of the decibel must be used, then extreme care should be taken that it is not used interchangeably or in conjunction with the classic decibel system. Figures 2-20 and 2-21 show the correct techniques when voltage readings are made and when dB readings are made, respectively.

Decibels and Voltage

The frequency-response curves of amplifiers, resonant circuits, and filters are often a plot of voltage ratio, expressed in dB versus frequency. This is legitimate because the same resistance is involved no matter what the frequency. (This is not exactly true, but tends to be essentially so within the passband of the device. Outside of the passband,

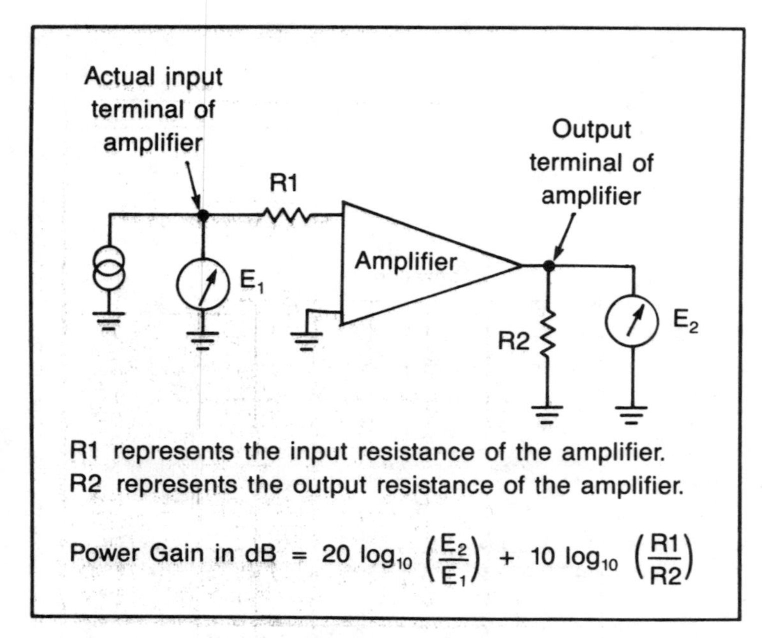

Fig. 2-20. Illustrated here is the correct use of the decibel system to indicate power gain when voltages are measured. R1/R2 can be ignored only in the singular case where R1 = R2. But when this is the case, power gain in dB = 20 $\log_{10}$ (E_2/E_1).

the general trend of attenuation as a function of frequency seems to be depicted accurately enough in most instances.)

Frequency-response curves of various filters are shown in Fig. 2-22. All of these curves represent relative amplitude response characteristics in terms of decibels, but we observe the following designations:

(A) Attenuation—dB
(B) Relative Voltage Response in dB
(C) Response—dB
(D) Amplitude Ratio (E_{out} E_{in})
(E) Voltage Gain (dB)
(F) Power Insertion Loss in dB

Additionally, there are other differences in the

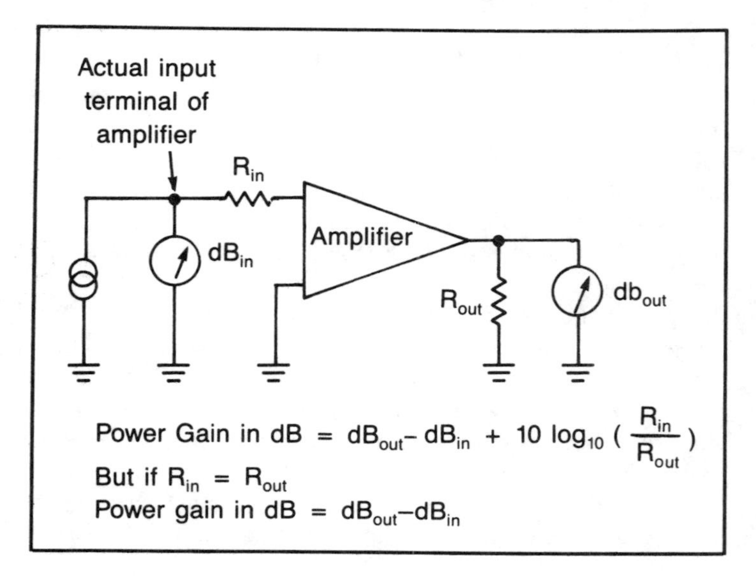

Fig. 2-21. Diagram showing the correct use of the decibel system to indicate power gain when decibels are measured.

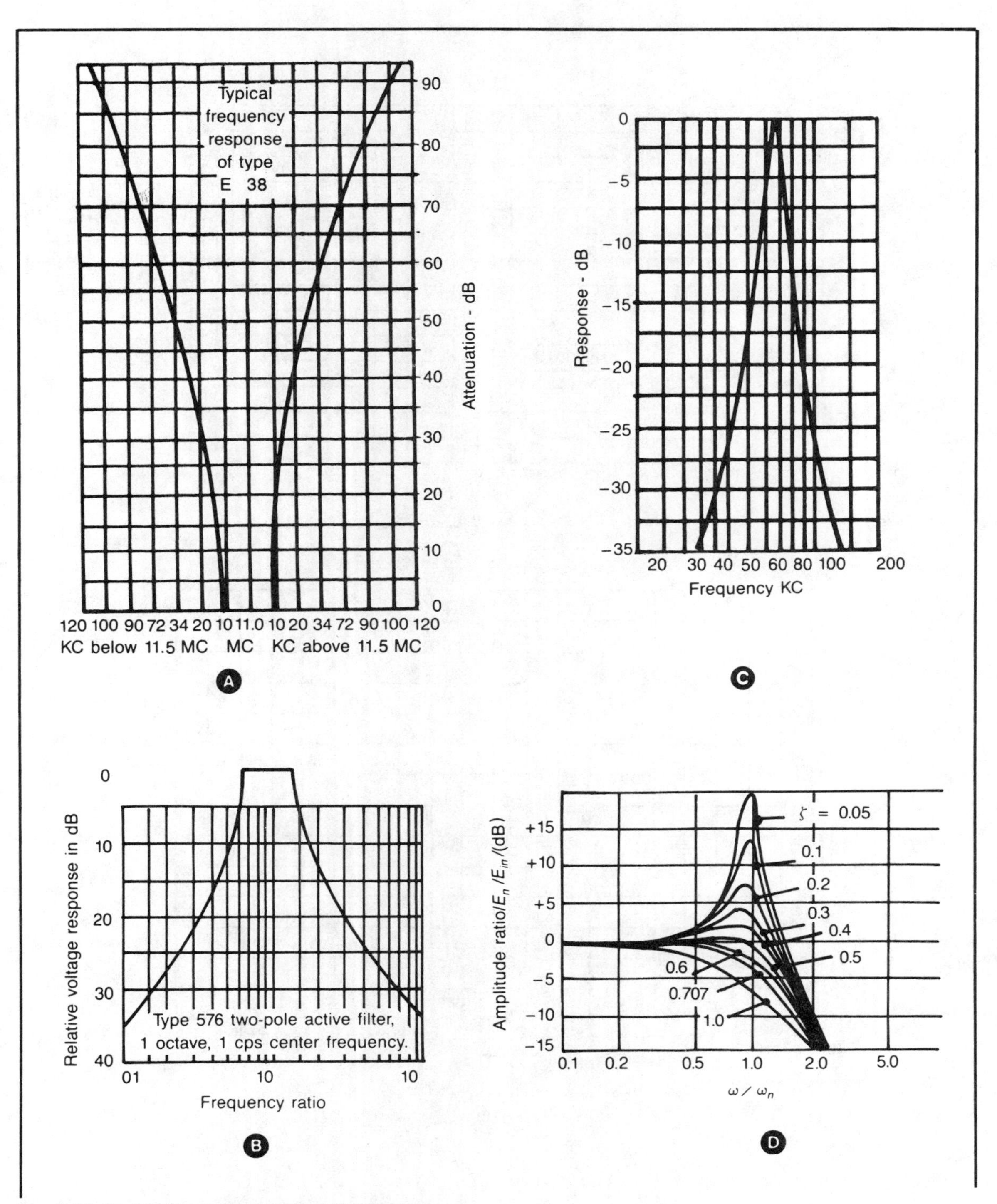

Fig. 2-22. Typical frequency response curves.

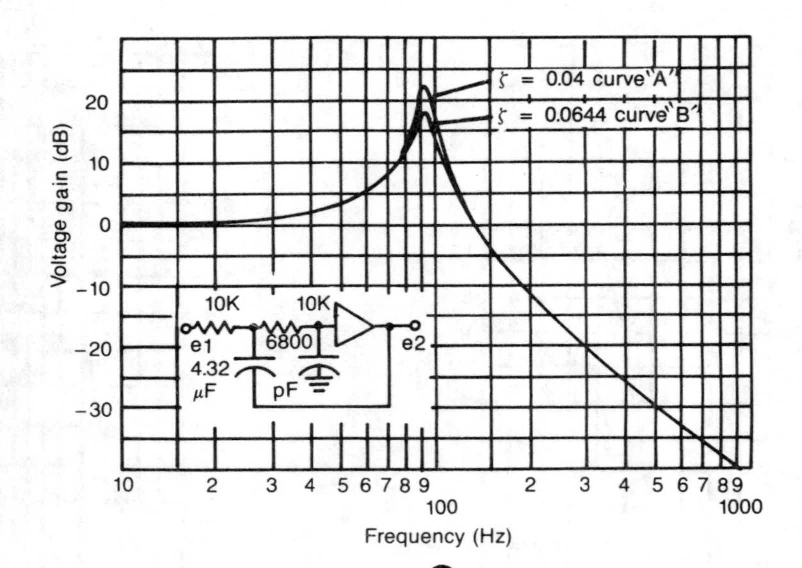

ζ = 0.04 curve "A"
ζ = 0.0644 curve "B"
Voltage gain (dB)
20
10
0
−10
−20
−30
10K
10K
e1
6800
e2
4.32
μF
pF
10
2
3
4
5
6
7
8
9
100
2
3
4
5
6
7
8
9
1000
Frequency (Hz)

E

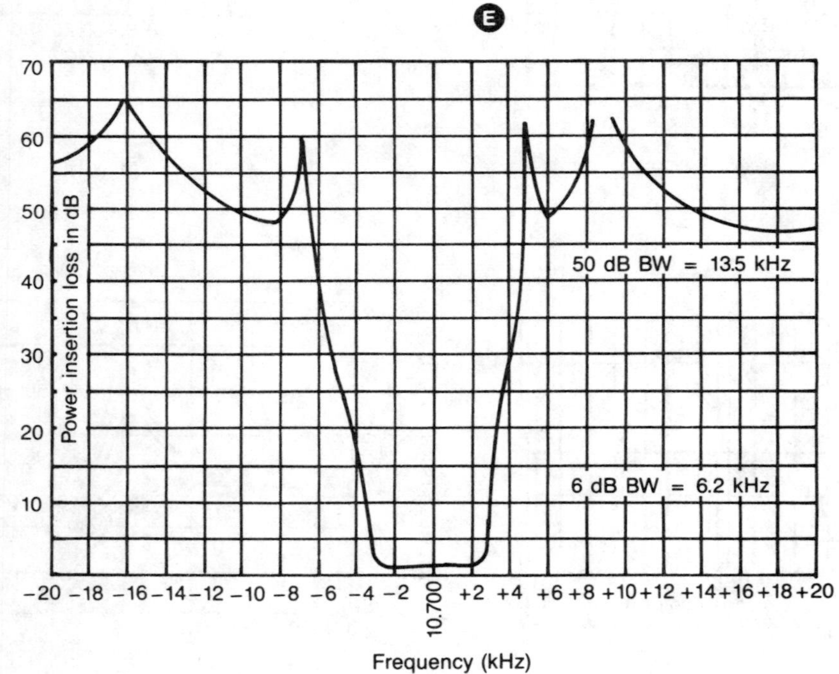

70
60
50
40
30
20
10
Power insertion loss in dB
50 dB BW = 13.5 kHz
6 dB BW = 6.2 kHz
−20 −18 −16 −14 −12 −10 −8 −6 −4 −2
10.700
+2 +4 +6 +8 +10 +12 +14 +16 +18 +20
Frequency (kHz)

F

manner in which the curves are plotted, some being vertically inverted with respect to others. Some use logarithmic scales for frequency, others do not. Let's try to resolve this confused state of affairs. At the very outset, as long as scales are properly labeled or described, there is no guessing. In the author's opinion, the response curve bearing the designation, "Power Insertion Loss in dB" (Fig. 2-22F), most faithfully carries out the classic intent of the decibel system. The other curves also are plots of relative power vs frequency whether or not the word or symbol for voltage is involved. In every one of these plots, there is a reference level. This reference is taken in the vicinity of the middle of the passband in bandpass filters and at a frequency as low as possible in low-pass filters. Therefore, outputs at other frequencies are related by certain ratios to the level at the reference frequency. If these ratios are expressed in decibels, it does not make any difference whether the initial measurement was in terms of voltage or power. Remember that response curves of this kind are based on readings which are supposed to occur across the same resistance or impedance; that is, the output of the filter, amplifier, or other device. Therefore, everything that has been said about decibel measurement with equal resistances applies here. In principle, the curves in Fig. 2-22 could have been derived from voltage or current measurements, from power measurements, or from readings directly from the dB scale of an electronic meter. (Current measurements are discussed in a later paragraph.) The reference level can be output power or input power. If it is input power, the insertion power gain or insertion power loss of the device is obtained. If the reference level is output power, the same response shape will be obtained, but the insertion gain or loss will not be seen on the plot. Of the five response-curves shown in Fig. 2-22 only curve F depicts an insertion loss. This response-curve represents the best engineering approach of the group because its deployment of the decibel concept is the most direct, and the least ambiguous. Note that the term "loss" implies that the dB indications are negative.

It is true that all the frequency response curves of Fig. 2-22 display useful information. With the possible exception of curve F, they can also lead to incorrect interpretations by technical people not privy to the impedance levels involved. Even curve F could be refined by the inclusion of the input and output impedances. (In the telephone industry, and with filter manufacturers, the terminology, "Power Insertion Loss in dB" implies that input and output impedances of the measured device are equal, or that the appropriate mathematical correction has been made.) Figure 2-23 shows the general setup used in measuring frequency response when input and output impedances are the same, as is often the case with passive filters. If these impedances are not equal, the more generalized decibel equations shown in Figs. 2-20 and 2-21 must be used.

The output impedance of the sine-wave oscillator should be low compared to the input impedance, R_{in}, of the measured device. Sometimes this requisite can be met by deliberately shunting a low resistance across the oscillator output terminals.

When frequency response plots are available for amplifiers or filters which are to be cascaded in a system, the overall response is easily obtained by simply adding the curves. This is due to the fact that the decibel, being an exponent, converts multiplication to addition. An example of this procedure is shown in Fig. 2-24. The −3 dB "break points," and the change-of-slope junctions are encircled in order to point out that some approximation is involved in these regions. For most practical purposes, this turns out to be more of an asset than a liability. Also, the straight-line depiction of the attenuation regions may not be precisely the way the actual measured response falls off. However, the simplified representation by a linear slope does have mathematical support from the equations used to determine attenuation. The attenuation of many devices is indicated with good accuracy by a linear slope of so many dB per octave, or so many dB per decade of frequency.

Notice that the response curves are plotted on log-log scales. It is true that the graph paper has logarithmically spaced divisions for frequency but linearly spaced divisions for response. However,

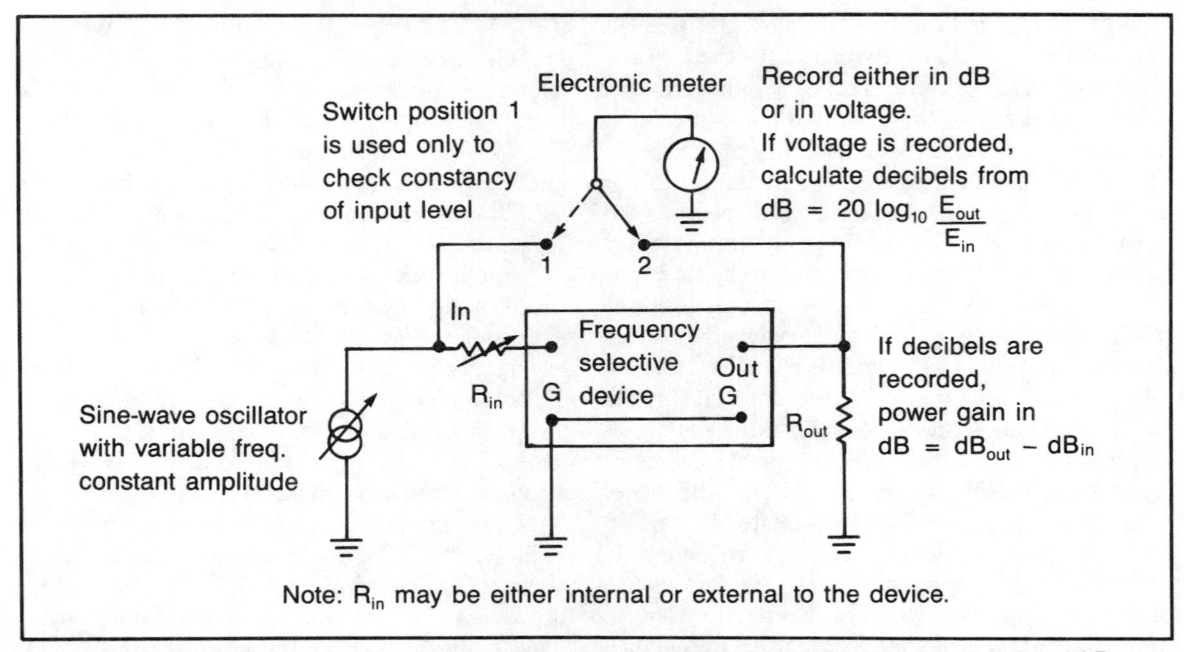

Fig. 2-23. Setup for measuring the frequency response of amplifiers, filters, tuned circuits, etc. when R_{IN} and R_{OUT} are equal. If these impedances are not equal, refer to Figs. 2-20 and 2-21 for appropriate corrective computations.

because response is plotted in decibels, the net result is that both parameters are actually plotted on logarithmic scales. An advantage of such plots is that individual response characteristics can be graphically or arithmetically combined to yield the overall response when the individual units are cascaded. For proper results, the response of an individual device should be the same as that developed when operating the cascade. For example, suppose amplifier 1 is to drive amplifier 2. Let the input impedance of amplifier 2 be 500 ohms. Then, the response of amplifier 1 should be derived with its output loaded by a 500-ohm resistance. In most instances, amplifier input and output impedances can be assumed to be resistive. In some cases, a more accurate simulation may be called for, particularly at higher frequencies. In Fig. 2-24, the slopes of the attenuating regions are most easily combined by first determining their individual decline rates in decibels per octave or in decibels per decade of frequency. These rates are arithmetically added to give a resultant slope. This makes the combining process relatively easy, for it is then only necessary to plot the resultant slope to be parallel to a line representing the summation of the individual slopes.

This use of the decibel system can save considerable mathematical analysis in determining some basic things about the overall situation. You know what frequency response to expect and can also reach valid conclusions concerning the stability of the combination from the resultant attenuation slopes.

Range Switching with Decibel Scales

When recording response, and it is necessary to switch, say, to a more sensitive range, do exactly what the range switch instructs. For example, if we are initially using the 0 dB range and soon find it necessary to switch to the more sensitive -10 dB range, continue to read the dB indication on the meter scale, but add minus 10 dB to each reading. Thus, a plus 1 dB indication is now actually minus 9 dB. Similarly, a minus 4 dB indication must be recorded as minus 14 dB. Obviously, you must be

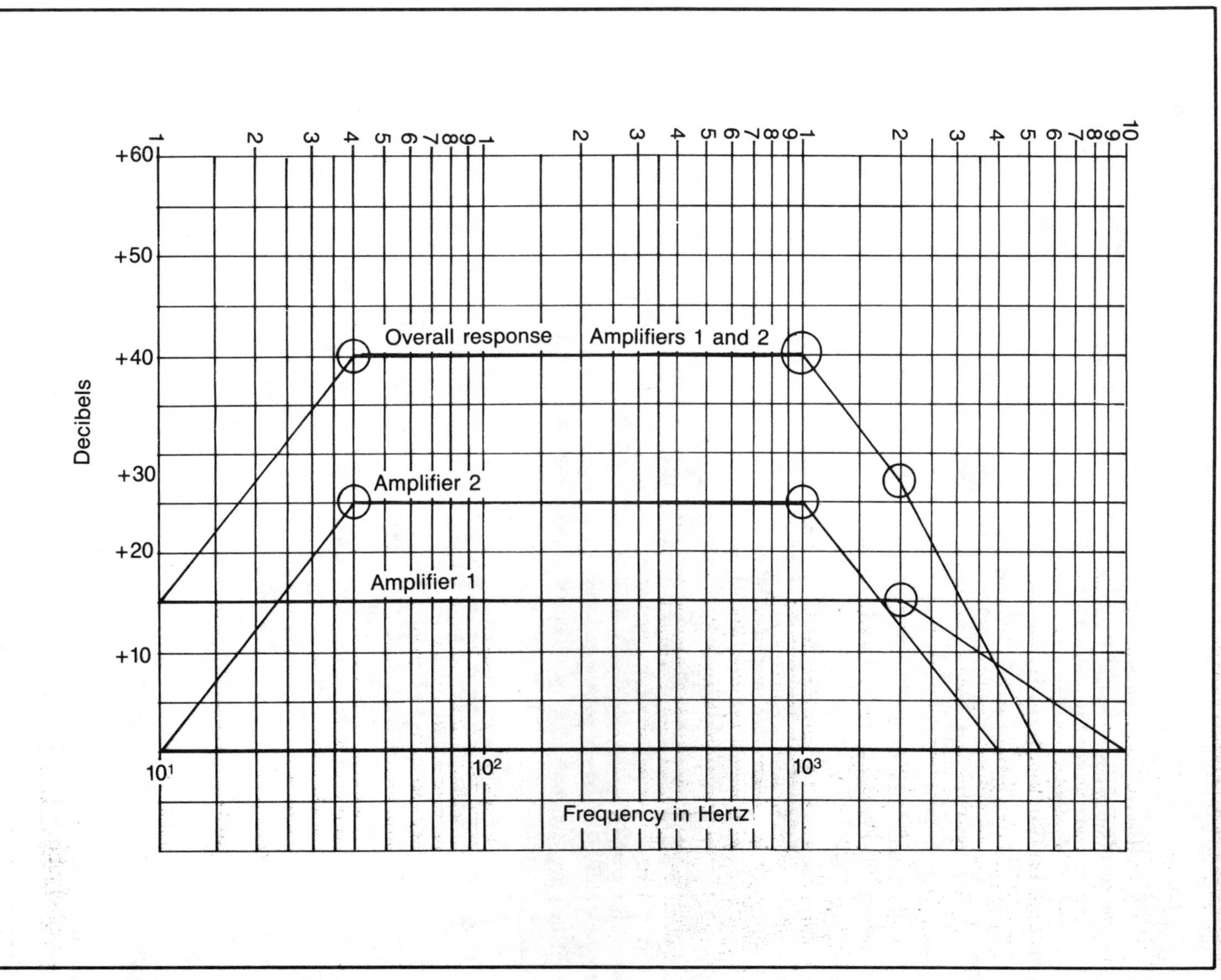

Fig. 2-24. Response curve of cascaded amplifiers.

careful when making decibel readings. The procedure is simplicity in itself, but it requires concentration. Otherwise, at a later time, we find ourselves asking, "now, shouldn't this reading of minus 19 dB really be minus 29 dB?"

Decibel Scale Changing on VOMs

VOMs and other meters usually do not have a range switch for the decibel scale. Rather, the relevant information is printed on the meter face and designates the number of dB which must be added to the actual reading when the range is changed. A dB range change is brought about in such meters by switching the voltmeter range. This is not as easy to use as is the multidecade dB selector switch on a meter designed specifically for decibel readings. The dB scale is sort of an afterthought, at least on the inexpensive VOMs. Nonetheless, a little practice will help you to make useful decibel measurements.

Use of the Decibel Scale for Measurement of Absolute Power Level

It may appear contradictory to say that the decibel system can be used to measure an absolute power level. We have repeatedly emphasized that the classic use of decibels was as a measure of the ratio of one power level to another; that is, of P2/P1. However, if the circumstances accompanying P1 are pinned down to a certain voltage developed across a certain resistance, then the power developed in R1 can serve as a reference power level. Thereafter, any P2/P1 ratio can be interpreted as so many times the power in P1, or in decibel notation, as a power level so many dB above or below P1. This use of the decibel system is discussed in detail in Chapter 3.

Decibels in Current Measurement

Almost everything said concerning voltage decibel relationships is likewise true of current measurements. The primary reason current has not been given as much attention is due to the fact that it is not used nearly as often as voltage in practical electronics measurements. However, when it comes to dealing with decibel relationships, some engineering texts consider current as the more fundamental parameter. When equal resistances are involved in the measurement of two currents, the calculation of decibel relationships is exactly analogous to that of voltage. Also, the use of the decibel table corresponds exactly to the procedure you would follow with voltage readings. However, for unequal resistances, the "corrective" part of the overall decibel formula is different from the voltage case. For clarity, the formulas involving current are:

$$dB = 10 \log_{10} (I2/I1)^2$$

when resistances R1 and R2 are equal

$$dB = 20 \log_{10} (I2/I1)$$

most commonly used form—again, only when R1 is equal to R2

$$dB = 20 \log (I2/I1) + \log_{10} (R2/R1)$$

when R1 and R2 are not equal

$$dB = dB_2 - dB_1$$

This is the way the number of decibels would be determined if a current meter had a decibel scale and we made two measurements involving equal resistances.

$$dB = dB_2 - dB_1 + 10 \log_{10} (R2/R1)$$

This is the way the number of decibels would be determined if a current meter had a decibel scale and we made two measurements involving different resistances.

The last two formulas tend to be of academic interest because a decibel current meter is not a commonly encountered instrument. It should be noted that in the third and in the fifth formulas, the corrective term utilizes R2/R1 rather than R1/R2 as in the case with voltages. For the inverse situation where we have the number of decibels, but wish to know the corresponding current ratio, the following formula may be used (with the provision that R1 and R2 are equal):

$$\frac{I_2}{I_1} = \text{antilog} \frac{(dB)}{(20)}$$

Of course, if the decibel value is a simple number, a quicker solution of the current ratio can be found in the decibel table.

Procedure for Evaluating Filter Response by Decibel Notation

Objective of Test: To use both the ac voltage and the decibel scales of an electronic meter to derive data for plotting the frequency response of a filter.

Test Equipment Required:

Inductor

Two .047 μF Mylar capacitors, plus or minus 10 percent. 200v

Two 10K composition resistors, plus or minus 10 percent, 1/2 watt

One 100-ohm composition resistor, plus or minus 10 percent, 1/2 watt

One electronic meter with RMS voltage scales and decibel scale

One sine-wave oscillator

One double-throw, single-pole switch.

Test Procedure: Connect the equipment as shown in Fig. 2-25. The LC network is a single pi section low-pass filter with a cutoff frequency of approximately 630 Hz. The 10K resistors must be considered to be part of the filter in this type of test. The 100-ohm resistor lowers the effective output impedance of the oscillator so that its behavior more closely approximates that of a voltage source. This technique restricts the available amplitude range from the oscillator, making it necessary to work at lower levels than would otherwise be necessary. This is acceptable, but the filter and its connections should be made with short leads, so that electrical noise does not become significant at low-level operation.

Initially, set the oscillator at its maximum output at a low frequency, say 90 Hz. With switch SW1 in its No. 1 position, find the least-sensitive dB scale where a minimum reduction of oscillator amplitude suffices for an indication of 0 dB. (Other dB indications can be used, but a zero setting is especially convenient.) Record the actual level by combining the dB designation of the range switch with the meter indication. For example, if the meter indication is zero, as suggested, and the dB range switch is positioned at – 10 dB, then the actual level is – 10 dB. This establishes the input reference level for the test procedure. If our filter is properly designed and implemented, the following conditions will prevail: (1) If switch SW1 is now placed in its No. 2 position, the output of the filter will be very nearly 6 dB lower than the input reference level. (Such image parameter filters exhibit a 6 dB insertion loss in the passband.) Call this power level the output

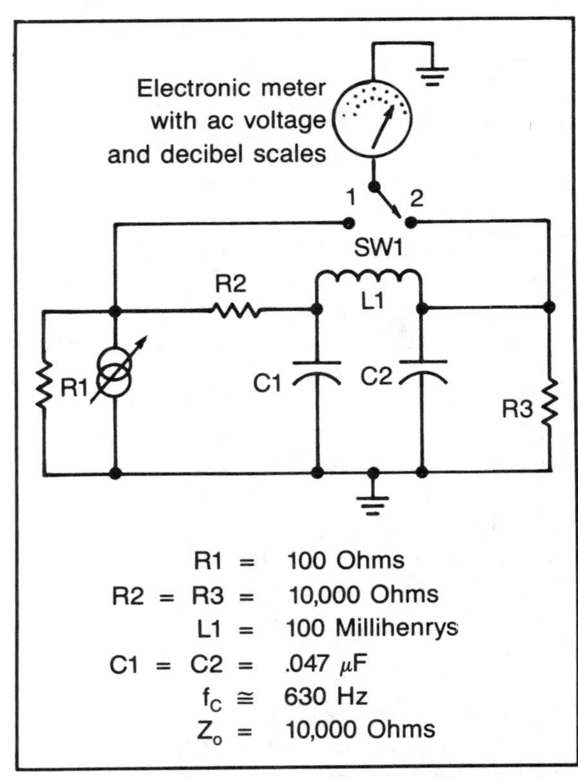

Fig. 2-25. Circuit for measuring the response of a low-pass filter.

reference level. (2) All subsequent dB values read from the meter when SW1 is in its No. 2 position will be no greater than, say, one dB higher than the output reference level. (3) As the frequency is increased past the cutoff frequency, the slope of the plotted response will be approximately 18 dB per octave.

In carrying out this test, switch SW1 should be placed in its No. 1 position frequently to make sure the input reference level remains constant. Usually, several or more adjustments of the oscillator output amplitude control will be necessary as the test proceeds.

The question always arises with regard to the number of measurements to make. A quick answer is the more the better. However, beyond a certain number, additional measurements become tedious to make, and are actually very little additional value. Some skill must be developed, though, if you are to get the maximum graphical information from the fewest readings. For example, if a rapid change of oscillator frequency shows the passband to be substantially flat, there is no need to take an abundance of measurements in this region. Generally, it is wise to make a series of fairly closely spaced measurements in the vicinity of the cutoff frequency. If the filter is "clean," the transition into the attenuation region will be substantially a constant number of dB per octave and it will be unnecessary to plot numerous points. However, it is wise to record down to 50 dB below the output reference level.

Repeat the measurements, using the same frequencies, but record ac volts instead of dB. The input voltage can be 0.1, 0.5, or 1.0 volt or other convenient value so that the ratio of output to input voltage can be computed mentally. Then, a second set of dB values can be computed from the formula,

$$dB = 20 \log_{10} (E2/E1)$$

Where convenient, use the dB table. The computed decibels should closely check the recorded decibels at each frequency. Plot the response of the filter on 5-cycle semilog paper. Label four of the decade log divisions from 10 Hz to 100 kHz. The linearly spaced horizontal divisions can each represent 5 dB intervals.

Comments on Test Results: The use of switch SW1 allows the test to be carried out with a single meter. Actually, the use of two meters is advantageous because the input reference level can be seen constantly. The input meter can be quite inferior to the output meter, since the sole function of the input meter is to make sure the input level is held constant. An oscilloscope can be used for this purpose, too.

A mistake commonly made by those making these kind of tests for the first time is to monitor the input level across capacitor C1, since it appears to be the input of the filter. This appears schematically logical, but image-parameter filters must be operated on a matched impedance basis; that is, the filter proper must "see" its own characteristic impedance both at the input and the output. That is why resistors R2 and R3 are associated with the filter. For practical purposes, R2 becomes part of the filter network and the input is measured across the oscillator as shown.

It would be instructive to experiment with curve-plotting techniques other than the one suggested. For example, you could draw a plot of output voltage vs frequency on simple nonlog graph paper. However, you will see that it is quite difficult to cover an extensive range, either in frequency or amplitude, without severe crowding. This is so even though we can start the frequency scale from zero, which we cannot do on a log scale. But, this is no real advantage for two reasons. It turns out that if the lowest frequency at which the first response is measured for a low-pass filter is a small fraction of the cutoff frequency, the amplitude corresponding to zero frequency (dc) will be substantially the same. Secondly, the test oscillator has its own minimum frequency.

In actual practice, there are two ways in which the E2/E1 ratio can be set up. If E1 is the input reference level, the plot of the response will show the approximately 6 dB insertion loss inherent in this type of filter. This method probably represents the best engineering approach. However, the sales department often prefers the situation where the response is plotted by using E1 as the output reference level. When this is done, substantially the same curve is derived as in the first method, but the passband coincides with the zero dB line on the graph. In other words, you do not see the 6 dB insertion loss. Obviously, this can be misleading. In some filters, the insertion loss, for various reasons, is much greater than 6 dB and the practice borders on the deceptive. On the other hand, you can show the passband at zero dB if the insertion loss is plainly labeled on the plot.

QUESTIONS

2-1. An ammeter was inserted in a circuit where it was known from other information that a 4-ampere current was flowing. However, the ammeter indicated about 3 1/2 amperes. The technician contacted the manufacturer and told them the 5-ampere full-scale meter was considerably in error. The manufacturer insisted that the meter was guaranteed to be accurate within plus or minus 2 percent of its full-scale indication, that it was a quality instrument with sapphire bearings, etc. However, in the interest of good customer relations, he made available another instrument in two days. The second ammeter also read approximately 3 1/2 amperes. An alert technician made a test with the two meters in series and it was then found that both meters indicated less than three amperes. What was the circuit difficulty that had been overlooked?

2-2. An electronic meter was not available to make ac voltage measurements on a high-impedance circuit. However, a high-sensitivity VOM (200,000 ohms per volt) was used because a quick calculation indicated that this meter had an internal resistance of ten megohms on the 50-volt scale. Therefore, it was felt that this meter should load the circuit about as little as an electronic meter. Was this conclusion justified?

2-3. A rectifier meter (ac voltage scale on a VOM) and an iron-vane meter are used to measure voltage in a low-impedance 60-Hz circuit. Both meters indicate substantially the same voltage. When a heavy load is placed across the ac power line, the two meters show reduced readings as might be expected, but now they also differ in their voltage indications. What does this suggest?

2-4. What instrument can measure the voltage from collector, plate, or drain of an ac amplifier with respect to ground and give you the most complete information about the operation of the amplifier?

2-5. Two electronic meters are connected to a function generator. Meter A presents the same voltage reading for sine, square, and triangular waves if, by means of an oscilloscope, the peak-to-peak amplitudes of these waves are maintained equal. Meter B changes its voltage reading so that the highest reading occurs with square waves and the lowest with triangular waves. What does this tell us about the meters?

2-6. Two students made voltage measurements related to the condition angle of a triac feeding a load resistance. Two voltmeters were available, one an RMS-responding electronic meter and the other the rectifier meter of a VOM. One student connected the electronic meter to the line side of the triac and the VOM to the load side. The second student reversed the roles of these two meters. The test results of the two students showed profound differences. Which student was likely to have recorded correct measurements? (Neither used corrections.)

2-7. A wattmeter is used to measure the input power, P1, and the output power, P2, of a low frequency attenuator. Is it correct to compute the decibel loss with the formula

$$dB = 10 \log_{10} \frac{P2}{P1}$$

if the input and output resistance of the attenuator are not the same?

2-8. An electronic voltmeter is used to measure the input voltage, E1, and the output voltage, E2, of an amplifier with unequal input and output resistances. Is the commonly encountered formula

$$dB = 20 \log_{10} \frac{E2}{E1}$$

valid in this case?

2-9. Decibel tables showing corresponding power and voltage ratios for given decibel values often stipulate that the two resistances associated with the voltage ratio must be equal. In what situation is a stipulation worded in this general manner apt to be misleading?

2-10. What basic distinction does the decibel system show between a voltage stepup transformer and a voltage amplifier of the same output-to-input voltage ratio?

ANSWERS

2-1. Since both meters tended toward the same indications under two different circumstances, it was very probable that neither instrument was giving an erroneous reading. Overlooked was the effect of the meter resistance on the circuit. The clincher here was the further reduction in current with the two meters in series.

2-2. No. The sensitivity of a VOM is based on the dc voltage ranges. The sensitivity of the ac voltage ranges, because of the rectifier arrangement used, is much lower. In this case the ac sensitivity would be 10,000 ohms per volt, a far cry from the anticipated ten megohms per volt.

2-3. Such a situation would result from distortion of the original sinusoidal waveshape. Since both meters are calibrated to read RMS voltages on a sine wave, they produce the same reading initially. With the equipment on the line, the RMS-responding iron-vane meter and the average-responding rectifier meter no longer sense the same voltage level. Under this condition, the iron-vane instrument remains essentially correct, but the rectifier meter produces an erroneous indication.

2-4. Dc oscilloscope because both dc and ac components can be identified and measured. Other effects such as wave distortion, parasitic oscillation, and noise can also be observed.

2-5. We know the parameter that these meters actually respond to despite their scale calibrations. Meter A is a peak-responding type, whereas meter B is either an average- or an RMS-responding instrument. (More information about the readings would be necessary to specifically determine the parameter of the wave to which meter B responds.)

2-6. The second student, because he was able to obtain correct RMS readings of the nonsinusoidal load voltage. Since the voltage on the line side of the triac remains sinusoidal, the VOM was adequate for this measurement. The first student would obtain wild readings from the VOM connected across the load.

2-7. Yes, this basic decibel formula is valid regardless of resistance. This illustrates the great advantage of making actual power measurements.

2-8. No. The corrective term

$$\log_{10} \frac{R1}{R2}$$

must be added to this formula. Here, R1 is the input resistance and R2 is the output resistance of the amplifier.

2-9. Such a stipulation is too restrictive. True, it does clearly exclude an entry of voltage ratios in which E2 and E1 were not measured across equal-value resistances. This is commendable, for a disregard of input and output resistances can make the intended use of the decibel system invalid. However, if the corrective term

$$10 \log_{10} \frac{R1}{R2}$$

is added to the basic

$$20 \log_{10} \frac{E2}{E1}$$

the error introduced by the unequal resistances is exactly compensated. Therefore, the entries and readings of the decibel table apply just as if resistances R1 and R2 had been equal.

2-10. Though an amplifier may be called a voltage amplifier, it is really a power amplifier. The decibel system specifies the power gain of such an amplifier, even though we find it convenient to view its operation as a voltage multiplier. The same viewpoint applied to a voltage stepup transformer emphatically discloses the lack of power gain, regardless of the ratio between its windings. Thus, if both transformer and "voltage" amplifier are designed to have the same input and output impedances, only the amplifier can provide true voltage gain! The only way voltage gain can exist when input and output impedances are equal is if a power gain is present. The 1:1 transformer, of course, does not meet this criterion.

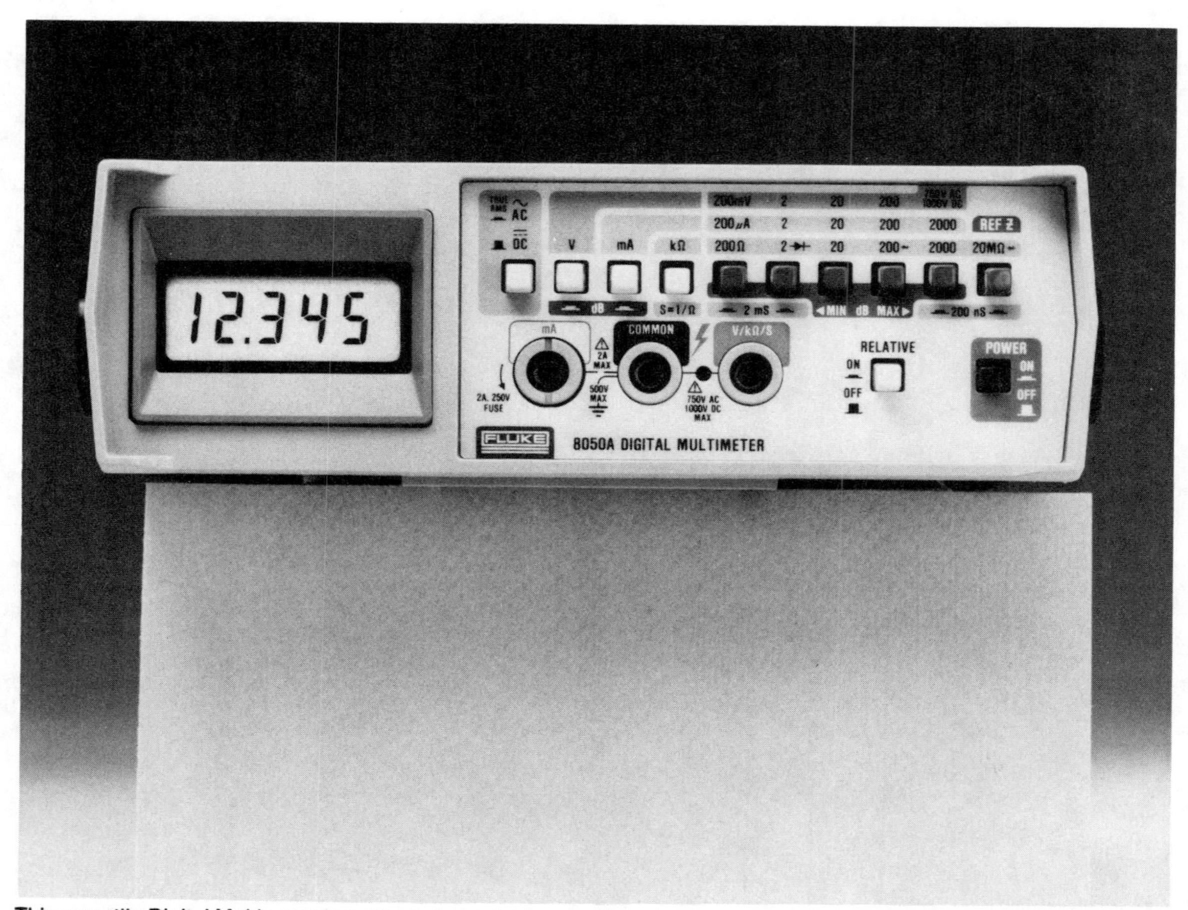

This versatile Digital Multimeter is particularly useful for a variety of power measurements. This stems from its true RMS capability and its display of dB and dBm units. By appropriate setting of the push-buttons, the readings can be referenced to a selected impedance level, thereby dispensing with computations. Courtesy of the John Fluke Mfg. Co., Inc.

Chapter 3
Power

The power level in electric and electronic equipment is not measured as often as voltage and current, but this is not an accurate indication of the importance of such measurements. Actually, we either knowingly or otherwise make many voltage and current tests which represent changes in the level of power. (If this were not so, the transformer could replace amplifiers and other apparatus, for it can provide almost any change in voltage or current desired.) If a radio receiver produces so many volts of audio signal in response to a few microvolts of rf at its input, we may be impressed with the tremendous overall voltage gain. However, basically this voltage-gain figure cannot give us a true comparison of one set with another. Only the overall power gain is truly meaningful, unless we take care to specify the input and output impedance levels of two sets being compared. But where we do that, we are actually comparing power gain even though we measure voltages.

Another example of "power behind the scene" is present in a transmitting installation where adjustments are made to provide maximum rf current into the antenna feeder line. It so happens that under proper conditions, maximum rf current at this point coincides with optimum performance of the station in terms of signal strength at distant radio receivers. However, we must not lose sight of the fact that the basic objective has been to cause the transmitting antenna to accept maximum power from the transmitter. The use of relative current measurements to determine this is only a practical convenience which is valid for a given installation, and then only on an "other-things-being-equal" basis. So many amperes of rf current flowing into the feeder line may not mean much if, for example, the physical length of the antenna changes on a very cold day from what it was on a very hot one, or if the transmitter frequency is changed appreciably. Thus, there can be conditions where a high current does not mean that an equally high power is being delivered to the antenna.

In more commonplace situations, the power delivered to equipment from the 60-Hz ac line may be very low despite inordinately high current. The measurement of current in ac equipment does not always lead to accurate information about its performance even though we also know, or have measured, the voltage also. In ac circuits, such data does not necessarily indicate the true power being consumed.

BASIC POWER DETERMINATION

The simplest approach to the measurement of power in electric circuits is illustrated by a dc circuit containing a resistance. Of course, we want to know how much power is being consumed by, or dissipated in, the particular resistance. Figure 3-1 shows three ways of determining the power dissipated in resistance R. In each instance, two of the three involved parameters must be known, but a single meter suffices if the value of R is known. This is the case in Fig. 3-1B and C. In Fig. 3-1A, notice that we deal with the current through resistance R and the voltage across resistance R. By this approach, only the power consumed by R is obtained; the power in other elements or portions of the circuit is automatically excluded.

It should also be realized that the dc source could be replaced by an ac source in the three circuits just considered, provided that appropriate ac instruments were used to measure I_r and E_r. However, this would be true only when resistance R is truly a resistive element. In the more general case where R is replaced by an impedance, Z, the formulas depicted in Fig. 3-1A, B, and C would not be applicable to the ac circuit. The measurement of power in ac circuits and systems will be taken up shortly.

The voltage and current measurements in Fig. 3-1D determine the total power delivered by the source. Here, we deal with the voltage across the source and the current flowing from (or into) it. The product of the voltage and current readings is the total power consumed by the three resistances associated with this circuit.

A common aspect of all four circuits is that the current may flow through elements other than the one which is being evaluated for power dissipation.

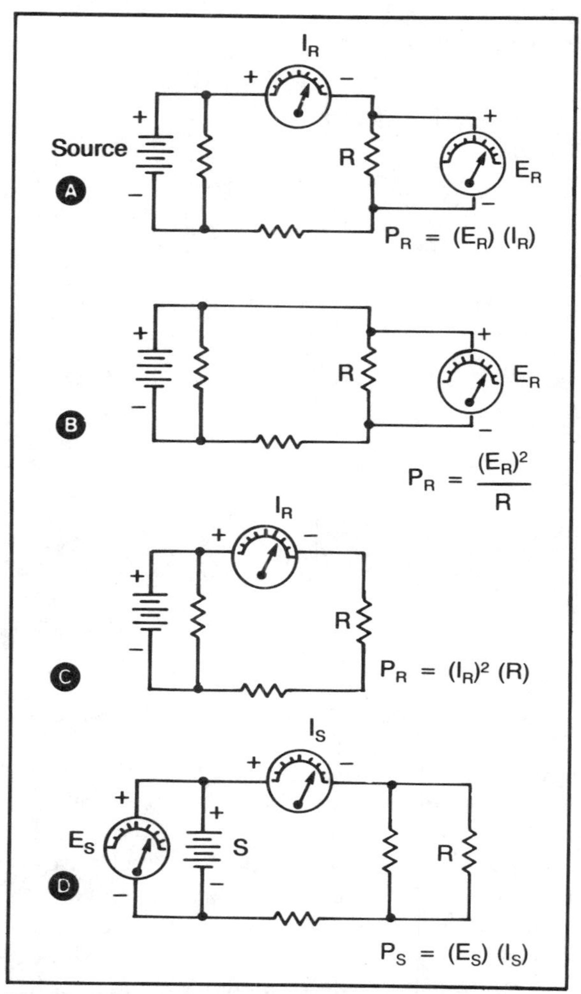

Fig. 3-1. Basic dc power relationships: A. Power consumed by resistance R; voltage across R and current through R are known. B. Power consumed by resistance R; voltage across R and R are known. C. Power consumed by resistance R; current through R and R are known. D. Power delivered by source S to the total circuit; voltage at the terminals of the source and the current supplied by the source are known.

However, the voltage measurement must be made directly across the desired element, such as R in Fig. 3-1A, B and C. It is surprising how often erroneous power determinations are made by ignoring this simple rule. Thus, had the voltmeter been connected across the battery in Fig. 3-1A, the computation would not have yielded the power dissipated in resistance, R.

MEASURING POWER IN AC CIRCUITS

Figure 3-2A shows a frequently encountered situation in ac circuits which requires essentially the same power measurement techniques discussed for the dc circuits in Fig. 3-1. The power dissipated in resistance R is determined with the same formulas cited for the dc circuits of Fig. 3-1. Although various phase displacements prevail between current and voltage throughout the circuit, it is always true that the current through a particular resistance, such as R, and the voltage across that resistance are in phase. Ac circuit power determinations using the three formulas in Fig. 3-1 are valid only when the current through the element in question and the voltage across it are in phase.

The measurement situations in Fig. 3-2A and Fig. 3-1 can be made identical by the use of ac-dc instruments such as electrodynamometer types. However, the power in the ac circuit in Fig. 3-2B cannot be measured in the same way as the power in Fig. 3-1D. Because of the previous discussion the reason may not be immediately obvious. When an ac source supplies current to a reactive circuit, the current is displaced in phase from the voltage across the terminals of the ac source. Under this condition, the simple product of source voltage and source current yields volt-amperes, a useful parameter, but not power in watts. This is why other instruments have been developed for determining the ac power into an impedance.

Power dissipation can be obtained from the volt-ampere product if we know the impedance angle, Θ, of the circuit. Thus, in ac circuits such as that in Fig. 3-2B, the power delivered by the source can be determined with the formula, P equals $E_s \times I_s \times \cos \Theta$, where $\cos \Theta$ is known as the **power factor**.

Practical Test Techniques

The test methods illustrated in Fig. 3-3 can be used to measure ac power. In Fig. 3-3A, resistance R is selected to be very low compared to the load resistance. Usually, you can get a rough idea of the load resistance from known circuit conditions. If R is then made one-fiftieth or a smaller fraction of the estimated load resistance, its disturbance of circuit operation will be negligible for most situations. The basic idea is to first obtain a reading across the load, then across R. To read the load voltage, E1, the electronic meter is set at an appropriate voltage range and the switch is placed in the No. 1 position. Next, place the switch in the No. 2 position and select an appropriate voltage range for a half- to full-scale deflection of the relatively small voltage. From this second measurement, E2, the current is computed as E2 divided by R. Notice that the arithmetic is simplified if R is 0.1 ohm, 1.0 ohms, 10 ohms, etc. The power in the load resistance is given by the following overall formula:

$$P = E1 \times E2/R$$

Although greater accuracy is possible if R is small, it should not be so low that noise voltages

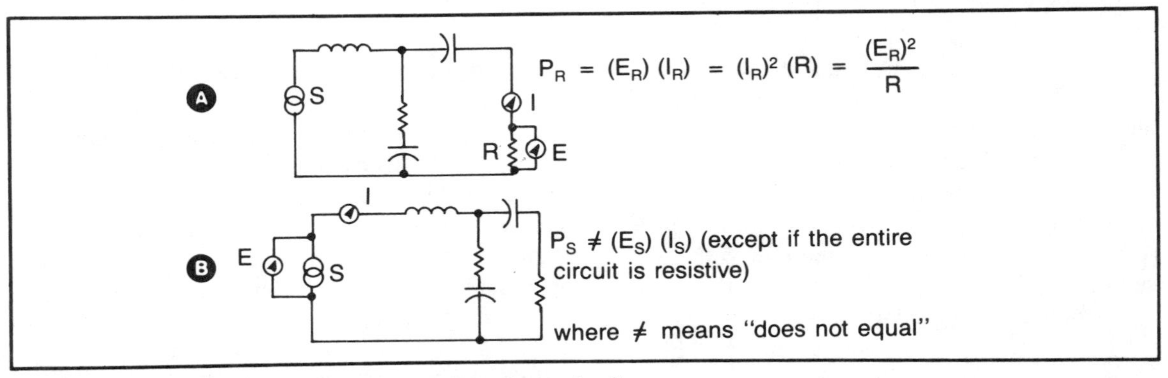

Fig. 3-2. Simple but important power relationships in ac circuits.

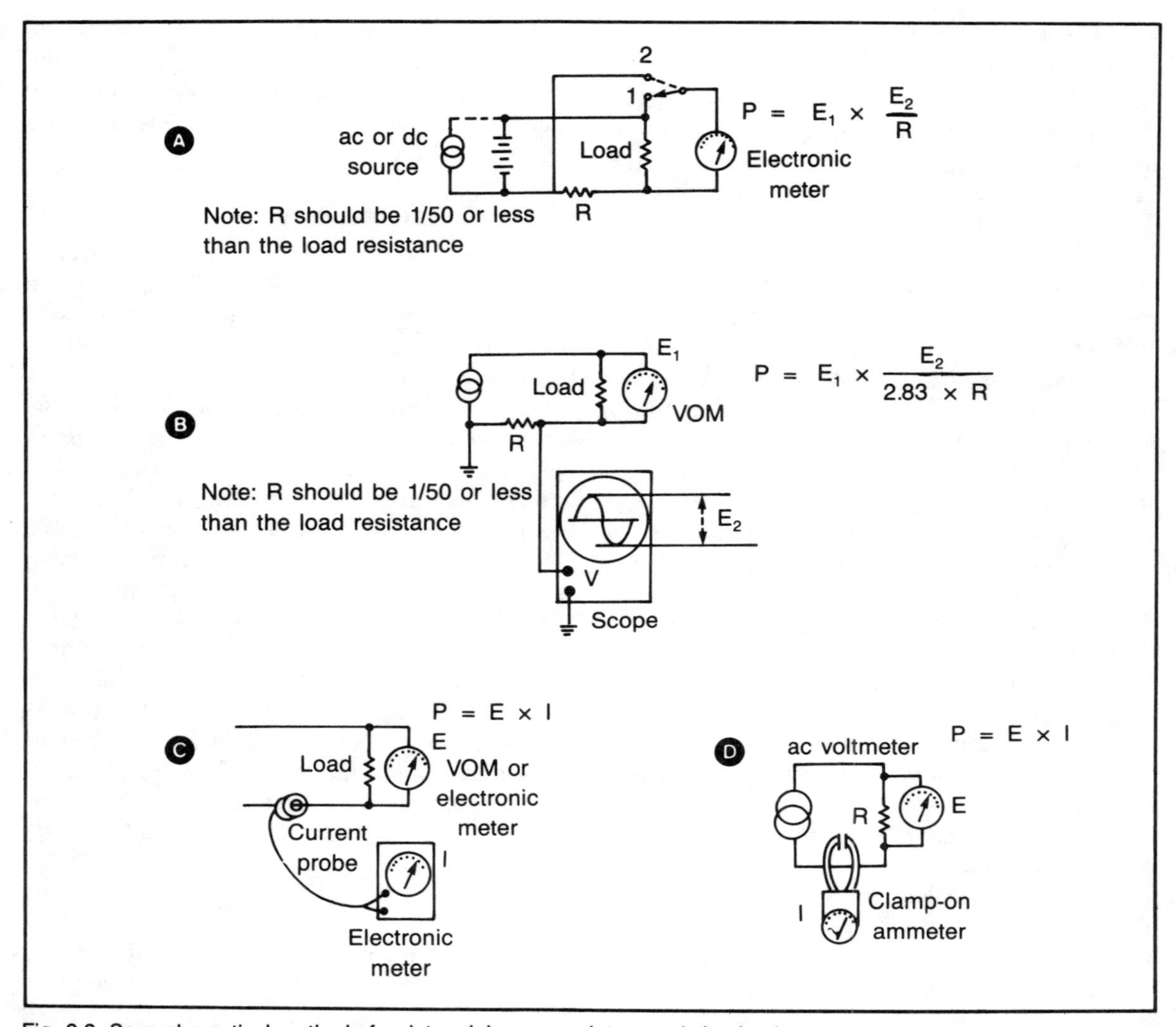

Fig. 3-3. Several practical methods for determining power into a resistive load.

can affect the reading of E2, nor should it be so low that the lowest voltage range of the electronic meter is not sufficiently sensitive to produce a half- to full-scale deflection. The lower the resistance of R, the less its power rating has to be. This usually poses no problem, for if R is one-fiftieth of the load resistance, the power dissipated in R will likewise be one-fiftieth of the power in the load resistance.

A similar test procedure is shown in Fig. 3-3B. Here, an oscilloscope is used to monitor the voltage across current-sensing resistance R. R can often be made quite low because, in addition to the sensitivity of the vertical amplifier in the scope, there is a nearly three-to-one advantage due to the peak-to-peak display in comparison to the RMS readouts of meters. Moreover, the scope provides visual evidence of noise voltages and other interferences which might become significant when R is made too low. In order to avoid ground conflicts, a VOM is shown instead of an electronic meter. The overall formula for load power is E1 × (E2 divided by 2.83R). The factor, 2.83, converts the peak-to-peak reading of E2 to the RMS value.

A third variation of these schemes is shown in Fig. 3-3C. Here, a current probe makes it unnecessary to interrupt the circuit. Also, the error introduced by the insertion of the sensing resistance, R, does not exist. Be sure that the current probe is used with its intended meter. You can get clamp-on meters capable of sensing the ac current by simple transformer action in ac circuits involving amperes through hundreds of amperes of current. This is shown in Fig. 3-3D. Make sure that the clamp is not kept from closing by dirt or foreign particles. Enclose only the conductor through which current is to be monitored unless it is definitely known that other conductors are not carrying current.

Procedure for Determining the Power Output of an Audio Amplifier

Objective of Test: To evaluate the power-output capability of an audio amplifier with commonly available test instruments.

Equipment Required:

Oscilloscope
Audio oscillator or function generator
Load resistance. This can be equal to any of the commonly encountered output impedance levels, such as 4,8,16,500, or 600 ohms. 50-watt wire-wound types are adequate for a wide variety of amplifiers used in public-address systems and entertainment equipment.
Electronic meter with RMS ac voltage scales

Test Procedure: Connect the equipment as shown in Fig. 3-4. Initially, set the audio source for a 1000 Hz sine-wave frequency and adjust the output control for zero or minimum amplitude. Tone or treble-bass controls on the audio amplifier should be set for flat response. The basic idea of this test procedure is to plot the maximum available output power versus frequency. However, a detailed plot of frequency is not always necessary. Often, the 1000-Hz determination, accompanied by one at a low frequency such as 100 Hz and at a high frequency such as 10,000 Hz, can reveal much about the basic operating condition or capability of the amplifier. Of course, the need for a more detailed analysis of the hi-fi characteristics of an amplifier will justify a number of power determinations, particularly in the regions of high- and low-frequency cutoff. In every measurement, the audio source output control should be slowly advanced until the overload point is reached, then backed down until

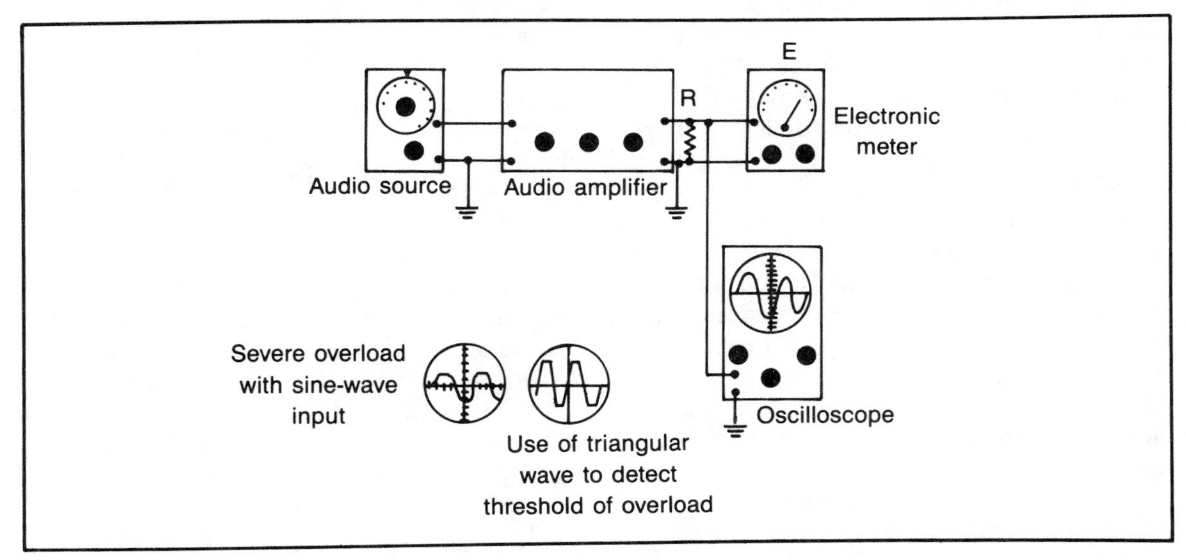

Fig. 3-4. Setup for measuring the power output of an audio amplifier.

a good visual indication of an undistorted wave appears on the scope. Record the RMS voltage from the electronic meter for this condition. Saturation and wave clipping should occur nearly simultaneously, and in equal proportion to both peaks of the wave. **Caution:** Do not operate the amplifier any longer than necessary at maximum output. An amplifier's thermal capacity may be considerably less than its electrical capability, particularly in solid-state equipment. The possibility of damage or destruction is somewhat confused because of varying methods employed by manufacturers in rating power output. However, this procedure is accurate because in all cases it defines a very real and practical performance parameter—the peak musical power capability. The power output is computed from the relationship:

$$P = \frac{(E)^2}{R}$$

Comments on Test Results: Obviously, some judgment must be exercised in evaluating the maximum undistorted output indicated by the scope waveform. This is particularly true in amplifiers where waveform clipping begins gradually. The use of a triangular wave is very helpful in making this "educated guess" more meaningful because distortion in the peaks of the triangular wave is much more obvious to the eye than is additional rounding of the tops of sine waves. (The triangular wave tends to be somewhat less effective for this purpose when the frequency is near the high-frequency cutoff of the amplifier's response. However, in most cases, it will still serve this purpose.). The triangular wave should be used only for the described test function. Voltage readings from the electronic meter should be recorded only under sine-wave operating conditions. Of course, more operational information can be gained from repeating the tests for different settings of the amplifier gain control and the bass-treble adjustment. In any event, make sure that the exact operating conditions are recorded.

Because of the non-ideal characteristics of output transformers, it is often found that performance at different impedance levels is not exactly the same. Be sure that the load resistance, R, matches the designated output impedance level.

If there is any evidence of parasitic oscillations on the scope pattern, the reason for this should be immediately determined. If such a situation is not the result of external feedback from unshielded connections, or the proximity of input and output test apparatus, the cause of the oscillation should be located and remedied: Sometimes you hear someone say that such parasitic oscillation is of no consequence. However, such a malfunction can limit the available power by upsetting biases, and by the useless dissipation of energy. Such a power limitation also tends to reduce the fidelity of the amplifier. "Birdies" and intermodulation distortion generally accompany parasitic oscillation, although detecting the presence of and the measurement of such degradation requires more sophisticated test procedures.

ELECTRODYNAMOMETER WATTMETER POWER MEASUREMENTS

Generally, voltage and current measurements are not enough to yield information about the power level in ac systems. The power in ac circuit is given by the equation:

$$P = E \times I \times \cos \Theta$$

where Θ is the impedance angle or the phase angle between the voltage, E, and the current, I. Thus, the determination of power in an ac circuit of only a slightly complex nature is quite involved.

The simplest and easiest way to measure ac power is with an electrodynamometer wattmeter. This meter indicates average power regardless of the circuit impedance or the phase displacement between voltage and current. Its torque is produced by a moving coil, the magnetic field of which reacts with that of a stationary coil. See Fig. 3-5. The moving coil senses the voltage and the stationary coil senses the current in the external circuit. Only effective power is capable of producing mechanical torque, so the meter ignores components of volt-

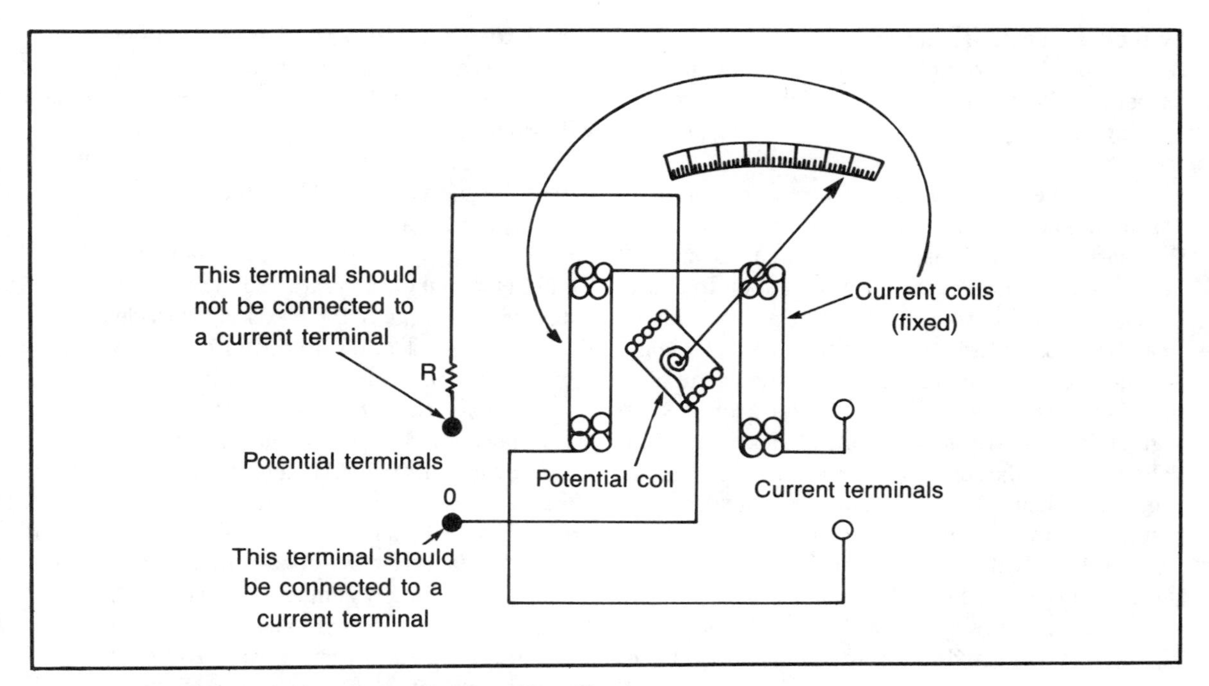

Fig. 3-5. Simplified circuit of an electrodynamometer wattmeter.

age and current which are not in phase. The scale indication of such a meter is exactly the power that would otherwise have to be worked out by E x I x cos Θ. Moreover, this power-indicating meter is suitable for use in dc circuits, too! Apparently, all that needs be done to measure power is to connect one of these meters into the circuit and record the reading. Well, not quite. As is the case with most measuring instruments, it is possible to both misuse and abuse the electrodynamometer wattmeter.

Figure 3-6 shows two correct methods for connecting the wattmeter into a circuit to measure the average power dissipated in or consumed by a load. The method depicted in Fig. 3-6A cannot be exact because the wattmeter must also measure the power consumed by its own potential circuit comprised of the voltage coil and its series resistance, R. (The function of R is similar to that of the multiplier resistance in a voltmeter.) Hopefully, this additional increment of power will be negligible. This can, indeed, be the case when measuring many hundreds of watts. However, in general, you should at least be aware of the possible need for correction. This is readily applied by the formula for true power:

$$P = W - E^2/R_p$$

where P is the true power, W is the power indicated by the wattmeter, E is the voltage across the load, and R_p is the internal resistance of the potential coil circuit of the meter.

Suppose, for example, that the wattmeter reading is 150 watts and it is determined by means of an accurate voltmeter that 100 volts appears across the load. The total resistance of the potential coil(s) together with multiplier resistance, R is 1000 ohms. These conditions lead to a corrected value of power, P equals 140 watts.

In Fig. 3-6B, the wattmeter is forced to indicate the power consumed by its current coil. Again, a correction must be applied in order to obtain true power. In this case, the formula for true power is:

$$P = W - I^2R_c$$

where P is the true power, W is the power indicated

by the wattmeter, R_c is the resistance of the current coil(s). For example, if R_c is 2 ohms and 1.5 amperes is being supplied to the load, the excess reading on the wattmeter will be $(1.5)^2$ (2), or 4.5 watts.

From the above, it might be inferred that a voltmeter and-or current meter should be used with the wattmeter. This is indeed the case, but not necessarily because of the excess reading. In many cases, just the fact that there can be an error leads to a choice of wattmeter connections which yields a sufficiently accurate reading so that no useful purpose is served by striving for greater accuracy. If a given load power is the result of relatively high voltage, but relatively low-current, the circuit in Fig. 3-6B should be used. But if the load power results from the opposite situation, the circuit in Fig. 3-6A is most likely to provide a reading with the least error. (A little thought will show that a given load power, say, 100 watts, can be the result of, say, 100 volts and one ampere, or 20 volts and 5 amperes, etc.).

The electrodynamic wattmeter lacks a feature inherent in common voltmeters and ammeters. In the latter instruments, a case of accidental overload is immediately evident because the meter needle is pinned. We all know that such an occurrence can stimulate remarkably rapid reflex actions in an otherwise calm, collected worker. However, in the wattmeter, a fivefold or tenfold current overload will produce no deflection of the pointer if there is no voltage across the potential coil. If the voltage across the potential coil is low, the resultant deflection may appear as a normal on-scale indication, despite the gross current overload condition. Similarly, a higher than rated voltage applied to the potential coil circuit of the meter is not evident from the on-scale deflection produced if the current is low. Thus, the rotor is vulnerable to damage or burn-out even though the load wattage is less than the full-scale indication of the meter.

In dc circuits such damage to the wattmeter is always possible. In ac circuits it is more of a probability. Why should this statement be made in face of the fact that the operating principle of the meter is substantially similar in both cases? The difference lies in the fact that so many amperes and volts in the dc circuit are always 100 percent effective in producing the mechanical torque required for pointer indication. This is not true, generally, in ac circuits. If the impedance angle, Θ, is a low value, a very high value of voltage or current may be applied to the wattmeter to produce only a nominal deflection of the pointer. The cosine of

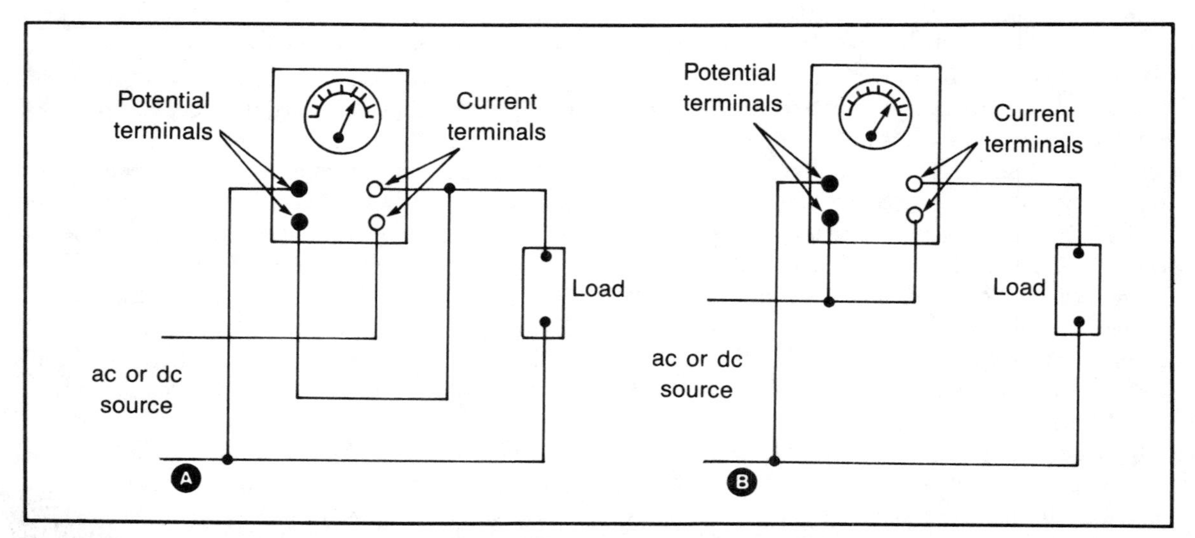

Fig. 3-6. The two correct connections for an electrodynamometer wattmeter. Connection A is used where current is relatively high and voltage is relatively low; B is used where current is relatively low and voltage is relatively high.

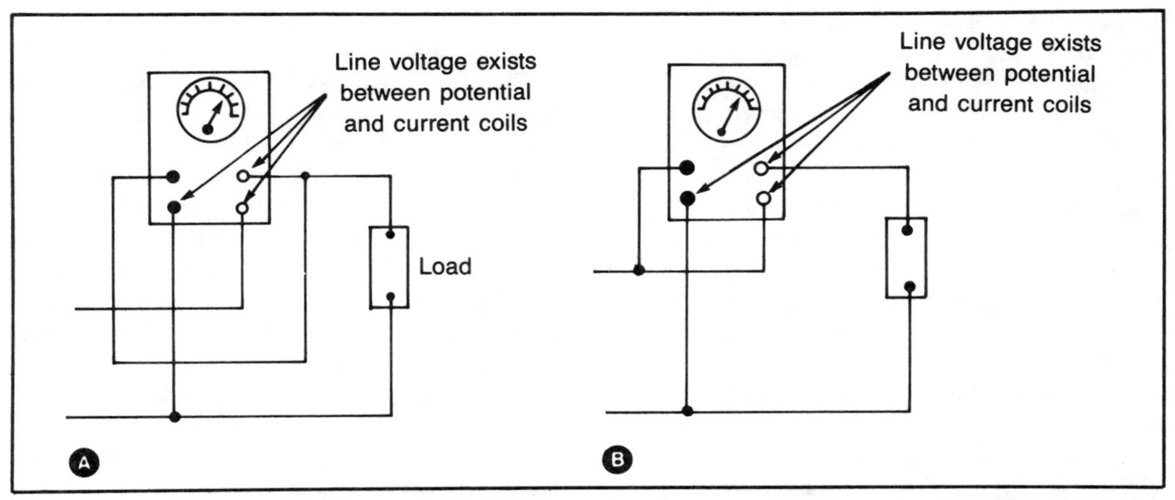

Fig. 3-7. The two incorrect connections for the electrodynamometer wattmeter. These are counterparts of A and B in Fig. 3-6, but in these arrangements error is likely from electrostatic force between current and potential coils. Also, the meter is endangered by insulation breakdown.

angle Θ is called the **power factor** and varies between zero and unity. Unity power factor corresponds to a resistive ac circuit where voltage and current are in phase; that is, Θ is zero (cos Θ equals 1.000). It is always desirable that power-consuming ac equipment operate as close to a unity power factor as possible because under that condition with a constant-voltage (or nearly so) supply, such as prevails with 60-Hz ac power lines, the current requirement for a given power is minimum.

From what has been said, it is important to know the current and voltage which will be applied to a wattmeter if we are to protect it from overload. Thus, even if you are not interested in correcting the power indication provided by the wattmeter, the voltmeter and ammeter should be used to determine that the wattmeter will not be damaged. It is not always necessary to leave the current meter and voltmeter in the circuit once this determination has been made; this can be a matter of judgment based upon overall circumstances.

The two connections shown in Fig. 3-6 do not exhaust all those possible. For example, Fig. 3-7 illustrates two additional wattmeter connections. Neither of these is desirable because the line potential is applied between the stationary and moving coils, giving rise to possible error because of electrostatic forces. Also, the stress on the insulation can lead to breakdown.

The wattmeters discussed so far have had four terminals. But since two terminals are always common, the manufacturer could make a three-terminal instrument. Indeed, many wattmeters do have only three terminals. Although this arrangement prevents the undesirable connection shown in Fig. 3-7 from being made, it also makes such a meter less flexible because you can no longer choose between the connections in Fig. 3-6A and B. In this book, we shall continue to deal with the 4-terminal meter because it more closely illustrates the classic use of the wattmeter. No confusion should result, whether the common connection is actually made internally or externally to the meter.

POWER FACTOR

The concept of power factor is closely associated with that of power when dealing with alternating-current circuits. Power factor is also useful because it is related to the impedance of the load in which the power measurements are made. Conversely, if the impedance is known, the power factor can be determined. Power factor values range from zero to unity. A unity power factor denotes the most efficient utilization of power. When the power factor is less than unity, more current must be supplied than is necessary to develop the power indicated on the wattmeter. The additional current flows through the resistance of power-line conductors and circulates in transformers and other devices, manifesting itself as heat and a correspondingly greater operating expense.

At the extreme of zero power factor, a perfect inductor or a perfect capacitor would consume current from the power line, despite the fact that ideal reactances would, in themselves, consume no power at all. Much effort in the design and installation of ac equipment is devoted to the achievement of power factors as close to unity as possible. Another way of stating this is that no matter what the circuit content of ac-operated equipment, its impedance should appear to the power line as being essentially resistive. Only a purely resistive load can present a unity power factor to the ac source. In a technique reminiscent of achieving resonance, reactive components in a load impedance are often deliberately cancelled by introducing a reactance of the opposite sign.

Figure 3-8 shows general setup employed to determine power factor. The wattmeter reads true power because it develops torque only from in-phase components of voltage and current. The product of voltage and current derived from the voltmeter and current meter would be misleading to the uninformed; such a product is appropriately termed apparent power. Except at unity power factor, apparent power is greater than true power. However, apparent power can be used to compute true power if we know the impedance angle.

Under such conditions, true power W is $E \times I \times \cos \Theta$. Conversely, if both true power W and apparent power $E \times I$ are measured as depicted in Fig. 3-8, the power factor, PF, is obtained from $W/E \times I$. Finally, tying these relationships together, we have, PF equals $\cos \Theta$ and Θ equals arc-cos $W/E \times I$.

No power factor consideration is needed in dc circuit. For dc, true power is indicated either by the wattmeter, or by the product of $E \times I$ read from

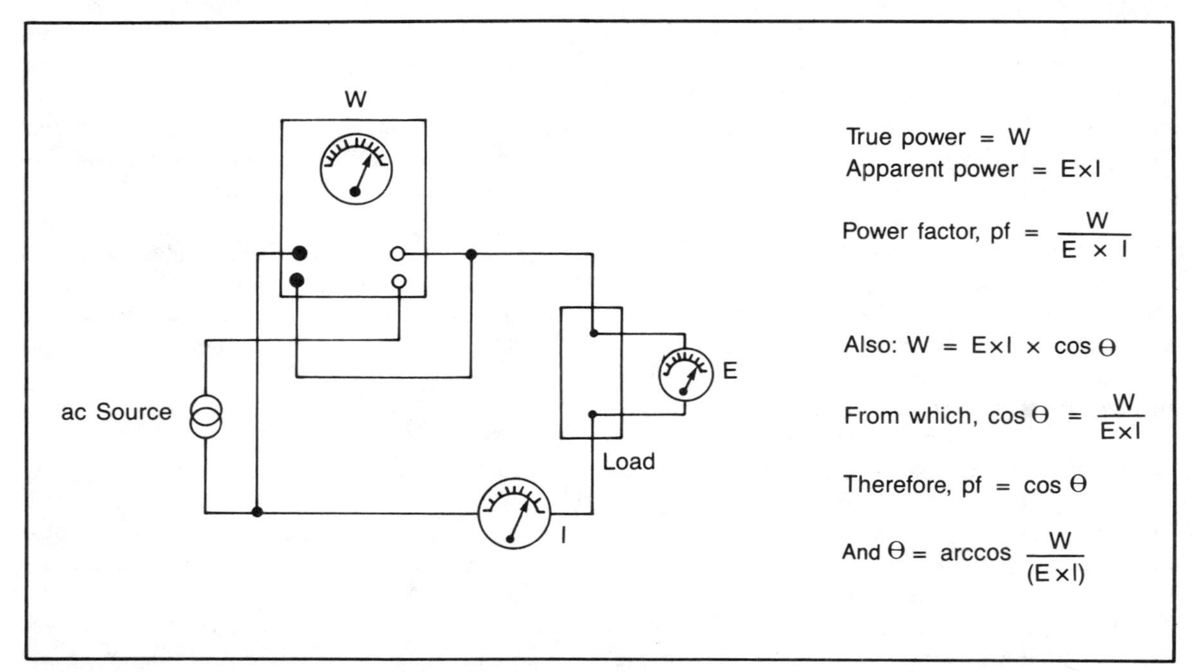

W
E
ac Source
Load
I
True power = W
Apparent power = E×I
Power factor, pf = $\frac{W}{E \times I}$
Also: W = E×I × cos Θ
From which, cos Θ = $\frac{W}{E \times I}$
Therefore, pf = cos Θ
And Θ = arccos $\frac{W}{(E \times I)}$

Fig. 3-8. Method of determining the power factor of an ac circuit or load.

a voltmeter and current meter. And because of the unity power factor in ac circuits which are resistive, true power may also be obtained from a wattmeter or from the product of voltmeter and current meter readings.

THREE-PHASE POWER MEASUREMENTS

Three-phase power systems are encountered in industry, in radio and TV stations, in laboratories, and elsewhere. Three-phase utilization of electrical energy is much more efficient than single-phase applications. There are other advantages, too. For example, the starting and running behavior of 3-phase motors is generally superior to that of their single-phase counterparts. Because of the widespread application of electronic control to power systems, it is a good idea for the worker in electronics to know the basic approaches in 3-phase power measurement. Such knowledge will also provide greater insight into 3-phase systems not associated with the 60-Hz power line, such as may be encountered in automotive, marine, aircraft, and various servo systems.

Voltmeter and ammeter tests such as shown in Fig. 3-9 provide the simplest approach to 3-phase power measurement. Notice that the computation for total load power is the same for the delta and Y connections. Actually, the formula, P equals 1.73 × E × I, is the general expression for volt-amperes. However, under the assumption that the load is resistive and the power factor is unity, the determination of volt-amperes is the same as that for true power. When the load is not purely resistive, the true power can be obtained only by the more general expression, P equals 1.73 × E × I × pF, where pF is the power factor. A common mistake made in 3-phase power measurement is to multiply the E × I product by three. This is not correct, although it is an understandable error resulting from a quick "common sense" evaluation. Notice that the multiplier, 1.732, is the square root of 3.

For the moment, we will not bring the power factor concept into consideration. Much useful information can be gained from the simple voltmeter-ammeter test under the assumption that the power factor is unity (the load is resistive). Even where it is known that such an assumption is not valid, the volt-ampere values obtained by this simple measurement technique can be very useful on a relative basis. For example, the volt-ampere consumption of a radio transmitter is often recorded daily or frequently in order to provide a general confirmation that, other things being equal, no departure from normal operation has occurred. Also, with such loads as dc power supplies for radio transmitters, you can often "ballpark" a power factor in the vicinity of 0.9 in order to arrive at a reasonable determination of true power from the voltage and current readings in Fig. 3-9. (A more useful approach to power factor will be shown later.)

Several ways of measuring 3-phase power are shown in Fig. 3-10. These techniques all make use of wattmeters based upon the 2-coil or dynamometer principle as discussed for single-phase power measurements. The most direct approach is the 3-wattmeter method depicted in Fig. 3-10A. The total power consumed by either a delta- or Y-connected load is simply the sum of the three readings. This is true regardless of load balance or power factor. If the three wattmeters do not have potential coils of the same resistance, the three wattmeter readings will not necessarily be equal, even though the load is balanced. However, the sum of the three readings will still yield true power for the total load.

The 2-wattmeter method of measuring 3-phase power, shown in Fig. 3-10B, must be used carefully because the total power can be either the sum or the difference of the two wattmeter readings. However, there need be no question involved in the test procedure if two identical instruments are employed and they are physically oriented just as shown. Under this condition it may be found that the deflection of one wattmeter is backwards. Before reversing terminal connections to this wattmeter, it would be marked, or otherwise kept in mind, as the negative-indicating meter. Thus, if W2 initially reads backwards, its indication is to be subtracted from the indication of W1. For unity power factor into balanced loads, both wattmeters will read alike. For the unique situation when the power

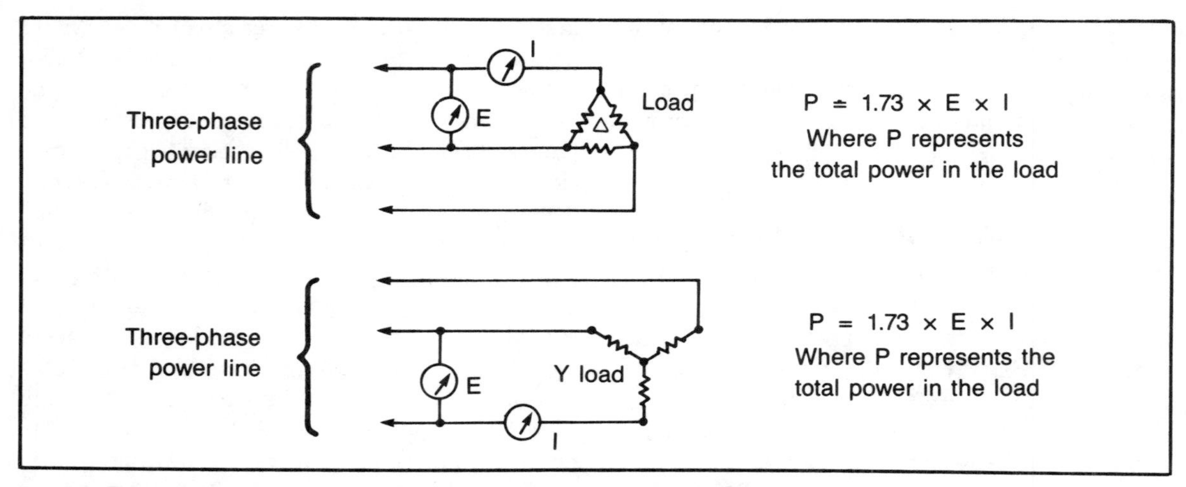

Fig. 3-9. Power measurement techniques for balanced and resistive 3-phase loads.

factor is 0.5, one meter will read zero and the total power must then be read from the other meter.

The measurement circuit in Fig. 3-10C is much simpler because only one wattmeter is necessary. However, its use is restricted to balanced loads only. If such is the case, true power is obtained by multiplying the reading by three, regardless of the power factor. Inasmuch as balanced loads are often encountered, this test procedure is a very useful one. The resistances (R) must be equal to the potential coil circuit of wattmeter W. This resistance can be determined by an ohmmeter measurement across the potential circuit terminals.

The single instrument method of measuring 3-phase power shown in Fig. 3-10D makes use of a polyphase wattmeter. This instrument combines in one physical unit the basic functions of the three meters in Fig. 3-10A. A study of the terminal connections will reveal a basic pattern which simplifies what might otherwise be a complex situation because of the large number of terminals. Once such a meter is properly connected into the 3-phase circuit, the true power is read directly from the scale.

The 2-wattmeter method of power measurement is particularly useful in that the power factor of the load can also be determined from the individual meter readings. This is done in two steps. First, the following formula is used:

$$\tan \Theta = (1.73) \frac{(W1 - W2)}{(W1 + W2)}$$

The algebraic signs of W1 and W2 must be carefully considered before inserting the values into this formula. Remember, if the connections to a wattmeter had to be reversed in order to produce a forward deflection, the reading from that meter is considered as being negative. After than Θ is obtained, look up the angle, Θ, corresponding to tan Θ in a table of trigonometric functions. Then record the value of cos Θ also shown therein. Cos Θ is the power factor, PF, of the load.

POWER DETERMINATION FROM DECIBEL INDICATIONS

In many cases, the most convenient way to measure power is with an electronic meter with a decibel scale. The power level of signals is most often below that which could be measured by ordinarily available wattmeters. Also, the 4-terminal connecting feature of the wattmeter imposes an inconvenience because of the necessity of interterupting a current-carrying lead. With an electronic meter you can measure throughout ranges embracing at least microwatts through hundreds of watts. On the other hand, commonly encountered wattmeters generally handle tens, hundreds, and thousands of watts. Another important advantage of the electronic meter is that its frequency response goes into the tens or hundreds of MHz compared to several hundred Hz for the wattmeter. What has been said about the electronic meter also applies to the VOM with a decibel scale, except that the VOM often has a frequency response not too far beyond the audio-frequency range. Also, the input impedance of many VOMs may be too low for high-impedance circuits. How can we derive power from decibel measurements when the decibel is not a unit of power, but rather a unit expressing the ratio between two power levels—and then only under certain impedance restrictions? Consider the basic decibel equation, dB equals 10 $\log_{10}$ (P2 divided by P1). We are not interested in two power levels, but rather only one. Let this power level be P2. Also, assume that all meter manufacturers agree to adopt a reference power level for P1. This being the case, decibel scales could be calibrated

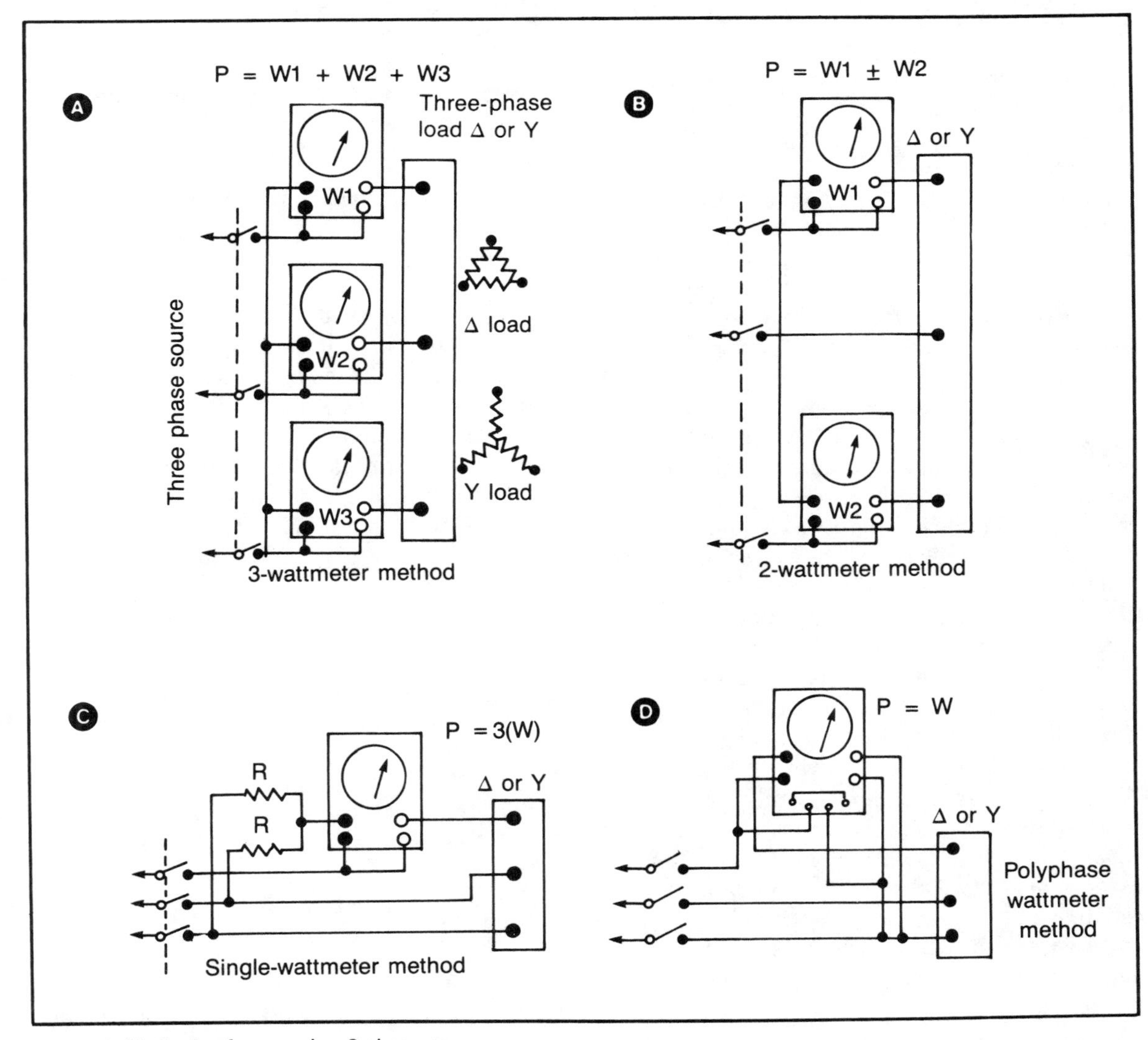

Fig. 3-10. Methods of measuring 3-phase power.

so that zero dB would represent P1. (Notice that when P1 equals P2, P2 divided by P1 equals 1. The log of 1 is zero, so the value of the right side of the decibel equation is likewise zero, confirming that zero dB represents reference power level P1.) Now, any dB value we read from the scale is so much above or so much below zero dB, the reference level. Another way of stating this is that a dB indication represents a certain power ratio to the reference level. Thus, the basic idea of the decibel as a power ratio is preserved. At the same time, the indicated ratio can be converted into an absolute value of power, so many microwatts, or so many milliwatts, etc. The conversion can be accomplished by charts, or with the decibel equation.

Some examples should illustrate that such power determinations are really quite straightforward. For the moment, assume that the meter is connected into circuits of a prescribed impedance level, say 500 ohms. (The manufacturer generally indicates the impedance level on the meter scale. The simple correction for other than prescribed impedance values will be dealt with shortly.) In the following examples, it will be assumed, unless otherwise stated, that the reference power level is six milliwatts, or .006 watt. (Other references, such as one milliwatt, are also used especially in audio work. As long as the proper reference is used, the procedures for converting decibels into watts, or vice-versa, are similar.) Figure 3-11 shows decibel scales with different power-level references. For the purpose at hand, let us first make use of the commonly encountered six-milliwatt reference; that is, P1 equals .006 watt. The decibel equation then becomes, dB equals 10 $\log_{10}$ (P2 divided by .006). As might be expected, we will also be interested in situations where power level P2 must be derived from dB readings. For this purpose, a simple transposition of the decibel equation yields, P2 equals 0.006 × antilog dB divided by 10. Now let us use numbers which may typify certain test procedures:

An audio amplifier delivers six watts of power into its rated load. How many decibels does this correspond to ?

$$\text{dB} = 10 \log_{10} (6/0.006) =$$

$$10 \log_{10} (1000) = 10 \times 3.000 = 30.0 \text{ dB}$$

If the meter with a dB scale were inserted into a circuit dissipating 6 watts into the resistance prescribed for the meter, the indication would be 30.0 dB. You can avoid confusion by saying, "30.0 dB referenced to 6 milliwatts."

If the circuit resistance was other than that prescribed for the meter, a simple correction would have to be applied to the meter indication. Upon doing this, we would again derive the value, 30.0 dB.

Suppose that the above amplifier requires 1.0 milliwatt at its input to deliver the rated output. In terms of dB, what is its required power input level?

$$\text{dB} = \log(0.001/0.006) = 10 \log(0.1666)$$

$$= 10 \times \bar{1}.223 = -10 + 2.23 = -7.77 \text{ dB}$$

Care must be observed in the manipulation of the minus characteristic of the logarithm. The power level which must be led into this amplifier to produce the rated output is −7.77 dB. Also, the power gain of this amplifier is 30.0 − (−7.77) equals 37.77 dB. Again, in order to avoid confusion with other references, you should describe this input power level as being minus 7.77 dB referenced to six milliwatts, or 7.77 dB below 6 milliwatts.

Now suppose that a low-level amplifier stage produces an output of −64 dB. What power does this represent? We solve this as follows.

$$\begin{aligned} P_2 &= 0.006 \times \text{antilog dB}/10 \\ &= 0.006 \times \text{antilog } -64/10 \end{aligned}$$

When the dB value isn't evenly divisible by ten, we make it so:

$$\begin{array}{rr} -64 & \\ -\ 6 & +6 \\ \hline -70 & +6 \end{array} \quad \text{from which dB/10} = \frac{-70}{10} + \frac{6}{10} = \bar{7}.6$$

Notice how the characteristic and mantissa are

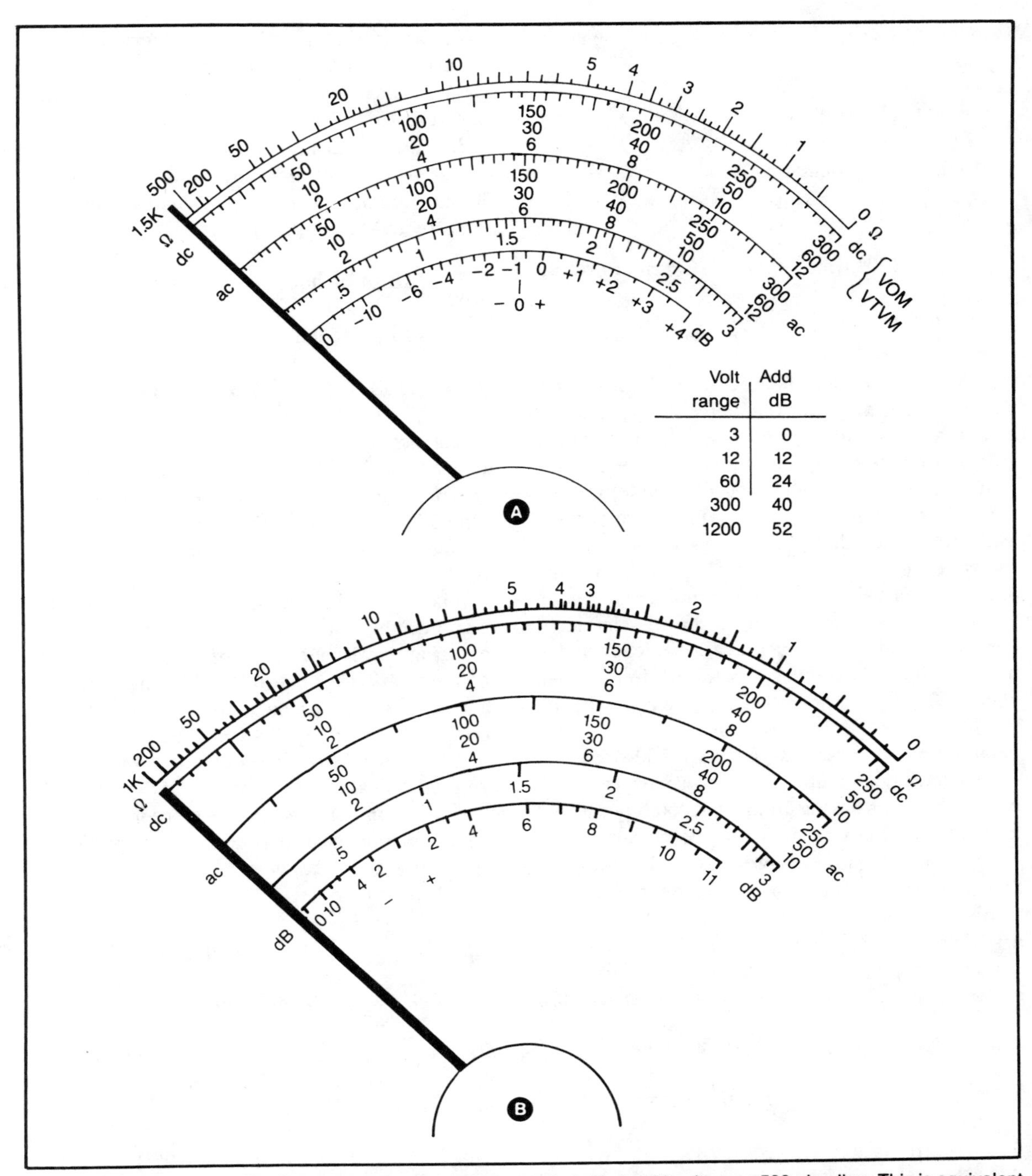

Volt range	Add dB
3	0
12	12
60	24
300	40
1200	52

Fig. 3-11. Two dB scales with different power-level references. A. 0 dB at 1.73 volts on a 500-ohm line. This is equivalent to 6 milliwatts. B. 0 dB at 0.774 volts on a 600-ohm line. This is equivalent to 1 milliwatt, or 1 dBm.

handled in this case. So now we rewrite the equation as:

$$P_2 = 0.006 \times \text{antilog } \overline{7}.6$$

$$= 0.006 \times 0.0000004 \text{ watt}$$
$$\text{or } 6\times10^{-3} \times 4\times10^{-7} \text{ watt}$$

$$= 24 \times 10^{-10} \text{ watt} = 24 \times 10^{-4} \text{ microwatts}$$

$$= 2.4 \text{ nanowatts}$$

What do we do if the circuit impedance is not the prescribed value for the meter? In practice, we cannot always enjoy the luxury of having the circuit impedance equal the impedance used to calibrate the meter. The best procedure under general conditions is to make an appropriate correction immediately. Then the dB value will be corrected for the effect of the impedance differences and eliminate any confusion in further evaluations or computations. The corrective procedure is simple. A certain number of dB must be algebraically added to the meter indication as follows: dB to be added equals 10 log Z_m divided by Z, where Z_m is the impedance for which the dB scale of the meter has been calibrated (This is generally 500 ohms for meters having a 6-milliwatt reference level and 600 ohms for meters having 1-milliwatt reference level.) Z represents the impedance of the circuit.

Let's investigate a couple of examples: Suppose that a meter calibrated for 500 ohms indicates + 10 dB in a 1000-ohm circuit. What is the corrected dB value representing the power level in this circuit? Making use of the formula, dB to be added equals 10 log Z_m divided by Z, we have:

$$\text{dB to be added} = 10 \log 500/1000$$

$$= 10 \log 0.5 = -10 \log 1/0.5$$

$$= -10 \log 2.0 = -10(0.301) = -3.01$$

Thus, the corrected dB is +10 −301 or approximately +7.0 dB. Notice the method used to avoid manipulating the negative characteristic which would have resulted from dealing directly with log 0.5. We make use of the equivalence, log x equals minus log 1 divided by x.

In another example, suppose that a meter calibrated for 600 ohms indicates −25 dB in a 300-ohm circuit. What dB value shall we record as the corrected reading? Again:

$$\text{dB to be added} = 10 \log Z_m/Z$$

$$\text{dB to be added} = 10 \log 600/300$$

$$= 10 \log 2 = 10(0.301) = 3.01$$

The corrected dB value is −25 +3.01, or approximately −22 dB. Notice in the first example that any dB indication on the meter is corrected by "adding" −3.01 dB. In the second example, any dB indication on the meter is corrected by adding +3.01 dB.

It is evident that the corrective procedure does not appreciably complicate the use of the meter for measuring power in terms of dB. In a given circuit, the computation for dB to be added need only be made once. Moreover, once dB indications are corrected, the corrected values may be converted into actual power levels as previously shown. The corrected dB values are treated as if they were the actual meter indications and no more involvement with the impedance situation is necessary. For example, the setup to convert the second example to power level would be:

$$P_2 = 0.006 \times \text{antilog dB/10 watts}$$

$$P_2 = 0.006 \times \text{antilog } -22/10$$

Note: If the meter had been calibrated for a 1-milliwatt reference, which is actually the usual case when Z_m is 600 ohms, the setup would be:

$$P_2 = 0.001 \times \text{antilog } -22/10$$

It is obviously very important to know the reference power level and the value of Z_m before using a decibel-calibrated meter for power determinations. Fortunately, this information generally appears on the meter. Once we start off on the right track, it is only a matter of practice and familiarity before the use of decibels for power measurements becomes second nature.

THE DBM

It has been shown that decibels can represent power levels if a reference power level is involved. Thus, a decibel indication, P2, is so many times greater or less than the reference power level, P1. Two commonly encountered references have been dealt with, six milliwatts and one milliwatt. The later reference is of particular interest because of its involvement in a scheme which leads to greater convenience in the use of the dB scale for power indication. Quite simply, when zero dB is referenced to one milliwatt in a circuit impedance of 600 ohms, the term dBm rather than dB is commonly used. Zero dBm corresponds to 0.774 volt across the 600 ohms. From P equals E^2 divided by R, we have $(0.774)^2$ divided by 600 equals 1×10^{-3} watt, or one milliwatt. By using the one milliwatt into 600-ohm system, power levels are generally dealt with in the actual dBm units, usually no attempt is made to convert into watts in the course of testing and aligning. Thus, you avoid the somewhat awkward logarithmic computations involved in converting to watts. Where desired, the conversion is essentially as previously given, P2 equals 0.001 × antilog dBm divided by 10 watts.

With this system, providing that the circuit impedance is 600 ohms, we simply read the meter and record the power level as being so many dBm. In audio work, and primarily in carrier-telephone circuits, 600 ohms is a common impedance level. In such circuits, it is very easy to evaluate the performance of complex installations by measuring and recording the dBm levels at strategic circuit junctions. The technique is just as easy as taking a voltage reading.

When testing the dBm level in a circuit which differs from 600 ohms, a correction must be applied to the indicated dBm value. This has been already discussed, but will be repeated here: A certain number of dB must be algebraically added to the meter indication as follows: dBm to be added equals 10 log 600 divided by Z, where Z is the circuit impedance.

When a power level is measured as so many dBm, it is not necessary to state that it is referenced to one milliwatt in 600-ohm circuit, since this is the specific implication of the term dBm. Often, but not necessarily so, power levels specified as so many dB are intended to be understood as being in dBm units. To avoid confusion or error, it is desirable that dBm be spelled out as such. Otherwise, there is nothing to infer that some other reference, such as six milliwatts, is not the intended one. In line with this suggestion, the manufacturer of the meter depicted in part in Fig. 3-11 would have served the cause of clarity by labeling the decibel scale, dBm.

MEASUREMENT OF RF POWER

Except for the use of appropriate instruments, the setup for rf power measurement shown in Fig. 3-12 does not differ from the basic techniques which can be used for dc and low-frequency power determination. Here, a thermocouple-type current meter is used to measure rf current, and an electronic meter in conjunction with an rf probe is used to monitor the rf voltage. What is not shown in Fig. 3-12 is the care necessary in setting up the equipment. It is essential that leads be kept as short as possible and that "hot" rf conductors do not get near other objects. If R is a physical resistor, the need to avoid the effects of distributed and stray reactance becomes more pressing with increased frequency. The measurement technique becomes almost an art. In general, wirewound resistances should not be used because of their inductance.

From the formulas for power, if R is known, only current or only voltage need be measured. However, in rf work, we can feel more secure when it is found that reasonably good agreement is obtained from E × I, I^2 R and E^2 divided by R. The weak point here is knowing what R really looks like to rf. Composition resistors can provide reasonable

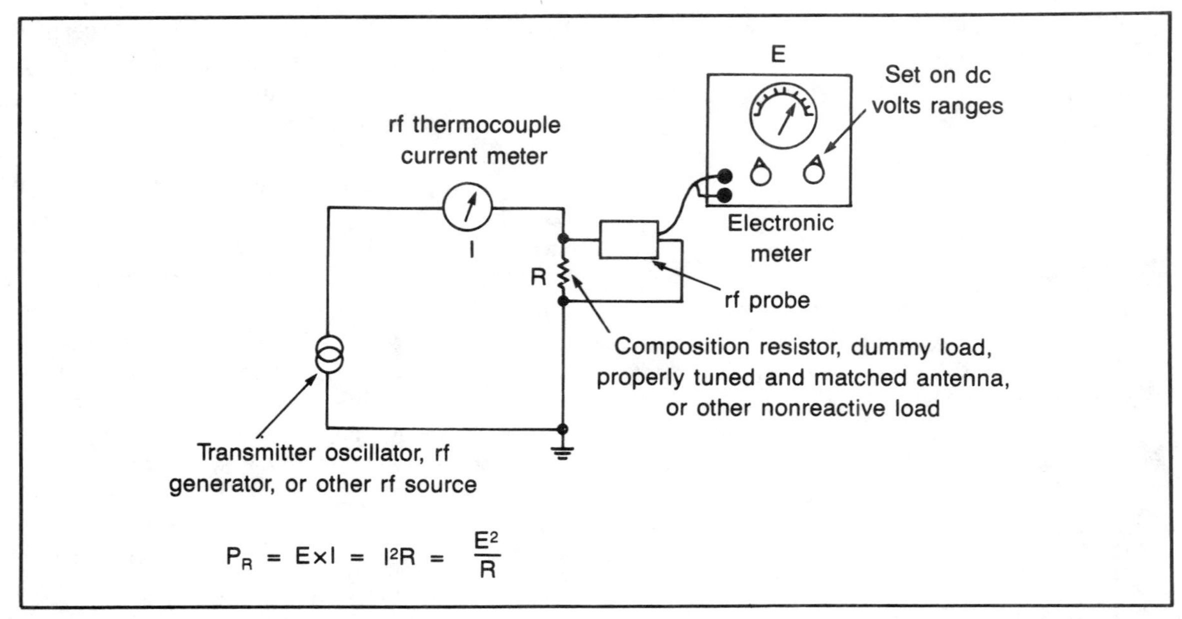

Fig. 3-12. Method of measuring rf power.

approximations through several tens of megahertz, but it is much better to work with special rf load resistances which are commercially available for such purposes. Open wiring should be avoided and the interconnections are best made by coaxial cable.

QUESTIONS

3-1. Two identical power transformers for TV sets are being tested side by side. The primary windings of both transformers are energized from the 60-Hz power line and all secondary windings are open. One transformer develops a considerable temperature rise. Both transformers produce about the same voltages across their secondaries. Also, ohmmeter checks when both transformers are cool show no significant difference. But when a wattmeter is inserted in the primary winding of the transformer which heats up, an appreciable power input is indicated. With the other transformer, the power input is relatively small. What would be a logical reason for this condition?

3-2. A wattmeter is inserted in the primary winding of the power transformer which supplies all operating power to an amateur AM transmitter. When the operator whistled into the microphone, the wattmeter increases its reading appreciably. However, there is very little change in the dc plate current to the final rf amplifier. Also, it is ascertained that the dc voltage regulation of the power supply for the final amplifier is very good. Since modulation adds power to the carrier, why isn't this added power seen as a current increase to the final amplifier?

3-3. An electrodynamometer wattmeter is used to measure the power input to an electric heater operating from a 120-volt, 60-Hz power line. If this same heater is operated from a 120-volt dc source and the same wattmeter is employed to measure the power consumed, how should the power reading compare to the value recorded for ac operation?

3-4. Voltage and current of one phase of a balanced 3-phase load are measured as E and I. The total load is computed from 3 × E × I. A subsequent check with three wattmeters shows a great discrepancy from the power determined with the voltmeter and ammeter. Assuming that the power factor is unity, why do the two power determinations not agree?

3-5. Because of inordinately heavy mechanical loads, a 3-phase motor consumes more power than had been anticipated. A second 3-phase source is available. It has been determined that this source provides the same balanced voltages as does the source connected to the motor. Also, the same alternator is supplying both sources of 3-phase power. Can the two 3-phase sources be connected in parallel on the basis of the above information?

3-6. An investigation in a factory showed that a number of large single-phase units were operating at a low power factor from the ac line. However, the argument was advanced that the units themselves displayed no signs of inordinate temperature rise. In the general case, can this be accepted as a valid justification for the operation of electrical equipment at a low power factor?

3-7. Automobile alternators are 3-phase machines. Since the phase connections are not brought out to terminals, how can the electrical power output be measured?

3-8. During three winter months, it is estimated that an electric heater is in use about 240 hours per month. The monthly statement from the utility company shows a total of about 720 kilowatt-hours of additional electrical energy consumed over the 3-month periods when the heater is in use. Assuming single-phase, 120-volt power circuits, can the power used by the heater be estimated?

3-9. The low power factor that had prevailed with certain large industrial equipment is brought close to unity by a special machine known as a synchronous condenser. This machine has the same effect as a bank of physical capacitors, and is employed in large installations because of the massive size that would be necessary if actual capacitors were used. After operation has been monitored for some time following correction of the power factor, it is noted that the equipment consumes more true power. How can this be explained?

3-10. During an interval while the starting motor is cranking an automobile engine, it is observed that the voltage across the 12-volt battery falls to 9.0 volts. Later, a dummy load of 0.05 ohm is momentarily held across the battery terminals and the terminal voltage is again observed to be about 9.0 volts. From this, it can be inferred that the dummy load simulated the load by the starting motor when the first measurement was recorded. From the power formula, P equals E^2 divided by R, we have, $(9)^2$ divided by 0.05 equals 81 divided by 0.05 equals 1620 watts. Inasmuch as there are 746 watts in one horsepower, the evidence shows that the ordinary 12-volt automobile battery delivered better than two horsepower during the measurement intervals! Is this reasonable, or is there a glaring fallacy in the test procedure or in its interpretation?

ANSWERS

3-1. The transformer which heats probably has a shorted turn. Often a single shorted turn, or a number of turns which do not short through and cause a substantially zero ohms resistance, can be quite difficult to identify from ohmmeter or voltage tests. However, the self-heating and the wattmeter checks definitely indicate abnormal loading. Such abnormal loading is readily evident by the wattmeter reading even when sufficient time is not allowed for the large mass of steel and copper to heat up. Another but less likely cause of such abnormal loading is sometimes due to the method of clamping or mounting the core of the transformer, which amounts to the equivalent of a shorted turn coupled to the core flux. Also, insufficient resistance between laminations can result in a high eddy current loss, which again is revealed by more than negligible power input to the primary.

3-2. The wattmeter indicates a true situation, for the modulation power added to the AM carrier must be supplied from the ac line. If a Class B modulator is used, the increased deflection of the wattmeter pointer can be quite pronounced because both modulator and final amplifier participate in the increased power demand from the ac line. However, the direct current to the final amplifier remains constant because the pointer of the current meter cannot respond to the audio frequency cycles. Thus, it appears that the dc power input to the final amplifier remains constant during modulation, when in fact it doesn't.

3-3. The reading should be the same in both instances. By definition, 120 volts RMS ac produces identical heating effect in a resistance as the same number of volts dc. Although RMS was not mentioned in the specification of line voltage, this is implied when working with ac circuits in general, and with sine-wave power circuits in particular. Additionally, the electrodynamometer wattmeter produces the same deflection for dc and ac if the same value of power is being metered.

3-4. Contrary to what may appear to be logical, the total power in a 3-phase system is not three times the individual phase power. Rather, it is $\sqrt{3 \times E} \times I$ where $E \times I$ represents the power in an individual phase of a balanced 3-phase system.

3-5. Not on the basis of the information given. One more fact must be determined. This is the sense of phase rotation. The sense of phase rotation can be reversed in a 3-phase source by transposing any two leads. Light bulbs may be connected between the phases of the two sources. When the right leads have been selected, none of the bulbs will glow.

3-6. No. The inefficiency resulting from low power-factor operation need not cause appreciable heat in the equipment itself. However, the excess line current associated with such operation can tax the safe current-carrying capacity of the wiring in the building. Such unnecessarily heavy line current produces a relatively high voltage drop in the supply conductors, thus adversely affecting lighting circuits and the operation of motors and other devices designed for operation over a limited voltage range. Finally, the electric utility company can penalize the use of low power-factor loads because such operation necessitates the supply of excessive current. The added current used by low power-factor loads does not actuate the watt-hour meters at the service junction of the building.

3-7. The rectified dc is brought out to a large output terminal. By simultaneously measuring the dc voltage at this terminal with respect to ground (car frame) and the direct current flowing out of (or into) this terminal, the needed data is obtained to compute the dc watts output from P equals $E \times I$. To measure the current, the load must be disconnected from the output terminal and an ammeter inserted in the circuit. The actual ac power generated will be somewhat greater due to the losses in the silicon rectifiers, but this is more academic than of practical value.

3-8. Yes, very readily, for if we divide kilowatt-hours by hours, the result is power in kilowatts. In this case, the heater would be in use for 3×240, or 720 hours over a 3-month period. Then, 720 kilowatt hours divided by 720 hours yields 1.0 kilowatt. As an alternative, notice that the average monthly increase in electrical energy is 720 divided by 3, or 240 kilowatt-hours. Again, from the concept that power can be obtained by dividing energy by time, we have 240 kilowatt-hours divided by 240 hours equals 1.0 kilowatt. (In computations of this kind, the same time unit should appear in both numerator and denominator so that they will cancel. Here, we have been dealing with hours.) Notice that the information about the phase system and voltage level of the electricity involved in operating the heater is of no importance.

3-9. With operation under conditions of low power factor, the inordinately high current demanded from the ac line results in a relatively high voltage drop in the supply conductors themselves. After the power factor is corrected, less line current is drawn and the voltage impressed directly across the equipment is then higher. Therefore, the true input power to the equipment tends to be higher.

3-10. The results are entirely reasonable: a lead-acid storage battery is capable of delivering large amounts of power for short intervals. It is not practical to look for too much accuracy in a power determination of this kind because many variable factors are involved. The battery heats internally and the chemical composition of the electrolyte and the plate surfaces change. The starter motor imposes an unstable load and it is difficult to make a consistent connection with the dummy load. However, such an investigation clearly establishes that 1 1/2 kilowatts of power or so can be drawn from the battery. Actually, considerably more power is often provided for short cranking periods by the automobile battery.

The autoranging feature of this Digital Multimeter greatly facilitates test and measurement procedures, freeing the user from the inconvenience of manually setting the range as the test prods are moved to different circuit positions in the equipment undergoing measurement. Courtesy of the John Fluke Mfg. Co., Inc.

Chapter 4

Impedance Concepts and Test Procedures

It is common knowledge that the behavior of an impedance in an ac circuit is somewhat similar to that of a resistance in a dc circuit. Both parameters limit the flow of current. Indeed, Ohm's Law for the ac circuit, I equals E/Z, in which Z represents impedance, emphasizes this comparison. Moreover, for many purposes, a knowledge of the magnitude of impedance is adequate for the circuit situation involved. But, there is more to impedance than such a comparison might indicate. One factor is that impedance tends to vary with frequency while resistance remains constant. (Except for skin effect and radiation, resistance behaves in most practical circuits as a frequency-independent parameter.)

Frequency dependency is not the only distinguishing feature of impedance. The understanding of the true nature of the parameter often glibly referred to as "impedance" provides the key to the behavior of alternating current in circuits. It is here that fact is separated from fancy. Without the insights that basic impedance mathematics makes possible, we can get by to a certain extent, but our innocence of the facts of ac life must ultimately invite trouble. For example, does knowing that two impedances have magnitudes of, say, 50 ohms each allow us to easily determine the value of the two impedances connected in series? Contrary to "intuition," or even a "common-sense" approach, the net impedance of such a simple arrangement can be, in principle, anywhere from zero to 100 ohms!

This does not mean that impedance magnitude is of no use. Rather, it should be appreciated that the magnitude designation of impedance is, in essence, an abbreviated identification. Thus, our 50-ohm impedances must have some other aspect to their natures. This we must realize because they do not, in general, add up arithmetically as would two series-connected resistances. The mystery appears to be deepened from the fact that 50 volts impressed across a 50-ohm impedance can produce the anticipated one ampere of current. (What appears to be undependable behavior was assumed to be just that before ac circuits were put on a firm mathematical basis.)

What has been neglected in our discussion of

impedance so far is that fact that impedance has not only magnitude but also an angle associated with it. Actually, impedance is a vectorial function. For the practical purposes at hand, we will simply consider it a vector. By this, we mean that impedance represents a quantity having both magnitude and direction. More specifically, it is useful to think of impedance as having a magnitude and an angle with respect to a reference. When impedance is completely described in this way, its symbol is dotted, thus Z. Now, enough of this academic hair splitting. Let's see what the idea of magnitude and associated angle leads to.

BASIC QUADRATURE RELATIONS AND THE RIGHT TRIANGLE

Before impedance tests can be meaningful, a few principles beyond those suitable for resistance tests must be considered. The concept of impedance revolves about the geometric and trigonometric relationships which exist in a right triangle. In Fig. 4-1A we see a diagram representing the ac voltages developed across a "pure" inductance (one with zero dissipation and no capacitance) and a series-connected resistance. The significant thing about this circuit is that the voltage developed across the inductive reactance is 90 degrees displaced from the voltage developed across the ohmic resistance. The displacement is a time or phase difference. For example, if we monitor the sine waves across the inductor and resistor with a dual-beam or double-trace oscilloscope, we do, indeed, see the 90-degree time displacement between the two waves (or very nearly so with an inductive reactance greatly exceeds the sum total of all dissipative losses). In Fig. 4-1B, the ac voltage across the inductor leads the ac voltage across the resistor. That is, E_l appears on the scope screen as being 90 degrees or one-quarter of a cycle ahead of E_r. By convention, the leading characteristic of E_l is depicted on the voltage vector diagram as a vertical line.

In Fig. 4-2, the voltage across the impedance, Z, is the vector resultant of the voltages across the

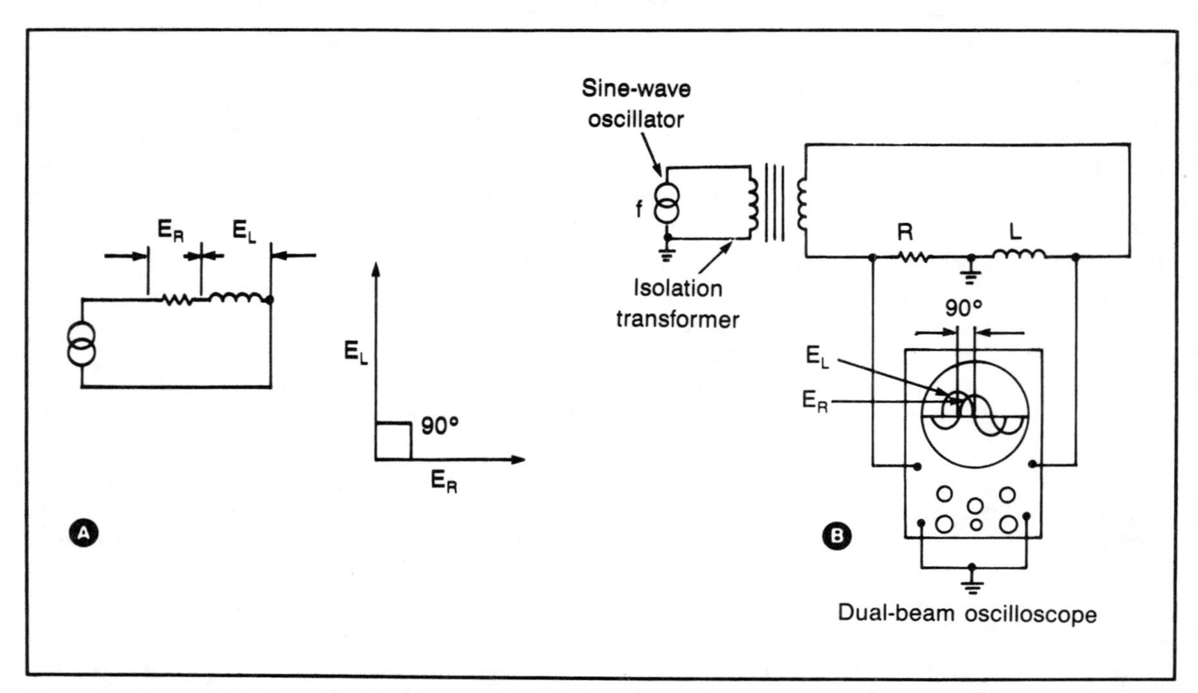

Fig. 4-1. Vector and waveform displays of the resistor and inductor voltages in a simple ac series circuit. Both, the scope display and the vector diagram show E_L, to be 90 degrees ahead in time with respect to E_R.

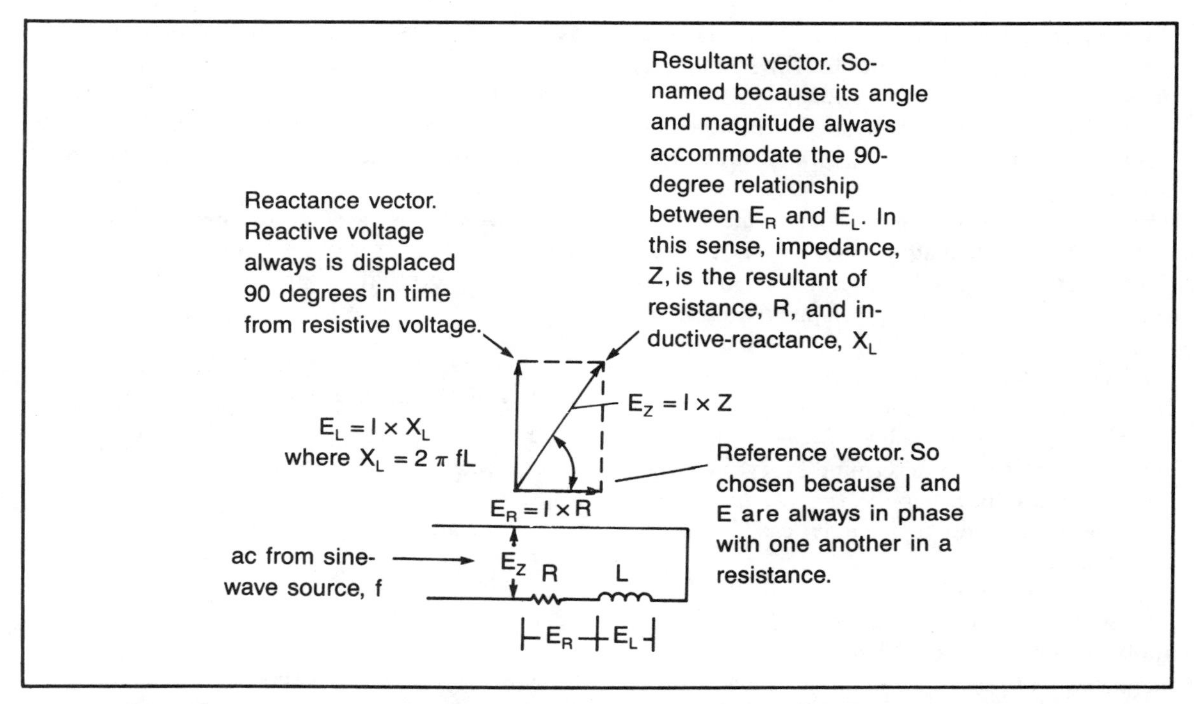

Resultant vector. So-named because its angle and magnitude always accommodate the 90-degree relationship between E_R and E_L. In this sense, impedance, Z, is the resultant of resistance, R, and inductive-reactance, X_L
Reactance vector. Reactive voltage always is displaced 90 degrees in time from resistive voltage.
$E_Z = I \times Z$
$E_L = I \times X_L$
where $X_L = 2\pi fL$
Reference vector. So chosen because I and E are always in phase with one another in a resistance.
$E_R = I \times R$
ac from sine-wave source, f
E_Z
R
L
E_R
E_L

Fig. 4-2. Three voltages are present in a simple ac series circuit. The resultant voltage is said to be developed (or generated) across the circuit impedance, Z.

resistance and the inductance. Thus, we would be in error if we arithmetically summed these two voltages in order to obtain the voltage across the parameter called impedance. Bear in mind that this diagram depicts the following conditions of a simple series circuit: (1) A common current, I, flows through the inductor-resistor combination. (2) The inductor presents an inductive reactance by which voltage E_l is 90 degrees time-displaced from voltage E_r. Specifically, E_l leads E_r in time by 90 degrees out of the full 360-degree sine-wave cycle. (3) A third voltage (the impressed voltage) exists across the two free terminals of the series circuit, and is obviously developed or impressed across a quantity comprising both inductive reactance and ohmic resistance. This quantity is the impedance of the series circuit. The angle Θ, associated with this third voltage is by convention measured with respect to the voltage across the resistance.

This convention tends to tie things down because the voltage and current associated with a resistance are always in time phase. Thus, no matter what the circuit or the frequency of the sine wave, the voltage across the resistance, or the current through it, is the reference from which the impedance angle Θ is measured. The basic idea of this reference prevails even if we devise a circuit in which there is no resistance. In such a case, the reactive voltages are still referred to a horizontal line as far as the impedance angle is concerned.

From Voltage Diagram to Impedance Diagram

Rather than always drawing the diagram with the three sides representing voltages, let's make life a bit easier for the moment and drop current from consideration. This should be feasible because, both electrically and schematically, current appears as a common parameter associated with resistance, reactance, and impedance to produce the voltage drops developed across these elements. Thus, instead of tediously dealing with $I \times R$, $I \times X_l$, and $I \times Z$, no damage will be done to the relationships if we deal simply with R, X_l, and Z. This is shown in Fig. 4-3A, where it is seen that the voltage relationships have not been violated.

We will call our right triangle an impedance diagram. Figure 4-3A shows a practical circuit arrangement for obtaining E_r, E_l, and I. This is of interest because even though it is convenient to deal with the impedance diagram for analysis, the quantities actually measured in a test situation are most likely to be the voltages developed across the resistance and the reactance(s). Also, convenience is best served by measuring the circuit current in order to complete the information. Such information, together with a knowledge of the frequency involved, enables us to determine everything about the behavior of the ac circuit. In particular, the impedance magnitude and angle can be readily computed.

Since the impedance vector in Fig. 4-3A is always completely identified by a magnitude and angle which are, in turn, governed by the magnitudes of resistance and reactance, it is only natural to investigate ways in which these quantities can be expressed in terms of one another. Because we are concerned with mathematics as a tool for dealing with impedance, we should make the tool as generally applicable as possible. To this end, we now consider reactance in general rather than specifically inductive or capacitive reactance. With this approach, we will already have covered the essential ground when the effect of the capacitor in the ac circuit is investigated. Let reactance (either inductive or capacitive) be designated by the letter X. Thus, X can be inductive reactance X_l or capacitive reactance X_c. For the time being the diagrams will continue to be drawn as they have been because we are not ready for the modification needed to represent a capacitor circuit.

Using Algebra and Trigonometry to Solve the Impedance Diagram

The name of the game now is to get the feel for the relationships shown with the impedance diagram in Fig. 4-3B.

We begin with the equation from Pythagoras' Theorem,

$$Z = \sqrt{R^2 + X^2}$$

As shown in Fig. 4-4, the impedance resulting from a resistance and reactance in series is the *square root of the sum of their squares.* And what about angle Θ? The best approach here is to list the trigonometric identities involving angle Θ, which appear in Fig. 4-5. Actually, we need deal with just three of these—the sine, cosine and tangent relationships. The beauty of these three identities is

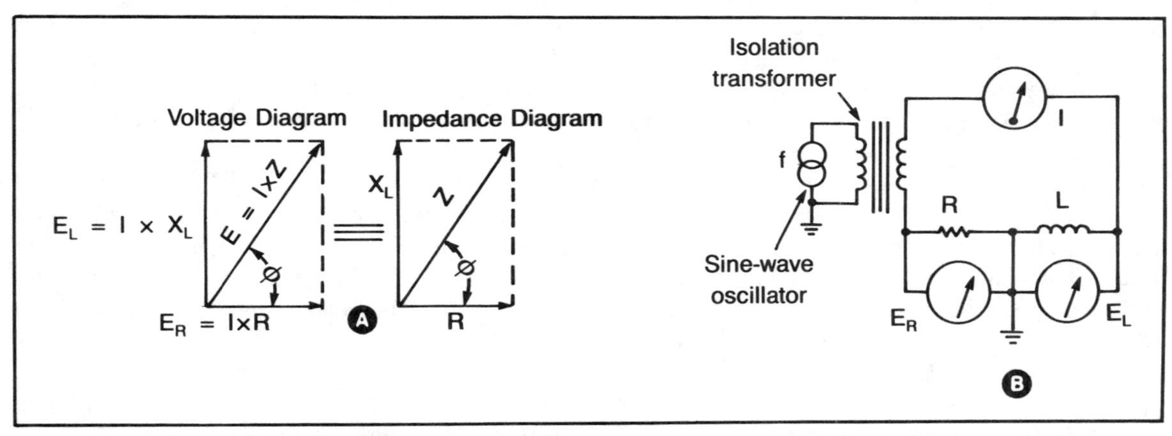

Fig. 4-3. Drawings showing the basic use of the impedance diagram in an ac series circuit. The impedance diagram (A) derives from the voltage diagram by deleting from consideration the common quantity, I. The circuit with the isolation transformer (B) avoids conflicts of grounds if electronic instruments are used for monitoring E_R and E_L.

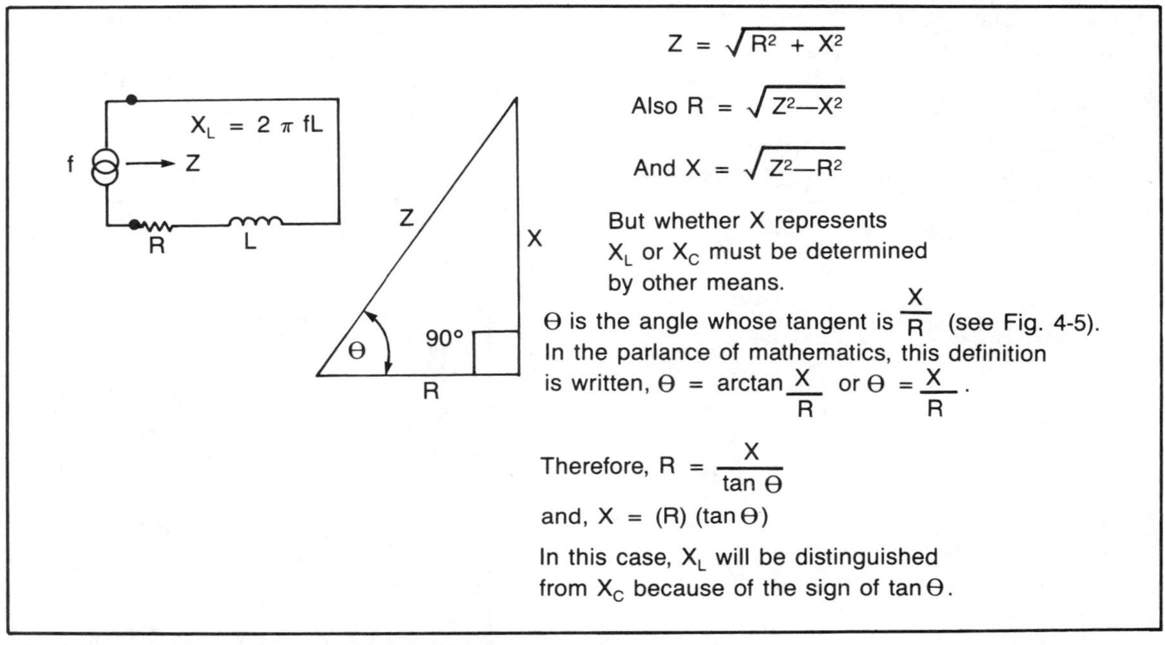

$Z = \sqrt{R^2 + X^2}$
Also $R = \sqrt{Z^2 - X^2}$
And $X = \sqrt{Z^2 - R^2}$
But whether X represents
X_L or X_C must be determined
by other means.
Θ is the angle whose tangent is $\frac{X}{R}$ (see Fig. 4-5).
In the parlance of mathematics, this definition
is written, $\Theta = \arctan \frac{X}{R}$ or $\Theta = \frac{X}{R}$.
Therefore, $R = \frac{X}{\tan \Theta}$
and, $X = (R)(\tan \Theta)$
In this case, X_L will be distinguished
from X_C because of the sign of $\tan \Theta$.
$X_L = 2\pi fL$
f
Z
R
L
Z
X
90°
Θ
R

Fig. 4-4. Impedance relationships illustrated by a right triangle.

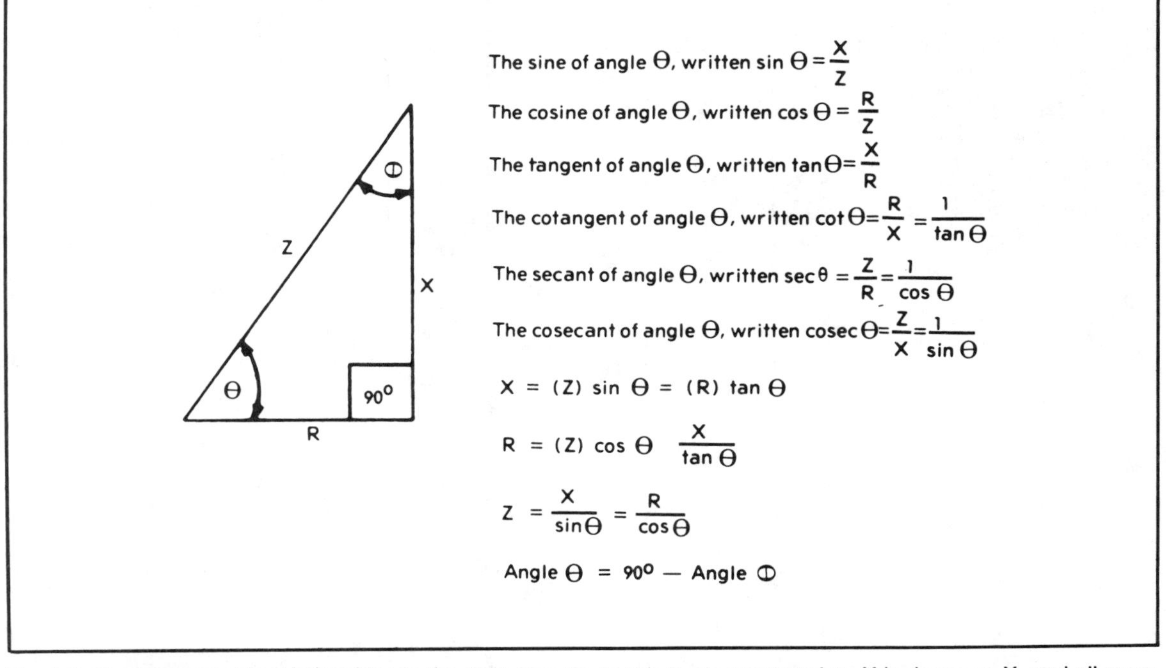

The sine of angle Θ, written $\sin \Theta = \frac{X}{Z}$
The cosine of angle Θ, written $\cos \Theta = \frac{R}{Z}$
The tangent of angle Θ, written $\tan \Theta = \frac{X}{R}$
The cotangent of angle Θ, written $\cot \Theta = \frac{R}{X} = \frac{1}{\tan \Theta}$
The secant of angle Θ, written $\sec \theta = \frac{Z}{R} = \frac{1}{\cos \Theta}$
The cosecant of angle Θ, written $\mathrm{cosec}\, \Theta = \frac{Z}{X} = \frac{1}{\sin \Theta}$
$X = (Z) \sin \Theta = (R) \tan \Theta$
$R = (Z) \cos \Theta \quad \frac{X}{\tan \Theta}$
$Z = \frac{X}{\sin \Theta} = \frac{R}{\cos \Theta}$
Angle $\Theta = 90^\circ -$ Angle Φ
Φ
Z
X
Θ
90°
R

Fig. 4-5. The trigonometric relationships in the right triangle. In this basic presentation, X is drawn as X_L and all quantities are positive.

that, knowing one side of the triangle and angle Θ, we can solve for the other sides. Conversely, with a knowledge of two sides we can quickly determine the other side (as an alternative to the use of Pythagoras' Theorem) as well as the angle Θ. All we need to speed our efforts is the trigonometric table such as appears in Table 4-1.

From what has been thus far discussed, we can clearly draw significance from the three impedance diagrams shown in Fig. 4-6. The three impedances represented all have the same magnitude but have different angles. If we are merely interested in limiting the flow of current to a certain value, one of these impedances is as good as another. Indeed, who cares about the angle? However, if any two of these impedances are connected in series (or in par-

Table 4-1. Trigonometric Functions. Only Sine, Cosine and Tangent Functions Are Listed, Since These Are the Most Useful of the Six Functions. (The Cosecant, Secant, and Cotangent Functions Are, Respectively, Reciprocals of the Listed Three.)

ANGLE φ	SIN.	COS.	TAN.	ANGLE φ	SIN.	COS.	TAN.
0	.000	1.000	.000				
1	.017	.999	.017	46	.719	.695	1.04
2	.035	.999	.035	47	.731	.682	1.07
3	.052	.999	.052	48	.743	.669	1.11
4	.070	.998	.070	49	.755	.656	1.15
5	.087	.996	.087	50	.766	.643	1.19
6	.105	.995	.105	51	.777	.629	1.23
7	.122	.993	.123	52	.789	.616	1.28
8	.139	.990	.141	53	.799	.602	1.33
9	.156	.988	.158	54	.809	.588	1.38
10	.174	.985	.176	55	.819	.574	1.43
11	.191	.982	.194	56	.829	.559	1.48
12	.208	.978	.213	57	.839	.545	1.54
13	.225	.974	.231	58	.848	.530	1.60
14	.242	.970	.249	59	.857	.515	1.66
15	.259	.966	.268	60	.866	.500	1.73
16	.276	.961	.287	61	.875	.485	1.80
17	.292	.956	.306	62	.883	.469	1.88
18	.309	.951	.325	63	.891	.454	1.96
19	.326	.946	.344	64	.898	.438	2.05
20	.342	.940	.364	65	.906	.423	2.14
21	.358	.934	.384	66	.914	.407	2.25
22	.375	.927	.404	67	.921	.391	2.36
23	.391	.921	.424	68	.927	.375	2.48
24	.407	.914	.445	69	.934	.358	2.61
25	.423	.906	.466	70	.940	.342	2.75
26	.438	.898	.488	71	.946	.326	2.90
27	.454	.891	.510	72	.951	.309	3.08
28	.469	.883	.532	73	.956	.292	3.27
29	.485	.875	.554	74	.961	.276	3.49
30	.500	.866	.577	75	.966	.259	3.73
31	.515	.857	.601	76	.970	.242	4.01
32	.530	.848	.625	77	.974	.225	4.33
33	.545	.839	.649	78	.978	.208	4.70
34	.559	.829	.675	79	.982	.191	5.14
35	.574	.819	.700	80	.985	.174	5.67
36	.588	.809	.727	81	.988	.156	6.31
37	.602	.799	.754	82	.990	.139	7.12
38	.616	.788	.781	83	.993	.122	8.14
39	.629	.777	.810	84	.995	.105	9.51
40	.643	.766	.839	85	.996	.087	11.43
41	.656	.755	.869	86	.998	.070	14.30
42	.669	.743	.900	87	.999	.052	19.08
43	.682	.731	.933	88	.999	.035	28.64
44	.695	.719	.966	89	.999	.017	57.28
45	.707	.707	1.000	90	1.000	.000	Infinity

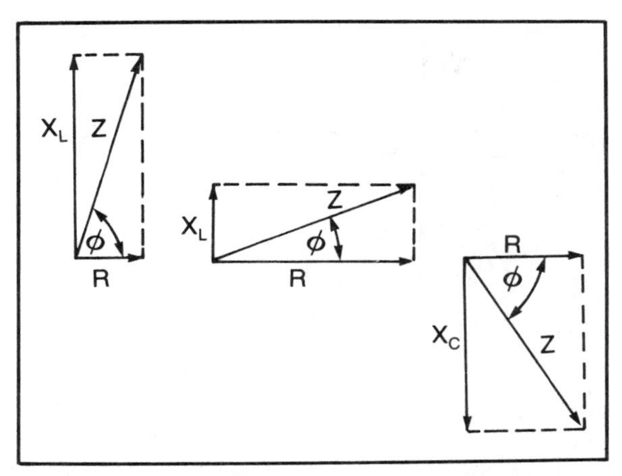

Fig. 4-6. The three impedances shown have the same magnitude but different angles.

allel) the resultant impedance will be determined by which two impedances are selected. And if there must be concern over the way in which any of these impedances influence the phase of voltage or current, as in a feedback network, then the circuit results will differ with each one despite their equal magnitudes.

Complex or R Plus or Minus jX Notation

Returning again to Fig. 3-4, the equation:

$$Z = \sqrt{R^2 + X^2}$$

can be written in another way which is uniquely useful to the situations encountered in ac circuitry. We refer to the expression, Z equals R plus or minus jX. Here, the letter j is an *operator*. In the operation to be performed, the terms R and X must each be square and the square root must then be extracted from the sum of these squares. In other words, we have deliberately created a relationship between two ways of arriving at Z. This implies that:

$$\sqrt{R^2 + X^2} = R \pm jX.$$

A reference to Fig. 4-7 will point out some of the features involved with the R plus or minus jX notation. The set of coordinates are labeled R and j and each can bear either the plus or minus sign. The impedance "vector" can occupy positions in any of the four quadrants. The components of impedance will be plus or minus R and plus or minus j, depending on which quadrant they are located in. In any event, j is always 90 degrees displaced from, or rotated from, the R axis. (Although Z must necessarily be depicted on paper as a stationary vector quantity, our imaginative powers can be extended to visualize vector Z as being in counterclockwise rotation at an angular speed determined by the frequency. The rotational aspect of vector Z is not closely related to our study of impedance, however.)

The essence of the foregoing development is that points anywhere in the four quadrants can be exactly specified by their R and j values. These points are said to represent impedances on the s plane. The s plane is the space encompassed by the quadrants. Notice that plus jX designates inductive reactance and minus jX designates capacitive reactance. Think of X or $-X$ as the value of j. If we were dealing with the geometry of the complex plane, the letter j would suffice for the labeling of the vertical axis. When we adopt this geometric system to the needs of the electric circuit, we bring in the reactance X to define the value of j. It can be seen that R plus or minus jX always occupies quadrant I and represents resistance and inductive reactance in series. R minus jX always occupies quadrant IV and represents resistance and capacitive reactance in series. (Quadrants II and III contain combinations of negative resistance with inductive or capacitive reactance. Such combinations are of importance in electronics, but are not typical of the simple test procedures that we shall mainly concern ourselves with.)

INDUCTIVE AND CAPACITIVE REACTANCE

In quadrants I and IV (Fig. 4-7), notice that the angle of the impedance vector is measured with respect to the line OR, which represents positive or "ordinary" resistance. Impedances in quadrant I have positive values of angle Θ, where as im-

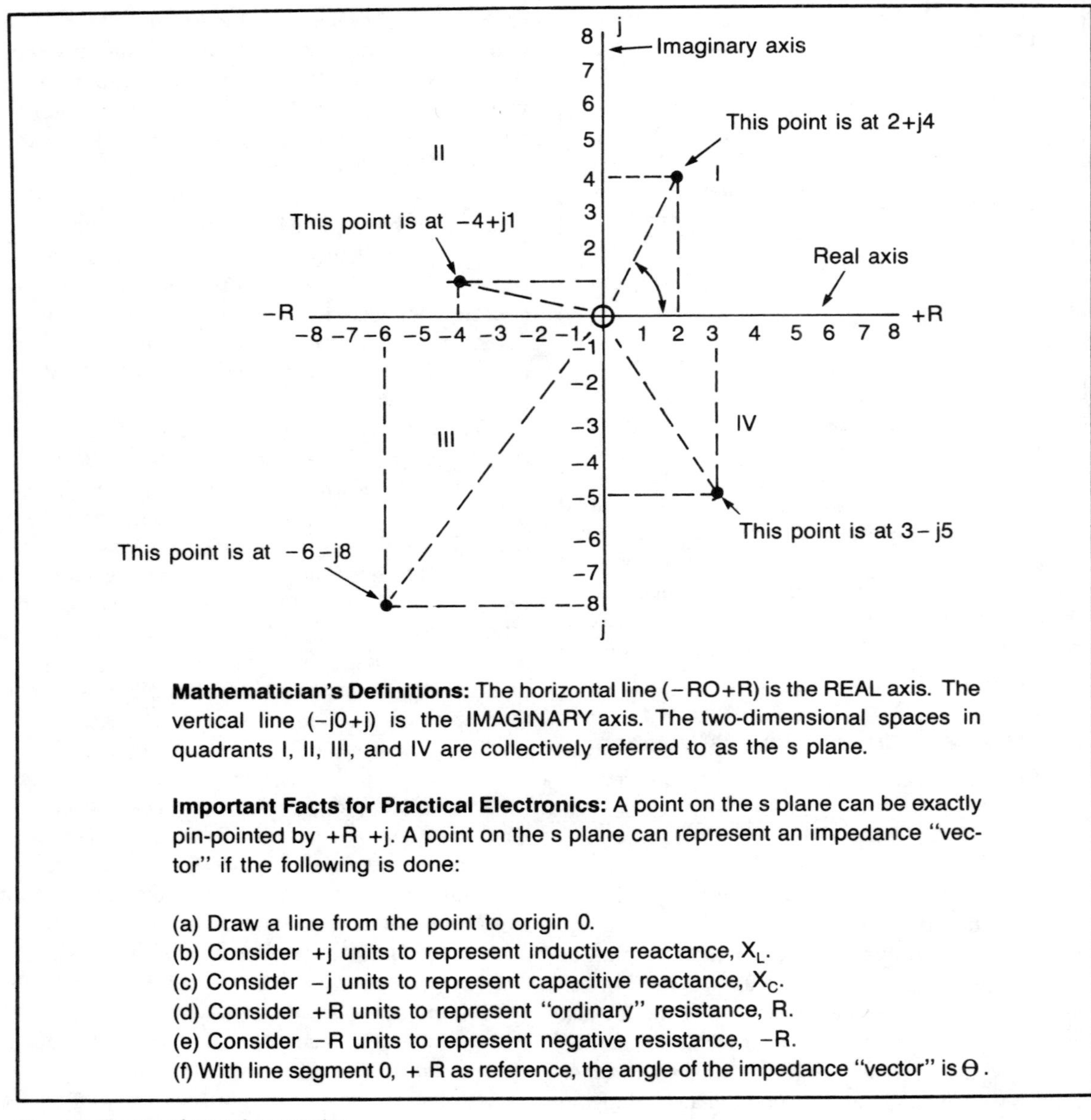

j
Imaginary axis
This point is at 2+j4
I
II
This point is at −4+j1
Real axis
−R
+R
−8 −7 −6 −5 −4 −3 −2 −1
1 2 3 4 5 6 7 8
8 7 6 5 4 3 2
−1 −2 −3 −4 −5 −6 −7 −8
III
IV
This point is at 3 − j5
This point is at −6 −j8
j
Mathematician's Definitions: The horizontal line (−RO+R) is the REAL axis. The vertical line (−j0+j) is the IMAGINARY axis. The two-dimensional spaces in quadrants I, II, III, and IV are collectively referred to as the s plane.
Important Facts for Practical Electronics: A point on the s plane can be exactly pin-pointed by +R +j. A point on the s plane can represent an impedance "vector" if the following is done:
(a) Draw a line from the point to origin 0.
(b) Consider +j units to represent inductive reactance, X_L.
(c) Consider −j units to represent capacitive reactance, X_C.
(d) Consider +R units to represent "ordinary" resistance, R.
(e) Consider −R units to represent negative resistance, −R.
(f) With line segment 0, + R as reference, the angle of the impedance "vector" is Θ.

Fig. 4-7. Basics of complex notation.

pedances in quadrant IV have negative values of angle Θ. The impedance itself is always positive in these two quadrants. The significance of this is that positive values of the angle Θ are always associated with the effect of inductance, whereas negative values of Θ are always associated with the effect of capacitance. All in all, R plus or minus jX contains information about the magnitude of the impedance, the magnitudes of its resistive and reactive components, the nature of the reactive component, and the value of the angle.

So much now for the s plane, the four

quadrants, and the real and imaginary axis. This is all part of the mathematical scheme of things and must be dealt with in advanced design and analysis. For our stated purposes, we will return to the simple triangle— the impedance diagram. However, we will continue using the R plus or minus jX notation.

Angle Information

We have already seen how the magnitude of Z is determined from R plus or minus jX, since we use the ordinary algebraic form of the complex notation.

$$\sqrt{R^2 + X^2}$$

Because of the squaring process, both R + jX and R − jX yield the same magnitude for Z. This is because both $(+X)^2$ and $(-X)^2$ become $+X^2$ according to the rules of simple algebra. Now, let us see how the angle information is contained in R plus or minus jX.

From trigonometry, angle Θ is defined as arctan X/R. This means that the angle Θ corresponds to the value of the tangent function given by the ratio of X to R. (Angle Θ equals arctan X/R may also be written, angle Θ equals $\tan^{-1}$ X/R. The −1 is not an exponent here, but a symbol denoting "the angle whose tangent is ..."). So, all we have to do is compute X/R and look up the result in the trigonometric table under Tangent. Then, the corresponding angle is Θ. Notice that angle Θ is positive in quadrant I, but is a negative angle in quadrant IV. When Θ is a positive angle, +X represents inductive reactance. When Θ is a negative angle, −X represents capacitive reactance. As you can see, the R plus or minus jX notation obviously contains all information needed to determine impedance. This is borne out by the two numerical examples in Fig. 4-8.

Now, recall that impedances in series generally do not produce a net impedance which is the arithmetic sum of the individual impedance magnitudes. Suppose that instead of being told that three impedances of such and such magnitudes are connected in series, we are given the information in complex notation; that is, R plus or minus jX. We can use some specific numbers for sake of both instruction and example. Referring to Fig. 4-9, the following impedances are connected in series: 5 + j7 ohms, 2 − j3 ohms, and 1 + j1 ohms. The resistive and reactive components of these impedances are added separately. Thus, we account for 8 ohms of resistance and +5 ohms of reactance. The overall or net impedance is, therefore, 8 + j5 ohms. The magnitude of Z is:

$$\sqrt{(8)^2 + (5)^2} = \sqrt{64 + 25} = \sqrt{89}, \text{ or } 9.5 \text{ ohms}$$

Now, if we are content with magnitude only, we can leave the result as is. It should be appreciated that the net impedance of the circuit could not have been calculated from impedance magnitude alone, however. The angle Θ of this overall impedance is given by arctan 5/8 and the trigonometric table shows this to correspond to 32 degrees.

Complex notation simplifies matters because we can add series-connected impedances without involved computations. This summing can usually be accomplished by quick inspection. However, close attention must be paid to plus and minus signs. Notice that the angle of each individual impedance has been taken into consideration without actually dealing with an angle of so many degrees. Of course, this follows from the fact that complex notation always carries angular information, whether or not such information is actually extracted for immediate use. The actual extraction of angular information may be made at any time by employing the relationship, angle Θ equals arctan X/R. When so doing, care should be taken to carry along the sign of X in order that the sign of Θ can be specified. Thus, arctan +X/R yields a positive value of angle Θ and relates to a series circuit with a predominantly inductive reactance. Conversely, arctan −X/R yields a negative value of angle Θ and refers to a predominantly capacitive reactance in the series circuit.

One of the advantages of the R plus or minus jX specification over the algebraic designation $\sqrt{R^2 + X^2}$ is that the sign of the reactance does not become lost. (The sign of X need not be lost when writing $\sqrt{R^2 + X^2}$, but more care is needed.) In compution, $\sqrt{R^2 + X^2}$ actually must be used to solve R plus or minus jX. However, specifying an impedance by the R plus or minus jX notation definitely reveals the sign of X.

When impedances are paralleled, we are tempted to combine them as resistances in dc circuits:

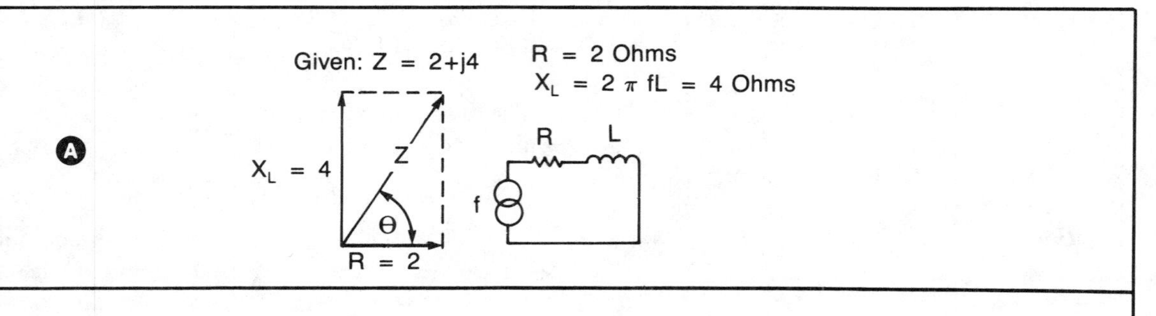

B

R = 3

Given; Z = 3−j5

$X_C = -5$

R

F

C

R = 3 Ohms

$X_C = \dfrac{1}{2\pi fC} = -5$ Ohms

In Fig. 4-7, the point at 2+j4 suggests a series circuit with 2 ohms of "ordinary" resistance and 4 ohms of inductive reactance.

The magnitude of Z is $\sqrt{(R)^2 + (X_L)^2} = \sqrt{(2)^2 + (4)^2}$

$$= \sqrt{4 + 16} = \sqrt{20} = 4.47 \text{ Ohms}$$

The angle, Θ is $\tan^{-1} \dfrac{X_L}{R}$ Where X is the j term.

$$\Theta = \tan^{-1} \frac{4}{2} \tan^{-1} 2 = 63.5°$$

In Fig. 4-7, the point 3−j5 suggests a series circuit with 3 ohms of "ordinary" resistance and 5 ohms of capacitive reactance. The minus sign should be used with −j quantities, so that the capacitive reactance is written as −5 ohms.

The magnitude of Z is $\sqrt{(R)^2 + (X_C)^2} = \sqrt{(3)^2 + (-5)^2}$

$$= \sqrt{9 + 25} = \sqrt{34 = 5.83} \text{ Ohms}$$

The angle, Θ, is $\tan^{-1} \dfrac{X_C}{R} = \dfrac{-5}{3} = \tan^{-1} -1.666$

$$\Theta = \tan^{-1} -1.666 = 59°$$

Fig. 4-8. Examples showing how to convert R ± jX to magnitude and angle of impedance, Z.

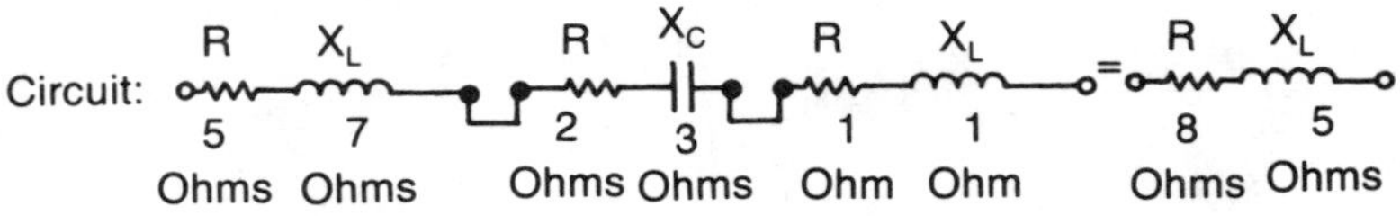

Circuit: R 5 Ohms, X_L 7 Ohms — R 2 Ohms, X_C 3 Ohms — R 1 Ohm, X_L 1 Ohm = R 8 Ohms, X_L 5 Ohms

Impedance complex notation: 5 + j7 Ohms   2 − j3 Ohms   1 + j1 Ohms   = 8 + j5 Ohms

Explanation: Summation of R components = 5 + 2 + 1 = 8 Summation of jX components = 7 − 3 + 1 = 5

Therefore net impedance is represented by 8 + j5 ohms. This corresponds to 8 ohms of resistance and 5 ohms of inductive reactance in series.

Fig. 4-9. Use of the complex notation, R + jX, in determining the net result of three series-connected impedances.

$$\frac{1}{Z} = \frac{1}{Z_1} + \frac{1}{Z_2} + \frac{1}{Z_3} \quad \ldots$$

For two elements, we have:

$$Z = \frac{Z_1 \times Z_2}{Z_1 + Z_2}$$

Will this approach work? As with the series connection, the answer is yes, if we take into account the angles associated with the impedances. Again, we cannot deal with magnitudes only. In the denominator of the second equation, we see Z1 + Z2. This is reminiscent of a series circuit comprising impedances Z1 and Z2. Indeed, the denominator can be handled in exactly the same way you would add the elements in a series circuit containing Z1 and Z2. Therefore, if the denominator is expressed in R plus or minus jX notation, the indicated summing operation is very easily performed. But, how do we perform the multiplying operation indicated in the numerator? To do this, we must use yet another notation, the so-called *polar form.* The polar form may be given, or it may be derived from complex notation or from trigonometric considerations. Actually, we have already encountered it without utilizing its most valuable property.

It has been shown that R plus or minus jX can be translated to the form Z at an angle Θ. (The opposite translation is also useful.) And, it has been shown that R plus or minus jX is uniquely adapted to the purpose of addition, such as occurs when dealing with series circuits. Now, it will be shown that the alternate form Z at an angle of Θ is uniquely adapted to the purpose of multiplication, as used in the equations associated with parallel circuits.

Let Z at an angle of Θ degrees be written $Z\angle\Theta°$ for quadrant I, and $Z \diagdown\!\underline{\Theta°}$ for quadrant IV. The later symbol indicates that angle Θ is negative. There is less chance of error with this symbology than is the case when minus signs are used with negative angles. $Z \angle\Theta°$ is, by this convention, the polar form of R + jX. And $Z \angle\Theta°$ is the polar form of R − jX. That is all there is to the notational aspect of Z in polar form. The translation of R plus or minus jX to its polar equivalent involves a magnitude determination from Z equals $\sqrt{R^2 + X^2}$ and an angle determination from Θ equals arctan X/R. As previously pointed out, we will confine our interest to the world of positive resistance which is represented by impedances acting in quadrants I and IV. For quadrant I, Θ varies between zero and +90 degrees. For quadrant IV, Θ varies between zero and −90 degrees.

Now, let's investigate the multiplying property of Z in polar form. It is very simple, indeed. The magnitude of the Zs are actually multiplied together, but their angles are added (with due regard

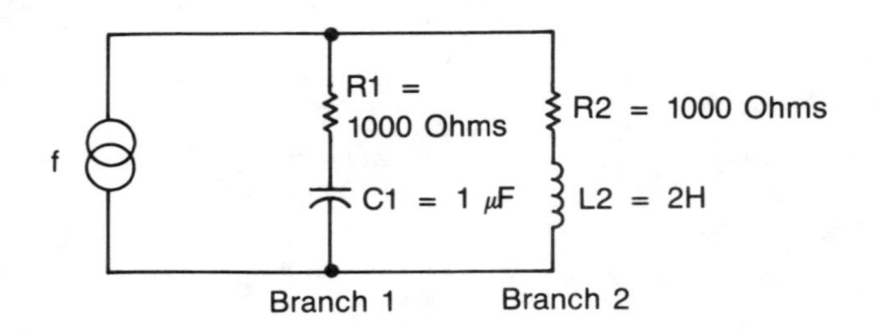

If f = 318.5 Hz, what is the impedance, $\underset{\bullet}{Z}$, of this circuit?

1. Basic formula for solution: $\underset{\bullet}{Z} = \dfrac{\underset{\bullet}{Z1} \times \underset{\bullet}{Z2}}{\underset{\bullet}{Z1} + \underset{\bullet}{Z2}}$ (The dots indicate that the Zs have both magnitude and angle.)

2. $\underset{\bullet}{Z1} = R1 + j(-X_{C1}) = \sqrt{(1000)^2 + (-X_C)^2}$

where $X_C = \dfrac{1}{2\pi fC} = \dfrac{1}{(2\pi)(318.5)(1\times10^{-6})} = \dfrac{1 \times 10^{+6}}{1 \times 10^{+6}} = 1000\Omega$

$\underset{\bullet}{Z1} = \sqrt{(1000)^2 + (-1000)^2} = \sqrt{2 \times 10^6} = 1414$ Ohms magnitude

Angle Θ of $\underset{\bullet}{Z1}$ = arctan $\dfrac{-X_C}{R1}$ = arctan $\dfrac{-1000}{1000}$ = arctan $-1 = -45^\circ$

Therefore $\underset{\bullet}{Z1} = 1000 - j1000$ is complex notation (useful in addition)
$\underset{\bullet}{Z1} = 1414\overline{|45^\circ}$ in polar notation (useful in multiplication, division)

We have solved the series circuit comprising R1 and C1 of BRANCH 1

3. $\underset{\bullet}{Z2} = R2 + j(X_L) = \sqrt{(1000)^2 + (X_L)^2}$

where $X_L = 2\pi fL = (2\pi)(318.5)(2) = (1000)(2) = 2000$ Ohms

$\underset{\bullet}{Z2} = \sqrt{(1000)^2 + (2000)^2} = \sqrt{1 \times 10^{-6} + 4 \times 10^6} = \sqrt{5 \times 10^6} = 2245$ Ohms

Angle Θ of $\underset{\bullet}{Z2}$ = arctan $\dfrac{X_L}{R2} = \dfrac{2245}{1000} = 2.245$

Arctan 2.245 = 66°

Therefore $\underset{\bullet}{Z2} = 1000 + j2000$ in complex notation

$\underset{\bullet}{Z2} = 2245\underline{|66^\circ}$ in polar notation

We have solved the series circuit comprising R2 and L2 of BRANCH 2

4. $\underset{\bullet}{Z1} \times \underset{\bullet}{Z2}$ in formula 1 = $1414\overline{|45^\circ} \times 2245\underline{|66^\circ} = 3{,}174{,}430\underline{|21^\circ}$

Fig. 4-10. Step-by-step solution of a circuit with two parallel branches.

5. $\underset{\bullet}{Z1} + \underset{\bullet}{Z2}$ in formula 1 =

$$\begin{array}{r} 1000 - j1000 \\ +1000 + j2000 \\ \hline 2000 + j1000 \end{array}$$

6. $\underset{\bullet}{Z1} + \underset{\bullet}{Z2}$ expressed polar notation $= \sqrt{(2000)^2 + (1000)^2} \text{ arctan } \dfrac{1000}{2000}$

$$= \sqrt{5 \times 10^6} \text{ arctan } \frac{1}{2}$$

$$\sqrt{5 \times 10^6} \text{ arctan } \frac{1}{2} = 2245\underline{|26.6^\circ}$$

7. $$\underset{\bullet}{Z} = \frac{\underset{\bullet}{Z1} \times \underset{\bullet}{Z2}}{\underset{\bullet}{Z1} + \underset{\bullet}{Z2}} = \frac{3{,}174{,}430\underline{|21^\circ}}{2245\ \underline{|26.6^\circ}} = 1414\underline{|21^\circ} - \underline{|26.6^\circ} = 1414\ \underline{|5.6^\circ}$$

Thus, the impedance of this parallel branch circuit has a magnitude of 1414 ohms and an angle of -5.6°

8. $1414\ \overline{|5.6^\circ}$ can also be expressed in complex notation:
The R component of $R - jX$ is given by $\underset{\bullet}{Z} \cos \Theta$.
$= 1414 \cos(-5.6^\circ) = 1414\ (0.9952) = 1407.2$ Ohms   1407.2 Ohms

The X component of $R - jX$ is given by $\underset{\bullet}{Z} \sin \Theta$
$= 1414 \sin(-5.6^\circ) = 1414\ (-0.0976) = -138.4$ Ohms
Therefore, $1414\ \overline{|5.6^\circ} = 1407 - j138.4$ Ohms
where $-j138$ denotes 138 ohms of capacitive reactance

Solving for C: $$C = \frac{1}{(2\pi f)X_C} = \frac{1}{(2\pi f)\ (138)} = \frac{1}{2\pi(318.5)\ (138)} = \frac{1}{138 \times 10^3}$$

$$= 00725 \times 10^{-3}\text{F} = 7.25\mu\text{F}$$

This shows that a series circuit consisting of 1407 ohms of resistance and a 7.25μF capacitor looks like the given parallel circuit. (But only at f = 318.5 Hz!)

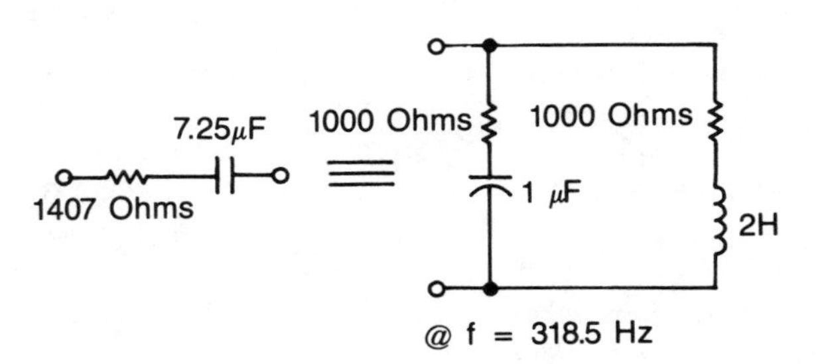

for plus and minus signs associated with these angles). That is:

$$\underset{\cdot}{Z}1 \times \underset{\cdot}{Z}2 = (Z1 \text{ at } \Theta)\ (Z2 \text{ at } \Theta 2)$$
$$= Z1 \times Z2 \text{ at } \Theta 1 + \Theta 2$$

The dot under the impedance symbol means that Z, so designated, involves both magnitude and angle. Although such symbols are often dispensed with, it helps clarify the situation presently under discussion. Let's look at some examples:

Let $\underset{\cdot}{Z}1$ equal 10 $\angle 30^\circ$ and $\underset{\cdot}{Z}2$ equal 5 $\angle 15^\circ$. What is the product Z1 × Z2?

10 × 5 equals 50 ohms (magnitude),
and 30 degrees plus 15 degrees
equals 45

Therefore, the product is 50 $\angle 45^\circ$, or 50 ohms at 45 degrees in quadrant I.

Let Z1 equal 3 $\angle 10^\circ$ and $\underset{\cdot}{Z}2$ equal 4 $\angle 2^\circ$. What is the product of $\underset{\cdot}{Z}1 \times \underset{\cdot}{Z}2$?

3 × 4 equals 12 ohms (magnitude),
and 10 degrees minus 2 degrees
equals eight degrees (angle)

Therefore, the product is 12 $\angle 8^\circ$, or 12 ohms at 8 degrees in quadrant I.

Finally, let $\underset{\cdot}{Z}1$ equal 7 $\angle 45^\circ$ and $\underset{\cdot}{Z}2$ equal 2 $\overline{\backslash 60^\circ}$. What is the product of $\underset{\cdot}{Z}1 \times \underset{\cdot}{Z}2$?

7 × 2 equals 14 ohms (magnitude),
and 45 degrees minus 60 degrees
equals minus 15 degrees (angle)

Therefore, the product is 14 $\overline{\backslash 15^\circ}$, or 14 ohms at minus 15 degrees in quadrant IV.

Once we get the gist of this multiplying process, we can return to the parallel circuit. Observe that the manipulations are very simple, but we must abide by the rules of algebra concerning the handling of plus and minus signs associated with both magnitudes and angles.

The equation for the impedance of a parallel circuit involving two parallel branches, $\underset{\cdot}{Z}1$ and $\underset{\cdot}{Z}2$, is:

$$\underset{\cdot}{Z} = \frac{\underset{\cdot}{Z}1 \times \underset{\cdot}{Z}2}{\underset{\cdot}{Z}1 + \underset{\cdot}{Z}2}$$

For a 3-branch parallel circuit:

$$\underset{\cdot}{Z} = \frac{\underset{\cdot}{Z}1 \times \underset{\cdot}{Z}2 \times \underset{\cdot}{Z}3}{(\underset{\cdot}{Z}1 \times \underset{\cdot}{Z}2) + (\underset{\cdot}{Z}1 \times \underset{\cdot}{Z}3) + (\underset{\cdot}{Z}2 \times \underset{\cdot}{Z}3)}$$

The 3-branch circuit is more involved, but it is essentially the same as the 2-branch circuit. Also, the impedance of any number of branches in parallel can be solved by similar equations. The main requirement is the ability to change back and forth from complex to polar notation. The dots under the Z tell us that we are dealing with impedance expressed as a magnitude and an angle (either in complex or polar notation).

A step-by-step numerical example of the solution to a 2-branch parallel circuit is given in Fig. 4-10. Notice that the solution of such a parallel circuit requires the ability to solve series circuits, and that we must be able to freely convert from complex to polar notation and vice versa. The ability to master such a problem as the one in Fig. 4-10 cannot fail to shed light on the behavior of ac circuits, and must make meaningful the true significance of test procedures involving impedance.

For example, when the solution to the circuit in Fig. 4-10 is finally converted back into R + jX notation, we obtain the constants for a series circuit which looks like the original parallel circuit—almost! If we had a black box with two terminals, we could not determine whether the contents consisted of the original parallel circuit or its derived series-circuit equivalent, at least as far as the impedance magnitude and angle are concerned. Does this imply that one circuit could be substituted for another in circuit applications? In some cases, perhaps, but generally speaking, probably not. The two "equivalents" exist at only at one frequency (318.5 Hz). For higher and lower frequencies, their magnitude and phase behavior become drastically different. For example, the parallel circuit will pass dc, whereas the series circuit will not.

IMPEDANCE MEASUREMENT BY THE EQUAL DEFLECTION METHOD

From the formula for impedance, Z equals $\sqrt{R^2 + X^2}$, it is obvious that a given numerical value of impedance can comprise endless combinations of R and X, and X can be inductive or capacitive. In a strict sense, an impedance is not fully defined by stating that it is, say, 10 ohms. In some applications, only a unique 10-ohm impedance would work properly, one in which the resistive and reactive parameters were specifically defined. However, there are many applications where a knowledge of the numerical, or absolute, value of impedance suffices. Here, the impedance may be made up of a number of components, but the overall impedance is treated as a 2-terminal device which influences an ac circuit in a way similar to the effect of resistance in a dc circuit. (Indeed, the impedance may even be 100 percent resistive.) No matter what the actual composition of the impedance is, its effect on the flow of ac current is described by the relationship I equals E divided by Z, which simulates Ohm's Law of the dc circuit.

The comparison with the dc circuit is instructive, but it ignores a significant fact of the ac behavior of impedance. That, of course, is its frequency dependency. For a given frequency, the absolute value of impedance is connected to an ac voltage and current by the Ohm's Law relationship. However, in the general case, the impedance will vary with frequency. The nature of this variation may be a significant part of the measurement. In such instances, the test procedure depends on data from which such variations can be plotted, or otherwise be shown. This means measuring Z over an appropriate range of frequencies. In other cases, where a single impedance measurement can provide useful information, the frequency of measurement should not be omitted from the report of the test.

The aforementioned circumstances lead to a simple impedance measurement technique. The general idea underlying the equal-deflection method of impedance measurement is shown in Fig. 4-11. A voltmeter is used to compare the voltage drop of the unknown impedance with that developed across a reference resistance. When these voltages are the same, the value of the unknown impedance equals that of the reference resistance in ohms. Observe that the actual value of the voltage involved is of no significance, except that it should be high enough to override noise or hum interference, but not so high as to overload the component undergoing test.

The accuracy of this scheme depends on the use of a high value of resistance for R1, a value at least of twenty and preferably one hundred times that of the impedance being measured. The function of R1 is to convert the oscillator to a constant-current source. Otherwise, the value of R1 does not enter into any computation involving the impedance value. It should always be determined, also, that the internal resistance of the voltmeter is a hundred times or more higher than the impedance being measured. Finally, from a basic concept of impedance, the oscillator should deliver a sine wave of good purity. The reference resistance can be a

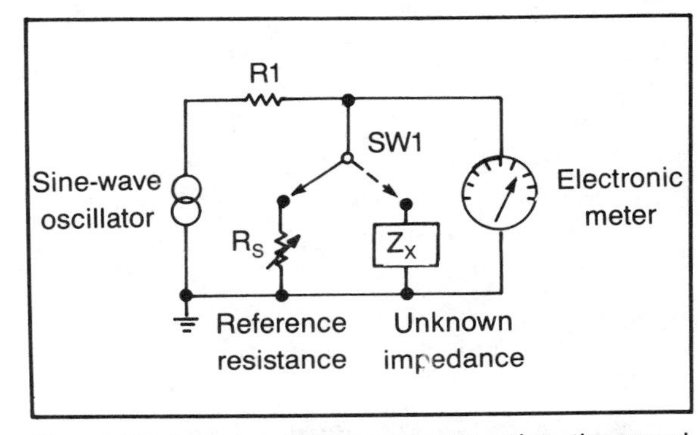

Fig. 4-11. Impedance measurement using the equal-deflection method. When the same indications are read for both positions of switch SW1, Z_x equals R_x.

decade resistance box, a linear potentiometer, or it can be a variable but uncalibrated resistance, in which case the resistance measurement is made after the equal-deflection test.

This method can be used to measure resistors, inductors, and capacitors, as well as general impedances comprised of two or all of these parameters. Measurements performed on inductors will be, in essence, impedance measurements because practical inductors do have some resistance. However, for many practical purposes, an inductor, particularly a high Q inductor, can be considered at its measured inductive reactance at the frequency of measurement. For all practical purposes, measurements on nonelectrolytic capacitors can be considered to be capacitive reactance at the frequency of measurement.

Inductance L is related to inductive reactance by the formula:

$$L = \frac{X_l}{2\pi f}$$

where L is in Henrys, X_l is in ohms, and f is in Hertz.

Capacitance C is related to capacitive reactance by the formula:

$$C = \frac{1}{(2\pi)(X_c)(f)}$$

where C is in farads, X_c is in ohms, and f is in hertz.

The accuracy of inductance and capacitance calculations depend on the accuracy of our knowledge of frequency. For many of the purposes of practical electronics where 5- or 10-percent tolerances are allowable, this need not be a barrier to the use of the equal-deflection method as a means of measuring inductors and capacitors.

Procedure for Determining Incremental Impedance and Inductance

Objective of Test: To use the equal-deflection method for measuring the impedance and inductance of an iron-core inductor carrying direct current.

Test Equipment Required: Constant-current dc power supply with an output impedance of 100K or higher at 60 Hz, a compliance voltage of 30 volts or higher, and a current capability of 20 milliamperes or higher.

- ac electronic voltmeter, with the emphasis on low ranges.
- 25K potentiometer and 50K resistor.
- Ohmmeter
- Sine-wave oscillator, with emphasis on high output voltage.
- 1.0 μF Mylar capacitor
- Double-throw single-pole switch, with no neutral position.
- Small power supply choke from one to ten henrys.

Test Procedure: Connect the components and equipment as shown in Fig. 4-12. Initially, the constant current supply should not be connected in the circuit. Select the desired frequency. This will often be 120 Hz, corresponding to the ripple frequency of a power supply energized from the 60-Hz line. With the switch SW1 in its No. 2 position, adjust the amplitude of oscillator for a half-scale reading on the electronic meter (set for low voltage range). Place SW1 in its No. 1 position and adjust R_s for the same meter reading. The impedance Z_x of the inductor now is equal to R_s. R_s can be measured with an ohmmeter when switch SW1 is again placed in its No. 2 position. The above procedure should work for a wide variety of iron-core inductors, with resistance R in the vicinity of 50K. If R must be lowered in order to obtain a sufficient ac voltage, it should be decreased no more than necessary.

The above procedure measures the impedance of Z_x and provides data for calculating L_x with no direct current in the winding of the choke. Repeat the above procedure with the desired direct current values by appropriate adjustment of the constant-current power supply (now connected as in Fig.

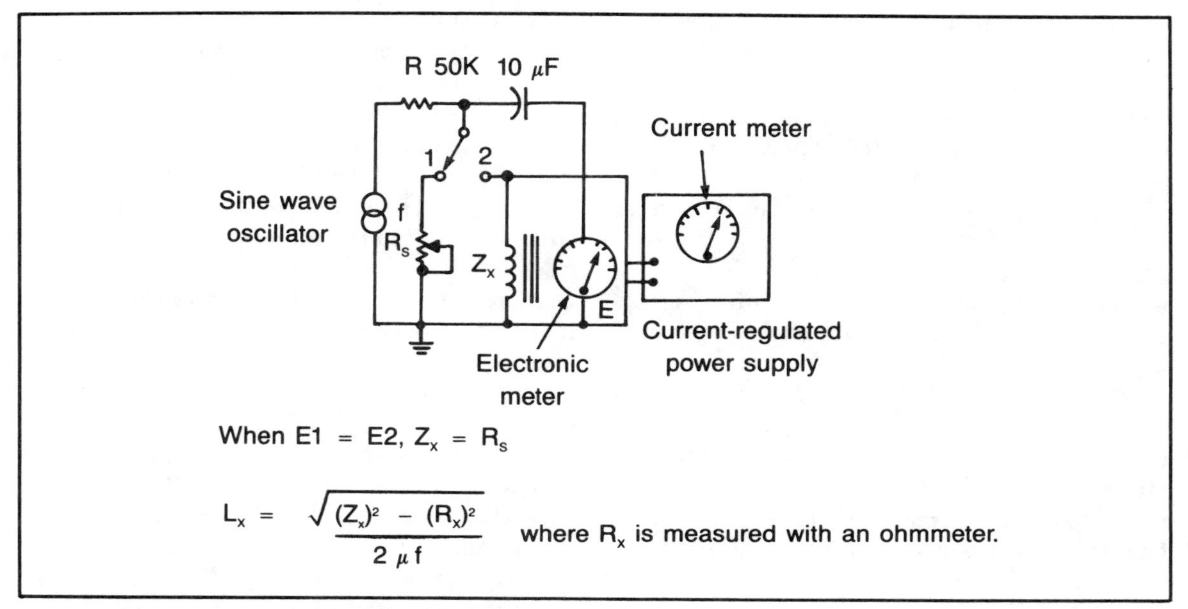

Fig. 4-12. Setup for measuring incremental inductance.

4-12). Except in rarely encountered instances where residual magnetism is present in the core, the introduction of even a small direct current will lower the impedance and inductance of the choke. Indeed, the need for a test of this nature is to determine whether or not an inductor still has sufficient inductance when it must carry a prescribed amount of direct current. The inductance (incremental inductance with dc in the winding) is calculated with the formula:

$$L_x = \frac{\sqrt{(Z_x{}^2 - R_x{}^2)}}{2 \pi f}$$

Comments on Test Results: One reason for using a current-regulated power supply is that the 60-Hz, or 120 Hz ac impedance of such a dc source can be very high. Therefore, the shunting effect of the constant-current supply on the inductor can be negligible. We use a constant-current supply, other than for its high ac impedance, because relatively low voltages can be used to cause several tens of milliamperes, or more current, to flow through inductors. If an ordinary power supply were used, ac isolation would have to be accomplished with a high series resistance. This would make necessary the use of a supply with a several hundred volt or even higher capability.

Some of the direct current is shared by the circuit comprising R and the oscillator. However, the high value of resistance R makes this leakage path relatively insignificant. The 1.0 μF capacitor keeps the dc voltage from entering the electronic meter, a precaution necessary with some meters. Because of the approximately 10-megohm internal resistance of such meters, this blocking capacitor does not introduce appreciable error. (This is especially so since the meter is used only in a relative way to indicate the same readings in both positions of the switch.)

The compliance voltage of a current-regulated supply is the highest output voltage at which it will regulate its output current. Compliance voltage usually pertains to one-half of the rated full-load current. Whatever the method of specifying, a higher compliance voltage is usually possible at lower current adjustments. In any event, it should be realized that regulation must be maintained because the supply no longer behaves as a high ac impedance if regulation ceases. (Those who use a current-regulated supply for the first time often are careful to prevent short-circuiting of the output terminals. However, this is a needless precaution, for unlike the voltage-regulated supply, there is no potential or actual danger involved.)

The reference resistance R_s can be a resistance box, in which case the accuracy will tend to be enhanced because the ohmmeter measurement will not be necessary. The polarity of the direct-current flowing through the inductor is not important ordinarily.

The ac impedance of a current-regulated supply decreases rapidly with frequency. This limits the use of this method. If frequencies higher than 60 Hz or 120 Hz are to be used, then the technique is limited to measurements of smaller inductances. Of course, much depends on the characteristics of the supply.

THE TRANSFORMER RATIO-ARM BRIDGE

Most impedance bridges are ac derivatives of the basic Wheatstone bridge. None of these instruments will tolerate any stray impedance between the component undergoing measurement and ground. Therefore, the component must be removed from its equipment and carefully connected to the impedance bridge with short leads. In sharp contrast, the transformer ratio-arm bridge is, for most practical purposes, immune to stray impedance between the test component and ground. Measurement leads on the order of 100 feet in length can be used to make capacity measurements as low as several picofarads! This makes measurements of components in a circuit and remotely located from the bridge practical. Such capability is very useful, but would not be feasible with the ordinary impedance bridge. The transformer ratio-arm bridge is generally used to measure the reactance of inductors and capacitors. It is also excellent for making precise resistance measurements. Impedance measurements can be made if an appropriate reference impedance is used.

The basic arrangement used in the transformer ratio-arm bridge is shown in Fig. 4-13. Balance occurs when the ampere-turns produced in T2 from the current through Z_x is equal and opposite to that from the current through Z_s. In order that a given reference impedance Z_s can help to balance the bridge over a wide range of unknown impedances, Z_x transformer T1 has multiple taps. Potentiometer R1 provides adjustment between taps. An important feature of this arrangement is that the secondary of T1 and the primary of T2 have very low impedances. This being so, the effect of stray impedance to ground or to the neutral terminal tends to be minimal. For example, stray capacitance to ground appears in the circuit in Fig. 4-14. Even many hundreds of picofarads do not ap-

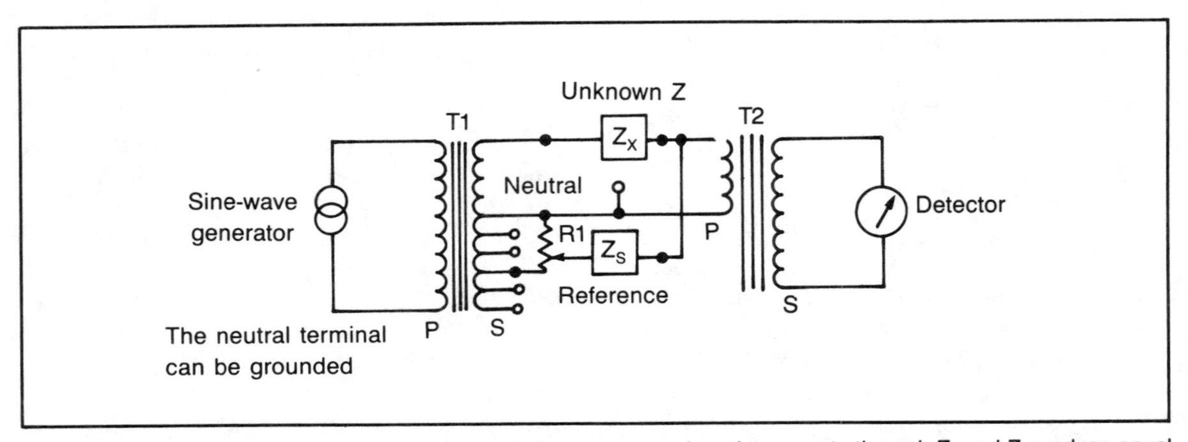

Fig. 4-13. Basic transformer ratio-arm bridge circuit. A null occurs when the currents through Z_s and Z_x produce equal and opposite ampere turns in the primary of T2.

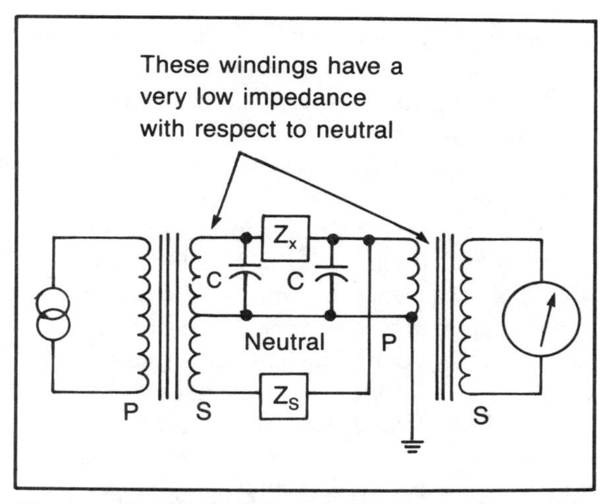

Fig. 4-14. Simplified ratio arm bridge circuit with stray capacitance C from test impedance Z_x to ground. The effect of C on the bridge balance tends to be negligible in most practical situations. Therefore, Z_x need not be removed from its circuit board. Also, components associated with Z_x need not be removed.

preciably affect the balance. The generator produces a relatively low frequency, usually in the vicinity of 1 kHz.

Suppose that a remote measurement is being made and the connecting leads introduce appreciable capacitance across the unknown impedance. The effect of such a capacitance can be exactly canceled by connecting a variable capacitor across the standard impedance and achieving a null with the connecting leads in place, but not connected to the unknown impedance. (See Fig. 4-15.) After this initial procedure, the leads are connected to the unknown impedance and a new null is established. From the settings of the transformer taps and the potentiometer, the magnitude of the unknown impedance is read in the conventional manner.

From the above, it is evident that a wide range of impedances can be measured despite the distance between the bridge and the equipment, and the impedance under test need not be removed from the equipment. This is very advantageous for testing printed-circuit boards and for making measurements inside of equipment where components are inaccessible. Such test procedures are completely beyond the capability of the Wheatstone bridge types.

Figure 4-16 shows the basic idea involved in adapting this test method to the measurement of the constituent parts of an impedance. This can be further extended to measure parallel networks, resonant circuits, quartz crystals, etc.

Procedure for Determining Impedance Transformation

Objective of Test: To demonstrate the use of the transformer in impedance transformation.

Test Equipment Required: Small radio-type power transformer with a center-tapped secondary. Total secondary voltage, 275 to 550. Secondary current rating, 15 to 100 milliamperes dc from a full-wave rectifier, 60-Hz primary winding.

Two electronic meters (test can be conducted with somewhat less convenience with a single meter).

Sine-wave oscillator.

Single-pole, single-throw switch.

Test Procedure: Connect the equipment as shown in Fig. 4-17. The resistance shown connected across the oscillator is used to make the in-

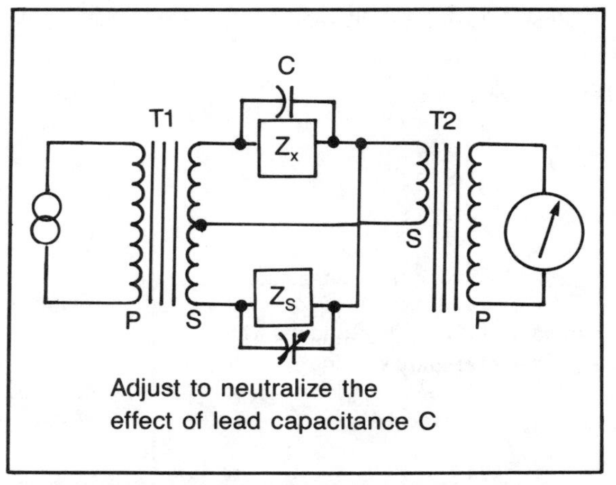

Fig. 4-15. Simplified ratio arm bridge circuit with capacitance C formed by the long connecting lead. For audio-frequency work, T1 and T2 can be filament transformers, with primary (120V) and secondary windings arranged as indicated.

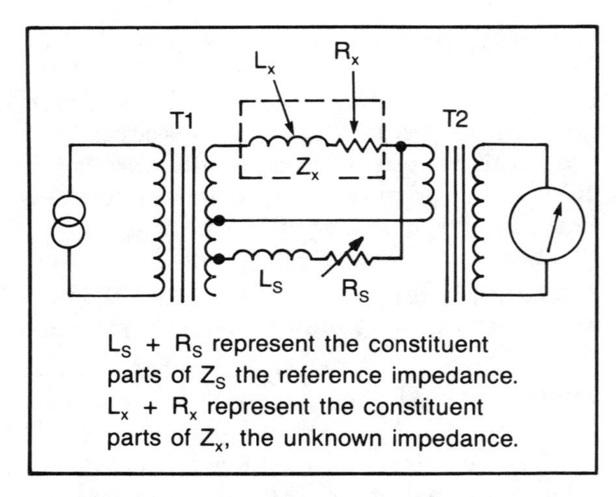

Fig. 4-16. Simplified ratio arm bridge circuit designed to measure the constituent parts of Z_x. k (L_s + R_s) equals L_x + R_x where k is determined by the position of the slider on the secondary of T1.

ternal impedance of the oscillator a negligible factor in the test, since it is considered more desirable to deal only with the 1000-ohm physical resistance in the primary circuit of transformer T1. (Some oscillators are immune to heavy loading; others are not. It would be better to make sure that the waveshape remains essentially sinusoidal with the 51-ohm shunting resistance. This is readily accomplished with an oscilloscope. Usually, distortion, even if severe, can be cleared up by operating the oscillator at a lower output level. With switch SW1 open, advance the output control of the oscillator so that both meters display a one-half to full-scale indication on appropriate ranges. The ratio of the voltage transformation of the transformer is E2/E1. The secondary impedance, Z_s is (Z_p) $(E2/E1)^2$, where Z_p is the impedance associated with the primary circuit. In our case, Z_p is 1000 ohms.

This relationship will now be confirmed from another approach. Close SW1 and adjust potentiometer R to produce one-half the original reading on meter M2. If meter M1 changes as a consequence of this adjustment, the output control of the oscillator should be adjusted to preserve a constant E1. When the half-deflection reading of meter M2 is achieved under the prescribed conditions, open switch SW1 and measure R with an ohmmeter. The resistance thereby measured should essentially confirm the value calculated from (Z_p) $(E2/E1)^2$.

Repeat the test procedure using the center-tap of the secondary and one outer lead. The impedance transformation of Z_p, should now be one-fourth of the value obtained with the full secondary winding. This new impedance value should again be calculated from (Z_p) $(E2/E1)^2$ and confirmed by the half-deflection or "6 dB drop method."

Reverse the circuit positions of the primary and secondary windings so that a voltage stepdown is obtained. Repeat the test procedure for the two

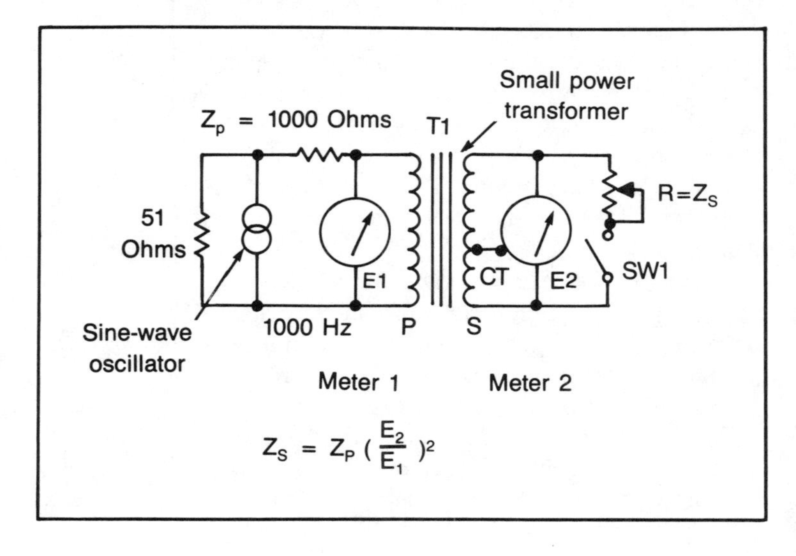

Fig. 4-17. Circuit for measuring impedance transformation. Z_p is transformed by the ratio $(E_2)^2E_1$ to the value, Z_s, in the secondary winding.

stepdown ratios resulting from use of the center tap and the full high-voltage winding (now the "primary"). Show that the impedance stepdown obeys the same rule as impedance stepup. It should be noted that in all tests, the impedance transformation ratio is $(E2/E1)^2$.

Comments on Test Results: Impedance transformation is commonly encountered in electronics. A transformer which couples the output of an audio amplifier to the speaker is an example of impedance stepdown. In wave filters, both stepup and stepdown of impedances are used. Examples of impedance stepup are often found in transistor circuits such as i-f amplifiers where the resonant tank operates at a higher impedance level than is presented to the transistor collector. Here, transformation is often accomplished by autotransformer action, but the basic idea remains the same as with 2-winding transformation. Impedance transformation between an antenna and the first stage of a radio receiver is usually stepup.

The quarter-impedance value available from the center-tap of a secondary winding is often puzzling when first encountered. Some would be inclined to suggest half rather than quarter impedance. However, the inductance of a winding such as is used on transformers is directly proportional to the square of the number of turns, not simply to the actual number of turns. This is the consequence of mutual induction. The flux linking a given turn links all other turns. Thus, one-half the number of turns produces $(1/2)^2$, or one-fourth the inductance. The reflected impedance from the primary then follows the same law as the secondary inductance and will be one-fourth the value corresponding to the impedance level for the full secondary-winding. Looking at it the other way, every time we double the number of secondary turns, we increase the impedance transformation ratio fourfold, even though the secondary voltage only doubles. Secondary voltage is proportional to the ratio of secondary turns to primary turns. Secondary impedance transformation with respect to the primary circuit is proportional to the square of the same ratio.

Procedure for Determining the Input and Output Impedances of Transfer Devices

Objective of Test: To devise a simple method for measuring the impedance of amplifiers, filters, attenuators, and other transfer devices.

Test Equipment Required:
Six 10K 1/2W 5 percent composition resistors
Two 15K 1/2W 5 percent composition resistors
The 5K resistance indicated in Fig. 4-18A is made from parallel 10K resistors. The 7.5K resistance is made from parallel 15K resistors.
Electronic meter
Oscilloscope (optional)
Ohmmeter
Two single-pole single-throw switches

Test Procedure: Make the 10,000-ohm attenuator as depicted in Fig. 4-18A, then wire the circuit as shown in the schematic. The sine-wave oscillator should be set to deliver a frequency in the vicinity of 1 kHz. Initially, switches SW1 and SW2 should be in their open positions. Adjust the amplitude control of the oscillator so that a reading of zero dB on the zero dB range is obtained on the dB scale of the electronic meter. Under this condition, the input impedance of the 10,000-ohm attenuator is matched, but not its output impedance. Close switch SW2 and note the new decibel indication. It should be very close to −6 dB with respect to the original 0 dB indication, because when a device is loaded with a resistance equal to its own internal impedance, the output voltage becomes one-half of its open-circuit value. One-half voltage corresponds to −6 dB.

Next, conduct a test with switch SW1 open and SW2 closed. Adjust the amplitude control of the oscillator to produce a −6 dB reading on the zero dB range of the electronic meter. Close switch SW1. The output level should rise to zero dB, or very close thereto. Here again the difference of 6 dB signifies that series resistance R1 "sees" its own value at the input of the attenuator.

We have measured the input and output impedances of the attenuator. Notice that when the

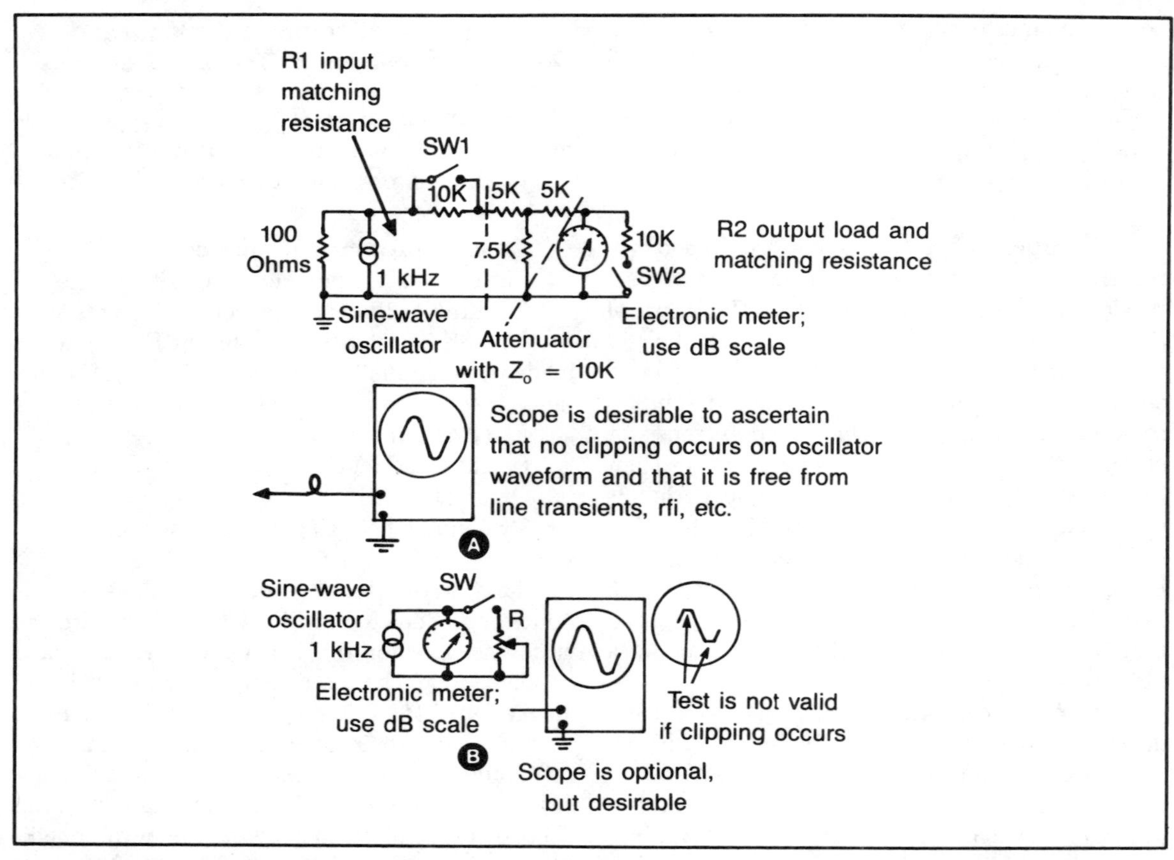

Fig. 4-18. Method for determining the input and output impedance (Z_o) of a T pad attenuator (A). Circuit B is used to determine the oscillator output Z.

output impedance was measured, the input of the attenuator was matched with its proper resistance, R1 equals 10,000 ohms. Likewise, when the input impedance of the attenuator was measured, the output impedance of the attenuator was matched with its proper resistance, R2 equals 10,000 ohms. The oscilloscope was not used, but its application with devices other than the simple attenuator is explained in the comments on the test results.

Another simple test is a measurement of the oscillator output impedance. In Fig. 4-18B, a variable resistance, R, is adjusted to cause a 6 dB drop in output level after switch SW is closed. The resistance of R is then measured with an ohmmeter. The resistance thus determined is equal to the oscillator output impedance at the frequency generated. The oscilloscope is in the picture. It reveals whether the generated wave is distorted by saturation, overloading, or clipping. If such is the case, the test must be made at a lower output level.

Comments on Test Results: The attenuator is representative of amplifiers, filters, and many active and passive devices which transfer energy from a pair of input terminals via some kind of processing to a pair of output terminals. It is often desirable to measure the impedance seen at a terminal pair of such devices; the "6 dB drop" or "half-voltage method" is simple and straightforward. Some precautions are in order, however. An active device such as an amplifier may not produce the same waveshape under the two test conditions (with and without an input or an output resistance). This can result if the voltage swing capability is exceeded with an open-circuited output or with zero resistance at the input. Other reasons can account for a change in waveshape such as non linearity, disturbance of feedback networks, and parasitic oscillation. These discrepancies can be quickly detected on the oscilloscope. When wave distortion accompanies one or both of the test conditions, the procedure is no longer valid. Either the test must be made at a lower level or other remedial measures must be taken. In most cases, a sine wave will be used, but occasionally a different waveshape is considered more relevant to the purposes of the device.

Where a meter with a decibel scale is not available, you can use an electronic meter with ac voltage scales. A reduction of a voltage level to one half of its previous value corresponds to −6 dB. Similarly, the doubling of a previous voltage level corresponds to +6 dB. Under some conditions, the ac voltage scale of a VOM or the accompanying dB scale can be used. However, due consideration should be given to the internal resistance of the meter and to its frequency response. If either of these do not fit in with the test conditions, the impedance determination is likely to be erroneous. The experiment with the oscillator alone shows that the general concept of the method is applicable to 2-terminal active devices also.

An interesting and useful impedance relationship applicable to attenuators (both T and pi) and to filters is as follows: Measure or compute the input impedance under two conditions, with the output open-circuited (Z_{Oc}) and with the output short circuited (Z_{sc}). Then, the characteristic impedance (Z_O) of the transfer device is,

$$Z_O - \sqrt{Z_{Oc} \times Z_{sc}}$$

For example, with the 10K T attenuator of this test procedure, Z_{Oc} equals 5K + 7.5K or 12.5K.

$$Z_{sc} = 5K + \frac{5K \times 7.5K}{5K + 7.5K} = 5K + \frac{37.5K}{12.5K}$$

$$= 5K + 3K = 8K$$

Then, Z_O equals $\sqrt{12.5K \times 8K}$ equals $\sqrt{100K}$ or 10K

which confirms the Z_o specification of this T pad.

OUTPUT IMPEDANCE OF VOLTAGE-REGULATED POWER SUPPLIES

Voltage-regulated power supplies have become commonplace both as laboratory instruments and as substitutes for brute-force supplies in a variety of electronic equipment. The use of voltage regulation generally upgrades the performance of electronic circuitry, but not necessarily for the reason often supposed. To be sure, the stabilization of the dc voltage level is almost universally desirable. However, the greatest benefit realized from a voltage-regulated supply is the low impedance offered to ac components. In other words, such a supply behaves, in essence, as a gigantic capacitor across the dc supply terminals of the powered equipment. This greatly decreases undesired mutual coupling between stages and between circuit sections in the equipment. Actually, although it is too infrequently emphasized, the power supply is not only part of the dc circuitry of single-ended (non-pushpull) amplifier stages, but is very much a part of the ac circuitry as well. A little thought will reveal that audio-frequency currents in such an amplifier must use the power supply as a return path. This being so, the supply should have as low an impedance as possible to the entire range of signal frequencies involved. In circuits other than amplifiers, it is equally important that the supply should approach a short circuit to a wide band of ac frequencies.

One of the best evaluations of a voltage-regulated supply which is intended to power any ac signal processing or generating circuits is a plot of the output impedance of the supply as a function of frequency. A representative plot is shown in Fig. 4-19. For simplicity, ordinary graph paper can be used. However, because of the way in which the uniformly spaced intervals are designated, the curve is actually a log-log plot. A feature of this plot is that in frequency regions where the response is

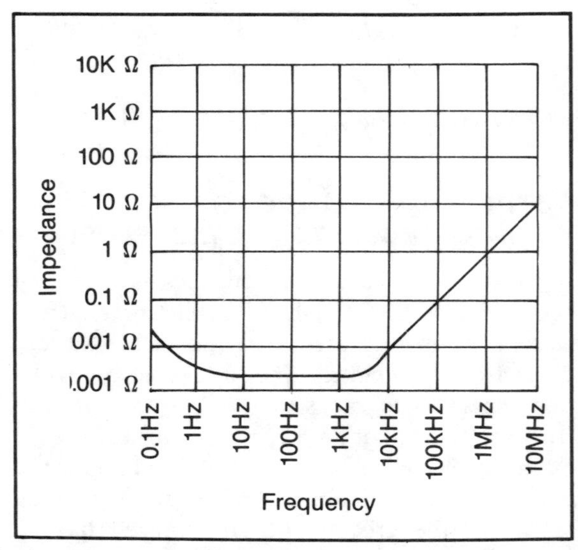

Fig. 4-19. Typical output impedance plot of a voltage-regulated power supply.

no longer flat and horizontal, the rate of change tends to be a straight-line slope. Therefore, once the slope is established and a quick investigation shows no irregularities, it becomes unnecessary to make a tedious series of closely spaced measurements; the impedance at the extended frequency regions can be obtained by extrapolation of the straight-line slope.

Figure 4-20 shows the basic setup for making such an impedance plot. The power supply is current-modulated by the output of a hi-fi or other audio power amplifier. The resultant voltage and current variations are observed on two oscilloscopes. Then, the output impedance is computed as the ratio of ac voltage to the ac current producing it. These voltage and current values are small compared to the dc capability of the supply and are known as incremental values. Both the incremental current and incremental voltage may be

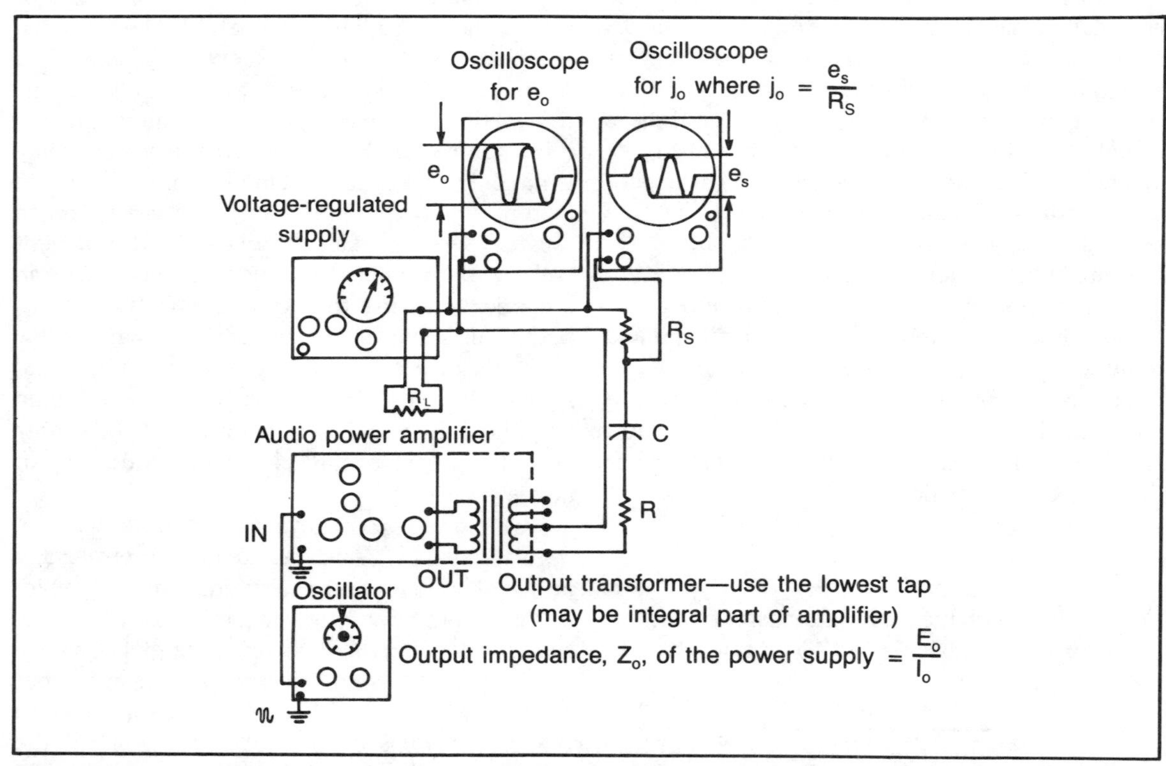

Fig. 4-20. Setup for a dynamic measurement of the power supply impedance (voltage-regulated). (Make sure there are no conflicts of grounds.)

read from the scope as peak-to-peak values. If available, a dual-beam oscilloscope may be used. A dual-trace oscilloscope can be used also, but sometimes with difficulty because of the small signals involved and unfortunate relationships between the switching frequency and the test frequencies. Meters could be used, but due consideration should be given to waveshape and frequency.

The audio amplifier is provided with a matched load by power resistor R. The dc-isolating capacitor C prevents the output transformer from short circuiting the power supply. This capacitor must be very large for low frequencies and of necessity must be an electrolytic type. Due regard must be given to its polarity. (Some transistor audio amplifiers of the "transformerless" variety already have the isolating capacitor, but be sure that its voltage rating is not being exceeded.) Notice that the current measurement is made by monitoring the voltage developed across a small resistance, R_s. A useful and convenient value for R_s is 0.1 ohm. This value will suffice for a wide range of measurement requirements. It is important that R_s be essentially noninductive, a requirement that can be met by paralleling many one-watt composition resistors. For example, a "sandwich" of thirty three 3.3-ohm resistors between two copper plates is a possibility. Commercial resistors are available for this and similar purposes.

The power supply is loaded to one-half of its rated current output by resistor R_l. The value of this resistor is determined by E divided by I, where E is the voltage output of the supply and I is one half the rated dc current of the supply. The wattage rating of R_l must be at least $E \times I$. (The half-maximum loading of the supply complies with general specification practice.)

How heavily should the power supply be modulated? Other matters being equal, the modulation should be as low as possible or just enough to give oscilloscope displays which are well out of the noise region. Otherwise, the modulation depth can vary over a fairly wide range without affecting the impedance determination. Of course, an excessively high modulating level will drive the supply out of its regulating range, or will produce wave distortion in the audio amplifier. Such operation is far removed from the mathematical ideal of incremental currents and voltages and must be avoided.

The output impedance, Z_O, from the setup in Fig. 4-20, is computed from the formula:

$$Z_O = \frac{E_O}{I_O}$$

where E_O and I_O represent incremental changes in the output voltage and output current of the power supply. The incremental current, I_O, is computed from E_s divided by R_s.

Notice that in this test a voltage variation occurs as the result of a current change. This cause-effect situation is opposite to that usually encountered. The audio amplifier thus simulates a variable load resistance and provides a convenient way to test the power supply dynamically.

DETERMINING INDUCTANCE AND CAPACITANCE FROM RESONANCE

If we know the resonant frequency of an inductance and capacitance, we can determine the value of one of the reactors if the other is known. Several simple test setups are shown in Fig. 4-21. The circuits in Fig. 4-21A and B are parallel resonant circuits, whereas circuits C and D are series resonant types. A common feature of all of the circuits is the location of the ground point. In each case, the sine-wave generator and the electronic meter or oscilloscope are grounded to the same point. This is very important. If the grounds or common terminals of these instruments are connected at different junctures of the circuit, proper resonance cannot be attained.

How do we decide whether to use parallel resonance (Fig. 4-21A and B) or series resonance (circuits C and D)? This decision must sometimes be based on an experimental evaluation to see which connection results in the sharpest response (highest Q) or best satisfies some other criteria. However, some general guidelines can be given. All of the circuits are based on the formula most often cited for resonance:

$$f_0 = \frac{1}{2\pi\sqrt{LC}}$$

This formula is always valid for series resonance, but involves error in low Q parallel resonant circuits. Where the Q factor for parallel resonance is less than ten, the very definition of resonance becomes somewhat uncertain and the slow rate of change of circuit impedance with respect to frequency makes an exact determination rather difficult. If the low Q at parallel resonance is due to high dissipative losses in the inductor or capacitor, it is doubtful whether series resonance will provide better measurement conditions. However, if low Q at parallel resonance is due to a low ratio of C to L, it may be found that reconnecting to achieve series resonance will produce a relatively high Q resonance. This may appear strange because both types of resonance would show infinite Q regardless of the ratio C to L, or L to C, if dissipative losses were zero. In actual practice, however, the Q pertaining to one type of resonance will be greater, often considerably so, than the other. The variation in dissipative losses (mostly in the inductor) is such that, for a given resonant frequency, the series LC circuit will generally be found to involve relatively high inductance and low capacitance compared to attainment or the same Q in the parallel LC circuit.

You can develop a fairly reliable sense about these matters. For example, in the mid-audio region of 1000 to 10,000 Hz, resonance involving capacitance of, say, 0.1 to several microfarads could be expected to perform best in the parallel circuit. On the other hand, resonance involving capacitance of several hundred picofarads to perhaps 0.01 μF could be appropriate candidates for the series LC circuit. This could be put on a more definite basis by actually taking into account the dissipative losses in the LC elements. Although this can be done, it certainly does not contribute to the simplicity of the test procedure. The experimental evaluation is quick and easy. Moreover, a fair portion of LC combinations can be expected to yield acceptable results for either parallel or series resonance.

In Fig. 4-21A the value of R must be high enough so that its presence imposes a negligible loading effect in the parallel resonant impedance of the LC circuit. A first trial of about 10,000 ohms

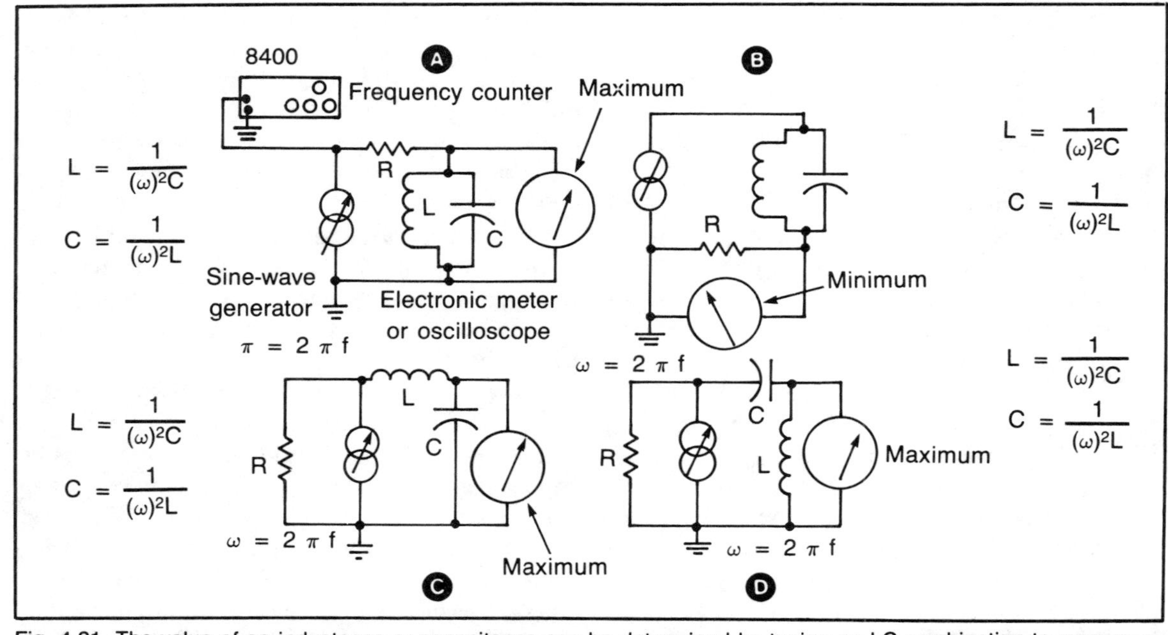

Fig. 4-21. The value of an inductance or capacitance can be determined by tuning an LC combination to resonance.

generally enables the establishment of resonance to be observed on the electronic meter or oscilloscope. However, R should be at least ten times the resonant impedance, R_0, of the parallel LC circuit. R_0 can be "ballparked" from the formula:

$$R_0 = \frac{L}{R_1 (C)}$$

where, R_1 is the dc resistance of the inductor. Adopted for L and C are the values corresponding to resonance. This use of the formula should not be for other purposes, since the dc resistance of L approximates the ac resistance only at low frequencies. Also, R_1 should include the dissipation losses in the core material (if any). Finally, at higher frequencies the assumption that R_c (the capacitor losses) is zero may result in appreciable error. However, for our purpose of making R ten or more times R_0, this approach often works very well.

A good practical way to ascertain that R is, indeed, high enough is to substitute a resistor of double or triple its value. If the Q of the resonant response remains substantially the same, we can conclude that R was OK. The Q of the resonant response is given by:

$$Q = \frac{f_0}{f_2 - f_1}$$

where f_0 is the resonant frequency, f_2 is a higher-than-resonant frequency where the voltage across the LC circuit is 70.7 percent of the resonant value (or −3 dB in power from the response at resonance), and f_1 is a lower-than-resonant frequency where the voltage across the LC circuit is 70.7 percent of the resonant value (or −3 dB in power from the response at resonance).

In the simplest use of this procedure, the frequency can be read from the dial of the sine-wave generator. For measurements of higher accuracy, a digital frequency counter can be used as shown in Fig. 4-21A. Also, many variations of the basic approach will suggest themselves. For example, any of the three parameters, f, L, or C may be variable if calibrated, or if a measurement can be made of it after resonance is established. Notice that in Fig. 4-21A, resonance corresponds to the maximum deflection of the detection instrument that can be obtained by tuning f, L, or C. In contrast, the circuit shown in Fig. 4-22 is resonant when the tuning of f, L, or C produces minimum deflection of the detection instrument. In this circuit, R should be high for the reasons described for the circuit in Fig. 4-21A. However, if R is too high in this case, the resonant dip will not be deep and may be difficult to observe and optimize. This is probably a more difficult test circuit to use. However, it can prove advantageous at higher frequencies because it is unnecessary to connect the capacitance of the detection instrument across the LC circuit. In any event, good results are obtainable from the circuit in Fig. 4-21A when the detection instrument is provided with a probe so that the LC circuit is not detuned by cable capacitance. (At very low frequencies, where C is relatively large, it can mask the effect of several hundred picofarads of cable capacitance.)

As previously stated, series resonance yields a higher Q in some LC circuits, usually those with relatively large inductors and small capacitors. Circuits C and D in Fig. 4-21 show the setups for series resonance. Resistance R now assumes a low value, on the order of perhaps 10 to 100 ohms. Its function is to lower the output impedance of the sine-wave generator, thus enabling the series LC circuit to more closely approach its inherent Q. Some generators are more tolerant than others when the output is shunted down in this manner. Therefore, an oscilloscope should be used to make sure that the waveform remains at least visibly sinusoidal. Also, some compromise with the lower limit of R may have to be made in order to obtain a reasonable output voltage. Most generators have output impedances between 500 and 1000 ohms. A more desirable way to lower the output impedance would be with the use of a stepdown transformer. For much of the audio spectrum, a filament transformer or, in some cases, a speaker output transformer can be used. There may or may not be any reason to prefer circuit C or D. Because practical capacitors tend to be "purer" reactances than practical inductors, the circuit in Fig. 4-21C might appear to more closely represent the basic idea of the series-resonance method. At higher frequencies, however, circuit C would tend to be more susceptible to detuning by the detection instrument.

In filter work, L and C are often the actual elements to be used in the filter. The objective here is to ascertain that resonance occurs at the calculated frequency. The selection of parallel or series resonance is determined by the actual type of resonance which will occur in the filter. Also, it is the general practice to make the LC circuit "see" a resistance equal to the characteristic impedance of the filter. Thus, if the characteristic impedance is 500 ohms for a parallel resonant LC "tank," circuit A could be used by shunting 33 ohms across the generator and making R approximately 470 ohms. Such a setup would closely simulate the situation experienced by the parallel resonant tank in the actual filter.

INDUCTANCE MEASUREMENTS WITH THE MAXWELL AND HAY BRIDGES

The Maxwell and Hay bridge circuits are shown respectively in Figs. 4-22 and 4-23. These bridges are capable of accurate inductance measurements because the phase displacement produced by the inductor is canceled by capacitor C_s in one arm of the bridge; the use of a capacitor in this fashion, rather than another inductor in an adjacent bridge arm, is highly advantageous because a capacitor can be a much more nearly an ideal reactance than a practical inductor. This arrangement also makes possible the balancing of the resistive component of the inductor being measured. This is brought about by means of R_s in the two bridge circuits. Thus, the bridges can be calibrated to read

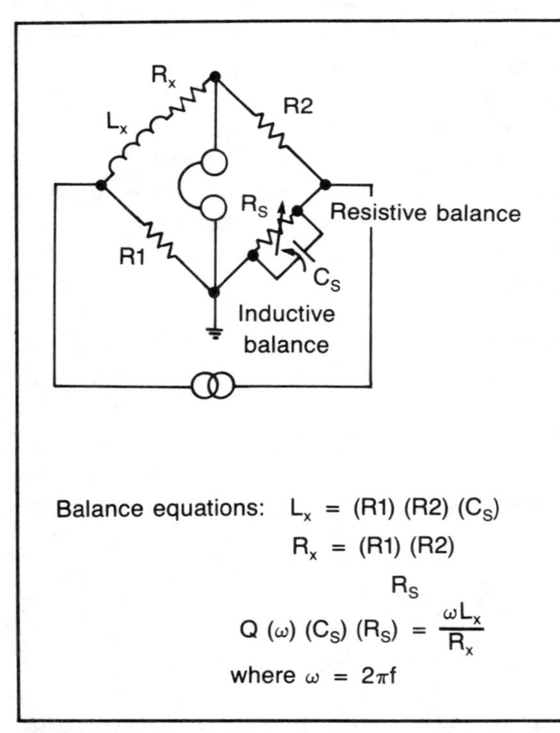

Fig. 4-22. Basic Maxwell bridge circuit. C_s can be fixed, in which case either R1 or R2 is made variable and becomes the "inductive balance" adjustment.

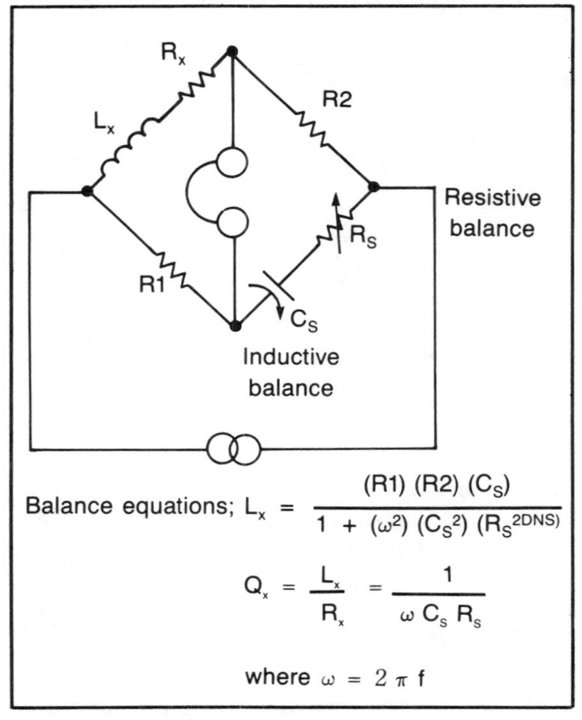

Fig. 4-23. Basic Hay bridge circuit. C_s can be fixed, in which case either R1 or R2 is made variable and becomes the "inductive balance" adjustment.

out both the true inductance and the resistive portion of the inductor under test.

The separation of L_x and its resistive component R_s is such that R_x includes the effect of core losses as well as ohmic resistance of the winding. This being so, the "pure" value of inductive reactance, and, therefore inductance, is separately balanced out in the nulling procedure. As a result, one dial can be calibrated to indicate inductance and another dial can be calibrated to indicate R_x, the equivalent series resistance which simulates the effect of all dissipative losses in the inductor. The reference capacitor, C_s, may be a fixed value, in which case resistive arms R1 or R2 can be variable and calibrated in terms of inductance. This applies to both bridges.

Inasmuch as both bridges have thus far been treated as similar circuits, it would be only natural to wonder about any differences in the performances of the two. There is, indeed, a profound difference, for it turns out that the Maxwell bridge is suitable for low Q inductors only, the opposite being true for the Hay bridge. That is why the Maxwell bridge is accompanied by an equation for R_x, which would have appreciably high values in low Q inductors. The Hay bridge can be used to determine R_x by a somewhat different equation, but in high Q inductors, R_x tends to be a negligibly small fraction of inductive reactance. Therefore, it is more usual to calculate or read out the value of Q itself, this being understood to pertain to the frequency of the bridge oscillator, usually 1000 Hz. On the other hand, when the Maxwell bridge is used, Q can also be calculated or read out for the bridge oscillator frequency. The Maxwell bridge is generally used when the Q of the inductor is 10 or less.

It is not always obvious which bridge will give the best balance, so most commercial bridges have a switch which permits selection of low Q or high Q measurements, which is to say, the Maxwell or Hay bridge circuits. A peculiarity in the balancing procedure of these bridges is the interplay between inductive and resistive balancing. Thus, in the circuits shown, you have to see-saw between the C_s and R_s adjustments in order to achieve an optimum null. This requires both patience and skill, but the final adjustment is rewarding because you have the feeling that you have used a measurement technique capable of great precision. These bridges can measure values from microhenrys to kilohenrys and can be used at high rf frequencies. In the forms shown, the aural and headphone response in the vicinity of 1000 Hz leads to extremely sharp nulling for a wide variety of inductors commonly encountered in everyday practice.

VOLTAGE-CURRENT METHOD FOR MEASURING AUDIO FREQUENCY INDUCTORS

The test circuit shown in Fig. 4-24 is based on the basic equation for inductance:

$$L = \frac{X_l}{2\pi f}$$

For a "pure" inductance, the inductive reactance, X_l, is simply the ratio of the voltage E impressed across the inductor to the resultant current I flowing through the inductor. Indeed, in many instances, the circuit in Fig. 4-24 can be used to

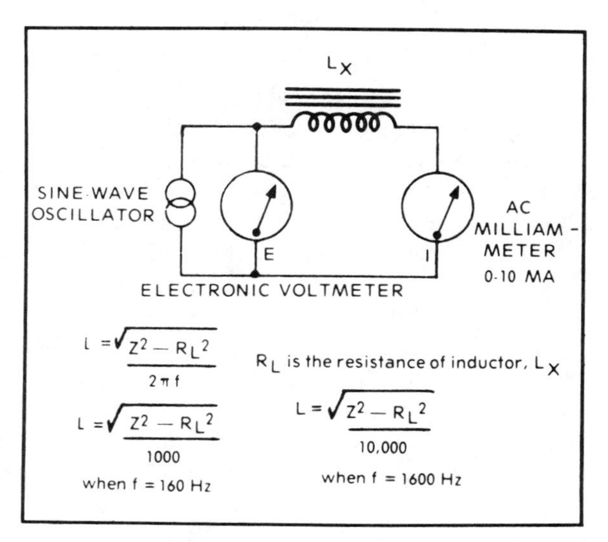

Fig. 4-24. Voltage current method for measuring audio-frequency inductors. The resistance of the inductor R_L is measured with an ohmmeter or a bridge.

determine quick but approximate inductance values by means of the following formula, which neglects the resistance of the inductor:

$$L = \frac{E/I}{2\pi f}$$

If f is adjusted to be 160 Hz, this formula becomes:

$$L = \frac{E}{I} \times 10^{-3}$$

Furthermore, if I is recorded in milliamperes, the 160-Hz formula reduces to the ultimate in simplicity, L equals E divided by I, where L is in henrys.

But since practical inductors also have resistance, the ratio of the impressed voltage to the resultant current is not inductive reactance, but rather impedance. The above was discussed because in certain applications such an approximation may be permissible. For example, a 15-henry choke has 5635 ohms of inductive reactance at 60 Hz. If such a choke has, say, 75 ohms resistance, the impedance,

$$Z = \sqrt{R^2 + X_l^2}$$

will be very nearly identical to the inductive reactance. However, in other situations, those in which the resistance is one fifth or a larger fraction of the inductive reactance, the impedance and the inductive reactance differ in value and gross error can result by using these magnitudes interchangeably. Therefore, the general test procedure advocated in conjunction with the circuit in Fig. 4-24 is to measure the resistance of the inductor with an ohmmeter or bridge. Then, E I is treated as the impedance due to the combined effects of inductive reactance and resistance. This gives rise to the general formula:

$$L = \frac{\sqrt{Z^2 - R_l^2}}{2\pi f}$$

This formula is an algebraic manipulation of the classic equation for impedance, Z equals

$$\sqrt{R^2 + X_l^2},$$

where X_l equals $2\pi fL$. Thus, this test procedure introduces no first-order approximations. To the extent that we can read the meters and determine the frequency accurately, very good results can be obtained from such a measurement technique. At higher than audio frequencies, second-order effects, such as errors from distributed capacitance and stray inductance, limit the use of this method.

It is a worthwhile precaution to use low values of voltage to avoid core saturation. Stepup and stepdown transformers can be inserted between the oscillator and the test circuit to extend the range of the inductance measurements.

CAPACITANCE MEASUREMENT BY PULSE-AVERAGING METER

The setup shown in Fig. 4-25 is the simplest practical version of a much used technique for measuring capacitance. When the scheme is built into a commercial test instrument, various sophistications may be incorporated, but the essential principle of operation remains the same. The deflection of a current meter is directly proportional to the rate of pulses of one polarity (unipolar) passing through it, providing the pulses are of constant amplitude, shape, and duty-cycle. The pulses are averaged by the mechanical inertia of the meter movement so that the pointer is stationary even down to relatively low repetition rates. The net effect of this is that a linear scale of capacitance results. In the arrangement in Fig. 4-25, the deflection produced in the current meter is directly proportional to the square-wave frequency and to the capacitance through which the square-wave current passes to the current meter. The shunt-connected diode clamps the baseline of the square wave, thereby making it have a unipolar rather than an ac characteristic.

It should be appreciated how this method differs from the use of sine wave for testing reactance. Whether the sine wave is associated with a simple or complex measuring technique, the reactance of

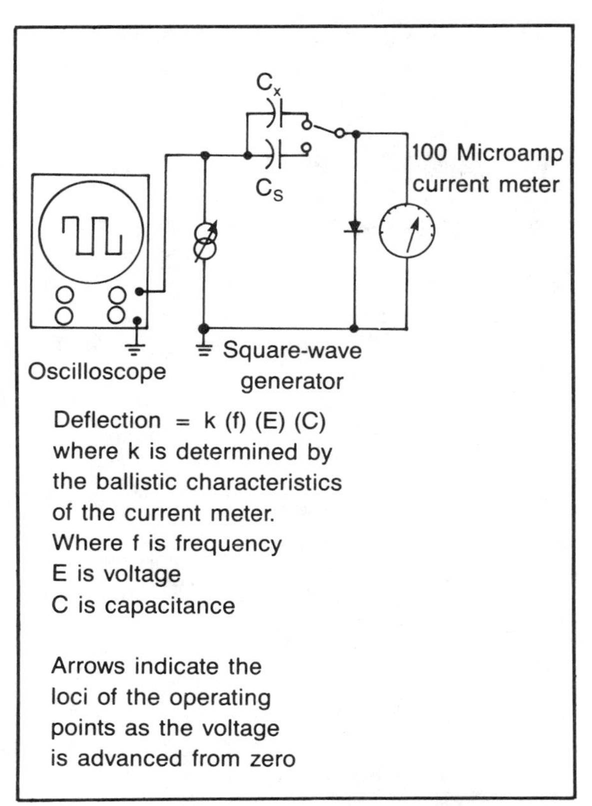

Fig. 4-25. Setup for capacitance measurement by averaging pulses.

a capacitor as a function of frequency is a nonlinear relationship. If a meter is calibrated to indicate capacitance as the result of the presence of a sine-wave current at a certain frequency, doubling that frequency does not produce twice the deflection, nor does halving the frequency result in one half the deflection.

The principal feature of a scheme utilizing pulses is that a single calibration suffices for a wide range of measurements. In Fig. 4-25, suppose that the generator is set to deliver a 1-kHz square wave. Using a reference capacitor of 0.01 μF, the amplitude of the square-wave generator is adjusted to produce a full-scale deflection on the current meter. The peak-to-peak amplitude on the oscilloscope is recorded. (It can be recorded as so many centimeters or divisions. We do not need to know the exact value of the square-wave amplitude; it is only necessary that it remain constant. Therefore, there is no need to accurately calibrate the scope.) Now, if the square-wave generator is set to deliver 100 Hz, a full-scale deflection of the current meter will correspond to 0.1 μF. At 10 kHz, it will correspond to 0.1 μF, and so on. And on any of these ranges, the capacitance will be a linear function of the deflection. Thus, if full-scale deflection represents 0.1 μF, a half-scale reading indicates a capacitance of .005 μF. Over a wide range of measurements, accuracy will depend, primarily, on the tolerance of the initial calibrating capacitor and the constancy of the square-wave amplitude.

A practical limitation is introduced by the fact that the diode is not an ideal rectifier. It does not begin to conduct until an appreciable fraction of a volt is applied, and its current vs voltage characteristic may depart considerably from a linear relationship. Therefore, the peak-to-peak square-wave voltage cannot be too low. Diodes vary greatly in their conduction characteristics. One should be selected which provides a linear characteristic at a relatively low voltage. Such a diode will enable a greater range of capacitance to be accurately measured. Although it is true that all junction diodes tend to "straighten out" with increased applied voltage, the square-wave generator has a limited output capability. So-called "high conduction" silicon diodes are well suited for this application.

Procedure For Measuring the Distributed Capacitance of an Inductor

Objective of Test: To use resonance as a means of determining the self-capacitance of an inductor.

Test Equipment Required:

Multirange oscillator with a frequency capability to 100 kHz or higher.

Oscilloscope

Several filament, power, or audio transformers.

Several 10 percent tolerance Mylar capacitors. Suggested capacities are: 0.047, 0.1, and 2.0 μF

Test Procedure: Connect the components as shown in Fig. 4-26A. Starting at a low audio frequency, say, several hundred hertz, slowly advance the frequency until a resonant rise in voltage amplitude is observed on the oscilloscope. More than one such increase may exist, but the resonance occurring at the lowest frequency is the important one for this test. (Although nonsinusoidal waves may sometimes be employed, the results tend to be more reliable with the use of a good sine wave.) Record the frequency, f_d, corresponding to the first resonant rise observed. Unless the Q of the inductor is inordinately low, or there are shorted turns, the priminence of this rise will be obvious. The initial test should be conducted with a low excitation voltage to avoid the possibility of magnetic saturation. This will not be a factor in filament and power transformers, but with certain audio inductors and filter solenoids, too much drive will invalidate the test.

Next, rearrange the apparatus as shown in Fig. 4-26B. C is chosen to cause parallel resonance at a low audio frequency, say, in the vicinity of several hundred hertz. The exact frequency is not important. The overall accuracy of the test will be governed by how closely frequencies can be read, as well as by the tolerance of capacitor C. The objective of this type of test is usually served by approximate information. Therefore, it is permissible to record frequencies from the oscillator dial. From the data derived from the setup in Fig. 4-26B, calculate L as follows:

$$L = \frac{1}{\omega^2 C}$$

where L is the inductance, C is the capacitance of the physical resonating capacitor, f is resonant frequency resulting from LC, and ω is $2\pi f$.

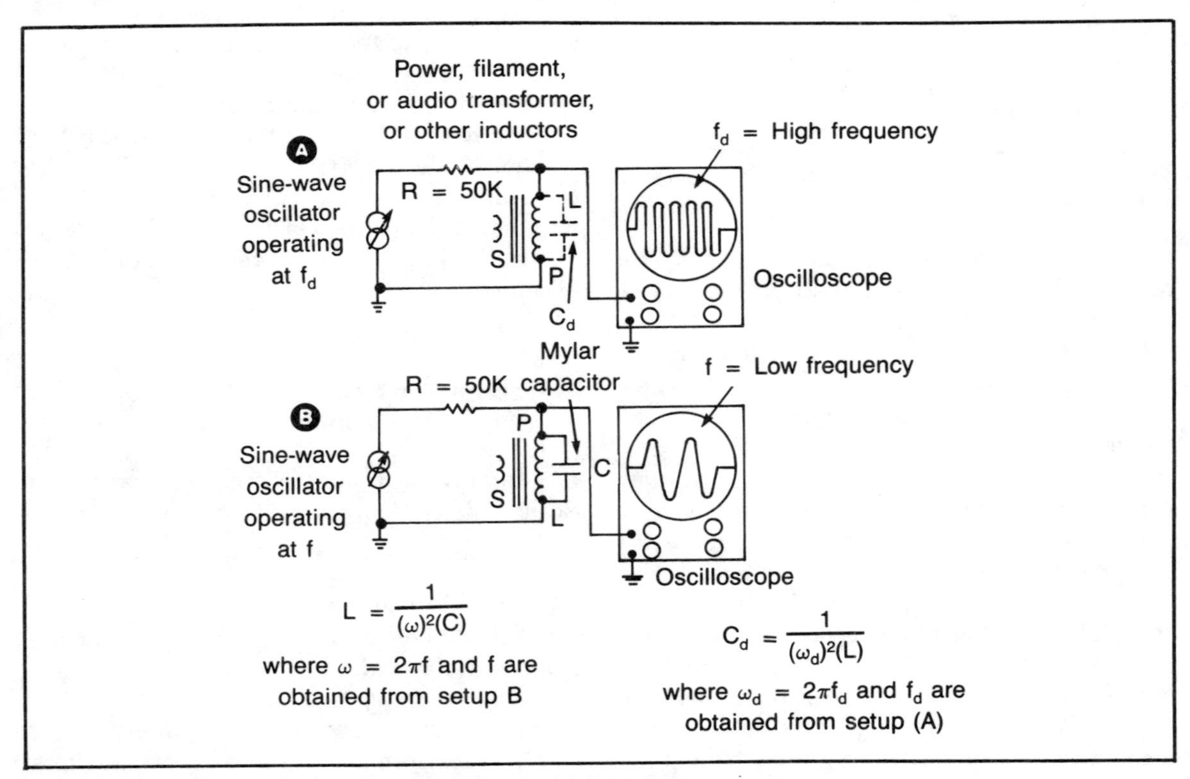

Fig. 4-26. Two methods of determining the distributed capacitance in an inductor.

Now, the distributed capacitance may be determined from a similar calculation:

$$C_d = \frac{1}{\omega_d^2 L}$$

where C_d is the distributed capacitance of the inductor, L is the inductance as determined from Fig. 4-26B, f_d is the resonant frequency resulting from LC_d as determined in Fig. 4-26A, and

$$\omega_d \text{ is } 2\pi f_d$$

Comments on Test Results: The sequence of events in this test procedure may appear opposed to logical procedure. However, the determination of the self-resonant frequency of the inductor is of prime importance. If this is not known, there is nothing to be gained from the remaining portion of the measurement technique. Although theoretically every practical inductor can be expected to exhibit parallel resonance with its distributed and stray capacity, this self-resonance may not always be readily found. As mentioned in the test procedure, low Q or shorted turns are possible reasons. Other difficulties may be involved. For example, because of low inductance as well as low distributed capacitance, the self-resonant frequency may be beyond the high-frequency limit of the oscillator. Also, at higher frequencies, the resonance may occur in the frequency region where the response of the oscilloscope drops off rapidly. It should be realized that self-resonance takes place with relatively large L and small C, just the opposite combination required for high Q in parallel resonance. Finally, in some instances, the distributed capacitance does not closely simulate the effect of a single parallel-connected capacitor. This is especially true for rf chokes.

The effect of distributed capacitance is of importance in certain applications. For example, suppose that an inductor in a bandpass filter is calculated for a specific inductance from filter equations, and it appears practical to construct such an inductor from considerations of wire size, window space, and core characteristics. Perhaps such an inductor is supposed to resonate at 50 kHz with a 0.001 μF capacitor. However, an actual test of the completed inductor shows self-resonance at 45 kHz. This inductor is useless because we can never reach resonance. There is already too large a resonating "capacitor" in the windings themselves. By measuring the distributed capacitance with this test procedure, we can decide whether a remedy is possible. Thus, if it appears that reducing the distributed capacitance to one half the first trial value would enable the 50-kHz resonance to be obtained this might be feasible through the choice of a wire with insulation having a lower dielectric constant, and/or with a different method of winding the wire on the core. Notice that in a case of this kind the desired resonance could exist with part physical capacitance and part distributed capacitance.

In rf tuners, particularly at higher frequencies, distributed capacitance greatly restricts the band coverage that would otherwise be attainable. In pulse circuits, such as the flyback circuit in TV receivers, the distributed capacitance of the flyback transformer must be reckoned with its initial design, and sometimes in servicing, for the effects of the self-resonance from distributed capacitance can produce shock-excited oscillations which find their way to the video circuitry, producing interferences in the picture.

QUESTIONS

4-1. An impedance consisting of an inductor and a resistor in series is used in a fixed-frequency application to limit the flow of current. A technician decides that circuit operation can be improved by further reducing the current through the impedance. Because of space restrictions, it appears that the most convenient way of accomplishing this might be to connect a capacitor in series with the described impedance, thereby making it a series LCR arrangement. A capacitor was tried and the results were encouraging in that the current was indeed decreased. The decrease was not enough, however, so a smaller capacitor was substituted. This time the current increased considerably. This was thought to be due to a faulty capacitor. A second try with a similar capacitor again gave the undesired and unexpected results. What had been overlooked in this attempt to manipulate impedance magnitude?

4-2. A parallel resonant circuit is contained in a "black box" with the two terminals of the LC network available. A high-impedance source generates one frequency, f_0; the resonant frequency of this network. The network behaves as if it had an ohmic resistance of R ohms at f_0. (Phase measurements with an oscilloscope show that the black box can be replaced with a resistor of R ohms.) Assuming that we are not allowed to inspect the contents of the black box, describe two ways in which it could be proved that the circuit comprises reactive elements and not simply a physical resistor of R ohms.

4-3. A bandpass filter has a shunt output arm consisting of a parallel resonant LC circuit. We want to increase the output impedance by four times. This means that a secondary winding with twice the number of turns on L must be added. (A two times voltage stepup corresponds to a four times impedance stepup.) With the finest wire that can be safely used there is not enough window space in the toroidal core to accommodate the desired secondary winding. The situation is particularly frustrating because only a little more window space would suffice, and a large order to an important customer hangs in the balance. However, the day was saved by the suggestion of an alert technician. What might have been the solution?

4-4. A low-frequency parallel resonant circuit from a piece of equipment was repaired by replacing the defective inductor. The new inductor was very similar to the old one, but because of improved core material, the resonant circuit had a Q of 14 instead of ten, as was the case with the original inductor. Both inductors can be assumed to be identical in inductance, distributed capacitance, current carrying ability, magnetic saturation characteristics, etc. Yet, it was found that the resonant frequency had shifted. It was determined that the resonant frequency with the new inductor was closer to the computed value of:

$$f_0 = \frac{1}{2\pi\sqrt{LC}}$$

Comment on this situation. Were the findings to be anticipated under the described conditions?

4-5. Intrigued by the possibility of using a variable resistance to shift a resonance frequency, a technician devised a scheme making use of a series LCR circuit with Q values somewhere in the 7-12 range, depending on the value of R. Resonance was detected by connecting a scope across either the inductor or the capacitor in order to observe the resonant rise in voltage. Although a digital frequency meter was available to check for shift of the resonant frequency, none could be observed. Further investigation appeared to indicate that resonance always occurred at:

$$f_0 .= \frac{1}{2\pi\sqrt{LC}}$$

What was the nature of the trouble here?

4-6. A student used an RMS-responding electronic meter for the first time. Voltage readings were recorded in an ac circuit consisting of a resistance and a large capacitance in series. It was found that the sum of the voltages monitored across the two circuit elements was considerably more than the measured voltage impressed across the series combination. The student tried to account for this discrepancy by taking into account the resistance of the inductor winding and the possible loading effect of the meter. It turned out that these effects were not appreciable and, therefore, could not be the explanation of the dilemma. Also, it was confirmed that the ac waveshape was a very good sinusoid. What had been overlooked?

4-7. Output impedance measurements are made

on a power supply with relatively poor dc voltage regulation. Yet the output impedance is found to be comparable to another supply with excellent dc voltage regulation. The output impedance is evaluated from frequencies below the range of interest to frequencies higher than the range of interest as determined by the powered circuitry. Does this appear to be a logical situation?

4-8. A Maxwell bridge is being used to make measurements on an audio inductor. The bridge oscillator is a self-contained 1000-Hz source and headphones are used as a null detector. When the optimum balance adjustment is attained, the 1000-Hz tone can no longer be heard, but all is not silence, for higher pitch tones remain and cannot be eliminated. Does this mean that the Hay bridge should be used in order to obtain a better null?

4-9. We want to identify various inductors for an audio-frequency filter. A variable frequency square-wave generator is available, together with an oscilloscope and an electronic voltmeter. Additionally, a number of 5-percent Mylar capacitors from 0.001 μF to 1.0 μF are on hand. Can the resonance method be used to measure the inductance of the unknown coils?

4-10. A low-power heater circuit in a thermostatically controlled oven makes use of a simple series circuit comprising a 1300-ohm heater element in series with a 2 μF capacitor, 120 volts at 60 Hz is applied across this combination. Since the capacitor has about the same capacitive reactance as the heater element has resistance, the voltages across both circuit components are nearly the same, approximately 84 volts. Because this simple network is a series circuit, the same current flows through the heater element and the capacitor. This being so, the product of voltage times current is the same for the capacitor as the heater element. Why then doesn't the capacitor develop the same amount of heat as the heater element?

ANSWERS

4-1. The evidence suggests that the value of the resistance was not high enough to mask or destroy the effects of series resonance. The size of the second capacitor must have been such that series resonance was closely approached, in which case the predominant part of the inductive reactance was effectively canceled out. Such reasoning clears up the apparent contradictory situation where a smaller series capacitor can cause the overall impedance of the series combination to permit increased current flow.

4-2. An ohmmeter could be used to show the relatively low dc resistance of L. From this information, it could be logically assumed that the inductance and capacitance must be resonating as a parallel "tank" circuit at frequency f_0. Secondly, and perhaps less simply, the presence of reactive, that is, energy-storage, elements could be demonstrated by the existence of a definite time interval for both the establishment of steady-state voltage and its decay to zero when the ac source is turned on and off. This could be done with an oscilloscope in conjunction with either an electrical or mechanical scheme for turning the ac on and off. Somewhat similarly, shock-excited oscillations at frequency f_0 could be observed when low duty cycle pulses are applied to the terminals of the black box.

4-3. Since there isn't enough space for a conventional secondary winding, the same impedance stepup of four could be accomplished by means of a tapped, or autotransformer, arrangement rather than that of a two-winding transformer. For example, the addition of the same number of turns as contained in L would transform the output voltage by two and the output impedance by the desired factor of four. The output tank inductor would then consist of a single tapped winding. The resonant portion of this winding would exist from tap to ground, and would be electrically unchanged from the original situation. The impedance stepup would be accomplished by the free terminal of the added turns with respect to ground. Another way of looking at this arrangement is to see it as a center-tapped inductor. The impedance between the outer ends of such an inductor is four times the impedance of the half-winding between center-tap and the common terminal or ground. From the description of the dilemma encountered with the attempted "straight" transformer approach, you can safely surmise that the autotransformer scheme saved the day!

4-4. The generally used formula for parallel resonance:

$$f_0 = \frac{1}{2\pi \sqrt{LC}}$$

is an approximation which, for high Q LC circuits, is very close to the exact frequency of resonance. However, when dealing with Qs of about ten or lower, this formula begins to show noticeable error. Because the Q was raised from 10 to 14 in the described situation, the true resonant frequency became closer to that computed from the generally used resonance formula. In most instances, this probably would not be of great consequence because the very definition of resonance at low Q becomes somewhat indistinct in the parallel LC circuit and the very broadness of the selectivity curve tends to make it unimportant whether the peak response is a few percent one way or another from the calculated f_0. On the other hand, there are applications where vernier tuning is accomplished in a moderate Q tank circuit by means of variable resistance associated with L, C, or both. And in very low Q circuits, f_0 would be much in error if calculated by the simplified formula.

4-5. The formula:

$$f_0 = \frac{1}{2\pi \sqrt{LC}}$$

is exact for series resonance regardless of the resistance or the resultant Q. Thus, whether the Q were 5 or 50, no tuning could be accomplished by means of variable series resistance. Second-order effects could change the series resonant frequency, such as excessive distributed capacity in the inductor, or a change of inductance with current, but these could be negligibly small. It is only in the parallel resonant tank circuit where vernier resistance tuning can be used, and then to an increasingly smaller percentage shift in frequency as the Q is made higher.

4-6. Overlooked was the very same phenomenon that puzzled no less than Thomas Edison. This illustrates the important fact that in ac circuits containing reactive elements, the voltages or currents do not combine arithmetically. Thus, if 10 volts is measured across a resistance and 10 volts is also measured across a series connected capacitance, the impressed voltage across the combination is not 20 volts, but rather the square root of the sum of their squares, or 14.1 volts.

4-7. Yes, a poorly regulated supply can, nonetheless, exhibit excellent ac characteristics; that is, low output impedance. This can be the result of a lack of a voltage reference source in an otherwise conventional voltage-regulated supply. Such a technique is sometimes encountered in TV sets and other equipment where the added cost of tight dc voltage stabilization is not considered justified. An even simpler approach is to use a very large capacitor across the output terminals of an unregulated power supply. Indeed, even in a voltage-regulated supply with very close regulation of the dc voltage level, an output capacitor often serves the purpose of reducing the output impedance at the higher frequencies where the internal amplifiers can no longer do this because of low gain.

4-8. Not in this case. If the 1000-Hz tone can be reduced greatly in sound level, it is an indication that the bridge circuit is very likely the best one to use for the particular inductor being measured. The inability to null out tones completely is commonplace and stems primarily from the harmonic content of the bridge oscillator. These harmonics encounter different phase conditions in the bridge than does the fundamental frequency; therefore, they do not cancel. The situation can be remedied by inserting low-pass filters between the oscillator and the bridge, but it is often difficult and costly to completely attenuate harmonic energy. Usually, the skill of the operator compensates for the presence of the residual tones, for the ear and brain can exercise considerable selectivity in concentrating on the fundamental tone.

4-9. Yes, both series and parallel resonance can be used. However, the accuracy may not be as good as with a sine-wave oscillator, and some precautions are necessary to avoid gross error. First, the oscilloscope, rather than the electronic meter, should be used because the condition of resonance will involve both wave amplitude and wave shape. A little practice will be necessary to identify resonance, but this is rather easy with the scope because the waveshape will be unique. It must be carefully ascertained that a resonant response is due to the fundamental and not to one of the harmonics of the square wave. The response to the fundamental is generally identifiable when it is found that setting the generator at one half and at one third the "resonant" frequency does not produce an even more obvious response.

4-10. A "pure" reactance consumes no power despite the fact that we can measure both current through and voltage across the reactance. What we do not observe on the meters is the time of occurrence of the voltage and current maximum and minimum values. These are such that the voltage is always 90 degrees out of phase with the current. A plot of such waves shows that the sign of power (the product of voltage and current) alternates from positive to negative at 90-degree intervals. What is the practical implication of this? Simply that power is consumed from the line for 90 degrees, then returned to the line during the next 90 degree interval. So over a full 360-degree cycle, no net power is consumed from the line or dissipated in the reactance. This theoretical situation can be quite closely realized with a capacitor designed to have low dissipation losses. Conversely, in the resistive heater element, voltage and current are always in phase, and the heat generated is proportionate to their product. Interestingly, the current that flows "through" the capacitor can generate heat in a resistance. Thus, if we connect a capacitor across the power line, power is consumed in the resistance of the wiring, even though the capacitor, itself, consumes no net power.

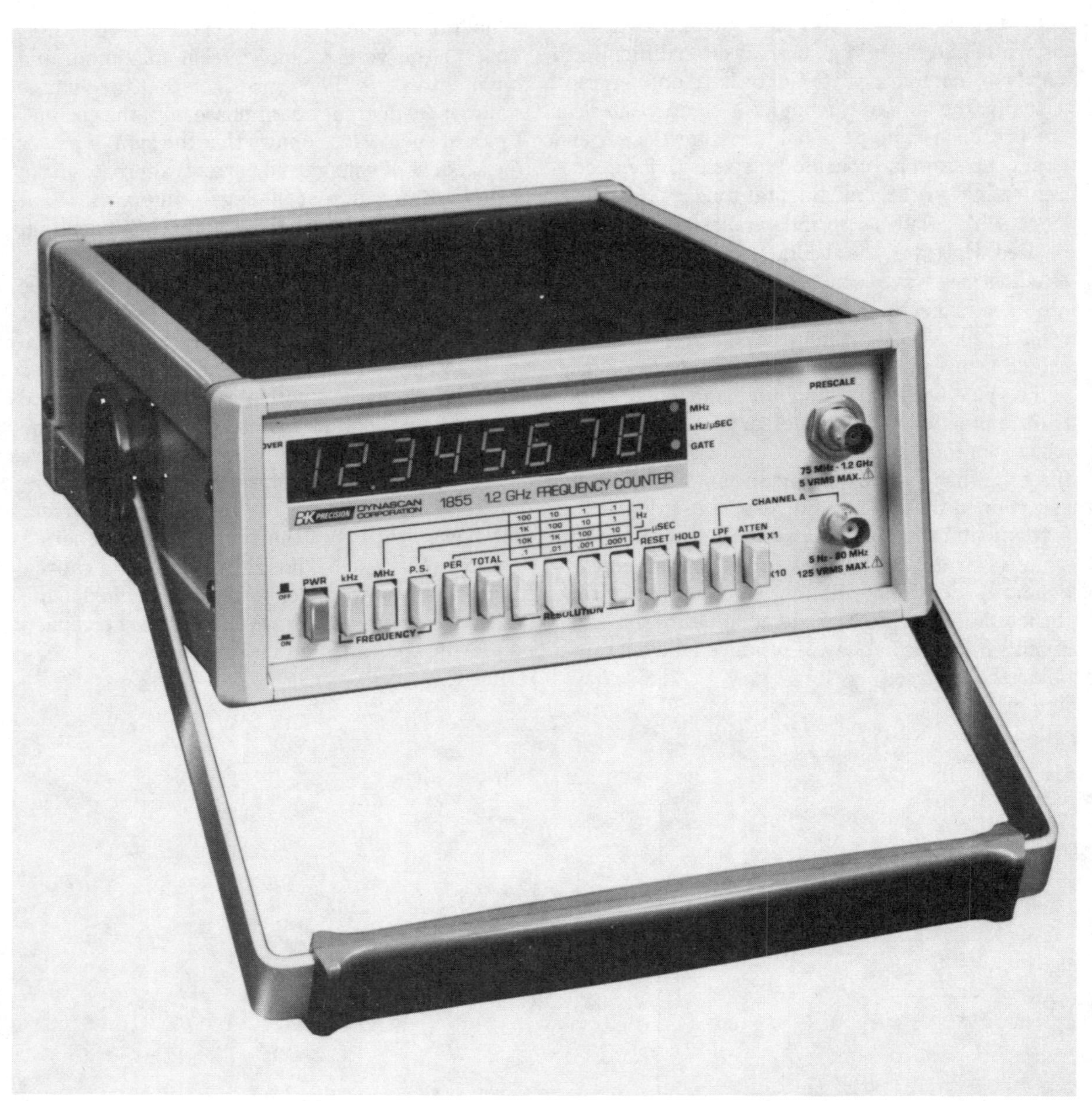

High-performance Digital Counters have made much progress since their inception. Relatively little skill is required in order to obtain an accurate readout and circuit disturbance is minimal. Courtesy of the Dynascan Corporation.

Chapter 5
Frequency and Phase

Much that goes on in ac circuitry depends on frequency and phase. This may be said to stem from the very nature of impedance in the ac circuit. If the magnitude and angle (another word for phase) of an impedance did not vary with frequency, and ac circuit would behave almost like a dc circuit an there would be but little to distinguish impedance from ordinary ohmic resistance. And if this were the case, radio communications would be deprived of the selectivity characteristic of tuned circuits, ac bypassing would not work, rectified waves could not be filtered, and many familiar and essential circuit techniques would not function. But since impedance and its specific constituents—inductive and capacitive reactance—do vary in magnitude with frequency, and do produce phase displacements which likewise are frequency dependent, the determination of frequency and phase are of prime importance to design, servicing, and maintenance of electronics equipment. There are other reasons why frequency is important: the radiation of a radio transmitter is not legal unless frequency stipulations are satisfied; the identification of interferences, such as parasitic oscillations, heterodynes, and crosstalk require measurement or determination of frequency; the frequency response of hi-fi amplifiers and other devices is meaningful only if frequency is properly measured. In other words, it is not easy to say something about an ac circuit without the mention of frequency.

Phase, too, is important, and it also has a bearing on the nature of ac impedance. At one time we could get by without a clear concept. But that day is but a memory of simpler times. Such developments as frequency modulation, stereo, color TV, and increasingly sophisticated instrumentation have made necessary the skill to investigate, measure, and interpret phase conditions in ac circuitry.

Although instruments are available to measure these quantities, it is not simply a matter of connecting the test instrument to the circuit and reading out the desired quantity. For example, one commonly encountered instrument which is very useful for both frequency and phase determinations is none other than the oscilloscope. But anyone with even a little practical experience with scopes can appreciate the need for various precautions if we are to derive trustworthy information from the display. Similarly, the digital frequency meter or electronic counter is quite capable of providing erroneous readouts unless the user is aware of certain facts connected with its operation.

USING THE DIGITAL COUNTER TO MEASURE FREQUENCY

The digital counter has come a long way since it first appeared on the commercial market. Even in its original form, it was recognized as perhaps the most useful measuring instrument to emerge from the laboratory since the oscilloscope. At first, this instrument was relatively expensive and did not appear destined to become a commonly encountered item. Also, earlier models were limited to submegahertz frequencies. With the advent of transistors and even more so with integrated circuits, the digital counter has become compact and inexpensive. Although it has not matched the almost universal value of the scope, it nonetheless is becoming an indispensable test instrument wherever a no-nonsense approach to frequency determination is a must. This includes not only the professional work of the engineer, but the endeavors of hobbyists, such as radio hams and audiophiles. In the service industry, the ability to quickly check a frequency is directly reflected in better profits.

Although the notion of connecting an instrument to a frequency source to get an instantaneous digital readout right down to the last hertz and beyond is intriguing, both the earlier instruments and sophisticated modern versions require care in the way they are used. Everything is fine if we are dealing with a repetitive wave of smooth form and high amplitude. However, erroneous readings can result if the wave has certain irregularities, involves transients, or is accompanied by noise, rfi or other disturbances. Moreover, many digital counters have an ill-defined region where insufficient amplitude can produce erratic counts or erroneous readouts which are not obvious from inspection. It is clear that stable operation requires that the amplitude of the frequency source be well above the threshold of sensitivity and well above the region of instability. Figure 5-1 shows some typical examples of good and bad signal test situations.

The safest way to use the digital counter is in conjunction with an oscilloscope (Fig. 5-2). The main function of the scope is to reveal the true nature of the signal presented to the digital counter. Not only does the scope tell us what we may logically expect in the way of readout accuracy, but a single glance at the display can quickly lead to remedial measures. For example, transients and rfi can often be sufficiently attenuated so that they no longer affect operation of the instrument. This may be accomplished by something as simple as connecting a capacitor across the input terminals of the counter. Sometimes other signal processing techniques can be advantageously used. Waves with interference on one polarity only can be rectified by a series or shunt diode. Where other frequencies are present, filter techniques ranging from simple to sophisticated may be necessary to remove or attenuate the offending frequencies.

With older style counters, difficulties may be experienced with very low frequencies, particularly with waveforms having slow rise and fall times. If sufficient amplitude is available, the remedy might be differentiation, waveshaping by nonlinear, or elements. Of course, regeneration of the basic signal by active circuits such as monostable multivibrators is a possibility, too. Fortunately, such monkey-business is not often necessary.

Often a give-away that the readout is not to be trusted is a variation of an indication with signal amplitude once the threshold of sensitivity has been exceeded. Now, this sounds a bit odd, but it is not difficult to spot in actual practice. If a clean wave is gradually increased in amplitude, you'll notice a certain amount of instability surrounding the threshold of sensitivity in the form of an erratic display of the digital readout. Increasing the amplitude further gets us through this region, where the readout becomes steady and remains so

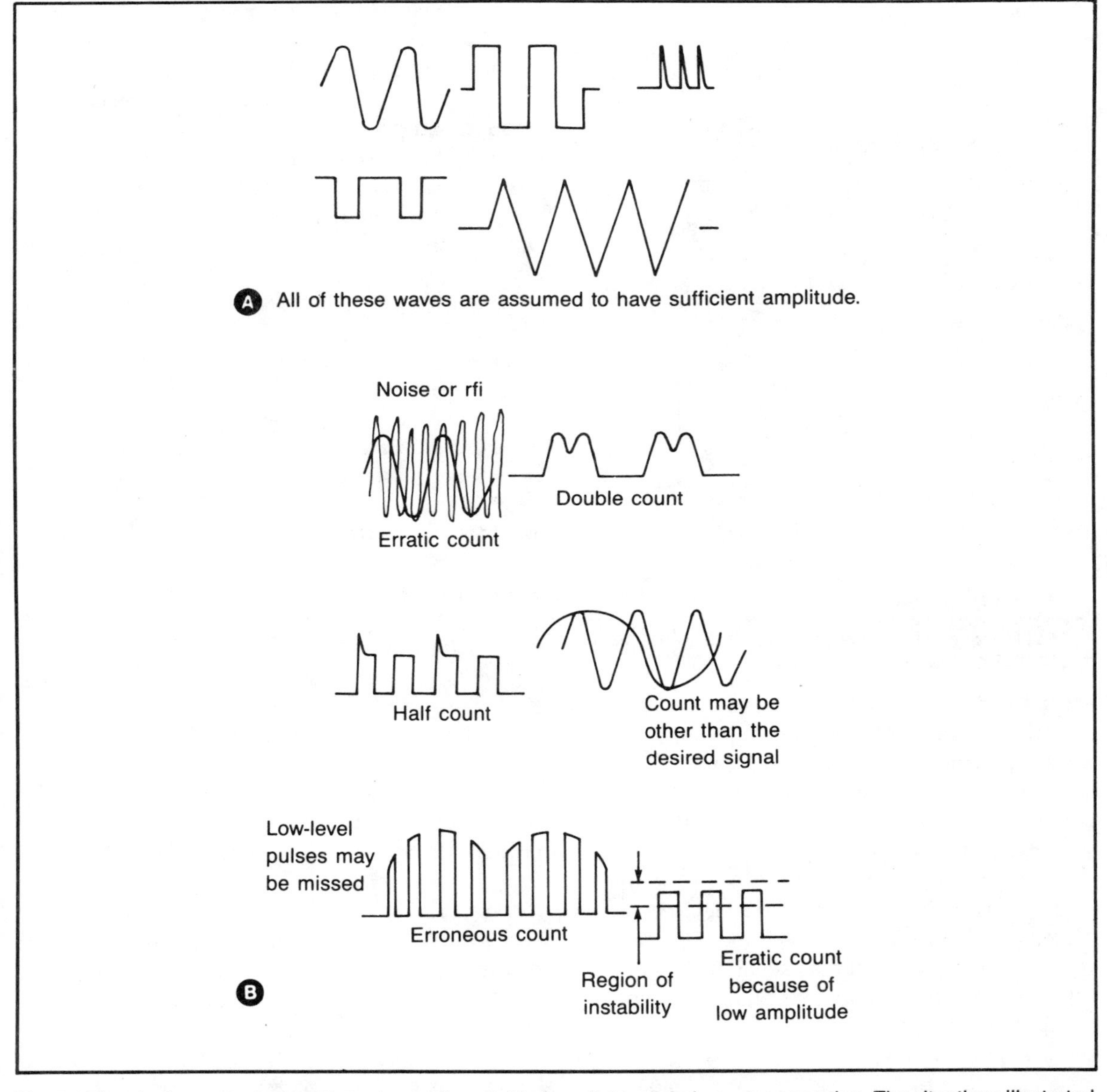

Fig. 5-1. Good, clean waveforms (A) are generally suitable for reliable digital counter operation. The situations illustrated at B may cause false indications.

over a very wide amplitude range. (Indeed, in most cases, instability due to excessive amplitude is not likely to occur.) Some allowances may have to be made for the fact that the frequency source itself may not be steady as a rock, or may change with amplitude. Also, the digital counters must be allowed a tolerance of plus or minus one count in the last digit in its count-scanning operation. Over longer periods, variations of several places in the last digit are likely because of various drifts within the counter circuitry itself. But for most practical work, there is no mistaking these effects with the

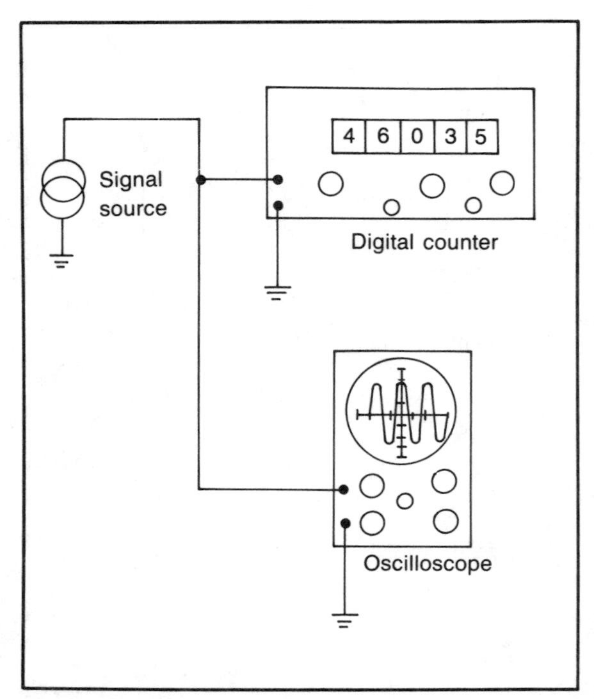

Fig. 5-2. A scope can be used to monitor the signal applied to a digital counter.

erratic or unstable readout caused by a signal with insufficient amplitude, or one accompanied by noise, transients, rfi, or other signals.

Wherever feasible, connections to the digital counter should be made with coaxial cable. Otherwise, what may originate as a clean wave could be contaminated by rfi and transients by the time it reaches the digital counter by unshielded wire. A notorious offender is the "hash" caused by fluorescent lighting systems. If the capacitance of coaxial cable is too high, then the unshielded connection should be as short as possible.

Fig. 5-3 shows several processing circuits which are useful for cleaning up or modifying signals, which otherwise might produce false readouts. With the exception of the rfi filter, these circuits generally require a signal of considerably more amplitude than would otherwise be necessary. This is particularly true for the differentiator circuit. All of these circuits should be physically situated close to the input connection of the digital counter.

If a frequency source provides signals which are of low amplitude, immersed in noise, accompanied by transients, or have waveshape distortions likely to cause a false readout from the counter, the setup shown in Fig. 5-4 can prove useful. Here, the oscilloscope is used as a wave regenerator; it converts the unsatisfactory signal to a smooth, high-amplitude sawtooth suitable for reliable operation of the counter. Additionally, the adjustment of the scope to produce a stationary display of one cycle of the wave implies that the sawtooth is synchronized to the frequency of the signal. The oscilloscope sweep is operated in its free-running mode. The apparatus should be physically arranged so that you can see both displays. When it has been determined that the single cycle on the scope screen is stationary, the readout on the digital counter is recorded.

Many oscilloscopes, even inexpensive ones, have a sawtooth output terminal. However, if this is not the case, it will be necessary to obtain a sample of the sweep voltage. This should be done with an isolating resistance, say, in the 100K to 1-megohm range, and a small series capacitor on the order of 150 pF. The exact values will depend upon the sweep circuitry and the point of sampling. The sawtooth probably will have considerably more amplitude than needed for the intended purpose and the sampling method is not critical. Perhaps the main thing to bear in mind is to be careful not to disturb the basic operation of the sweep. A good way to tell that all is well is to check the normal operation of the scope even with the sampled sawtooth grounded.

The accuracy of this method does not depend on the linearity, frequency response, or fidelity of the oscilloscope, provided that a single cycle can be stabilized on the screen by synchronizing its sweep with the sampled signal. Therefore, a relatively simple and inexpensive instrument should suffice.

Procedure for Calibrating a Variable-Frequency Source

Objective of Test: To use a digital counter to calibrate the dial of a variable-frequency source.

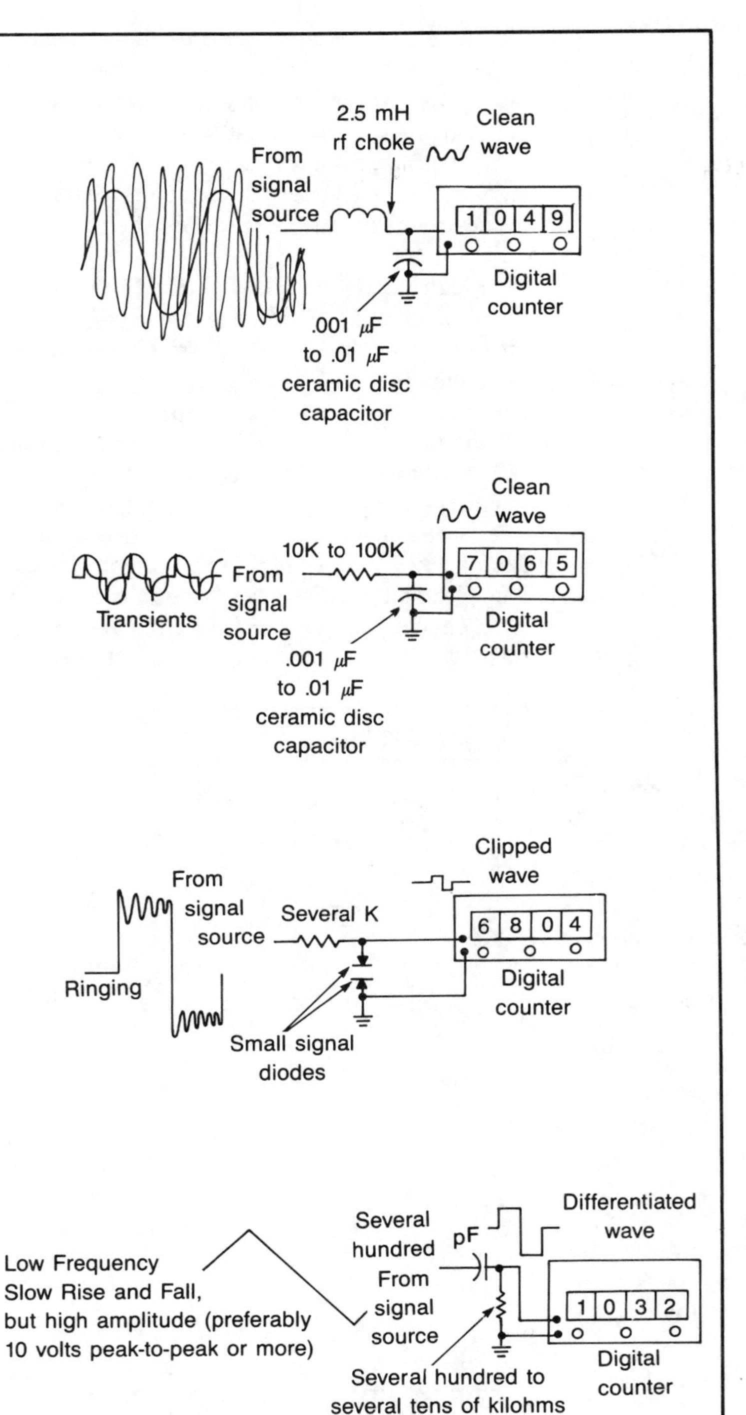

Fig. 5-3. Processing circuits for signals likely to cause false readouts.

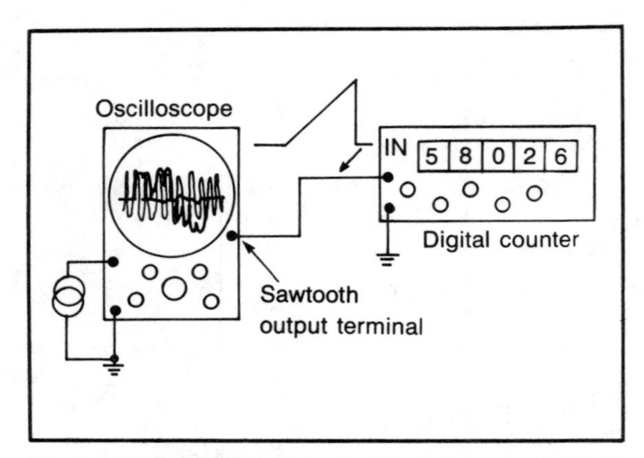

Fig. 5-4. Technique for measuring the frequency of signals unsuitable for the digital counter.

Equipment Required:

One of the following: a laboratory sine-wave oscillator, a function generator, or a signal generator.
Oscilloscope
Digital counter

Test Procedure: Connect the equipment as shown in Fig. 5-5. If a sine-wave oscillator or function generator is used, select a range in the audio spectrum, say, from several hundred Hz to 10 kHz. (Set the function generator for sine waves.) If a signal generator is used, select the lowest frequency range, but make certain that the digital counter is capable of covering that range. Allow all instruments about ten minutes to warm up. If the counter has a provision for self-testing, use it to determine that counting operation is basically OK. Then return the counter to its normal external frequency-measuring mode. Become acquainted with the amplitude threshold level of the digital counter and notice the scope peak-to-peak amplitude which corresponds to the level needed to cause repetitive cycling of the digital counter. About one and one-half times this amplitude should be used when calibrating.

Beginning at the low-frequency end of the dial, calibrate the major divisions by recording dial indications vs frequency. A study of the results will reveal whether or not an overall improvement can be effected by mechanical positioning of the dial on the shaft. If some readings are too high and some are too low, a better average situation can be achieved by such mechanical adjustment. (Some signal sources have electrical adjustments for the low end and for the high end of the dial.)

Comments on Test Results: This simple test procedure acquaints you with a very useful instrument combination—the variable-signal source in conjunction with the digital counter. It is difficult and expensive to make a precision source of selectable frequency. And even when this is done, the inevitable weak links are the analog dials from which the frequency must be read. Because of the effects of aging, temperature, and other influences, such a source is not generally suitable for everyday electronics. However, even an inexpensive "workhorse" source becomes a laboratory instrument when monitored with a digital counter. In

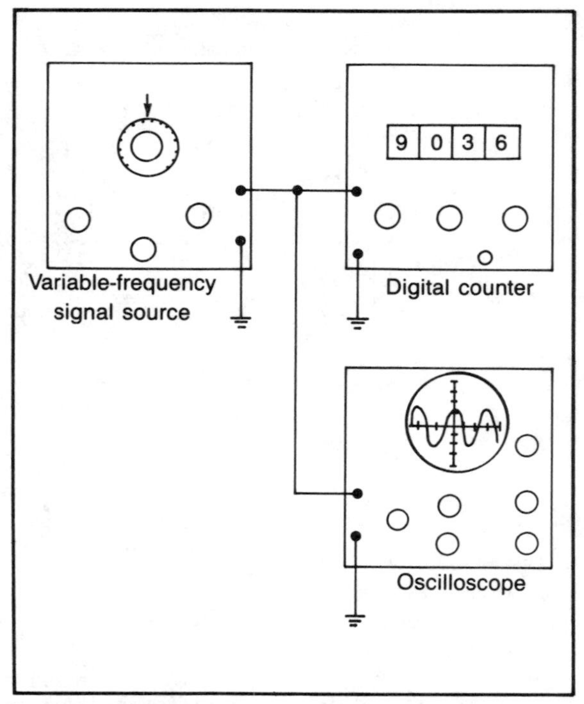

Fig. 5-5. Setup for calibrating the dial of a variable-frequency signal source.

most cases, short-time frequency stability of such a source will at least be good enough for a number of repetitive readouts to be observed on the digital counter.

Self-Testing

Most digital counters have a means for self-testing. Thus, if the crystal oscillator in the time base operates at one megahertz, this frequency and decade submultiples of it can be injected into the input terminal. Then, you should obtain such readouts as 1,000,000; 100,000; 10,000, etc. The fact that the readout is, say ,100,000 in the test mode does not mean that the frequency is necessarily exactly as indicated. What we look for in the test mode are gross departures from decade division. This used to be more common in older instruments than in later versions using integrated circuits. Insofar as concerns the absolute accuracy of the readout, only a calibration test with Station WWV or with a secondary frequency standard can show whether a vernier adjustment of the crystal oscillator is needed. However, for most of the ordinary applications in practical electronics, the digital counter is likely to provide more resolution than is necessary, even though the last place or two may require correction. Compared to the relatively inexact techniques long used for frequency determination, the digital counter, despite the worst condition of internally accumulated drifts, greatly exceeds the accuracy of the former methods.

PERIOD MEASUREMENT

Sometimes it is more useful to measure period, rather than frequency. In any event, period and frequency have reciprocal relationships; that is, period equals 1 divided by the frequency and frequency equals 1 divided by the period. Period measurements are often made for the lower frequencies, say, below 100 Hz. Most digital counters can be placed in the period mode by properly positioning a panel-mounted function switch. (When in the period mode of operation, the internal time base becomes the input signal. The actual input signal is used to time the gate which, in the frequency-measuring mode, is controlled by the time base.) For low frequencies, the period measurement can provide higher resolution (more decimal places). It should be determined that the input impedance is still high enough for the measurement being made, however; on some counters the input impedance is high for frequency measurements, but lower for period measurements. For various reasons, sometimes it will be found that a wave which is unsatisfactory for frequency measurement will produce a stable and accurate period readout.

USE OF THE OSCILLOSCOPE IN FREQUENCY MEASUREMENTS

There are several ways in which the oscilloscope is uniquely adapted to the task of frequency measurement. As a voltmeter (or simulated current meter) it can be used to determine maximum or minimum levels associated with resonant circuits. If the scope has a calibrated sweep, frequency can be measured by actually counting the number of cycles displayed over a span of time-calibrated divisions on the horizontal axis. Yet another way of using the oscilloscope for frequency measurement is as a comparing instrument. When thus used, a display is stationary when the frequency under test bears a harmonic or subharmonic relationship to a reference frequency. The slightest deviation from such a relationship results in a non-stationary pattern. This method can provide very high accuracy with even an inexpensive oscilloscope. Stationary displays include Lissajous figures, "gear wheels," and "slotted wheel" patterns. The use of the scope in these various methods of frequency measurement is discussed in the following paragraphs.

Scope-Indicated Resonance

Fig. 5-6 shows several circuits in which the oscilloscope indicates maximum or minimum levels at resonance. Frequency is then computed from the equation:

$$f_0 = \frac{1}{2\pi\sqrt{LC}}$$

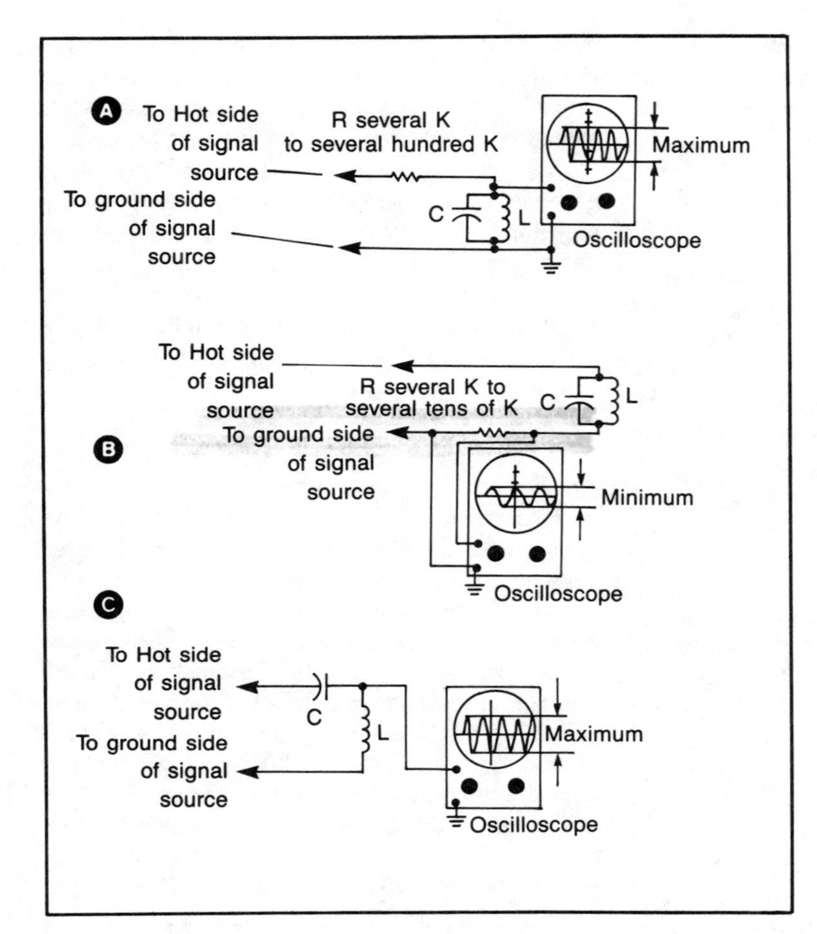

Fig. 5-6. A scope can be used to indicate maximum and minimum signal points while measuring frequency by the resonance method.

Of course, a disadvantage of this approach is that we must somehow know the values of L and C. Yet, in many instances, the circumstances may be such that this method will be convenient. For example, there are likely to be a few known inductors and many known capacitors. From various parallel or series combinations, the condition of resonance can be ascertained and we can at least arrive at a good "ball-park" notion of frequency. Nor is this method necessarily restricted to sine waves, for with a bit of logical reasoning, the resonance of the fundamental frequency contained in waves of various shapes can be detected. Remember, the fundamental frequency is the same as the pulse-repetition rate in pulse wavetrains. The maximum and minimum points of non-sinusoidal waves are not as clean-cut as with sine waves, but the waveshape displayed on the oscilloscope screen will also undergo a unique change at resonance. With a bit of practice, this condition is easy to recognize.

Phase Indication

The frequency measuring setups shown in Fig. 5-7 again require resonance, but are somewhat more sophisticated than the circuits in Fig. 5-6. Here, resonance is detected not as maximum or minimum levels, but as a unique phase condition. Specifically, at resonance (either series or parallel) the LC combination no longer behaves reactively. In series resonance, current flowing through the circuit encounters resistance only. In parallel resonance, the same is true, although here the resistive component is "stepped-up" to a relatively high value. In both cases, the current supplied to the LC circuit and the voltage impressed across the LC circuit are in phase. This being so, a unique Lissajous figure will be displayed by an appropriately connected oscilloscope, namely a straight line. It turns out to be more practical to use parallel resonance in implementing this scheme. In series resonance, the voltage across the LC elements may be too low for practical use. However, with the aid of an isolation transformer to remove ground conflicts, a modified approach can provide good results with series resonance. This is shown in Fig. 5-7B. Here, the straight-line display is produced by the 180-degree phase displacement between inductive and capacitive voltages at resonance.

Lissajous Figures

When the horizontal sweep circuit of an oscilloscope is disabled or disconnected and both vertical and horizontal deflection plates are driven from sine-wave sources, the resultant display will be in motion, except when the frequencies of the two sources are integrally related. By this we mean such frequency relationships as 5:1, 3:1, 2:1, 1:1, 1:2, 1:5, etc. Also included are such fractional ratios as 7/5, 3/2, 2/3, 5/2, etc. All of these ratios involve either whole numbers or integral fractions. Thus, ratio such as 4.23:1, or 1.76 : 1.91 will not result in a stationary pattern. This being the case, frequency determination is possible if we can interpret a stationary pattern and if one of the two frequencies is known. The general arrangement for producing Lissajous figures is shown in Fig. 5-8.

The easiest pattern to interpret is that which results from a 1:1 ratio between 90 degree out-of-phase sine-wave frequencies. If the vertical and horizontal amplitudes on the screen are equal, a perfect circle will be displayed. However, in the general employment of this scheme, it is not necessary that such relationships prevail between the signals reaching the vertical and horizontal deflection plates. Rather than a circle, an ellipse will be seen when the two amplitudes differ and/or the phase difference is other than 90 degrees. Notice however, that whether a circle or an ellipse is formed, the display of a simple closed loop with no crossovers indicates a 1:1 ratio. So when we see any such stationary pattern, whether it is a relatively flat ellipse, a closed loop with or without wiggles or inflections, or a full-blown circle, the unknown frequency is then equal to the known frequency.

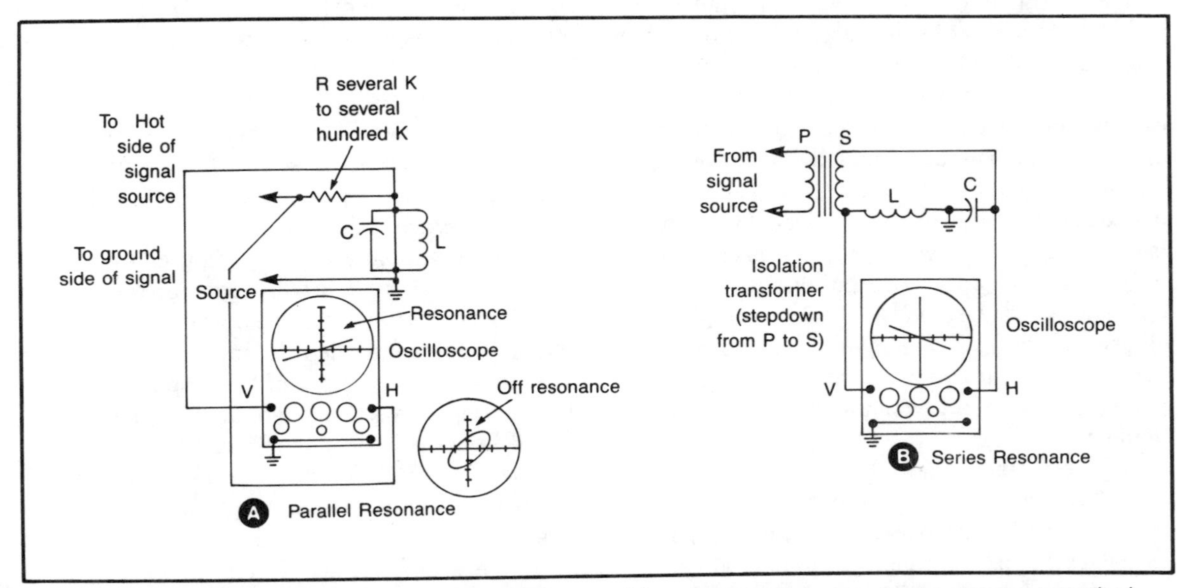

Fig. 5-7. An oscilloscope can be used as a phase indicator while measuring frequency by the resonance method.

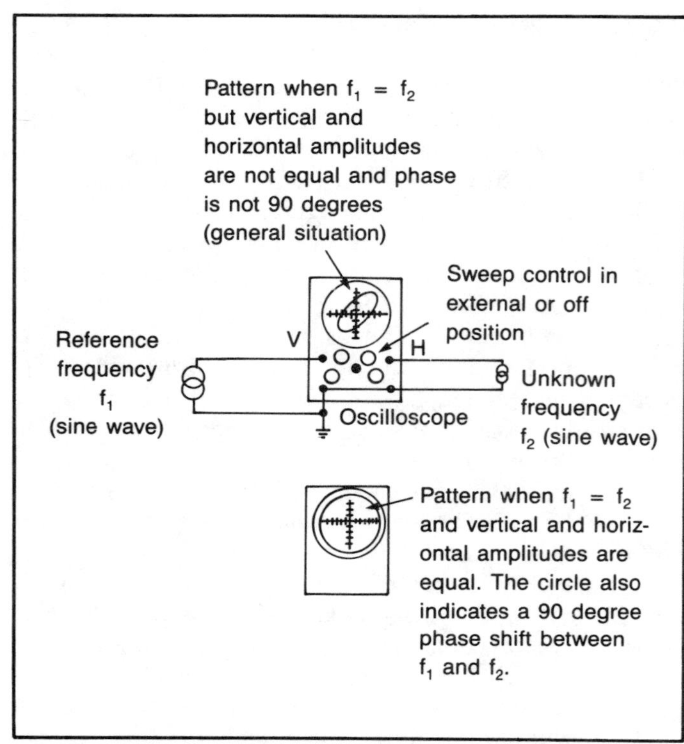
Pattern when $f_1 = f_2$ but vertical and horizontal amplitudes are not equal and phase is not 90 degrees (general situation)
Sweep control in external or off position
Reference frequency f_1 (sine wave)
V
H
Unknown frequency f_2 (sine wave)
Oscilloscope
Pattern when $f_1 = f_2$ and vertical and horizontal amplitudes are equal. The circle also indicates a 90 degree phase shift between f_1 and f_2.

Fig. 5-8. Basic setup for a display of Lissajous figures. There is generally no need to have the vertical and horizontal amplitudes equal. Also, the phase difference between vertical and horizontal signals need not be 90 degrees for this test procedure to be useful.

Wiggles and other distortions are caused by harmonics or other non-sinusoidal aspects of one or both frequency sources. Assuming again that both frequencies are sinusoidal, the tilt or inclination of an elliptical pattern is determined by phase shifts both external to and internal to the oscilloscope. One of the particularly useful aspects of this method is that, although sine-wave signals tend to be easier to interpret at ratios other than one to one, the basic precision of the determination is not affected by waveshape, by phase difference, or by amplitude differences, providing that the interpretation can be made at all. For example, if our ellipse resembles a rectangle, the frequency ratio between vertical and horizontal signal sources is precisely one to one when such a pattern is stationary.

If the pattern is seen to rotate at the rate of, say, 10 revolutions per minute, this represents the frequency difference between the two signals; that is, one-sixth Hz. Obviously, this method can provide great sensitivity in attaining exact synchronization between two frequencies. By varying the unknown frequency slightly above and below that resulting in a stationary pattern, you can learn to differentiate between too high a frequency and too low a frequency by observing the sense of rotation. When making use of pattern rotation, an ellipse or any irregular open figure which closes upon itself without crossovers is suitable. (A true circle undergoes cyclic shape variations above and below frequency synchronism, but may not be as easy to count as the initially flattened figures.)

Ratios other than unity: For the sake of simplicity, let us first confine our discussion to the arrangement where the reference frequency is applied to the vertical input of the scope and the unknown frequency is applied to the horizontal input. It will be assumed that the internal vertical and horizontal amplifiers of the scope are used, although this is not strictly necessary. (At higher frequencies, a connection is sometimes made to the deflection plates directly. Also, when you want to avoid the phase shifts of the internal amplifiers, the direct

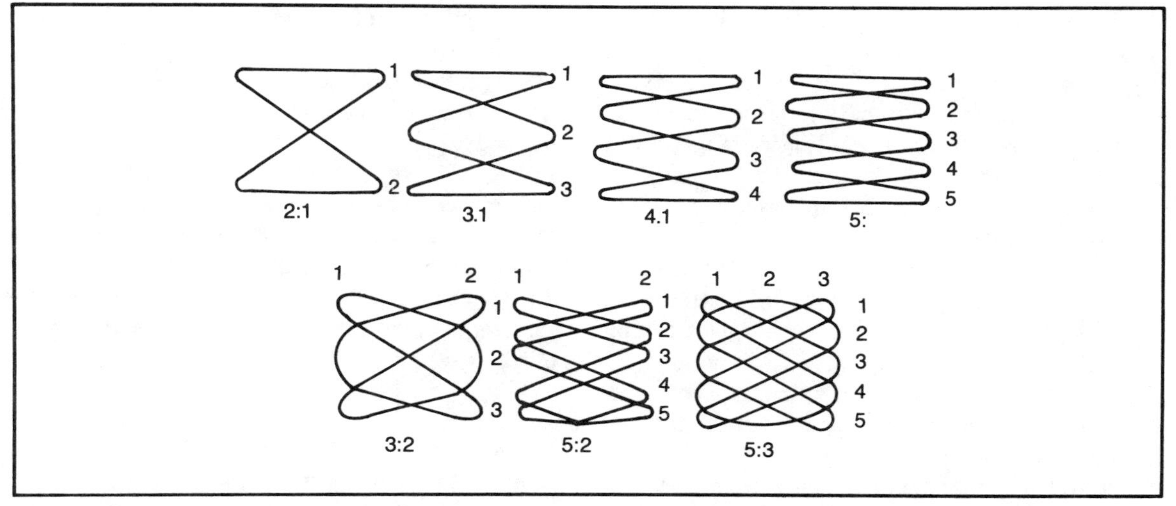

Fig. 5-9. Lissajous figures for several ratios when f_2-f_1 exceeds unity.

connection is useful. Obviously greater amplitude is then required from the signal sources to produce a reasonably sized pattern.) Figure 5-9 shows the Lissajous figures resulting from several ratios greater than unity. That is, the unknown frequency is greater than the reference frequency by either a simple whole number ratio or by an integral fraction. Again, for simplicity, it will be assumed that both frequencies are essentially sinusoidal.

A study of the patterns should convey the general idea involved in comparing the unknown to the reference frequency. Notice that the number of horizontal loops exceed the vertical loops for frequency ratios in excess of unity. Figure 5-10 shows Lissajous figures corresponding to several ratios less than unity. That is, the unknown frequency is lower than the reference frequency, again by ratios expressed as integral functions. The situation is similar, but now the greatest number of loops will always be seen as vertical loops.

Actually, some skill must be developed in counting and positioning the pattern. High or low ratios, or close ratios such as 8:7 or 5:6 can be difficult to interpret. Sometimes by deliberately causing the pattern to rotate slowly (by a slight readjustment of the frequency ratio), it becomes easier to count obscure loops. The rotating patterns help make prominent those loops which are actually separate, but cover one another because of unfavorable phase conditions. Sometimes, changing the vertical and horizontal size of the pattern makes the loops easier to count. And adjustment of the brightness, focus, and astigmatism controls can be helpful, too. For example, the attainment of fine focus in the outer edges of the scope screen, even though at the expense of the central area, can provide the extra resolution needed to separate closely spaced loops. The opertor can develop other techniques which are helpful. Of course, fewer loops or the simpler patterns are easier to work with.

Misinterpretation of Lissajous figures is quite easy when there are many loops to count. Not only is it possible to miscount the loops which are displayed, but under unfavorable phase conditions, many loops can be obscured by being closed,

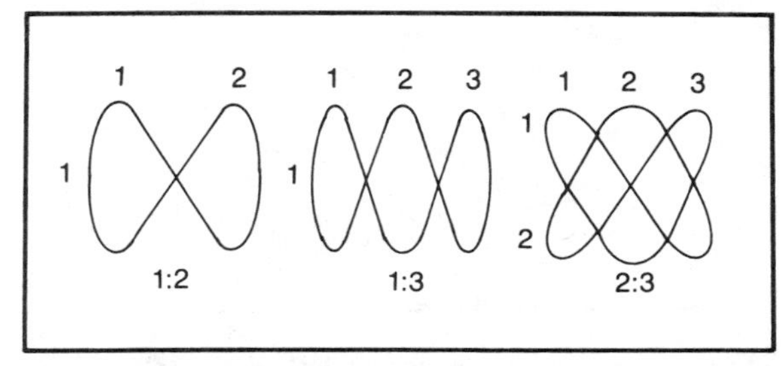

Fig. 5-10. Patterns showing several Lissajous figures when the ratio f_2-f_1 is less than unity.

thereby becoming overlapping lines. Such a pattern, though stationary, is useless because you can't see its ratio-indicating features. If you are not on the alert for such a possibility, you will come up with an incorrect frequency indication. Figure 5-11 shows an example of such a pattern. Actually, this is a 6:5 Lissajous figure, but this fact is not clear without a detailed analysis, as well as some lucky guesses based on much experience with these patterns. Fortunately, such indeterminate patterns are quite noticeable if you look carefully. Notice that the right and left sides in Fig. 5-11 do not contain the same number of loops. In other such patterns, the same situation may exist between the top and bottom loops. When this kind of pattern is encountered, a slight change in phase conditions will restore the display to a usable Lissajous figure. This may be brought about by a change in any circuitry associated with the two frequencies which can influence phase. Often a capacitor connected across either the vertical or horizontal input terminals of the scope can accomplish this. Conversely, a very slight change in one of the frequencies will open up the closed loops; although, in this case, the pattern will slowly rotate.

When dealing with ratios which cannot be expressed as a whole number with respect to unity, such as 3:2, 6:5, 9:7, etc., special care must be exercised. In these instances, the corner loops are counted in both the horizontal and the vertical directions, seemingly serving double service. Experience and practice are needed to avoid error in interpreting such complex patterns. Often, a very complex Lissajous figure can be simplified by a slight change in the reference frequency. When the reference frequency is provided by a variable oscillator, this is easily demonstrated. For example, a ratio of 20:9, which would be exceedingly difficult to deal with, can become a 2:1 ratio by causing the relatively small change to 20:10. Much smaller percentage changes sometimes will accomplish the same result. If a crystal oscillator is used as the reference source, it can prove worthwhile in this respect to make use of even the vernier frequency adjustment possible with such a source.

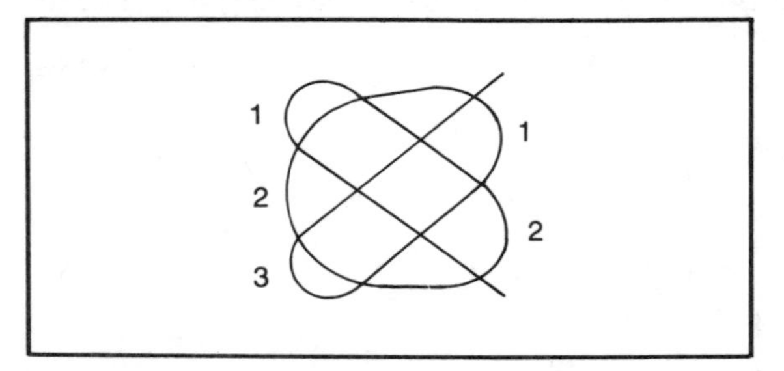

Fig. 5-11. Indefinite Lissajous figure. An unfavorable phase condition is indicated by the unequal number of loops in the left and right sides.

The Gear-Wheel Pattern

A method bearing both similarity to and differences from the Lissajous technique is shown in Fig. 5-12. The pattern produced by the setup shown resembles a gear wheel, although the "teeth" tend to be something less than acceptable to the mechanical designer. The basic circular configuration of this gear wheel is produced by the reference frequency in conjunction with a 90-degree phase-shifting network. We know that the one-to-one ratio Lissajous figure is a perfect circle if the signals applied to the deflection plates of the oscilloscope are equal in magnitude (assuming equal deflection sensitivities for both sets of plates) and differ in phase by 90 degrees. The same logic applies here, except that the vertical signal and the horizontal signal both come from a common source, f_1. This is a desirable situation because a circle (or ellipse) so produced is likely to remain stationary, or very nearly so. Source f_1 would have to be quite unstable for any appreciable change of the basic circle (or ellipse) to occur during ordinary viewing time. This is not to say that f_1 can have sloppy drift characteristics; the more stable f_1 is, the better, as we shall now see.

If the magnitude of f_2 is zero, f_1 will produce a circle or ellipse. The introduction of frequency f_2 into the horizontal channel of the scope will modulate the basic circle or ellipse to produce the "teeth" of our gear wheel. The teeth will be stationary only when f_2 is a whole-number multiple of the reference frequency, f_1. It sometimes happens

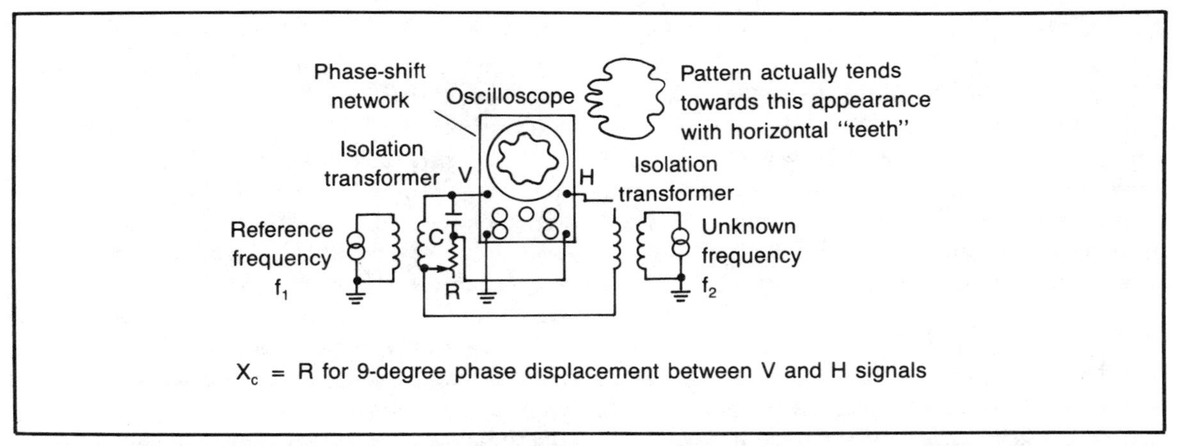

Fig. 5-12. Frequency measurement setup using a gear-wheel pattern. $f_2 = n \times f_1$. (f_2 must be higher in frequency than f_1 when using this method.)

that a particular ellipse makes it easier to count the teeth. (In practice, any multiple greater than unity is useful; thus, f_2 may be $2 \times f_1$, $3 \times f_1$, $6 \times f_1$, $15 \times f_1$, etc.) The use of the term "modulate" is not strictly accurate from the engineering viewpoint. What we observe is a mixture of waves rather than a true modulation process.

The gear-wheel method of frequency determination is particularly useful for high ratios of unknown-to-reference frequency because in such situations the identification of loops or bumps in a Lissajous figure often becomes rather difficult. The gear teeth tend to be relatively easy to count even when present in great number.

The unknown frequency, f_2, is simply $n \times f_1$, where n represents the number of teeth and f_1 is the reference frequency. This method is not applicable to the measurement of frequencies lower than the reference frequency, f_1, in the sense that the Lissajous figure method is. However, a technique which is useful in certain instances involves shifting the phase of the unknown frequency, f_2, in order to produce the basic circle or ellipse (see Fig. 5-13). This is accomplished with variable R and/or C elements. WIth this setup the stationary teeth are caused to appear by introducing reference-frequency f_1, which is variable. Adjusting f_1 produces a gear-wheel or toothed pattern from which the frequency of f_2 is computed by f_2 equals f_1 divided by n where n, as before, represents the number of teeth.

In ordinary use of this modified scheme, frequency f_1 would be read from the dial of a variable oscillator. Of course, a digital counter could be used with the variable oscillator in order to measure f_1 with greater accuracy. However, there could be several reasons why a direct application of the digital counter to measure f_2 might not be practical. For example, frequency f_2 could be associated with other frequencies and accompanied by noise or transients which could make operation of the digital counter difficult. A scope pattern under such conditions would generally be easier to interpret. Also, the amplitude of f_2 might be insufficient to operate the counter, but of suitable level

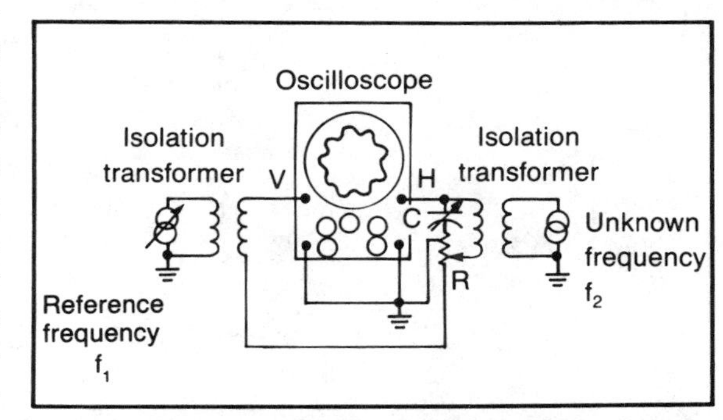

Fig. 5-13. Setup for use of the gear-wheel pattern when the unknown frequency is lower than the reference frequency.

for use with the internal amplifier of a high-gain oscilloscope.

The particular circuits shown in Figs. 5-12 and 5-13 need not be used if the basic objectives are kept in mind. Other phase-shifting networks, such as resonant circuits, can be utilized. The isolation transformers are not always necessary, but in the arrangements shown, they are used in order to prevent conflicting ground connections.

The Slotted Wheel Pattern

The test setup shown in Fig. 5-14 represents a somewhat more sophisticated version of the idea used in the gear-wheel pattern. Here, again, the unknown frequency must produce an integral number of interruptions or modifications around a circle or ellipse. In this case, the modifications are alternate interruptions and enhancements of beam brightness. The unknown frequency is again n times the reference frequency, where n is either the number of spaces or line segments of the stationary slotted-wheel pattern. As with the gear-wheel pattern, the unknown frequency must be a whole-number multiple of the reference frequency. The technique of bringing this about also calls for a variation of the reference frequency (unless it is desired to adjust the unknown frequency so that it is a known multiple of the reference frequency).

The basic idea is, in part, similar to that of the gear-wheel method. A phase-shifting circuit is incorporated in the reference frequency, f_1, signal input circuit for the purpose of producing a circular or elliptical pattern prior to chopping by the unknown frequency. The chopping is brought about by feeding the unknown frequency, f_2, into the Z axis of the scope. Many oscilloscopes have a provision on the rear panel whereby the cathode-first grid circuit of the CRT can be interrupted just for such a purpose as this. The unknown frequency can be a wide variety of waveshapes, although those with equal duty cycles are particularly suitable. Square waves are excellent because of the clean chopping accompanying such signals. In any event, the unknown frequency must be of sufficient amplitude to cutoff the beam. If trouble is encountered here, there is some advantage in operating the scope in a dark room using patient adjustment of the brightness and focus controls. Under good conditions, this method is probably even better than the gear-wheel method where high ratios of frequency must be determined. There can be some confusion in the counting of gear teeth in the top and bottom portions of the gear pattern, but the counting of dashes or spaces tends to be easy if clean chopping and fine focusing is attained.

The slotted-wheel methods of frequency determination is not suitable for measuring frequencies lower than the reference frequency, f_1. However, frequencies below that of f_1 can be measured by the procedure modification described for the gear

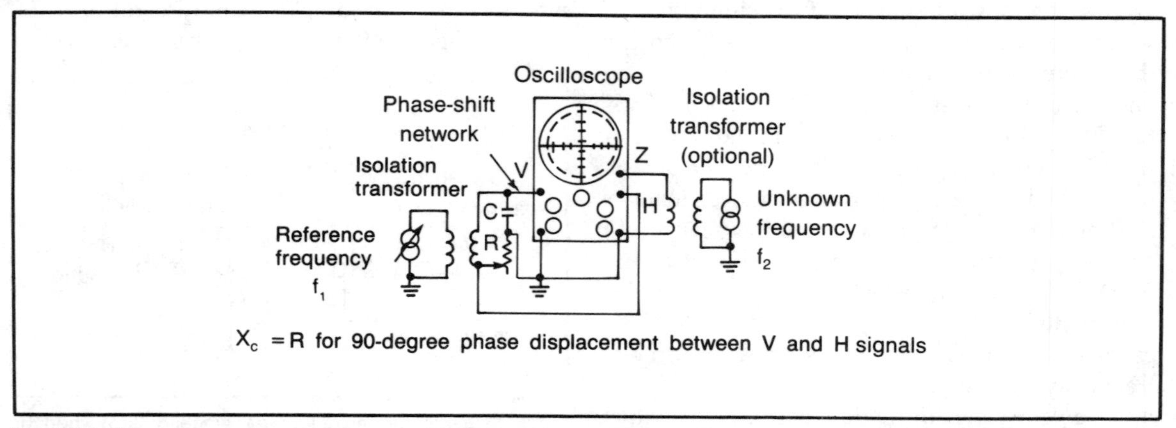

Fig. 5-14. Frequency measurement setup using the slotted-wheel pattern. $f_2 = n \times f_1$. f_2 must be higher in frequency than f_1 when using this method. Also, f_2 must be a whole-number multiple of f_1 to produce a stationary pattern.

wheel method. Because of the essential similarity involved, the modification will not be repeated here.

Again, other circuit approaches are possible if we keep the objectives in mind. It is to be noted that the isolation transformer shown in conjunction with source f_2 is optional here. That is, no ground conflict will result from its omission. However, in some cases, a stepup or stepdown in voltage might prove advantageous to the particular situation.

FREQUENCY DETERMINATION FROM RC NETWORKS

Two RC networks which are useful for the measurement of an unknown frequency are shown in Fig. 5-15. The main feature of these networks is that a sharp null is obtained without the use of inductors. A disadvantage is that network components must be ganged together if the setup for reading frequency from a calibrated dial is to be convenient to use. However, once the component procurement and mechanical headaches have been overcome, unknown frequencies can be determined very quickly. For many situations encountered in practical electronics,these networks can be used for "ballparking" an unknown frequency. However, almost any accuracy can be approached, since the accuracy directly depends on the tolerances of the components and the care exercised in nulling and calibrating. In both networks, an attempt should be made to use resistors in the several kilohm to several tens of kilohm range. There is really nothing sacred about this suggestion, but if you use resistors, say, in the several hundred kilohm to several megohm range, stray capacitances can present a problem in attaining a sharp null. On the other hand, the use of resistors in the several tens or several hundreds of ohms can result in excessive loading of the frequency source.

These statements may have to be tempered somewhat when making measurements outside of the audio frequency range. With regard to the capacitors, you must expect practical difficulties when less than several tens of pF are involved. In fact, it is advisable to try to keep capacities in excess of several hundred pF. At the other extreme, it is wise to limit maximum capacitance to the vicinity of several or several tens of μF. Although electrolytics, particularly solid-state tantalum units, have been successfully used, by far the best results are likely to be achieved with mylar or other types with a low dissipation factor.

The Twin-T Network

The twin-T or parallel-T circuit shown in Fig. 5-15A provides a common ground and operates well over a wide range of input and output conditions.

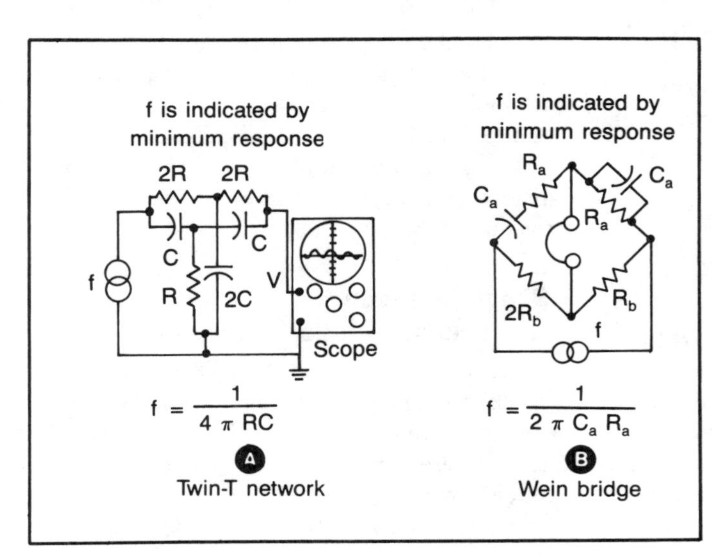

Fig. 5-15. These RC networks and a scope can be used to determine an unknown frequency.

Ideally, it should see a zero impedance source and an infinite impedance null detector, but in practice these matters do not ordinarily have to be given much consideration. In order to make an easy-to-use frequency-measuring device from the twin-T network, either the three resistors or the three capacitors should be ganged. Gross changes in range, such as by decade steps, can be accomplished by switching in different sets of the unganged components. The basic idea is that the indicated relationships between the C and R values must be maintained at all times. For example, you cannot use different values for the two series-connected capacitors. Each of these must be C. Nor, can the value of the shunt capacitor be selected arbitrarily; rather, its value must be 2C. Similar restrictions apply to the selection of the R values. The behavior of the twin-T network is similar to a series resonant LC circuit because maximum attenuation is provided at frequency f.

A number of small, quickly constructed twin-T networks can be made up and used at appropriate times to check the frequency at various dial settings of oscillators, function-generators, and other signal sources. The twin-T network is likely to retain its original accuracy, whereas variable-frequency sources are very much more vulnerable to the effects of aging, mechanical changes, etc. Twin-T networks assembled for this purpose should use 1 percent or better precision resistors and 5 percent mylar capacitors.

Invariably, it will be found that the null can be greatly improved by vernier adjustments of the elements. To make this practical, the shunt capacitor, 2C, is chosen like a woman's shoe, one size too small. Then small padding capacitors are selected for the best null. With the tolerances suggested, no modification of the other components need be undertaken. Notice that because we can't obtain the exact component values needed for a chosen frequency, it becomes quite difficult to attain an exact null frequency. In this situation, we can often accept what we get, say 992 Hz, or 1014 Hz instead of 1 kHz. On the other hand, if the null is deep, say, 50 dB or more, we nonetheless have a very good frequency marker or, in essence, a passive frequency reference. Use as pure a sine-wave source as possible when initially adjusting the network. "Rock" the frequency of this calibration source while padding shunt-arm 2C for optimum null.

The Wien Bridge

The Wien Bridge circuit (Fig. 5-15B) also indicates the "resonant" frequency by means of a null. It has the constructional advantage in that only two components need be mechanically ganged together. These are either the resistors designated as R_a or the capacitors designated as C_a. Notice that the arms, R_b and $2R_b$ bear no relationship to the other components of the network. For these reasons, the selection of components and the mechanical construction of the Wien Bridge tend to involve fewer conflicts than in the twin-T network. However, there can also be a basic disadvantage involved in the practical implement of the Wien Bridge. This is not readily apparent from the simple circuit in Fig. 5-15B, but in many practical situations there is likely to be a conflict in grounds between the null detector and the signal source. If headphones are used, as shown, such trouble may not be directly encountered, but at higher audio frequencies, it may be difficult to find a good null because of stray capacitance to ground which tends to unbalance the bridge. An isolation transformer at the signal source, and/or the null detector is a step in the right direction, but the required isolation still may not be readily obtained at higher frequencies. Throughout the audio-frequency range, good results can generally be had if the isolation transformer(s) have grounded electrostatic shields between windings.

A Simple Frequency Transducer for Audio Frequencies

The arrangement shown in Fig. 5-16 will provide accurate and reliable frequency indications in the audio range and somewhat beyond. A toroid of square hysteresis loop material carries primary and secondary windings, but differs in operation from a conventional transformer in that magnetic satura-

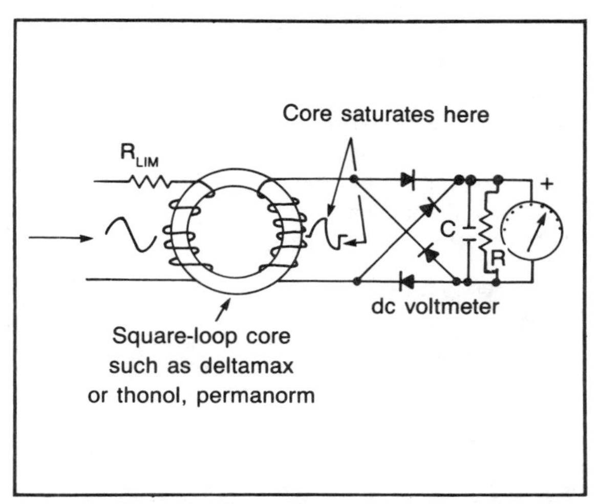

Fig. 5-16. Simple frequency meter circuit using a saturating transformer.

tion is the desired feature. The volt-seconds involved in saturation are independent of frequency and amplitude, but the number of equal-area pulses developed in the secondary increases in direct proportion to frequency. The rectified and filtered dc voltage is, therefore, a linear function of frequency and a dc voltmeter may be used to provide the frequency readout. The scheme is passive and is not periodic in the sense that no tuned circuits are involved. Because the hysteresis loop of the core does not vary in any way, the overall precision is probably as good as can be obtained from the meter itself. A digital meter could provide an additional refinement.

A possible disadvantage for some test situations is the relatively low input impedance with easily constructed windings (by hand). This can be partially offset by using a high-value primary limiting resistance, provided that considerable amplitude is available from the frequency source. Of course, a transistor stage of amplification ahead of the primary winding can be used to achieve higher input impedance. Load resistor R is not absolutely necessary if filter capacitor C is not too large; these matters, in turn, will vary with the type of meter used. Capacitor C is necessary to preserve linearity and steadiness of the meter response down to very low frequencies.

DIRECT OBSERVATION OF PHASE DISPLACEMENT ANGLE

The observation of phase as the actual displacement between two waves is a useful and instructive test procedure. Such a display may be produced with either a dual-trace oscilloscope or with a conventional oscilloscope in conjunction with an electronic switch. Electronic switches are made specifically for such purposes as this and are becoming a common test item. The major manufacturers of oscilloscopes often supply electronic switches as auxiliary items, and in some instances the electronic switch is designed as an optional plug-in module. An electronic switch enables the ordinary single-trace oscilloscope to duplicate display functions which would otherwise require the use of the relatively expensive dual-trace oscilloscope.

The setups for monitoring phase angle with both single- and dual-trace scopes are shown in Fig. 5-17. The switch, SW, provides a self-check for the test procedures. When this switch is in the No. 1 position, the oscilloscope should be capable of displaying a single wave because the two displayed waves come from signal source No. 1 and can overlay one another. This does not mean that the two waves cannot be vertically separated by adjustments of the vertical amplitude or position controls. However, there should be no horizontal displacement between like zero crossings of the two waves, since this would indicate a differential phase shift somewhere in the setup. In the event that the two waves are not exactly in phase, an appropriate correction will have to be applied when switch SW is placed in the No. 2 position. If the necessary correction is more than a few degrees, it is best to determine the reason for it and correct it. With the dual-trace oscilloscope, the trouble would most likely be due to some defect in one of the vertical amplifiers. With the single-trace scope, the trouble is more likely to be found in the electronic switch.

Suppose that one full cycle is adjusted to occupy six major horizontal divisions of the graticule. This corresponds to 360 degrees divided by 6 or 60 degrees per division. If each major division is divided into five minor divisions, then each minor

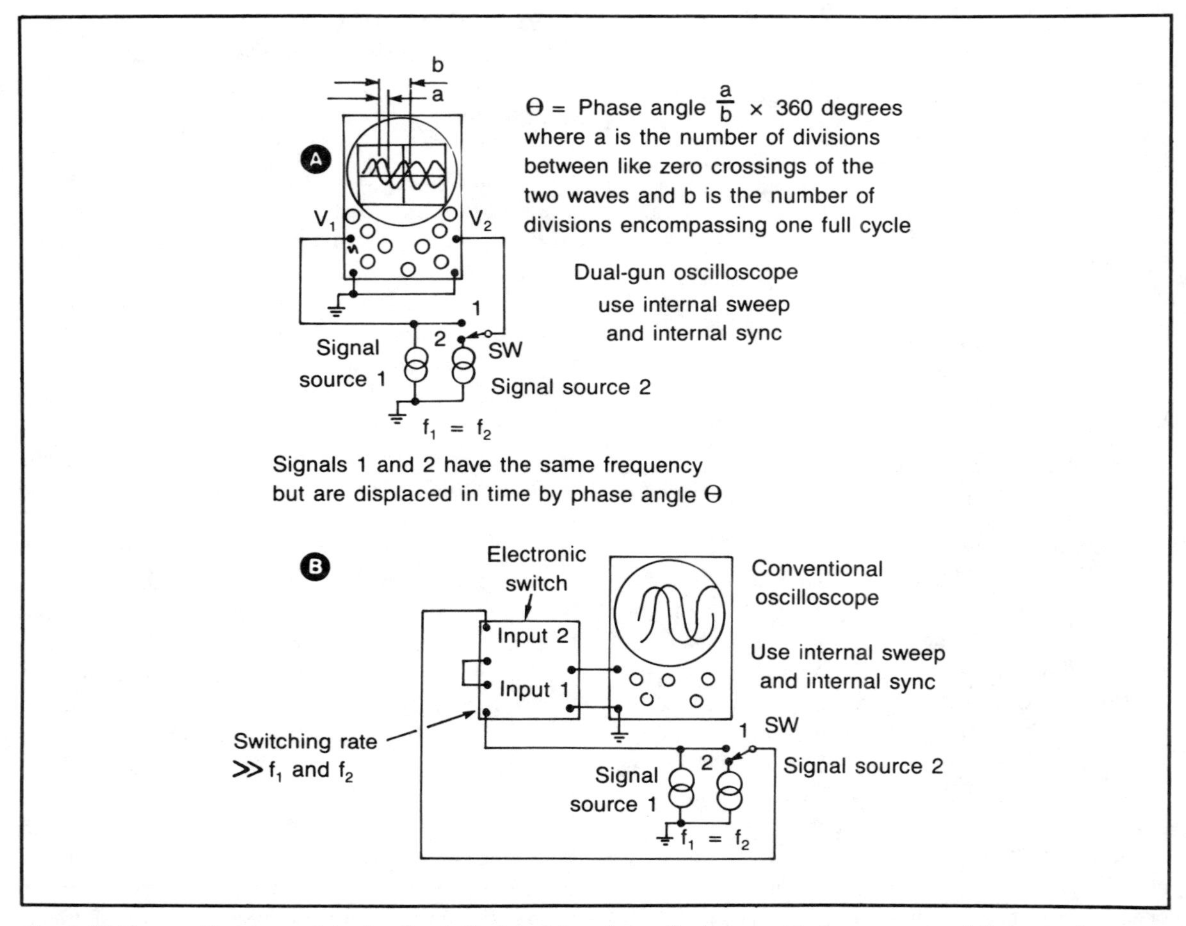

Fig. 5-17. Two oscilloscope setups for direct display of the phase displacement between waves of the same frequency.

division represents 12 degrees of electrical time. If the horizontal displacement between like zero crossings of two displayed waves is one major division and one minor division, then, Θ, the phase angle of one wave with respect to the other is 60 degrees + 12 degrees, or 72 degrees. Since the sweep proceeds from left to right, the measurement can be refined by assigning a plus or minus sign to the phase angle. Thus, the wave which lags in its time of occurrence is so many minus degrees with respect to the wave which has already attained a specific point on its cycle. For convenience, the "specific point" is generally taken to be a zero crossing. Notice that a wave has an "up-headed" and a "down-headed" zero crossing. The phase between two waves pertains to like zero crossings. Single-trace scopes in which the electronic switch is an integral part of the instrument or in which the switching function is provided by a plug-in module usually produce the dual-trace pattern by either of two switching modes selectable from the panel. In the "chopped" mode, tiny segments of the two waves are alternately displayed on the screen at rates between 100 kHz and 1 MHz in most instruments. If the switching rate is much greater than the frequencies of the chopped waves, the eye perceives the waves as continuous line displays. The other switching mode is usually designated as "alternate" switching, but here the implication is that the two complete waves can be alternately swept across the screen. Again the eye sees steady wave patterns on the screen. Notice that the terms "chopped" and "alternate" could, strictly speaking, be construed to describe both modes, but it is a matter of how rapidly wave information from the two sources is selected for presentation on the screen. Both modes are useful for direct observation of phase angle. Under certain conditions, various interference problems exist between the two waves and it is a good idea to experiment with the two modes.

Determining Phase Angle From Lissajous Figures

In the simple one-to-one Lissajous figure, there is a wealth of phase information in a generalized elliptical pattern. Figure 5-18 shows the setup and the measurement technique involved. Although two frequency sources are depicted, in most practical situations the two generators represent two different circuit junctions, both of which are energized from a single source. It is assumed that a difference of phase exists in the voltages monitored at the two junctions, or that you want to determine whether such a phase difference exists. Thus, the two generators are symbolic of two ac voltage sources which can actually originate from a single

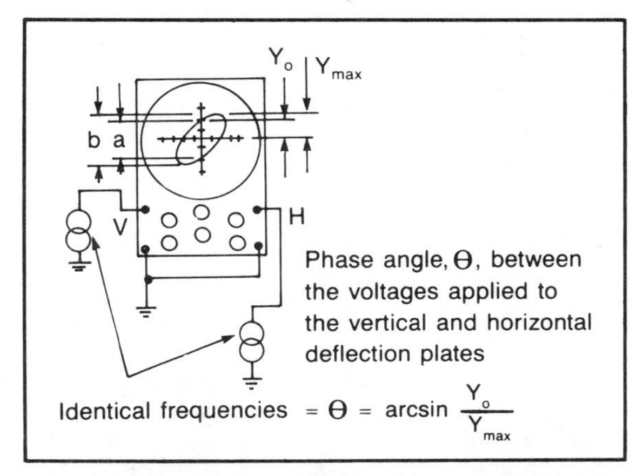

Fig. 5-18. Phase measurement setup using a Lissajous figure. The alternative measurement technique can also be used: $\Theta = \arcsin a/b$.

generator. (On the other hand, it is well to understand that the phase angle between two physically separate generators might be of interest; the basic stipulation here being that the two frequencies must be identical.)

If the voltages impressed on the vertical and horizontal deflection plates are equal, the pattern becomes a circle when there is a 90-degree phase difference between the two waves. The other geometric extreme (straight-line patterns) occurs when the relative phase angle between the two waves is zero or 180 degrees. The validity of the trigonometric conversion of measurement data to phase angle does not depend on equal vertical and horizontal deflection voltages. However, the actual measurements may not be easy to make if the elliptical patterns are too flat. Therefore, the ratio between vertical and horizontal signal amplitudes (from the viewpoint of the deflection plates in the cathode-ray tube) should not be too great nor too small. When the conditions are right to produce a perfect circle, it will also be found that, provided the amplitudes of the two waves (at the deflection plates) are maintained constant, the straight lines produced at zero or 180-degrees phase displacement will be inclined at an angle of 45 degrees.

The general procedure is to first make sure that the vertical and horizontal channels of the oscilloscope do not produce an appreciable amount of differential phase shift. This is done by connecting the vertical and horizontal input terminals together and driving the two from a source providing about the same frequency at which the phase measurement is to be made. The pattern should be a straight line with an angular inclination determined by the amount of gains developed by the vertical and horizontal amplifiers. If an ellipse is displayed, the scope will be in error when making phase determinations because of an internal difference in phase shift between the two amplifier channels. This internal phase "offset" can be measured (see the test procedure for internal phase differential) and the appropriate correction can be applied to tests on external circuits. It is also possible to develop some method of equalizing the phase shifts in the two amplifiers.

When making the pattern measurements as illustrated in Fig. 5-18, either the Y_0, Y_{max}, or the b, a, dimensions may be recorded. In either case, the ellipse should be studied carefully in order to center it as precisely as possible. Compute the appropriate ratio to three or four decimal places and look up the result in the sine-function column of a trigonometric table. The angle corresponding to this value is Θ, the phase angle between the two waves.

Useful phase information is shown in Fig. 5-19. You can quickly estimate phase situations by making use of the illustrated patterns. Phase angles equally displaced from 90 degrees, such as 60 and 120 degrees, have numerically equal sine functions. However, the inclinations of the phase patterns quickly reveal the different phase situations. Confusion with regard to which wave leads or lags the other can often be resolved by carefully looking at the patterns. The column of patterns containing the perfect circle shows the various Lissajous figures which result from having equal deflection voltages applied to the X-Y plates of the oscilloscope. The two columns at the left depict the modifications of the circular and straight-line patterns when the deflection voltages are not equal. In the left column of patterns, the X deflection voltage exceeds the Y deflection voltage. In the next column of patterns, the opposite situation exists. Intermediate ellipses are not shown for unequal X-Y deflection because their phase-angle indication is not especially easy to determine by looking at the display. However, the mathematical computation of the phase angle as indicated in Fig. 5-18, is still valid.

Test Procedure for Investigating Oscilloscope Phase Shift

Objective of Test: To determine the degree of phase shift between the vertical and horizontal channels of an oscilloscope.

Equipment Required:

Oscilloscope

Source of variable-frequency sine waves

Test Procedure: Connect the equipment as shown in Fig. 5-20. Disable the scope internal sweep by appropriately setting the panel function switch. The basic idea is to produce a diagonal line on the screen of the oscilloscope. Adjust the X-Y positioning controls so that the trace line crosses the X-Y graticule lines at the center. If the scope does not have a graticule, the results will not be as

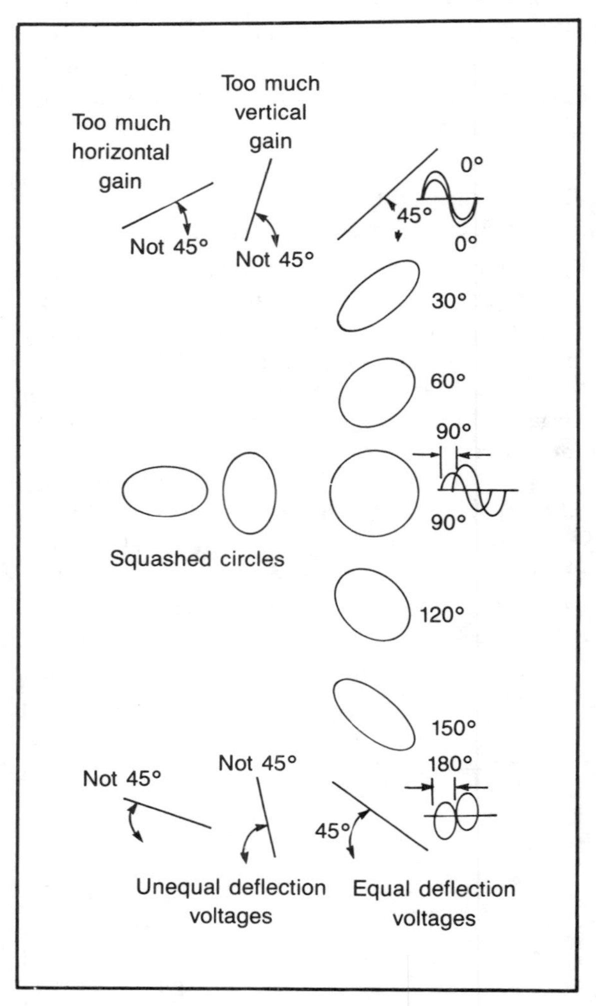

Fig. 5-19. Quick-look phase patterns. Amplitude variations change the inclination of the ellipses (and lines). Phase variations change the eccentricity of the ellipses and open the lines when the phase difference deviates from 0 or 180 degrees.

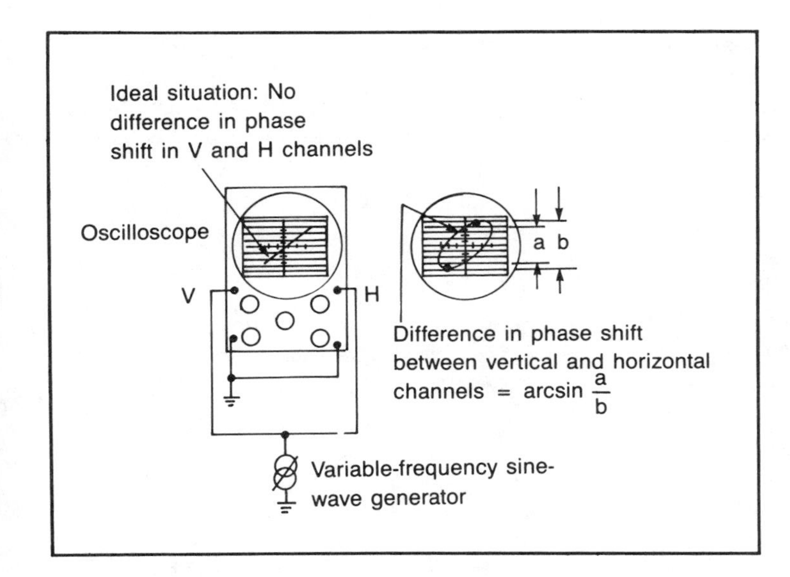

Fig. 5-20. Method of determining the internal differential phase shift in the vertical and horizontal amplifiers in an oscilloscope.

precise but will still yield useful information. In any event, the ends of the trace line should not be too close to the edges of the screen.

Initial adjustments should be made at a low frequency in the vicinity of 100 Hz. Increase the frequency while observing the inclined line trace. Depending upon the oscilloscope and its operating condition, the inclined line may ultimately open into an ellipse. This may occur at several to several tens of kHz in some cases and at tens of megahertz in others. In some scopes, the ellipse may open considerably, whereas in others only a slight departure from the inclined line will be seen. Whatever the situation, record the results as a function of frequency. With this information you can calculate the phase difference as shown in Fig. 5-18. When so doing, the pattern should be made as large as the ability to read the graticule markings will allow. If the oscilloscope does not respond to dc, the above test procedure should be extended to low frequencies. Considerable phase shift between the two channels often occurs in the 10- to 100-Hz region.

Comments on Test Results: Even a high-quality oscilloscope with identical vertical and horizontal amplifiers can have, or develop differential phase shift. The existence of such phase shift directly affects the accuracy of phase measurements. Therefore, the above described test procedure should be applied to an oscilloscope before it is used to make phase measurements.

There are two ways to deal with the problem. First, determine the reason for the differential phase shift, especially where the shift occurs at relatively low frequencies. Possible causes are different amplifier circuits for vertical and horizontal channels, defective components such as leaky coupling capacitors, changed value of resistors, or defective tubes. The problem is likely to occur where transistors have been replaced. Because of the wide tolerance in the characteristics of transistors of a given type, a differential phase shift between two identical amplifier channels can exist even though both channels have a constant amplitude vs frequency response. If at all feasible, an oscilloscope should be serviced to produce a minimum phase differential from very low to very high frequencies. In scopes with dc capability, there should be no problem at low frequencies. However, in ac oscilloscopes, phase difference will be found at low as well as high frequencies. If this region is near the low-frequency cutoff of the amplifiers, and the vertical and horizontal amplifiers are not identical, not much can be done without redesign or by some modification such as altering the size of coupling capacitors. Larger coupling capacitors can extend the region of low-frequency phase shift to lower frequencies, often a worthwhile accomplishment.

The second way to deal with differential phase shift is to accept it and make the necessary corrections when using the oscilloscope for phase measurements. In this case, precise data must be available and this calls for a graticule on the scope screen. For example, by consulting such data, you should know whether to add X degrees for 400-kHz measurements, subtract so many degrees at 10 Hz, or whether phase measurements can be made over a wide frequency band with no corrections.

Although the phase characteristics of the oscilloscope must be known to ensure the accuracy of phase measurements, it is still possible to accurately determine frequency ratios from Lissajous patterns regardless of phase differences in the oscilloscope. When the frequency ratio is very great, or is a large fraction, a small phase differential near the measurement frequency could conceivably make the loops of the pattern more discernible, but in most practical situations this is at best insignificant.

If, by adjustment or repair, a substantial gain is made in minimizing differential phase shift, the overall performance of the oscilloscope will often be improved. For example, it may reproduce square waves more faithfully. This is particularly true if a defect is found in the vertical amplifier. The correction of differential phase shift also is likely to improve linearity in one or both of the amplifiers, although this does not necessarily follow.

THE COLOR TV VECTORGRAM

A powerful color TV receiver servicing technique is the ability to evaluate signal phase conditions at the color tube. Although it is not our intent to deal with specific steps involved in the diagnosis and repair of equipment, some aspects of the subject under discussion are particularly important because of the way they relate to phase and Lissajous figures. Indeed, the vectorgram, which is a Lissajous figure, can reveal some worthwhile information.

Figure 5-21 shows the pattern obtained under ideal conditions when a color TV receiver is driven by a keyed rainbow generator and the signals at two of the three grids at the color tube are monitored by an oscilloscope. The keyed rainbow generator test signal produces a display of ten discrete vertical color bars on the TV screen. (This test signal

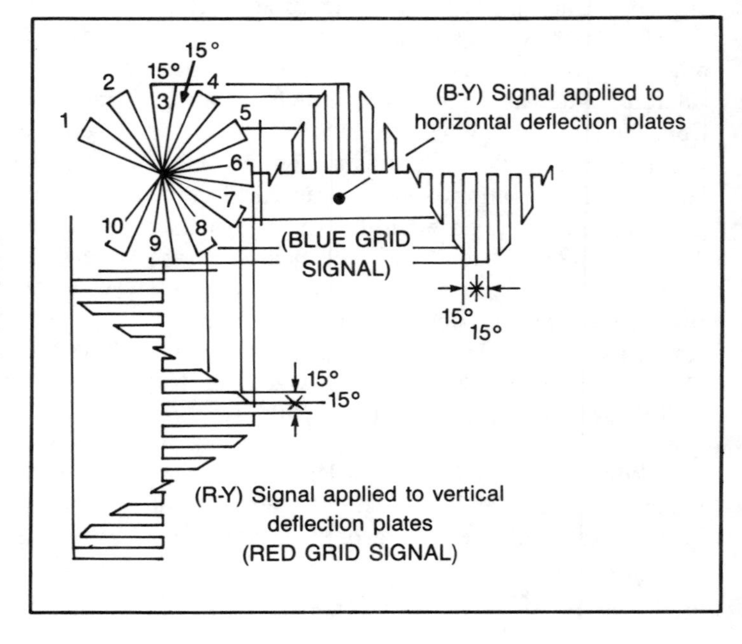

Fig. 5-21. Idealized Lissajous figure resulting from signals sampled at the red and blue grids of a color picture tube when the receiver is driven by a keyed rainbow generator. Two "petals" are missing because of blanking interval.

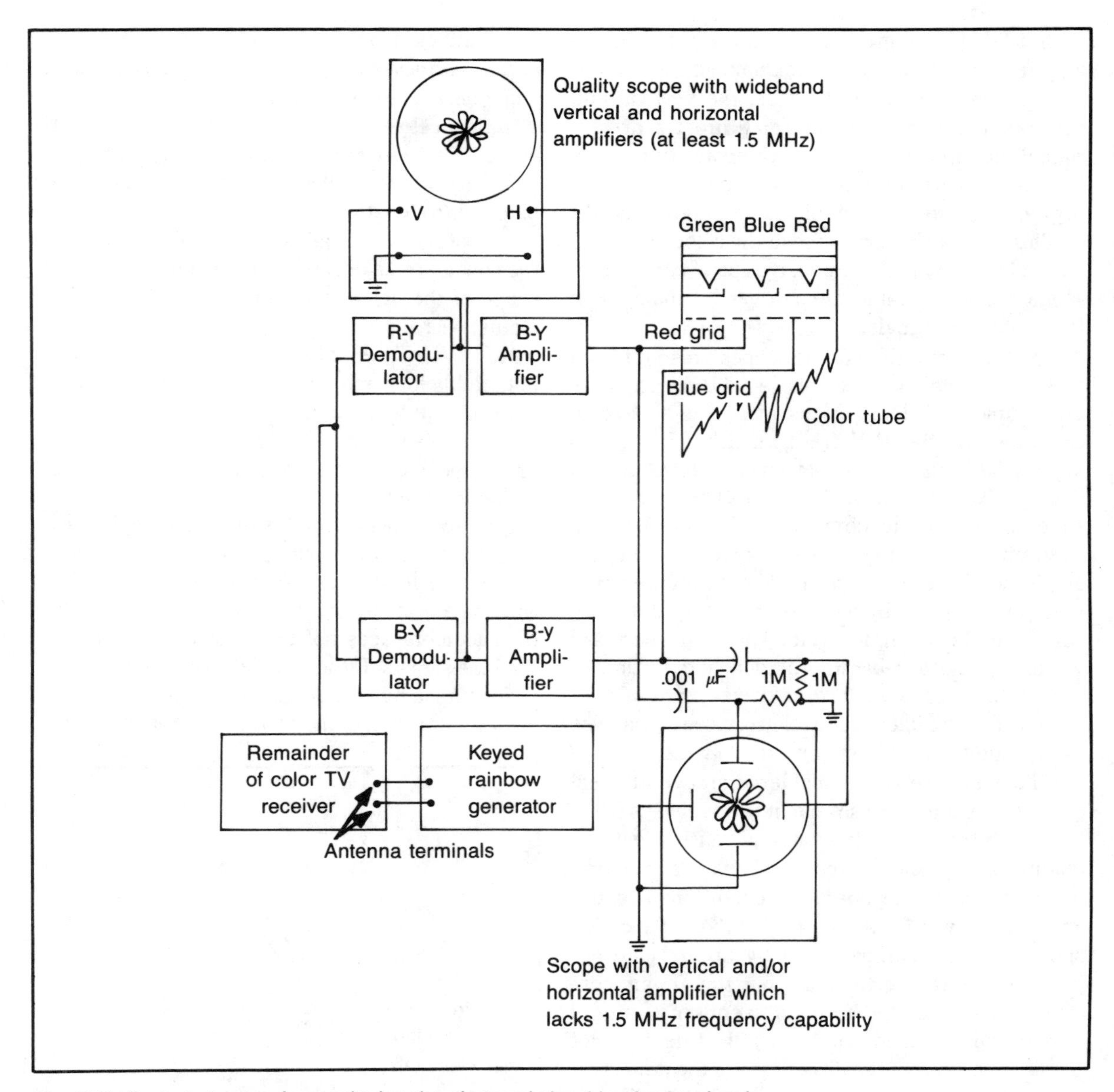

Fig. 5-22. Basic test setup for monitoring the phase relationship of color signals.

is actually in the form of modulation applied to one of the VHF channels so that the signal can be processed by the receiver just as would be a signal from a color TV station.) When the ten color bars do not appear in their normal hue, intensity, or sequence, the next logical step in locating the cause of the trouble is to study an oscilloscopic display of the signals producing the color bars. When such a display is in the form of a Lissajous figure (commonly called a vectorgram in TV service parlance) the vital aspects of the electrical signals responsible for the rendition of color are revealed.

Figure 5-22 shows how to connect a scope for a color vectorgram display. Either a lab-type scope with 1.5-MHz or greater vertical and horizontal amplifier response or an inexpensive instrument can be used. In the latter case, the vertical and horizontal deflection plates are directly driven. Signal pickup then occurs at the color-tube grids because this is where the color signals are at their highest amplitude level. The high quality oscilloscope can be connected at the color tube grids also. However, the inexpensive scope cannot present a display of usable size if driven by unamplified R-Y and B-Y signals.

Notice that although the scopes present Lissajous figures, no 90-degree phase-splitting networks are employed. This is because the sine-wave envelopes of the R-Y (red) and the B-Y (blue) signals are already displaced in phase by 90 degrees. This is depicted in Fig. 5-21 and is responsible for the basic circular outline of the vectorgram. In fact, if these signals were not keyed, and if the TV receiver, itself, did not obscure a segment of the figure by its blanking interval, the result would be a simple circle. The actual serrated waves convey color-bar information, and the flower-like pattern of the vectorgram reveals a wealth of clues which the knowledgeable serviceman can use as a guide to trouble spots.

Thus far, we have considered an overall ideal situation. Although desirable, it is not always feasible to obtain a circular pattern. With the high-quality scope, a circular pattern can always be produced under proper operating conditions because any inequality in the amplitudes of the red and blue signals can be compensated by appropriate adjustments of the vertical and horizontal gain controls. However, when an economy model oscilloscope is used, such an adjustment is not possible because of the direct feed to the deflection plates. Instead of a circular pattern, you will then have to work with an ellipse, such as shown in Fig. 5-23, or in some cases, the horizontally constricted counterpart of this pattern. The ellipse still contains useful information, however. In particular, notice that unequal amplitudes of the red and blue signals cannot produce inclined ellipses, but only ellipses with vertical and horizontal major and minor axes. You can quickly identify the inclined ellipse with the high quality scope when you are in doubt because no adjustment of vertical or horizontal gain will convert such an ellipse into a circle. The significance of the inclined ellipse is that the demodulator circuits are not producing the required 90-degree phase displacement between the red and blue color signals.

Another non-ideal aspect of the vectorgram is the considerable difference between the geometric shape of the petals shown in Figs. 5-21 and 5-23. Actual vectorgrams can assume some rather weird patterns. However, the basic phase information is not readily lost, and the very shapes of the "petals" are often quite meaningful. Notice the numbering of the petals in Fig. 5-21. If we use the hue control of the receiver to position petal 3 vertically, petal 6 should then be horizontal. If this is not the case, the pattern must be that of an inclined ellipse even if this is not readily evident from an inspection of the overall vectorgram. As previously pointed out, this means that the required 90-degree phase displacement does not exist between the red and the blue signals. Either the demodulators are defective or misadjusted or the subcarrier voltages injected into the demodulators are not correctly

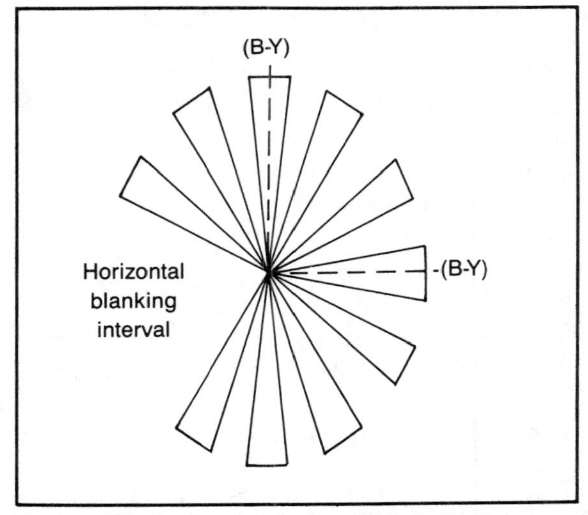

Fig. 5-23. Idealized non-circular pattern resulting from unequal vertical and horizontal deflection voltages applied to a test scope. In this case, the basic shape is that of a vertical ellipse. Where gain controls are available, such a pattern can be circularized.

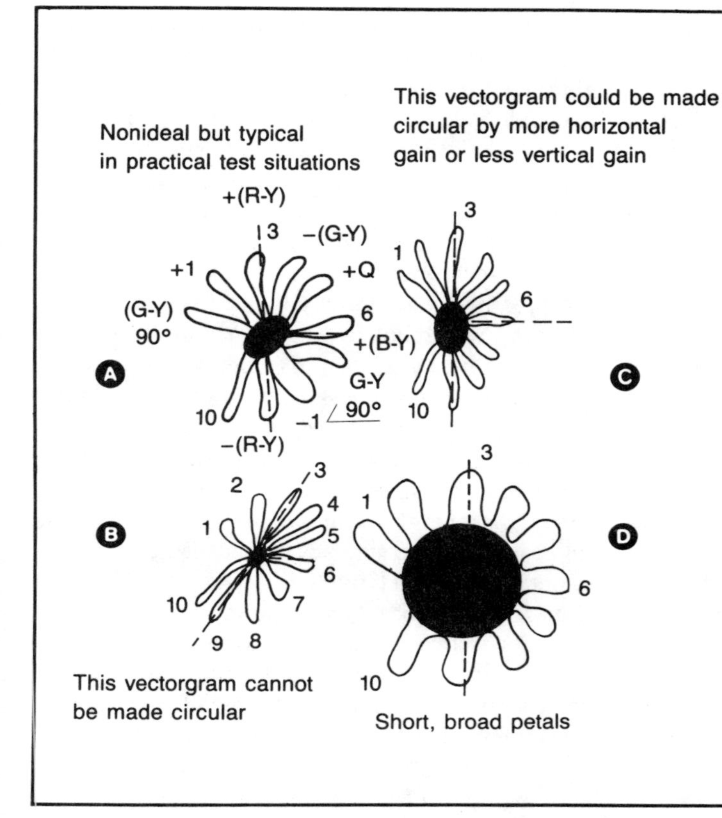

Fig. 5-24. Typical vectorgrams. A—Phase relationships and other aspects OK. B—Elliptical pattern; basically OK. C—Inclined ellipse; phase error in demodulation. D—Phase relationships OK; malfunction in the bandpass amplifier.

phased. If the third petal cannot be made to stand up vertically, adjustment of the color burst circuit is probably needed. A compressed vectorgram with short and usually broad petals is generally indicative of trouble in the bandpass amplifier. Vibrating petals indicate contamination by interference, often a 4.5-MHz signal from a mistuned trap. Figure 5-24 shows several vectorgrams and what they indicate.

Although not always necessary, it is usually desirable to use low-capacitance high-impedance probes with an oscilloscope. In any event, it should always be ascertained that there is negligible differential phase shift between the vertical and horizontal scope circuits. Oscilloscopes manufactured specifically for this type of test are called vectorscopes. These are equipped with phase-angle calibrated graticules and have many other features particularly useful for the display of vectorgrams.

QUESTIONS

5-1. An oscillator is constructed using a tank circuit which had been parallel-resonated from a square-wave frequency source. The oscillator appears to operate well, but when its sine-wave output is checked with a digital counter, the frequency is found to be about three times that intended. What might have gone wrong?

5-2. A frequency divider, driven by a crystal oscillator, produces a stable Lissajous type pattern on an oscilloscope. The dividing factor is clearly observed to be 1:2. However, a digital counter behaves erratically when an attempt is made to use it to confirm the information derived from the scope. The digital counter appears to be in proper operating condition as determined from self-checks and from other tests. What might be causing the contradictory results with the two test instruments?

5-3. A frequency source is to be adjusted as closely as possible to the frequency of a very stable crystal oscillator by observing a 1:1 Lissajous figure on an oscilloscope. However, the precision oscilloscope is not available because it has not been returned at the promised time by the calibration laboratory and only an inexpensive scope with a notoriously unstable horizontal sweep circuit is available. Suggest what might be done, assuming there is not time to wait for the better instrument.

5-4. If the vertical and horizontal channels of an oscilloscope do not each have a net phase shift of zero degrees, could we accurately measure phase shift using a basic Lissajous pattern such as an ellipse?

5-5. A digital counter is self-tested against its own time base and divider chain. On every range, all digits show correct readouts. When this instrument is used to check a crystal oscillator which has been in service a long time, small but undesirable departure from the desired frequency exists. Should immediate attempts be made to "pull" the oscillator back to its original frequency, or should something more basic be done first?

5-6. Two sine waves of the same frequency but 90 degrees displaced in phase are fed to the vertical and horizontal input terminals of an oscilloscope. Amplitude and gain controls are adjusted to produce a circular pattern needed for a data display technique. Elsewhere in the system are two triangular waves which are derived from these two sine waves. The argument is presented that, since the two triangular waves will follow the respective sine waves in the zero-crossing times, and since the use of triangular waves would affect the X and Y axis of the scope in the same manner, a circular display could also be produced with the triangular waves. Is this a correct deduction?

5-7. An unwanted signal in a system has been identified as the cause of disturbances. It is suspected that this frequency is a harmonic of the 60-Hz power line. At a certain circuit junction, the interfering frequency has a peak-to-peak amplitude of several volts. How can the suspicion be quickly checked for validity if the only test instrument available is a simple oscilloscope?

5-8. During the course of test procedures on a system, it is decided to make a single amplifier stage responsive to frequency, f, but not to frequencies appreciably above and below f. This is attempted by the seemingly logical use of a twin-T network, which has been designed and proven to reject frequency, f. It is reasoned that if this network is connected between the output and input terminals of the amplifier, progressively higher negative feedback will occur as the departure from frequency f becomes greater. Therefore, according to this reasoning, the amplifier should provide the highest gain at frequency f, because at this frequency there is relatively little signal voltage fed back from the output to the input of the amplifier. At other frequencies, the opposite is the case and the amplifier should provide drastically reduced gain. Upon trying out this scheme, it is found that the amplifier has become an oscillator at frequency, f. What was lacking in the test approach and what can be done to save the situation?

5-9. In a phase-controlled SCR circuit, the gate pulses are essentially of rectangular shape and the anode-cathode voltage is sinusoidal. We want to make a test which shows the relationship of the phase difference between gate and anode voltages to output current. Can a dual-trace oscilloscope be used to display meaningful phase information when the waves are of different shapes?

5-10. A full-wave 500-volt dc power supply appears to be basically operative, but it is no longer a satisfactory source for the low-power transmitter it powers. Some of the defects evident are poor regulation and reduced output, inordinately high temperature rise in the power transformer, and high ripple. A sample of the output viewed on an oscilloscope shows 60 Hz to be the predominant ripple frequency. The supply, operated from the 60-Hz power line, utilizes a silicon bridge rectifier and a single electrolytic filter capacitor. Can the filter capacitor be defective, or would it be wise to make other tests before diagnosing the trouble?

ANSWERS

5-1. When the LC circuit was measured, apparently the resonance observed corresponded to the third harmonic of the square wave's basic repetition rate. This mistake could have been avoided by using a sine wave instead of a square wave, or by carefully ascertaining that the resonant response observed with the square-wave source could not be duplicated with even greater reaction on the scope or voltmeter at a higher square-wave repetition rate. Thus, if an LC circuit is being resonated at 1000 Hz, there is the possibililty that only sufficient capacitance will be added to the inductance to resonate at 3000 Hz, the third harmonic of a 1000 Hz square wave. But if the dial of the square-wave source is quickly swept through the 3000-Hz region, the error can be easily detected. In the instance cited, since the resonance must be established at one third of the wrongly chosen resonance, a capacitor of about nine times that initially used is needed in the LC circuit. Similar mistakes are also possible at the fifth and seventh harmonics, although these are less likely to occur. Another safeguard is to have in mind a general value of the capacitor needed to resonate the inductor at the desired frequency.

5-2. The waveforms of frequency dividers are often considerably contaminated with switching transients. These can be caused by actual feed-through from the oscillator or driving source, or by shock excited oscillation originating in stray circuit parameters. Such transients can disturb the operation of digital counters. The remedy could be an RC low-pass filter, or just a capacitor connected across the input of the digital counter. In such cases, a reasonable amount of attenuation of the transients rather than their complete removal is usually adequate.

5-3. When observing Lissajous figures, the horizontal sweep is replaced by one of the two signals being compared (usually the unknown source). Moreover, since the scope is being used only to show the frequency relationship between two external signals by means of a stationary pattern, such ordinary defects as poor calibration, nonlinearity, and even poor frequency response need not detract from the potential accuracy which can be achieved. The simple answer here is to use the inexpensive scope and with full confidence.

5-4. If the vertical and horizontal channels of the oscilloscope provide the same phase shift, no measurement error will be introduced.

5-5. The self-test of the digital counter does not necessarily mean that its internal time base is precise. Rather, it shows that the divider chain in the time base is operating properly and that the instrument is generally functional. This suffices for a majority of test procedures. However, where real precision is involved, the digital counter itself must be calibrated with respect to Station WWV or a secondary frequency standard.

5-6. Two triangular waves of the same frequency and with a 90-degree phase difference will produce an open Lissajous type pattern which will be stationary. Unfortunately, however, the pattern thereby produced (for equal deflection voltages at the X-Y plates) will be a diamond (a square rotated 90 degrees) rather than the desired circle.

5-7. The interfering signal and a sample of the 60-Hz line voltage can be connected to the vertical and horizontal input terminals of the oscilloscope in order to produce a Lissajous pattern. If the pattern is stationary, the harmonic relationship between the two sources is confirmed. If the pattern is not stationary, the interfering signal is not harmonically related to 60 Hz. However, various frequencies, other than those harmonically related, can be superimposed on the power line and can thereby be introduced into a system. Usually, a test of the kind suggested is intended to find harmonics of 60 Hz which are generated after the ac power enters the system. This is usually brought about by rectifers or other nonlinear elements.

5-8. Apparently, it was not realized that the twin-T network, in addition to its amplitude response, has a 180-degree phase-shift characteristic at frequency f. Such behavior produces positive feedback at frequency f, thereby converting the amplifier into an oscillator! However, in order to do so, the amplifier must provide enough gain to overcome the high attentuation of the twin-T network at frequency f. Herein lies the solution to the oscillation problem, for if the gain of the amplifier is gradually reduced, a point will be reached where oscillation will not take place. Then the scheme becomes workable! Notice that if the network had provided infinite attenuation at frequency f, the problem would not have existed. Practical twin-T networks are usually found to produce attentuation at rejection frequency f, in the vicinity of 45 to perhaps 60 dB, in terms of the input to output voltage ratio.

5-9. Yes, the meaning of phase displacement is not altered by the fact that the waveforms are different. The interval between the zero-crossing time of each waveform can be measured as with two sine waves. And as with sine waves, one full cycle of the rectangular wave constitutes 360 degrees. The important factor here is that both waves have the same frequency.

5-10. In a properly operating full-wave supply, the predominant ripple frequency should be twice the line frequency, or in the above case, 120 Hz. The fact that 60 Hz was found to be the predominant ripple frequency suggests that it is operating as a half-wave power supply. Such operation can be caused by a single open-circuited rectifier in the bridge.

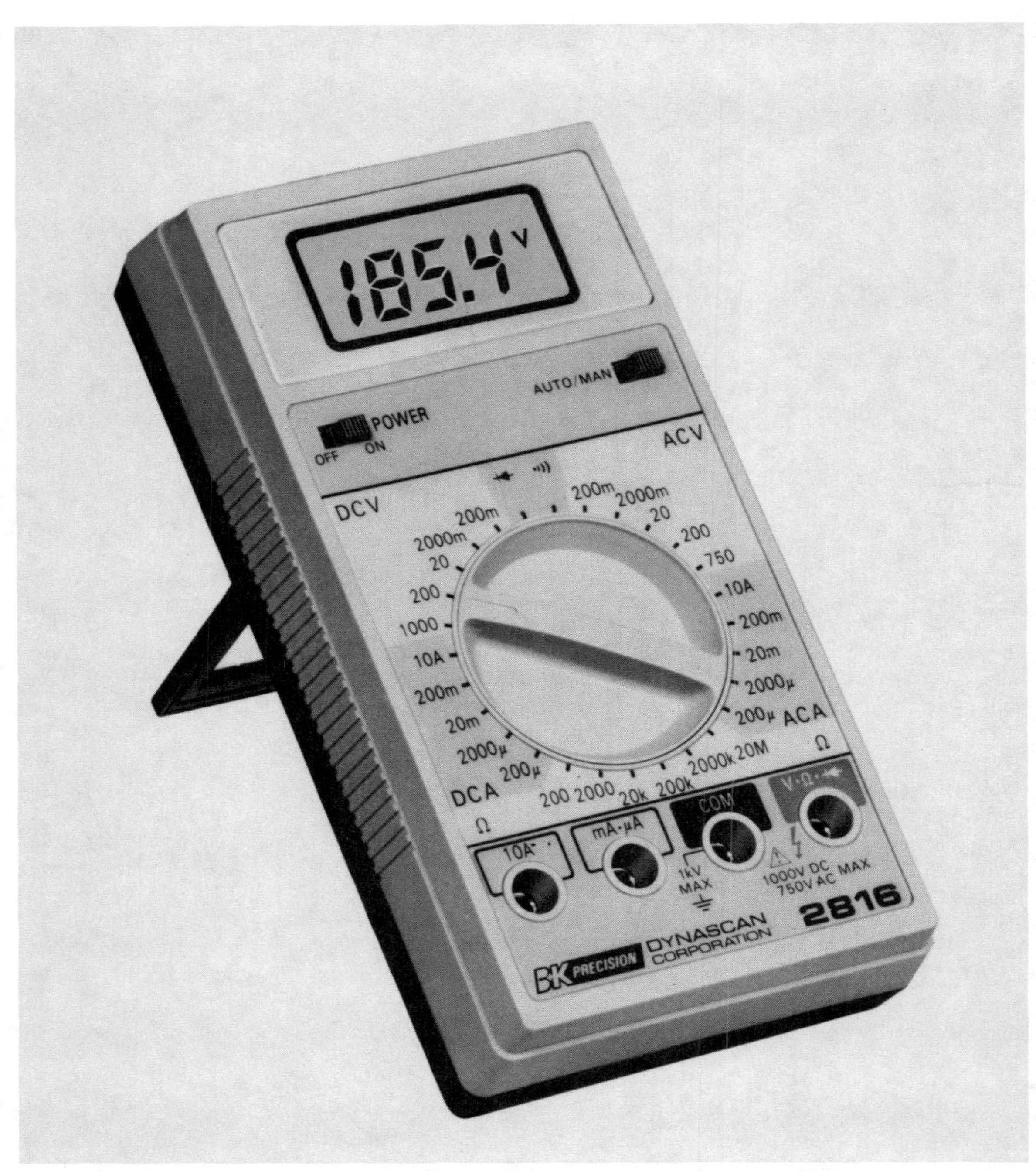

This general-purpose Digital Multimeter is easy to use, yet it provides extensive coverage in its various measurement modes. Courtesy of the Dynascan Corporation.

Chapter 6

Electron Tubes

There are basically two types of tube testers—the emission tester and the transconductance tester. Both may provide indicating meters with GOOD-FAIR-BAD segments and the operator may not be aware of the type he is using. The emission tester involves simpler circuitry and is generally the type used in the do-it-yourself version and in inexpensive service testers. It is generally satisfactory as a tester of the relative "goodness" of tubes, although it is somewhat more likely to indicate as GOOD a tube which might not perform in certain functions than is the transconductance tester.

The transconductance tester is, from the technically oriented person's vantage point, superior to the emission tester because the tubes are operated during the test in a manner simulating the basic conditions of actual circuit operation. This is particularly true because most transconductance testers are dynamic types; that is, they subject the tube to an ac signal and provide a real-time evaluation of its performance as a voltage-to-current transducer. Transconductance testers most certainly belong in the developmental laboratory, but they also provide the serviceman who wishes to invest beyond the cost of economy equipment with a superior test instrument.

The term *transconductance*, denotes the ratio of plate current change to a small grid-voltage change under the condition that plate (and other electrode) voltage is maintained constant. Transconductance is basically the important parameter of tube operation; it involves tube action of the kind generally experienced in circuit operation and is simultaneously dependent upon electron emission. If a tube tester were designed to measure the voltage amplification factor of a tube, good indications could be obtained even with nearly exhausted cathodes.

Transconductance (g_m) testers are labeled as such, or, sometimes more descriptively, they are designated as dynamic transconductance testers. "Mutual conductance" has also been used to describe testers which essentially measure g_m. However, some "conductance" testers actually measure the quantity known as *perveance* of a diode. In other words such testers are in essence emission testers. And some "dynamic conductance" testers are, again, emission testers, but with ac applied to the electrodes.

It is little short of amazing that commercial testers can accommodate even a reasonable portion of the tremendous varieties of tubes in use. To begin with, there are diodes, triodes, tetrodes, pentodes, heptodes, and various combinations of these inside one envelope. Socket arrangements include octal, Loctal™, 9-pin miniature, 7-pin miniature, 4-prong, etc. Multifunction tubes run the gamut from Nuvistor™ to compactron. There are innumberable combinations of filament and heater voltage and current requirements. Some tubes have anode or grid connections at the top. There are thyratrons, voltage regulators and electron-ray indicators, to say nothing of remote cutoff tubes, beam-power amplifiers, and gated-beam discriminators. Plate currents and voltages for these varieties of tubes range from tenfold or a hundredfold in magnitude. If it weren't for the luxury of actually seeing tube testers in existence, a logical conclusion would be that the combinations and sheer numbers of tube types in use would require a tester comparable in size to a large office building.

The function of the tube tester is to check tubes for behavior common to all or to most types. For example, most electron tubes have filaments or heater-cathode systems which emit electrons. One of the predominant reasons tubes age, grow weak, or malfunction is because electronic emission diminishes. Therefore, many types and varieties can be tested for relative emission which is based on the current that can be collected by the various elements tied together to form an overall anode. Another behavior characteristic common to most, but not all tubes, is the ability to amplify. Some tubes can be individually evaluated in terms of output current produced by a fixed input voltage. Tubes fail for reasons other than weak emission. The presence of gas, shorted or open elements, intermittents, leakage, and microphonics are typical of those often encountered. Many testers can test for all or most of these defects.

Sometimes even the most sophisticated tester will not give us any idea why a tube will not oscillate throughout the VHF band, or why another tube causes hum in the preamplifier of a hi-fi system, or why a particular VR tube generates relaxation oscillations, or why a TV horizontal output amplifier produces parasitics. It is often found that judgment based on experience is quite valuable in assessing the GOOD-FAIR-BAD indication on the tube tester. Sometimes the best test of all is the substitution test, where a questionable tube is replaced by a new or known good one.

Cathode-ray and TV picture tube testing has become a separate art. Here again, the number and types of tubes in use stagger the imagination. Not only do the testers provide an indication of the relative "goodness" of these tubes, but often incorporate provisions for "rejuvenation" of emission, the welding of open electrodes, and the clearing of shorted electrodes. Adapters are available which enable ordinary tube testers to test TV picture tubes.

EMISSION TESTER

The basic circuit of a typical emission tester is shown in Fig. 6-1, where all electrodes, other than cathode and heater of a multielement tube (a pentode in this illustration), are tied together to operate as a single anode for collecting electrons emitted by the cathode. In the case of a triode, the grid and plate are tied together. And a diode is tested as is. Notice that with the pentode shown, even the suppressor grid, an element not ordinarily polarized positive with respect to cathode, becomes part of the electron collecting structure.

The applied voltage is dc in some testers, but for reasons of economy, is often raw ac. This is feasible because the tube, allowing only unilateral conduction, functions as a half-wave rectifier while being tested and, therefore, unidirectional pulses flow through the dc current meter. The inertia of the meter movement provides the "filtering" needed for a steady average value of deflection. The current meter deflection is a relative quantity and must be determined by the designer for the ex-

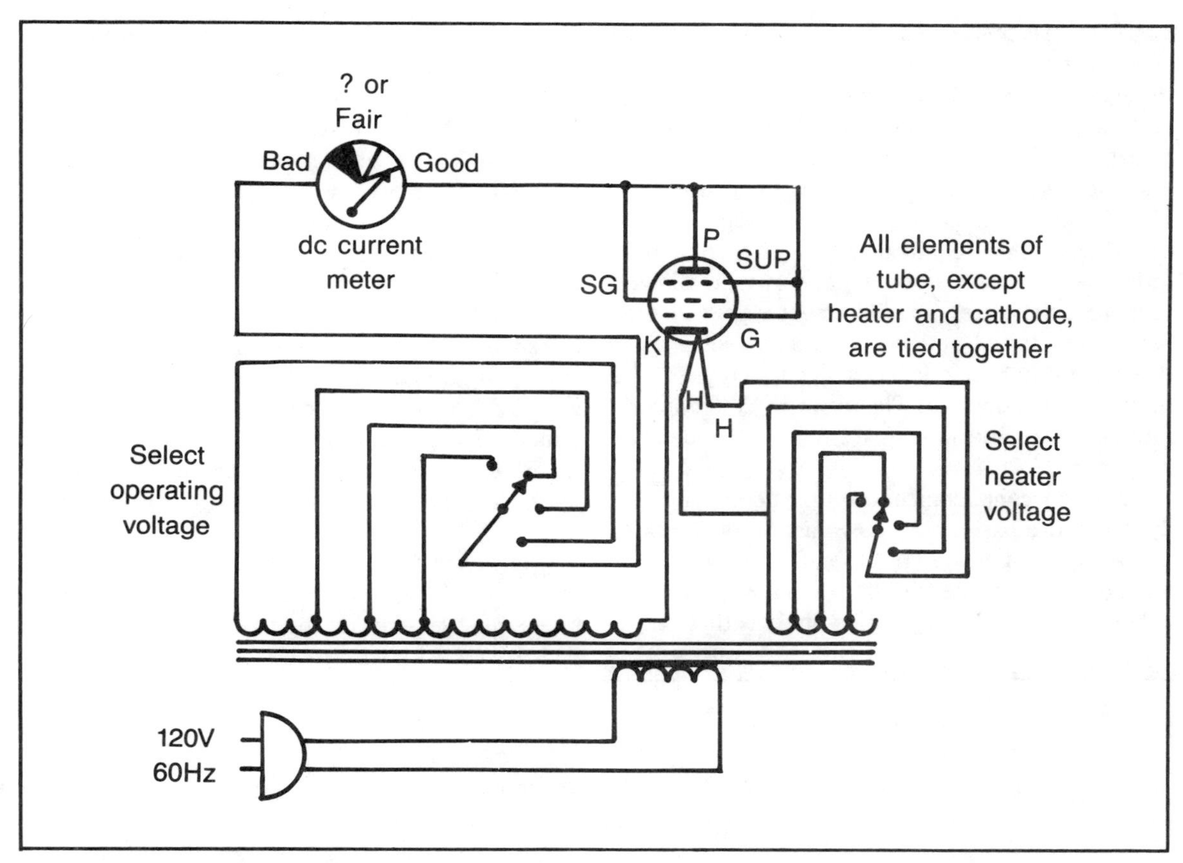

Fig. 6-1. Basic emission tester circuit.

pected emission range of new tubes in order to set the GOOD-FAIR-BAD limits. Actually, a predetermined voltage is applied according to tube type. The meter scale may have red, yellow, and green segments and/or a numbered scale from which the user refers to a chart to see what the prescribed deflection range is for a particular tube type. Small testers often have the basic socket types, such as octal, 9-pin miniature, 7-pin miniature, etc., and a number of rotary tap switches for selecting the appropriate heater and "anode" voltages. Large testers commonly have many sockets and it is necessary to consult instructions to determine which socket to plug a given tube into. Usually, the operator must set one or two knobs in order to provide the appropriate element connections and voltage for the tube under test.

DYNAMIC TRANSCONDUCTANCE TESTER

All transconductance testers, except a few specialized laboratory types, utilize the dynamic test principle to achieve transconductance readouts. The term "dynamic" simply means that an ac input signal is used for instantaneous readings. Similar to dynamic testing is static testing in which the same data is obtained by manually applying two dc input signal voltages and computing the transconductance from the two output indications thereby produced. Obviously, the static test technique is not suitable for rapid service work, and certainly not for technically untutored persons.

The basic idea involved in the operation of a dynamic transconductance tester is shown in Fig. 6-2. Transconductance is indicated by the ac plate current in response to a small ac grid voltage. In the scheme shown, a 1000-Hz source is used, but as may be guessed, 60 Hz is also used in some testers in the interest of economy. Another design variation makes use of an electrodynamometer-type current meter in the plate circuit, rather than the transformer-ac milliammeter combination shown. In that case, the dc component of plate current must either be nulled out in some manner, or, if allowed to actuate the meter, an adjustment is provided for setting the initial deflection on a reference line. Then a pushbutton is depressed which connects the input signal to the grid. The resulting deflection indicates the quality of the tube as a transconductance amplifier.

All transconductance testers, whether static or dynamic, are based on the evaluation of the ratio of plate-current change resulting from a small grid-voltage change while the plate (and screen) voltage is held constant. It may not be obvious that this is the basis of the test from the ac circuit in Fig. 6-2. However, this is indeed the case, and this fact will be borne out in the following test procedure dealing with the static test for determining transconductance. The static test technique may be viewed as a slow-motion version of the dynamic test. As such, it provides a more direct insight into the operation of a tube as a so-called transconductance amplifier.

Pentodes and other multielement tubes are tested for transconductance in the same way as the triode, but it is essential that their screen grids be supplied from a dc source with good voltage regulation. Also, the screen grid must be at ground potential in relation to the ac signal; this can be ensured by means of a large bypass capacitor from screen to cathode.

The transconductance test has no significance for diodes. Therefore, rectifiers and other diode types must be checked by the emission test. The

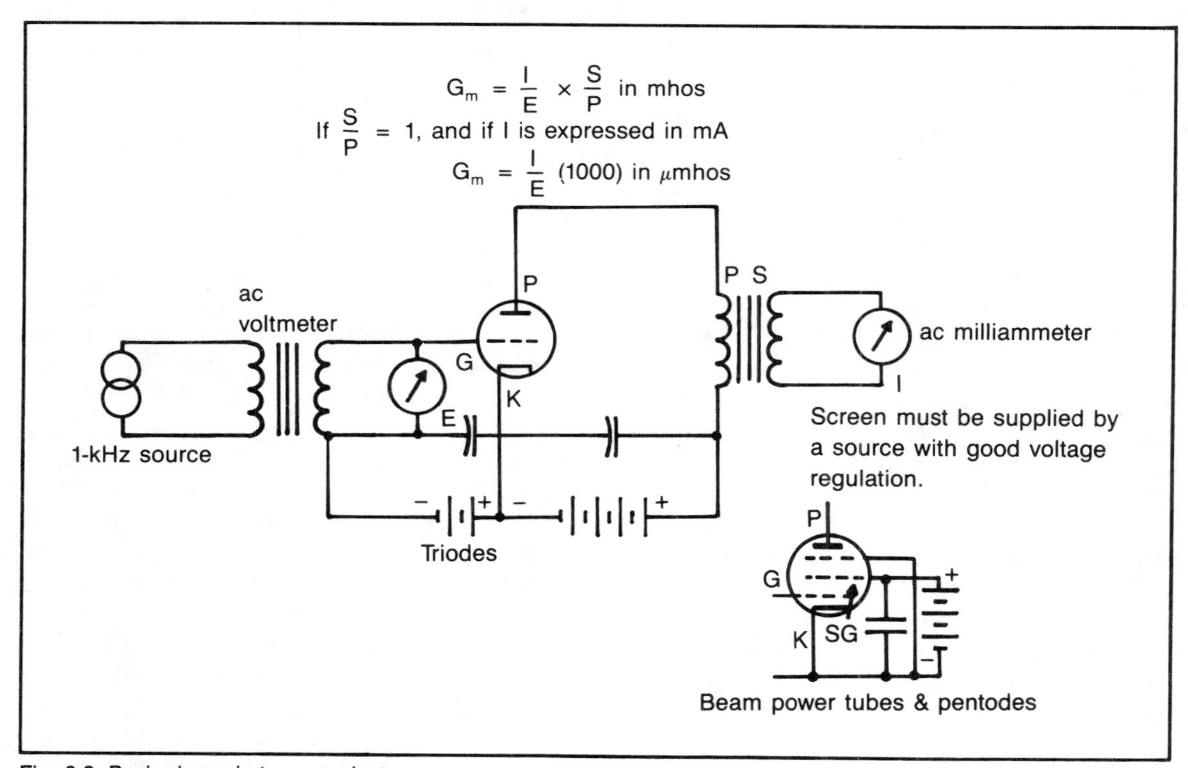

Fig. 6-2. Basic dynamic transconductance tester circuit.

means of conducting such tests is incorporated in the majority of transconductance tube testers.

"Transconductance" testers sometimes come in another form in which a fixed dc bias is applied to the grid and the quality of the tube is evaluated by the amount of dc plate current thereby resulting. If a multielement tube is being tested, the screen grid and other elements are polarized with voltages similar to those that might be encountered in a typical amplifier circuit. Notice that the cause and effect relationship does not actually abide by the mathematical description of transconductance. Yet, neither is this type of test a simple emission test, for the controlling action of the grid certainly enters the picture. Figure 6-3 shows a basic circuit which is representative of such testers. It appears somewhat as a blend of a dynamic and a static transconductance test, and yet is neither. But for practical purposes, it turns out that no harm is done by referring to these testers as transconductance types, and experience has shown that despite the simplifications and economics involved, such testers do generally simulate the test results provided by true transconductance testers.

Procedure for Determining Transconductance by Two Static Test Methods

Objective of Tests: To demonstrate simply ways to derive meaningful tube-performance data from which transconductance evaluations can be made.

Test Equipment Required:

Medium-mu triode such as 6J5, 6C4, one-half of a 12AU7, or one-half of a 6SN7

6.3 volt filament transformer

Three VOM's

0-300 volt variable dc power supply: 20 milliampere current capability is suitable.

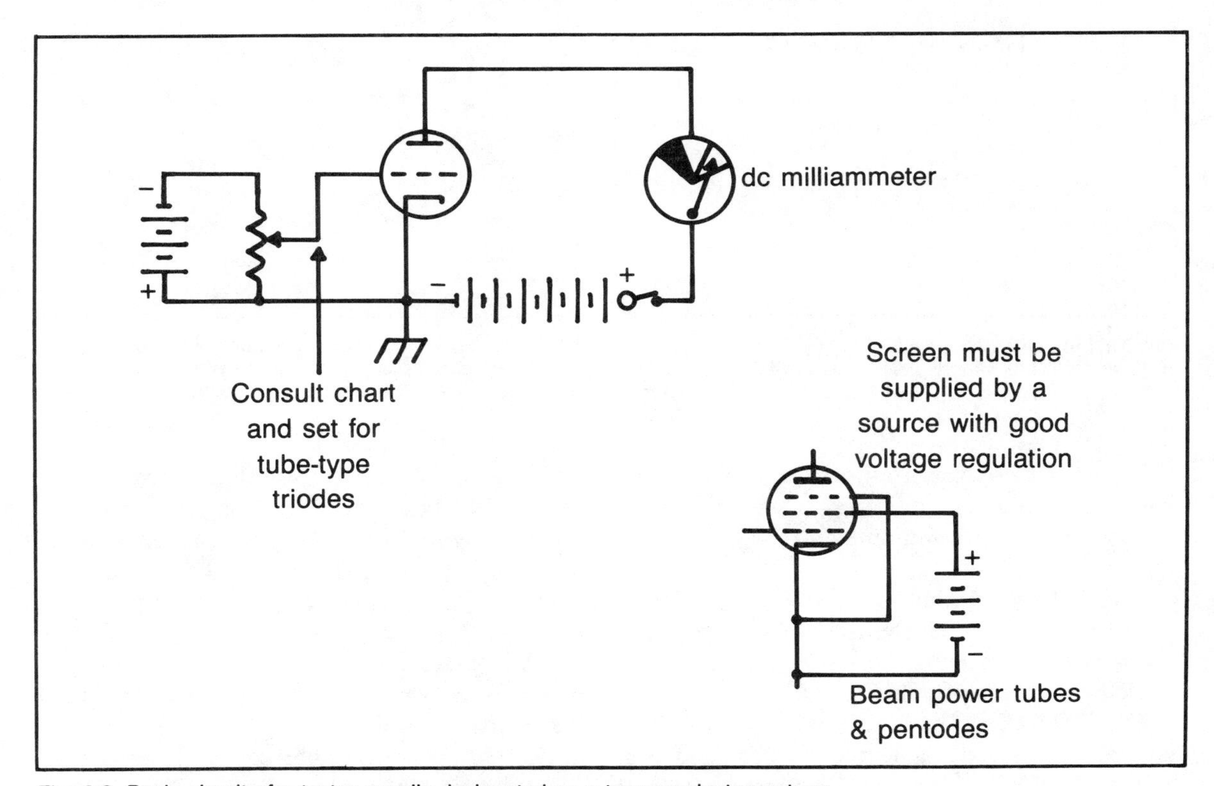

dc milliammeter
Consult chart and set for tube-type triodes
Screen must be supplied by a source with good voltage regulation
Beam power tubes & pentodes

Fig. 6-3. Basic circuit of a tester usually designated as a transconductance type.

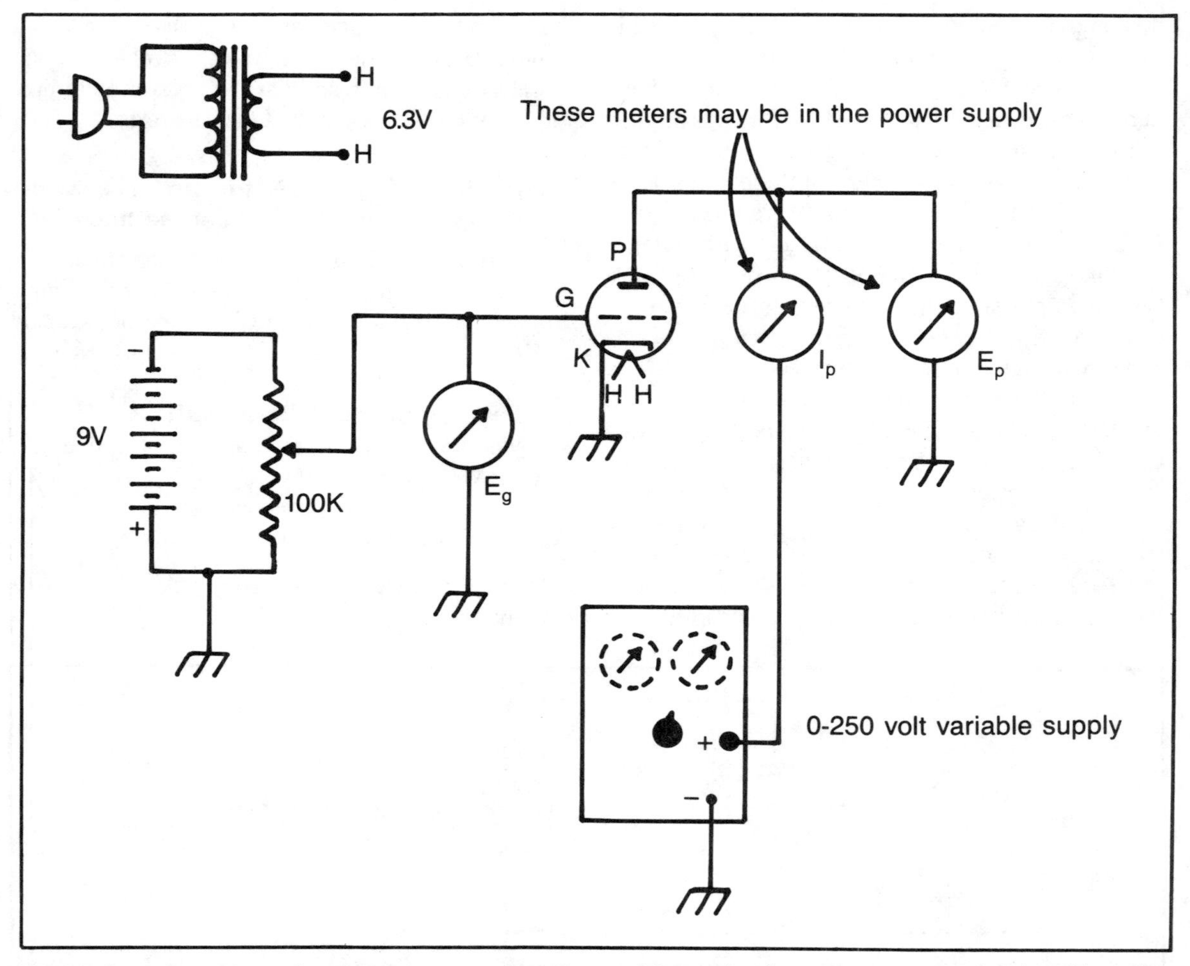

Fig. 6-4. Setup for static test for tube performance data. Two useful sets of performance curves may be plotted from data obtained from this simple arrangement. Plate current, I_p, may be recorded and plotted as a function of grid voltage, E_g, with plate voltage, E_p, maintained constant. Conversely, I_r can be measured as a function of E_p with E held constant.

Two nine-volt batteries (transistor radio type). These are connected in series.

100,000 ohm potentiometer

Test Procedure:

Connect the circuit shown in Fig. 6-4. Refer to the sample curves in Fig. 6-5AB. Note the two ways of plotting these curves. In Fig. 6-5A, the plate voltage is varied while the grid voltage is maintained constant. By selecting a number of such fixed grid-voltages, sufficient data may be recorded to plot a family of plate current vs. plate voltage curves. In Fig. 6-5B, the grid voltage is varied while the plate voltage is maintained constant. By selecting a number of such fixed plate voltages, sufficient data may be recorded to plot a family of plate-current vs. grid voltage curves.

An interesting aspect of these two families of performance curves is that they convey essentially the same information. Practical matters and graphical convenience often dictate the preference. Moreover, one is readily convertible into the other.

Of more importance, either may be used to derive the transconductance curve shown in Fig. 6-5C. Note that the single transconductance curve combines the interacting effects of grid voltage, plate current, and emission capability. A family of such transconductance curves could be plotted for a range of plate voltages. Also, it is often desirable to extend performance into the positive grid-voltage region. For sake of simplicity, this has not been done in our test setup. Some tubes, low-mu types are fairly linear well into their positive grid region.

For both, Fig. 6-5A and Fig. 6-5B, transconductance is derived from the same basic concept. Namely, one divides a small change in plate current by the small change in grid voltage responsible for it, but with plate voltage maintained constant. The smaller the changes, the more accurate is the derivation. In any event, four quantities are always involved—two grid voltages and two plate currents. If E_{g1} is the initial grid voltage, then I_{p1} is the corresponding plate current. A small change in grid voltage would give us E_{g2} and the

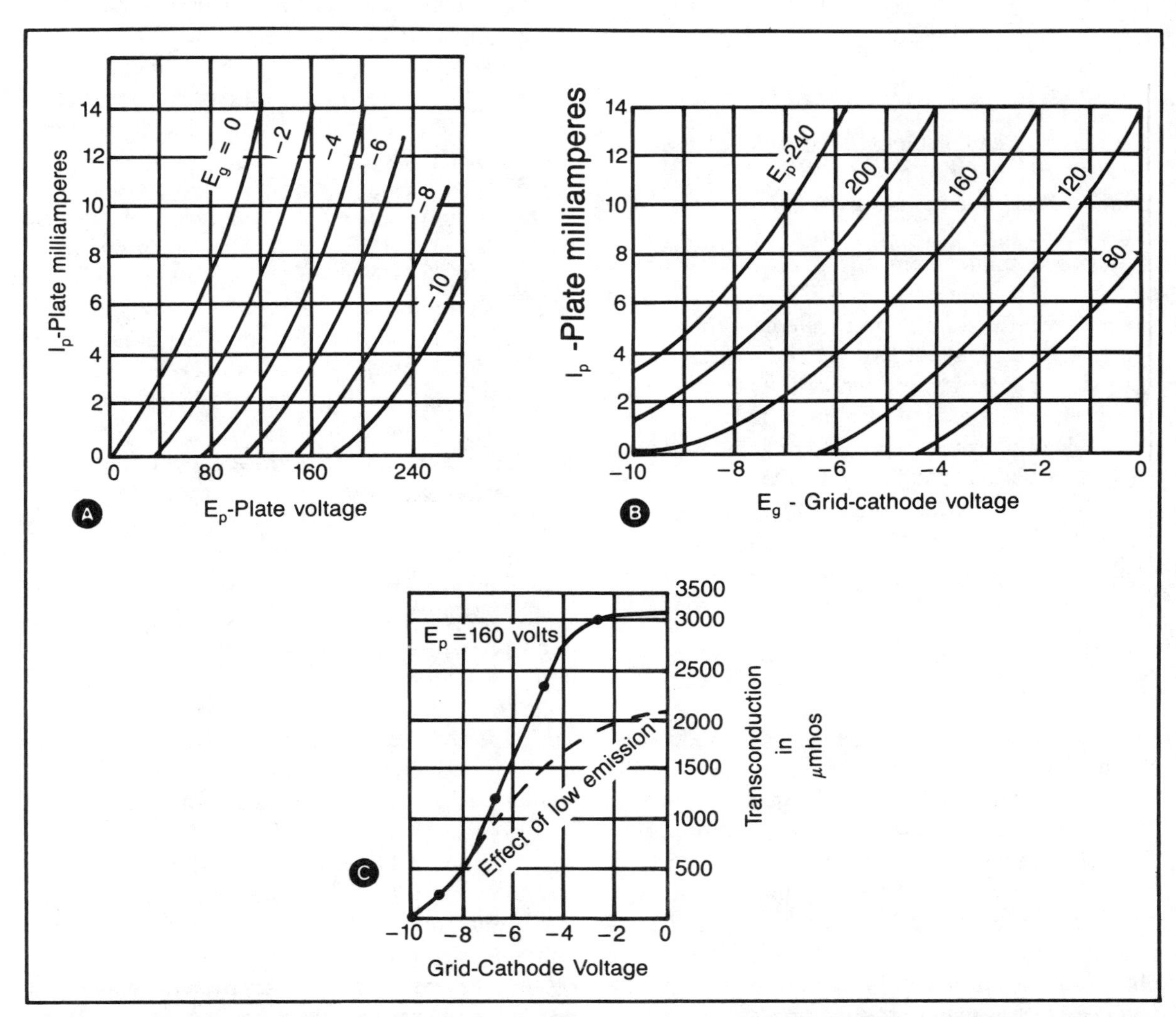

Fig. 6-5. Performance curves of typical medium-mu triodes. (A) Plate current vs plate voltage—constant grid voltage. (B) Plate current vs grid voltage—constant plate voltage. (C) Transconductance vs grid voltage—constant plate voltage.

corresponding new plate current would then be I_{p2}. This data can be read from either Fig. 6-5A or Fig. 6-5B and is directly useful from the mathematical expression for transconductance, g_m. Specifically, we have:

$$g_m = \frac{I_{p2} - I_{p1}}{E_{g2} - E_{g1}}$$

If I_{p1} is larger than I_{p2}, the expression can be written:

$$g_m = \frac{I_{p1} - I_{p2}}{E_{g1} - E_{g2}}$$

If the measurement units are volts and amperes, the transconductance given by these equations is in mhos. It is important to keep in mind that these expressions for g_m are valid only if the plate voltage is maintained constant over the region of change, even though it is small. The basic idea is to see how influential a control element the grid is. Thus, any plate-current change produced by changing plate-voltage would invalidate the result.

If plate currents are expressed in milliamperes and grid voltages are expressed in volts—which is the usual situation encountered—a very convenient way to use the above expressions for g_m is to multiply them by one-thousand and to express the result in micromhos. For example:

$$g_m \text{ in micromhos} = (1000)\,\frac{I_{p2} - I_{p1}}{E_{g1} - E_{g2}}$$

This convenience is due to the fact that the transconductance of most receiving-type tubes falls within the range of 1500 to 15,000 micromhos. Moreover, grid-bias voltages tend to be from several to several-tens of volts and plate currents are nearly always dealt with in milliamperes.

Tetrode (screen-grid), pentode, and beam-power tubes can be measured for their transfer characteristics in much the same way as depicted for triodes. However, a stable source of screen voltage is needed. This is best obtained from a regulated power supply or a battery. Good test results are not forthcoming from schemes using resistance networks to lower or divide down the voltage from the plate-voltage supply. This is because plate current is very dependent upon the screen-grid current, which in turn, is largely controlled by screen-grid voltage. For many of these tubes, the screen-grid voltage is between one-third and one-half of the plate voltage. The generalized plate transfer characteristics of these tubes is shown in Fig. 6-6. They feature near-independence of plate current on plate voltage, as revealed in the almost horizontal slope of their usable operating regions. The beam-power tube is the most "ideal" of the three, whereas the tetrode suffers from the effects of secondary emission over a substantial portion of its operating range.

These tubes have amplification factors much higher than ordinarily encountered in triodes. Measurement accuracy is accordingly more vulnerable to misreading meters, and to failure to keep the required voltages constant. Also, graphical derivations such as the plotting of the transconductance curve(s) from the transfer characteristics, requires special care. The use of a digital voltmeter for measuring the control-grid voltage is a good idea if one is interested in accurately evaluating these high-performance tubes. Interestingly, the pentode and beam-power tubes display plate-transfer characteristics resembling the output characteristics of bipolar and field-effect transistors in the sense that all of these devices tend to be "constant-current" amplifiers, i.e., plate, collector, and drain currents are nearly independent of the voltage applied to these electrodes.

As with triodes, many of these tubes have extended operation well into their positive control-grid regions. Although the test setup of Fig. 6-6 does not provide for this, the behaviour in the zero and negative control-grid voltage region suffices for many test and measurement purposes.

In maintenance and service work, we should cultivate the art of quickly assessing the condition of tubes before removing them from the equipment and subjecting them to testing. Much time can often be saved in this way. The most obvious evalua-

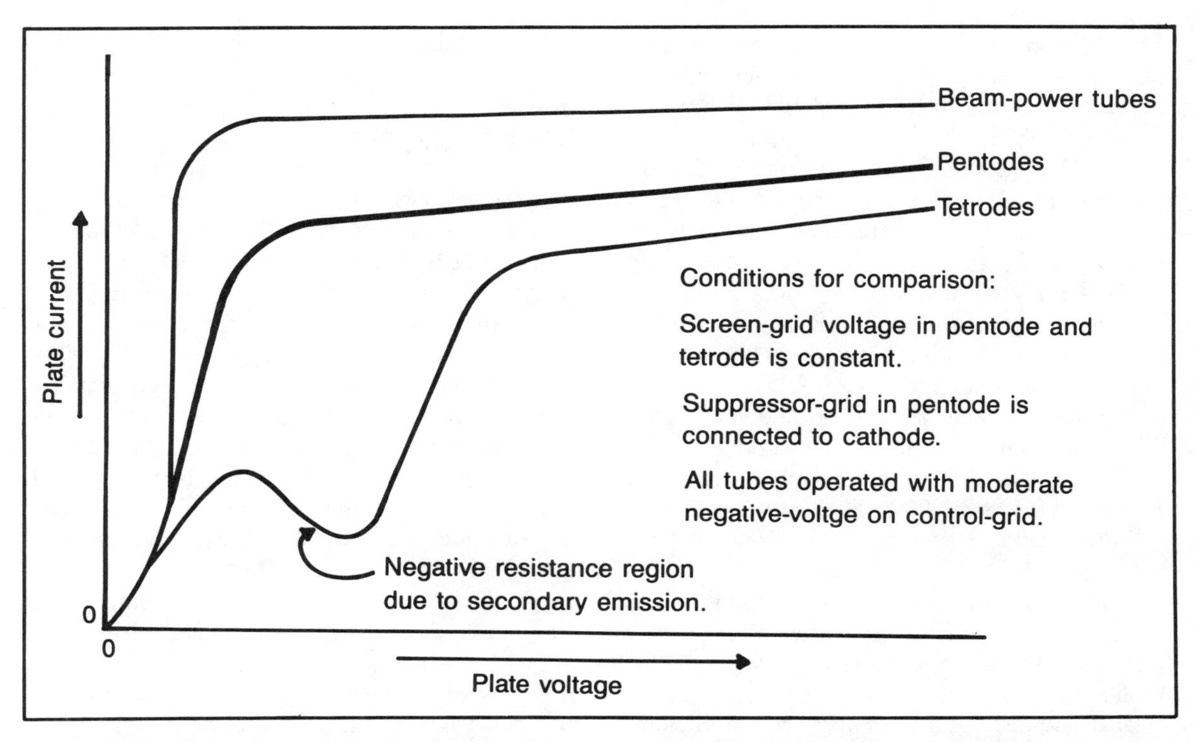

Fig. 6-6. Generalized plate-transfer characteristics of tetrodes, pentodes and beam-power tubes. The constructional features of pentodes and beam-power tubes eliminates the negative resistance region and extends the useable operating range.

tion of glass envelope tubes is to see whether their filaments or heaters are on. But just because a glow is observed from each tube does not necessarily mean they pass this test. Some tubes, such as double triodes, have two heaters—one for each triode section. These heaters can be connected to operate either in series or in parallel. Most often, the socket connections are such that the heaters operate in parallel. Therefore, if one heater opens, the other will still glow. When inspecting tubes with more than one heater, make certain that both heaters glow.

Sometimes, we cannot readily see the glow of a heater because the glass envelope has become opaque from a deposit of "getter" material and from an evaporated film produced by the cathode or filament. (The "getter" is a quantity of magnesium, barium, phosphorus, or other substances which is ignited after the tube has been evacuated in the factory. This removes residual gas.) Of course, with metal tubes and Nuvistors, we cannot see the heater glow. In all of these cases, we can tell if the heater is operating by feeling the tube. Usually, a casual touch will reveal the warmth from heater operation. A few minutes of operation are often required for the temperature to rise enough to feel. Some tubes have quite powerful heaters and can become quite hot from heater operation alone.

If air gets into a tube due to a fracture or from a broken seal, it often produces a characteristic white deposit. Filaments tend to burn out quickly after such an event, but heaters may continue to glow for some time. Such a tube cannot work, however. The slower development of internal gas, say from heated metallic parts, is often indicated by a blue or purple glow. This is particularly true in larger tubes which dissipate relatively high power or operate at higher voltages. Included here are audio-output amplifiers, TV horizontal-output tubes,and transmitting tubes. If we look at these tubes in subdued light while they're operating, the difference between a grossly gassy tube and a normal one is not too difficult to distinguish.

Tubes which have become loose in their bases are candidates for intermittent or continuous shorts or opens, if this is not already the case. These can usually be restored to reliable operation by appropriate cementing and soldering.

When TV picture tubes become gassy, spot focusing is degraded and a blurred display results. It is impossible to attain sharp screen lines with the focus control in such a tube.

VR tubes should have a soft and stable glow. The glow should not be concentrated at a small area, but should be uniformly distributed throughout the length of the elements. Erratic shifts in the intensity or pattern of the glow should be cause for concern.

Tubes often indicate the conditions in a circuit. The most obvious is the situation where several tubes with series-connected heaters are cold. This implies that either the ac circuit or the heater in at least one of the tubes is open. Sometimes in a pentode or beam-power tube you may see the screen grid glowing incandescently. This is an indication that there is an open in the plate circuit. When plate current is absent, virtually all of the available electron emission is diverted to the screen grid. A tube which has operated in this condition for any appreciable time may still test good, but the probability is also good that its life span has been shortened. Tubes, particularly power types, which have lost a substantial part or all of the applied negative grid bias tend to operate inordinately hot. Sometimes adjacent components or circuit-board areas become charred or discolored from the excessive heat. Discoloration of the socket, itself, is often a tell-tale sign of this circuit condition.

Most tubes, and particularly power types, operate noticeably hotter when supplying plate and screen grid current than when warmed by heater operation alone. With experience a tube deprived of its power supply voltages can often be identified by the touch test.

An intermittent circuit trouble is often the most time-consuming type of defect to remedy. The first thing that must be known is whether the intermittent is in the circuit or in the tube. By lightly tapping the tube while it is operating we can often determine whether or not the trouble is in the tube or at least localize the trouble. Tapping a tube will often cause sounds in the speaker, disturbances on the picture tube screen, or fluctuations of meters. A particularly disturbing tube defect that can be quickly identified by this test is a microphonic condition. In a microphonic tube there are elements that will vibrate when the tube is tapped. The vibration produces a mechanical oscillation which is amplified by the tube. Such tubes are not intermittent in the true sense, but cause objectionable interferences in audio, video, and other equipment. Sometimes, the slightest jar—a passing automobile or the footsteps of a person walking in the house—is enough to trigger the condition.

A not uncommon trouble in tube equipment is the failure of the tube pins to make reliable contact in the socket. (This, of course, can be the fault of the socket.) Sometimes, oxidation forms on the tube pins which results in intermittent, or poor contact. The pins should be carefully cleaned with a very fine grade of sandpaper where this is suspected. Using a brush and a good solvent, all abrasive particles and metallic dust should be removed before returning the tube to its socket. Such maintenance can prove quite effective in restoring old equipment to good operation, particularly where high-resistance contacts in the heater pins have developed.

OHMMETER TESTS FOR HEATER CONTINUITY

The most obvious use of the ohmmeter in tube testing is checking the continuity of the filament of heater. Although they involve little mental effort, such tests are quite important because of the wide-spread use of series-string heater circuits in entertainment equipment. When the heater of one tube in such an arrangement burns out, all of the remaining tubes in the series are deprived of heater current. An ohmmeter check of the heaters may be made at the tube sockets, provided that the line

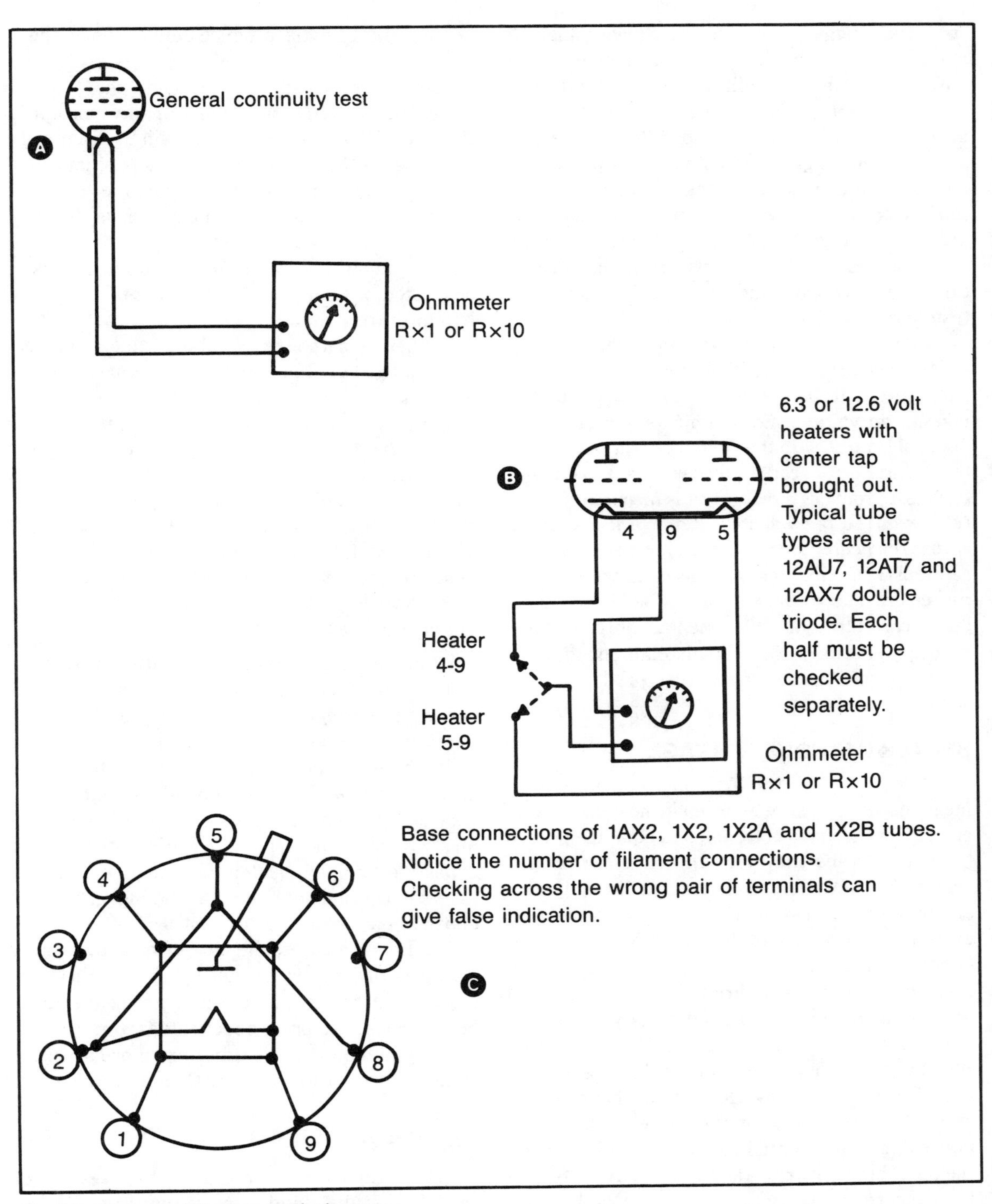
General continuity test
A
Ohmmeter
R×1 or R×10
6.3 or 12.6 volt heaters with center tap brought out. Typical tube types are the 12AU7, 12AT7 and 12AX7 double triode. Each half must be checked separately.
B
4
9
5
Heater 4-9
Heater 5-9
Ohmmeter
R×1 or R×10
Base connections of 1AX2, 1X2, 1X2A and 1X2B tubes.
Notice the number of filament connections.
Checking across the wrong pair of terminals can give false indication.
C
1
2
3
4
5
6
7
8
9
Fig. 6-7. Heater and filament continuity tests.

cord is unplugged from the 60-Hz power line. (Do not make tests while the unit is connected to the power line even though the power-line switch is in the off position.) It is all too easy to inadvertently burn out the ohmmeter or short out the line voltage while probing around with the test prods. The best way to conduct such tests is to remove the tubes from the sockets and apply the ohmmeter prods directly to the tube pins. The R × 1 and R × 10 ranges are the most useful for heater continuity tests. Obvious and not-so-obvious continuity tests are shown in Fig. 6-7.

In order to avoid the statistical game of hoping to discover the defective tube before checking a half dozen of them, the bad one can be found by making ac voltage measurements across the individual heaters with the equipment turned on. (Set the voltmeter to read ac line voltage.) When one such test produces a reading, it is most likely at the tube with the open heater. Remove it and check across the heater pins with the ohmmeter. Be very cautious while making voltage tests with the set turned on. The voltmeter will not be in danger but guard carefully against accidental shorts with the prods. It is often wise to insulate all but the very ends of either needle or clip-type prods.

HEATER-CATHODE LEAKAGE

A tube with less than many hundreds of megohms of resistance between its heater and cathode can cause various troubles in equipment even though the tube may test OK on emission and transconductance testers. Hum in audio equipment and 60-Hz bars on a TV screen are two of the most common indications of heater-cathode leakage. Other troubles may be more elusive, such as interference with synchronizing circuits and waveform contamination of laboratory test equipment. TV picture tubes often develop this trouble, but it need not necessitate replacement of the tube. Some judgment is needed to evaluate heater-cathode isolation in a circuit. In general, when a noticeable ohmmeter indication can be obtained on the megohm resistance range, this may be reason for concern in some circuits. When cathode-to-heater leakage paths produce deflections on the R × 10,000 or the R × 1000 ohms ranges, such tubes should be replaced even though there is no noticeable degradation of performance (except in the case of TV picture tubes, as will be discussed). Leakage all the way down to a short circuit may be also encountered. In some cases, the results of an ohmmeter test on a cold tube, such as shown in Fig. 6-8A, may not indicate the conditions which exist when the heater is in operation. If this is suspected, the simple setup in Fig. 6-8B can prove helpful in this situation.

When a cathode-ray tube develops heater-cathode leakage, the expensive tube may still exhibit like-new behavior with regard to screen brightness, focal qualities, etc. A simple way to keep such a tube in service is to supply its heater current from a separate filament transformer as shown in Fig. 6-9. This technique provides the required electrical isolation so that a complete heater-cathode short will not inject ac into the circuit nor upset the picture tube biase. An extra dividend often derived from this technique is the brightening of previously washed-out displays.

Anemic-looking screen illumination is often the result of the gradual drop in cathode emission with use. By operating the heater at a higher than rated temperature you can again increase the emission and thereby extend the useful life of the tube. This is accomplished with special "brightener" transformers which supply a slightly higher filament voltage. For example, 7.5 or 8 volts might be applied to a 6.3-volt heater. The length of time the tube will continue operating at the higher voltage depends on circumstances, but it is often worthwhile. In some cases, the lack of brightness is at least partially due to a low line voltage and this technique need hardly provide more than the rated heater current to bring about satisfactory results. Even some new picture tubes are more sensitive to the effects of low current than are others.

GAS TESTS

Other things remaining equal, the presence of gas in a tube tends to change its control

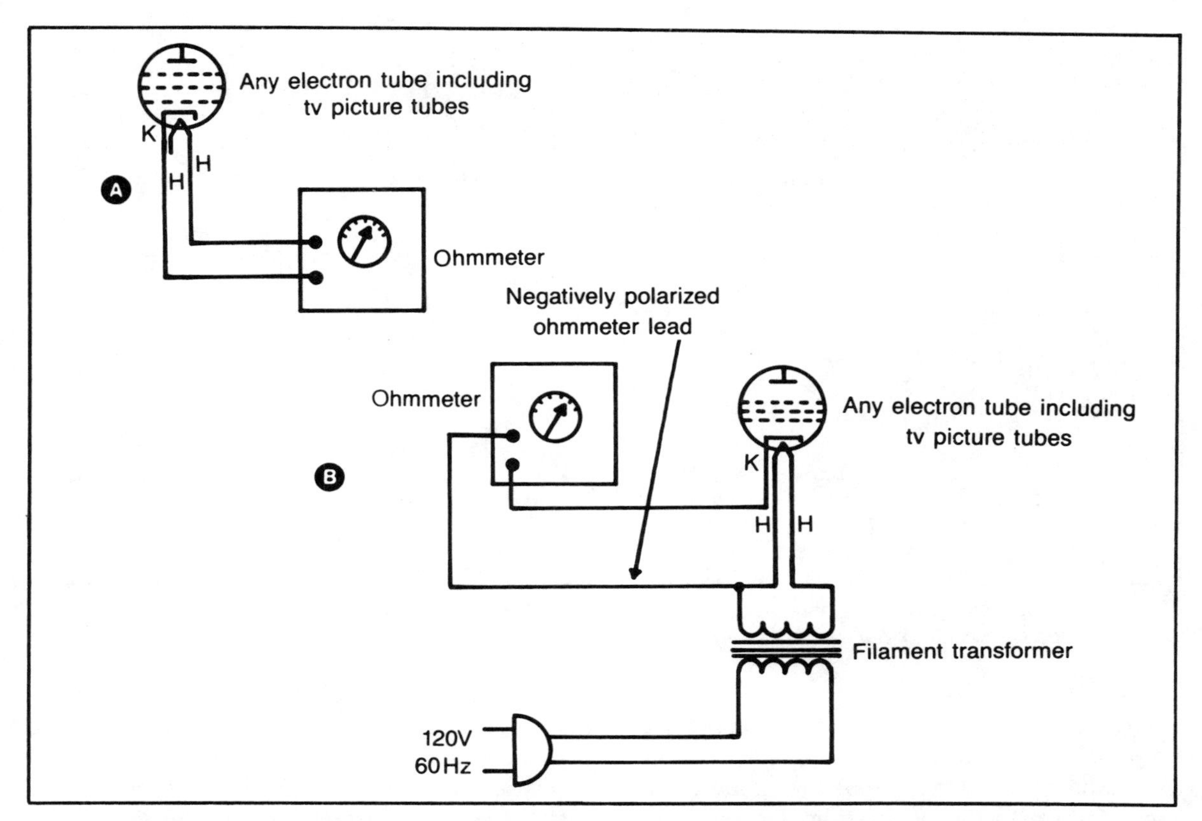

Fig. 6-8. Tests for heater-cathode leakage for shorts. A—Cold heater test; OK. In many situations. B—Hot heater test; more likely to catch intermittent shorts or leakage conditions which are worse when the heater-cathode structure is at operating temperature.

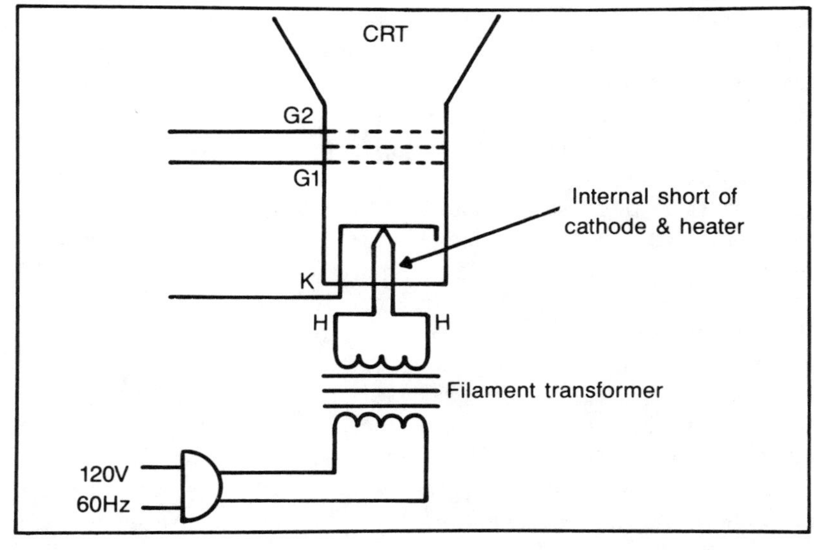

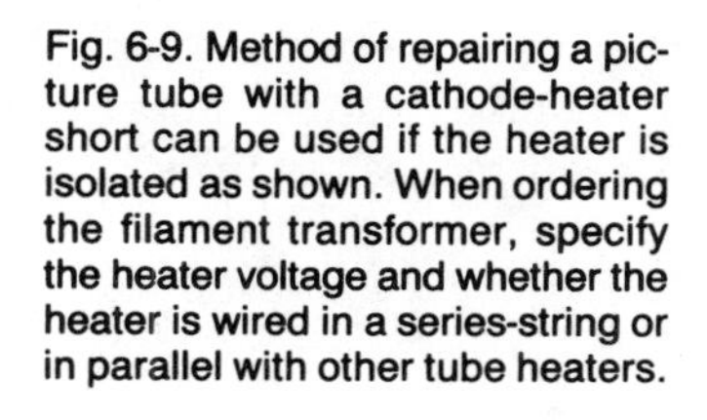
Fig. 6-9. Method of repairing a picture tube with a cathode-heater short can be used if the heater is isolated as shown. When ordering the filament transformer, specify the heater voltage and whether the heater is wired in a series-string or in parallel with other tube heaters.

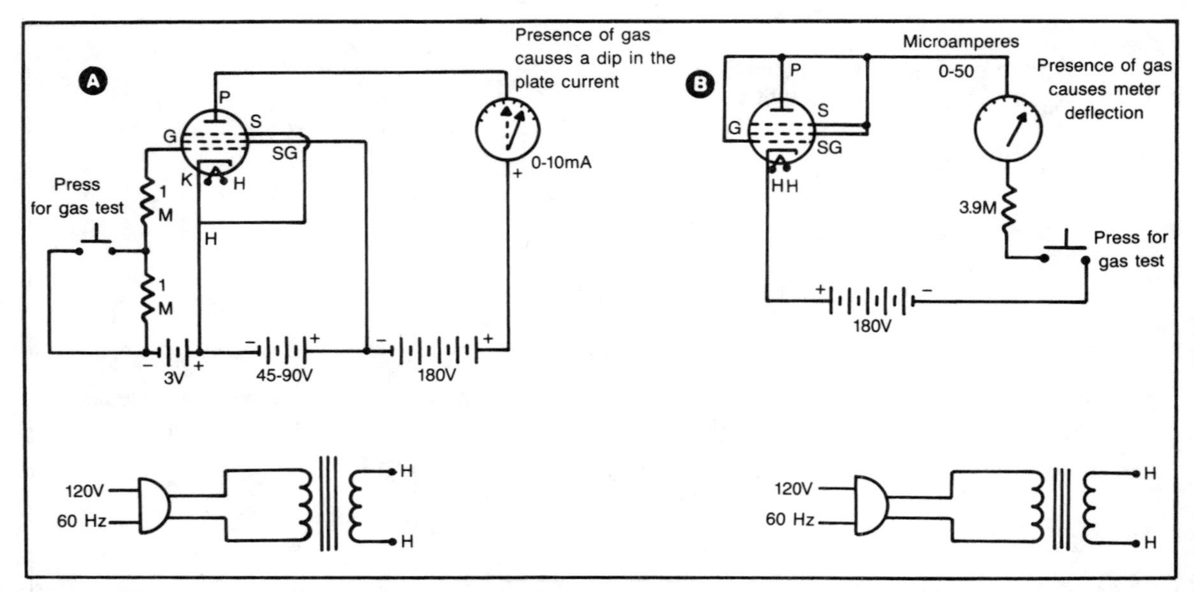

Fig. 6-10. Test circuits for revealing the presence of gas in tubes. Circuit A is used in both emission and transconductance tube testers. Circuit B is suitable for a relative indication of gas.

characteristics. Gas causes distortion in amplifier stages, runaway tendencies where appreciable power is involved, increased susceptibility to internal flashover or voltage breakdown, and a more rapid drop in cathode emission. In cathode-ray tubes, the result may be poor focusing, smeared display outlines, reduced contrast, and abnormal disturbances in the picture. The presence of gas is less likely to show on emission testers than on transconductance testers where a loss of grid control may result in a reduction of transconductance. However, long before a tube develops this much gas, there may still be enough to produce large shifts in dc grid bias in many circuits.

Because neither emission nor transconductance tests will reliably detect gas in a tube, both types of testers generally incorporate other circuits for gas tests. One of these is shown in Fig. 6-10A. If there is no gas present, shorting out of part of the grid-return resistance will have no effect on the plate current; there is no grid current and, therefore, no voltage-drop across the lower grid-return resistance. However, if the tube is gassy, an ionization current flows in the grid-return resistances and the lower one does develop an IR drop, thereby decreasing the negative bias on the grid. When the lower resistance is shorted during the gas test, the grid becomes more negative and this is seen as a reduction in plate current. Notice that the transconductance of the tube serves the purpose of the test by amplifying the effect.

The scheme shown in Fig. 6-10B measures ionization current directly. The various electrodes, other than heater and cathode, are tied together as in emission testing but polarized negatively with respect to the cathode. This test is best made with normal heater current, even though the electron emission from the cathode is not collected by a positive electrode. The hot cathode helps to ionize the residual gas within the tube; therefore, such a test tends to be more realistic than with a cold cathode.

TV PICTURE TUBES AND OTHER CATHODE-RAY TUBES

TV picture tubes which are evaluated primarily in terms of their emission capability, would be difficult to test if each one had to be removed from the receiver. So in-equipment testing is accomplished by means of an extension harness with an appropriate CRT socket. This socket is substituted for the regular in-circuit socket and the other end of the extension harness connects either to a conventional tube tester or to a specially designed CRT tester. The latter arrangement has advantages, but the former is generally adequate to confirm low emission and to make interelectrode short and leakage tests.

The concept of measuring the transconductance of cathode-ray tubes has not gained commercial popularity. Although it is conceivable that such a test might be put to use, it is of little importance in the way CRTs are used.

The basic test performed on both color and black-and-white tubes measures the cathode emission by using the screen grid, G2, as the plate, with control grid, G1, at zero bias voltage with respect to cathode. By using the triode thus formed, it becomes unnecessary to provide high voltage in order to make meaningful tests. It is commonplace to apply raw ac to the screen grid, allowing the tube, itself, to perform the rectification. About 300 microamperes of screen grid current generally indicates that the tube will have normal brightness capability. The voltage applied to the screen grid when making this test varies with different tubes. Testers provide selectable ranges to accommodate this variation. Figure 6-11 shows how the basic emission test is performed.

A word of caution is in order with regard to the voltage applied to G2. The G2 voltage specified for TV picture tubes ranges at least from 50 to 500 volts. If a considerable overvoltage is applied to a tube with a low-voltage screen grid, the tube can be damaged. Even if there is no damage, at the very least you could get a false reading. The latter situation applies where insufficient G2 voltage is applied, although in this case the tube cannot be damaged.

An alternate arrangement for testing emission in cathode-ray tubes is shown in Fig. 6-12. When you first glance at the circuit, you may think that the use of dc is the significant difference between this test setup and that shown in Fig. 6-11. But of greater importance, notice in Fig. 6-12 that the control grid, G1, and the screen grid, or first anode, G2, are tied together, just as is done in emission tests of ordinary electron tubes. The current

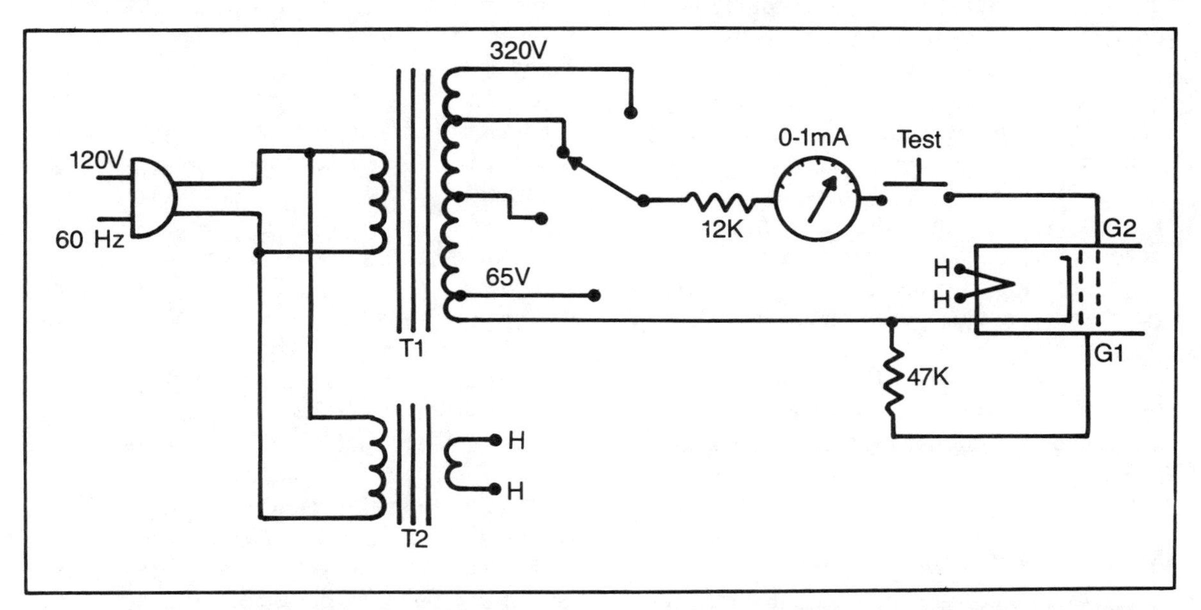

Fig. 6-11. Basic test setup for determining the emission capability of cathode-ray tubes. In order to protect the dc current meter from the effect of internal shorts in the CRT, transformer T1 should have a low current capacity and-or the meter should be protected by silicon diodes shunted across its terminals.

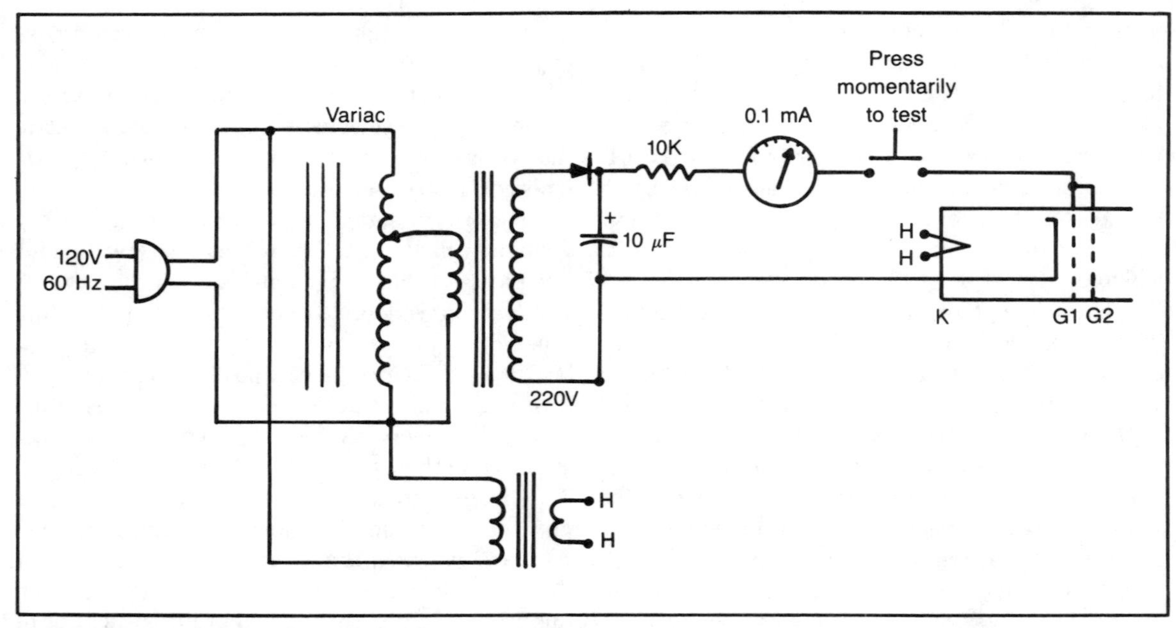

Fig. 6-12. Alternate CRT emission test circuit.

reading is relative and depends on the tube type, applied voltage, and emission capability. This testing technique is likely to be encountered with adapters used with ordinary emission-type tube testers. It is also a convenient tester to set up on the bench. The best way of "calibrating" and evaluating the meter indication is to connect it to a tube known to be good.

Do not allow current to flow any longer than is necessary to read the meter when using this test, because prolonged and excessive control grid current can be damaging. Also, tubes with low-voltage screen grids (around 50 volts) should be tested with 50 volts or less.

The tester in Fig. 6-11 probably provides a more realistic test for TV picture tubes because zero bias on the control grid, G1, corresponds to TV set operation when the video drive is "white peak" and the brightness control is fully advanced.

Continuity and Short Tests

Tests for element continuity and interelement shorts in cathode-ray tubes may reveal some surprising results. Defects such as open-circuited elements are not common in conventional tubes, but they are frequently found in TV picture tubes. Also, even elements other than those physically adjacent will frequently short together. This is generally caused by flakes of conductive material which lodge between structures, their supports, and their leads. The flakes are usually from the cathode, and you may or may not be able to remove them by physical or electrical means.

When a simple symmetrical-electrode neon bulb is connected in a cathode-ray tube test circuit using ac voltage, the manner in which the bulb glows will reveal what is happening in the circuit. When a CRT element behaves as a diode with respect to the cathode, only one electrode of the neon bulb will be illuminated because this is its glow pattern on dc or rectified ac. If the tube element or the cathode is open, there will be no glow. Similarly, if the cathode emission is very low, the neon bulb will not glow. If there is a short between the element and the cathode, both cycles of the ac will pass through the bulb and both electrodes will glow.

Figure 6-13 shows a test setup in which neon bulbs are inserted in the heater, control-grid, and screen-grid circuits of a CRT test setup. A glance at the bulbs indicates the existence of any of the defects shown in Fig. 6-14. Thus, the use of inexpensive neon bulbs eliminates a more conventional test approach, which could take much more time and motion, as well as involve vague meter readings.

Cutoff Testing

The basic idea in cutoff testing is described for color tubes because an additional evaluation—tracking—must be considered. The cutoff test procedure would be applied in the same way as for black-and-white tubes, except that tracking is not important for single-gun tubes. Cutoff testing, as the term implies, is the measurement of the negative control-grid voltage needed to cut off "plate," that is, G2 current. A natural question here is that once emission has been confirmed to be satisfactory, why test for a "designed-in" characteristic which depends on such geometrical and physical factors as cathode-grid spacing, grid electrode fabrication, etc. The need for such a test arises from the very close spacing between cathode and control grid. In response to thermal and mechanical effects, this spacing can change with time. A very small change in spacing can produce a relatively large change in the control characteristics of the grid. The contrast range of pictures displayed on the tube screen is critically related to the grid control characteristics—a compressed control range leads to greater contrast, whereas an extended range of grid control may deprive the picture of contrast which cannot be restored by circuit adjustments. The latter situation is identified by a higher than normal grid cutoff voltage. The former situation is less frequently encountered because the grid will ultimately short with the cathode.

The setup shown in Fig. 6-15 can be used to

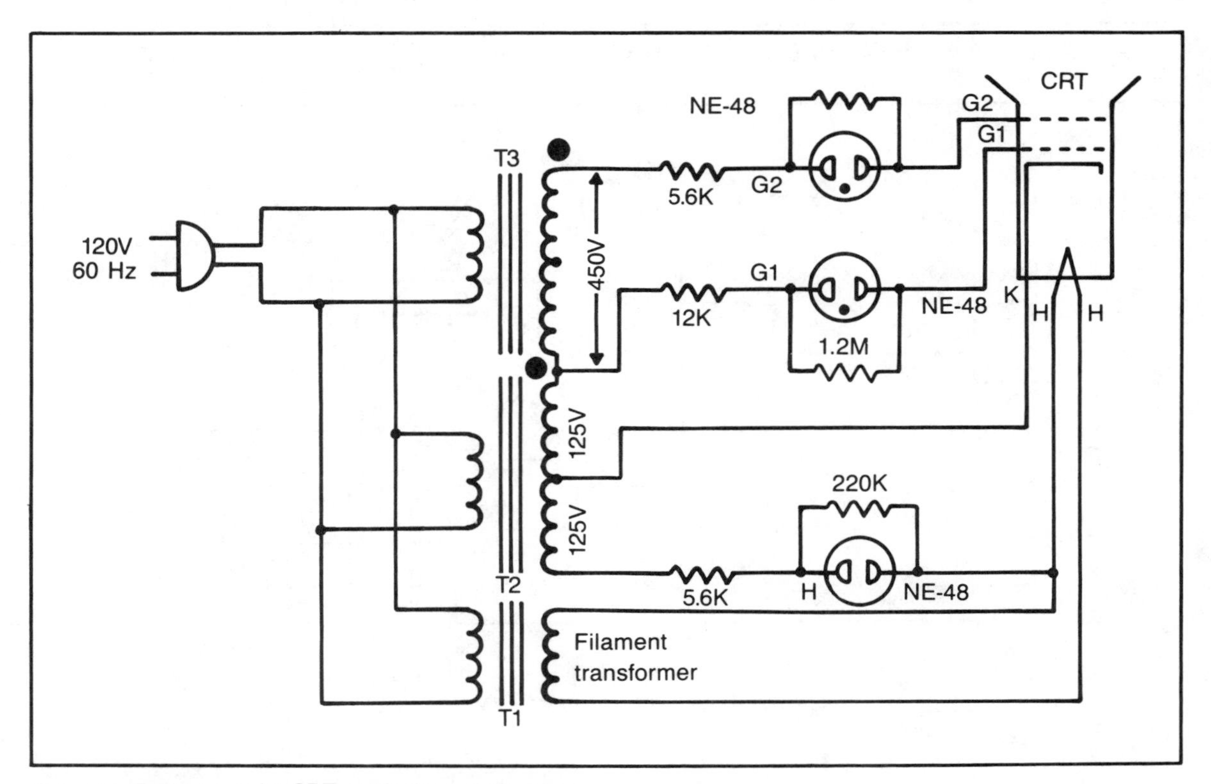

Fig. 6-13. Basic setup for CRT continuity and short tests.

Electrode glow			Interpretation
H	G1	G2	
○	◐	◐	**Good Tube**—No shorts or opens in K, H, G1, G2 structures
○	◐	○	**Bad Tube**—G2 is open
○	○	◐	**Bad Tube**—G1 is open
◐	◐	◐	**Bad Tube**—H-K short or high leakage
●	◐	◐	**Bad Tube**—Also H-K short
●	●	◐	**Bad Tube**—H-G1 short
●	◐	●	**Bad Tube**—H-G2 short
○	●	◐	**Bad Tube**—G1-K short
○	◐	●	**Bad Tube**—G2-K short or G1-G2 short
○	○	○	**Bad Tube**—K is open, or emission is very low

Fig. 6-14. Meaning of the neon-bulb indications in test setup in Fig. 6-13.

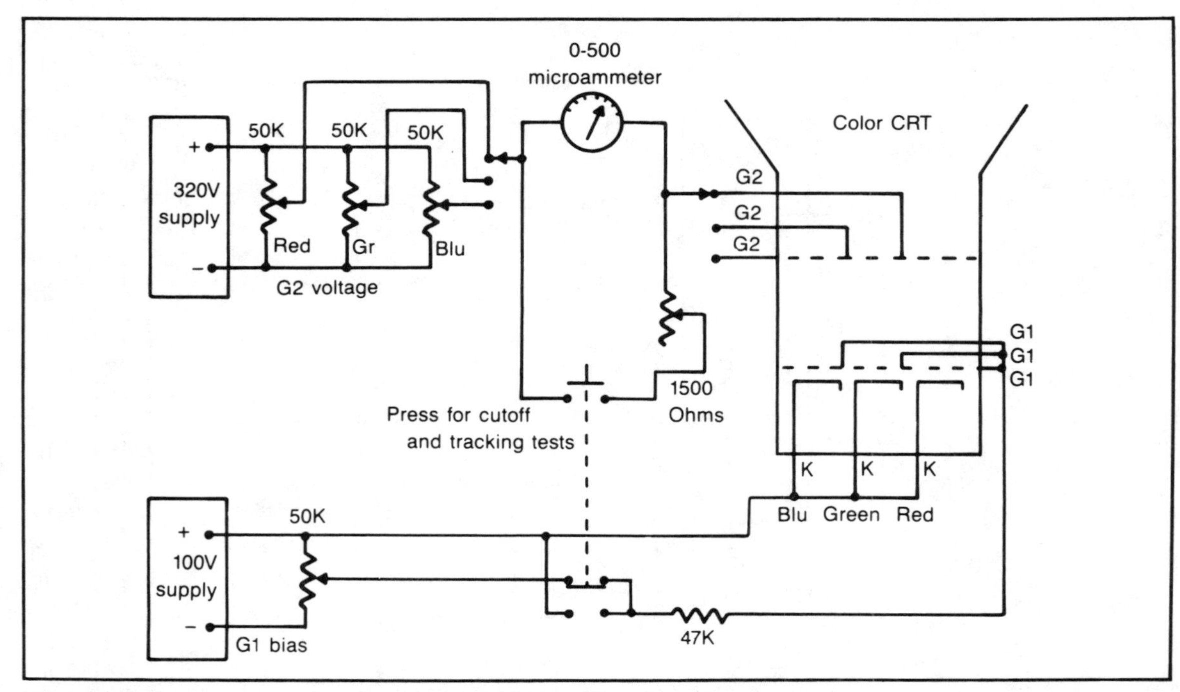

Fig. 6-15. Basic test setup designed to determine CRT cutoff and tracking characteristics.

see if a negative bias on the control grids will cut off the screen-grid (G2) current at nearly the same screen-grid voltage for all three guns. A test procedure which relates to the way in which color tubes are operated in TV sets follows: Set the G1 bias at about −45 volts. Adjust the G2 voltage to produce a G2 current of about 15 microamperes. (This operating condition corresponds to a just visible trace on the screen.) Next, reduce the G1 bias to zero and measure the G2 current. (This operating condition corresponds to the application of maximum video drive to the tube.) After this has been done for all three guns, the tube can generally be considered acceptable if the highest G2 current does not exceed the lowest by more than 150 percent. By this criterion, the three guns are assumed to track sufficiently well in their control-grid cutoff characteristics.

For single-gun tubes, the grid bias should reduce the G2 current to the vicinity of about 15 microamperes. For such a test, the G2 voltage should be the nominal value applied in the set. In any event, when comparing tubes, the same G2 age should be used. Notice that a cutoff evaluation is of no practical significance unless it is also determined that the tube's emission is OK. This can be readily checked by observing whether a 300-microampere or more current is available from G2 (at the nominal G2 voltage) when the bias on G1 is zero volts.

QUESTIONS

6-1. A tube tested in an emission-type tester produces a good indication, but this tube performs very poorly as a driver stage to Class B output amplifiers. Later, it is found that when the tubes are tested in a transconductance-type tester, the meter indication is in the reject region. How might the discrepancy between the two testers be explained?

6-2. Two tubes produce the same reading on an emission tester. However, it is observed that one warms up quickly, whereas the other requires several minutes to reach its maximum reading on the meter. Other things being equal, what might we consider to be the expected lifespans of the tubes?

6-3. Trouble is encountered with marginal B+ voltage in a TV set. It is suspected that an inordinately high voltage drop is taking place in the rectifier tube. Should this tube be tested in an emission tube tester or in a transconductance type?

6-4. A tube is found to be OK insofar as emission is concerned. However, it produces considerable distortion as an audio amplifier. Assuming that the cathode is in good condition, and there are no internal shorts or leakage paths, what particular test would help establish the nature of its defective performance?

6-5. A tube which just passes a transconductance test performs well as an oscillator in a UHF TV tuner. Another tube, made by a different manufacturer, shows a very high transconductance but does not oscillate throughout the UHF range. Since high transconductance should lead to vigorous oscillation, should the tube tester be held suspect?

6-6. A TV set has an "instant on" feature, where a small amount of ac is continually applied to the tube heaters. Power supply voltages are applied only when the set is switched on. The owner of this set decides to extend tube life by removing the ac line plug from its receptacle when the set is not in service. Is this a valid decision?

6-7. A cathode-ray tube produces displays with fuzzy outlines. The focusing circuit appears to be all right, judging from the fact that the focusing control does make the raster lines appear sharp. The astigmatism control is somewhat less effective, but does produce the general action expected from it. Emission, cutoff, leakage, and short-continuity tests yield acceptable indications. Assuming the CRT circuitry is OK, what might be the nature of the defect in the CRT?

6-8. The injection of 60-Hz interference into the picture tube display would not occur in the event of a heater-cathode short if the heater were operated from direct current. Aside from the added cost of doing this, cite a reason why such a technique would not be likely to increase the average operating time of picture tubes.

6-9. A cathode ray oscilloscope is constructed from surplus components. Initial operating tests appear encouraging. In subdued light, the blue trace of the electron beam within the cathode-ray tube can be seen, but at this stage further tests are not feasible because of trouble in the sweep circuit. What other troubles can we expect?

6-10. It is decided to test a tube used as the middle stage of 3-stage resistance-capacitance coupled amplifier. The tube is to be left in the equipment, but the B+ lead feeding its plate load resistance is interrupted to accommodate a dc milliammeter. Dc voltage from an external supply is applied to the grid. In this way when a small change in grid voltage is imposed, the resultant change in plate current is seen on the milliammeter. Is this a valid procedure for testing the transconductance of the tube? (Assume that the cathode connects directly to ground and that the grid connects to ground through a resistance of several megohms. Grid bias is zero volts.)

ANSWERS

6-1. This is not an uncommon occurrence. Cathodes sometimes develop "hot" spots—small areas from which high emission takes place. However, the remainder of the cathode emission capability may be virtually exhausted. Under such a condition, the control grid cannot exercise normal control of the space current and this is revealed by the transconductance tester.

6-2. The slow warm-up tube probably will not continue to render usable service as long as the other tube. This test has other variations. Sometimes, the probable lifespan is indicated by the length of time the emission holds up after the heater voltage is turned off. Yet another approach is to make an evaluation of the minimum heater voltage required for, say, an 80 percent meter deflection relative to the normal heater voltage. Such tests are more of an art than a science and must be tempered by judgment based on experience. However, these tests do tend to show up tubes which probably have already seen their best days.

6-3. It would not make any difference because the concept of transconductance measurement has no significance for diodes. If the tube were tested in a transconductance tester, it would be subjected to the same test as in an emission-type tester.

6-4. This tube should be tested for gas. It is likely that a portion of the "emission" current indicated by the meter of the tester is actually ionization current from gas.

6-5. No. Oscillation capability, particularly at high frequencies, is influenced by many factors not checked by the tester which subjects the tube to a 60-Hz, or at best, an audio frequency. The slight constructional differences in tubes of different brands result in different interelectrode capacitances, or different lead inductances for the various elements, especially the cathode. This is a function where good performance is often hard to predict. Often, the best performer must be determined by actual trial. It also is not uncommon to encounter circuits which appear to favor the very tubes which fail to perform satisfactorily in other circuits. In spite of all this, the transconductance test does tend to be a more reliable indicator of portable performance than emission tests for such cases.

6-6. Experience seems to show that whether tubes are continually operated at a below-normal heater current or at the normal heater current, their lifespan is not shortened. Conversely, the evidence of precision instrument manufacturers suggests that the actual tube life is prolonged when they are so operated. Sparing the heaters the thermal shock of repetitive turn-ons, and keeping the tube at a near constant temperature probably contributes favorable conditions which override the effect of mere "mileage."

6-7. The occurrence of defeats which tend to sneak past test procedures is one of the things which makes electronics an art as well as science. In this case, a burned grid aperture was responsible for the fuzzy outlines. Such a defect can (but does not necessarily) show up on the cutoff test.

6-8. Actually, a heater-cathode short is disabling for two reasons. It is true that dc operation of the heater would eliminate the possibility that 60-Hz bars and other line-frequency interferences would appear in the picture. However, a heater-cathode short also upsets the grid bias, a critical operating parameter of the tube. If the heater is operated from an isolated transformer winding, the heater-cathode short neither causes 60-Hz effects nor does it upset grid bias. Thus, under this condition it is not even necessary to operate the heater from dc. Although installing and isolation transformer will clear up the effect of a heater-cathode short and save an otherwise good tube, manufacturers invaribly use series string or other inexpensive heater circuits in which such a defect causes problems.

6-9. It will likely be found that focus is poor and that the brightness control does not work well. Displays will tend to be fuzzy rather than sharply outlined. These symptoms point to a gassy cathode-ray tube! The blue trace of the electron beam which is visible within the tube is actually a path of ionized gas. In a tube with a good vacuum, there is no visible evidence of such ionization phenomena.

6-10. No, it is not. The plate current must be governed by the plate of the tube alone. If a resistance is present in the plate circuit, the B+ must then be adjusted to maintain a constant plate-to-cathode voltage for the two plate current readings. Operating the tube at zero grid voltage is acceptable if the grid signal is very small and if an appropriate tube and plate load resistance are used. This test could be made valid by injecting the signal in such a way that a few volts of negative grid bias becomes the grid operating point, and by shorting out the plate load resistance. For example, if the grid bias is varied from −2 volts to −3 volts, the transconductance can be computed for the grid operating point of −2.5 volts.

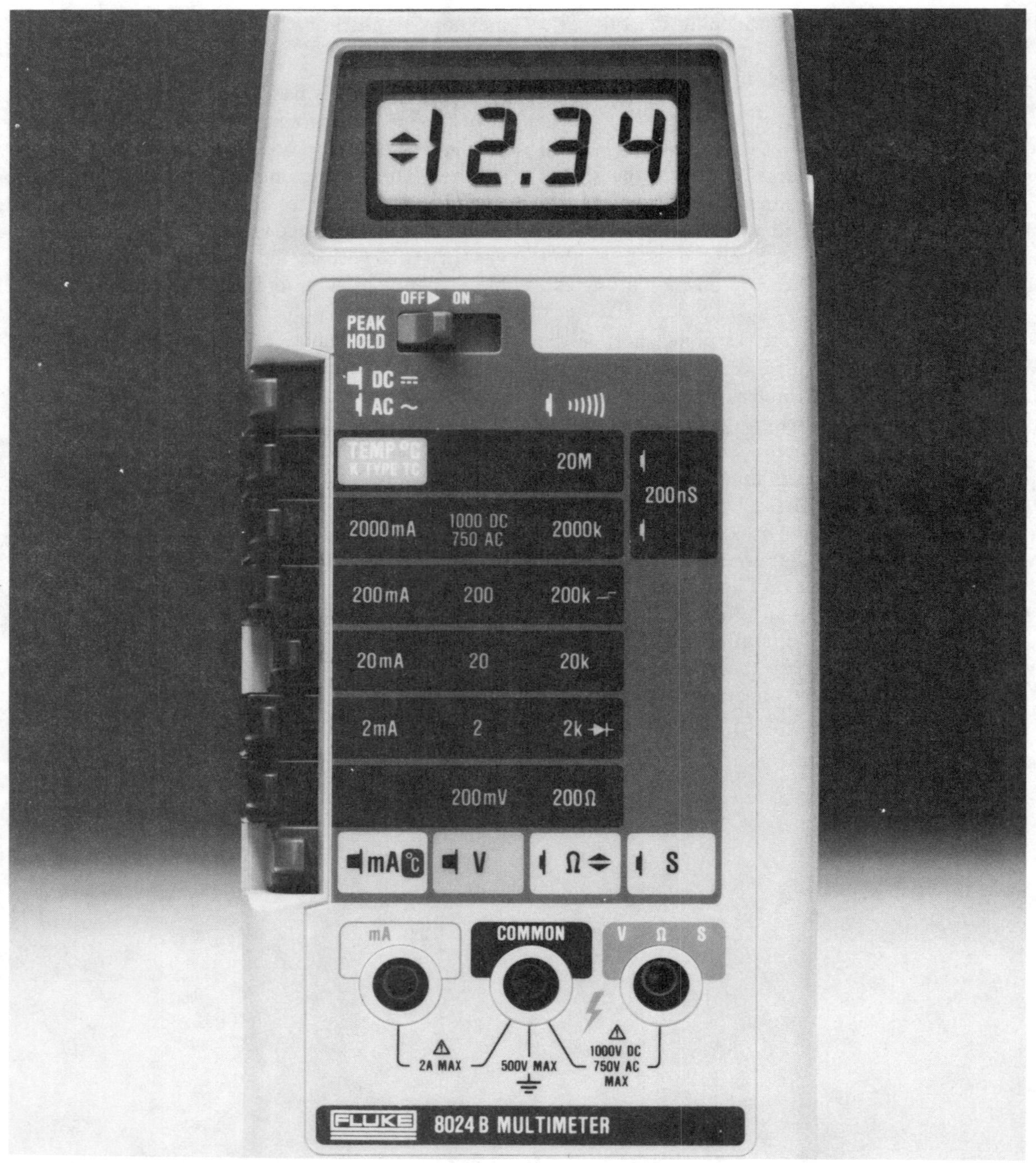

This Digital Multimeter has a diode test mode. The usual implication of the diode symbol is that two measurement conditions are provided for silicon pn junction-devices. In the first, or ordinary, the test voltage is 0.3 volt or less. This enables testing components on PC boards without turning on semiconductors. The second, or special, is a constant-current and with 0.7 or more volts at the test prods. This enables polarity tests and the monitoring of the junction voltage. Courtesy of the John Fluke Mfg. Co., Inc.

Chapter 7

Testing Semiconductor Devices

If we always had access to expensive test equipment and complex methods of testing semiconductors, it would be a relatively simple matter. It is, indeed, worthwhile to learn how to use such equipment and test approaches even if they are not available to us in our everyday pursuits, because of the insights that consciously, or otherwise, tend to carry over to simpler test procedures. In a very small sense, "sophistication" in semiconductor testing is to be found in the simplicity, economy, and the time it takes to carry out the procedure. And, in compliance with the purpose of this book, this is the path we will follow. It is surprising what the knowledgeable user of just an ohmmeter can do in troubleshooting a board full of diodes, transistors, and perhaps other types of semiconductor devices. We do not intend to play down the real importance of the beta tester, the curve tracer, the in-circuit tester, and various test equipment which allows us to measure the very parameters called out in the specification sheet supplied by the maker of the semiconductor device. But human skill and intuition aided by knowledge and verified by various cross-checks is hard to beat. Even when the more involved test equipment is available, the simple tests with more commonly encountered instruments are often employed to confirm the results.

THE SEMICONDUCTOR JUNCTION

A common characteristic of many different semiconductor devices is the pn junction. From a test standpoint, the pn junction has an "energy gap." But since we are going to look at the pn junction as a 2-terminal diode without regard to the internal physics of the device, it is more practical to refer to the pn junction as a voltage gap. Although not always emphasized, the voltage gap manifests itself in two ways. First, when the pn diode is polarized in a forward direction (minus to the N side of the junction) actual conduction does not occur until a certain voltage difference is reached. Below this value, the ideal junction diode can behave very much like a device with two electrodes extending

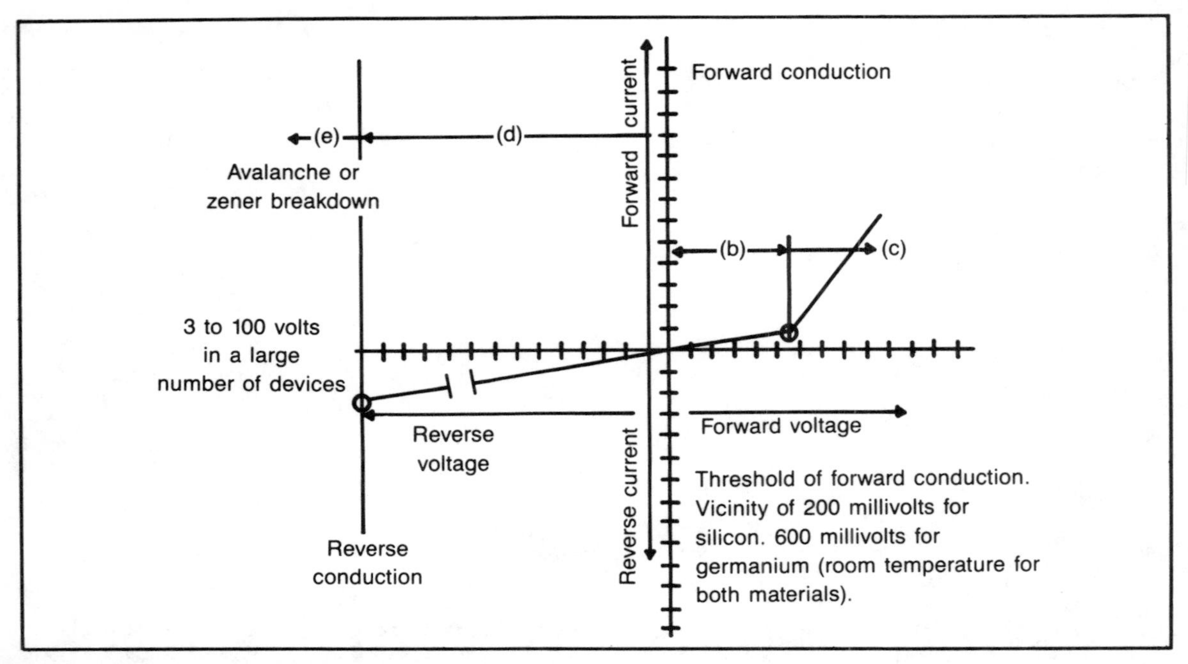

Fig. 7-1. Generalized junction-diode E1 curve.

into a vacuum—that is, no current passes through the device. Real diodes may approach this ideal to various degrees. The diode material, fabrication, and temperature all play a part determining how close an actual diode comes to the ideal diode. An ideal diode behaves as a perfect insulator up to a certain forward junction voltage, then it abruptly starts conducting. Most germanium junction diodes begin to conduct in the neighborhood of 0.2 volt at room temperature (21 degrees C). The transition between nonconduction and conduction tends to be relatively gradual, with a good portion of the nonconducting region behaving somewhat as a leaky conductor. Nevertheless, there is no mistaking the conductive region, once the voltage across the junction is raised sufficiently.

The silicon pn junction has a higher threshold voltage of approximately 0.6 volt for most diodes at room temperature. Moreover, the silicon pn junction is usually more ideal because most of the voltage region below the conduction point is characterized by very low leakage. On a graph, the current vs voltage characteristic curve of the silicon pn junction is more like we would draw for an ideal rectifier with no conduction until a certain voltage, and then a relatively abrupt change into the conduction region. In other words, the "knee" of the curve has a small radius, even approaching an angular change from the horizontal to a nearly straight line slope in some units. (Infinite conduction or zero resistance, which would correspond to a vertical slope, might be assumed to represent an even more "ideal" diode or rectifier.) For practical test purposes, the generalized pn behavior shown in Fig. 7-1 is useful for a large number of semiconductor devices.

SIMPLE TESTS ON PN JUNCTIONS

Many of the simple diode tests are performed to answer the following questions: Does the pn junction display high or extremely high resistance when polarized for forward conduction but with a below-threshold voltage? And does the pn junction display good conductivity when the forward-polarized junction voltage exceeds the threshold of conduction? High conduction at very low junction voltages indicates a shorted junction. Low or zero conduction after the threshold voltage is exceeded by a comfortable margin indicates an open-circuited junction. Both defects disable the device.

Although most pn junctions which have been subjected to electrical abuse tend to fall definitely in one or the other of these classifications of damage, in-between situations are also encountered. Thus, a pn junction may show high leakage current well below the threshold of conduction; conversely, the conductivity well above the threshold of conduction may not be as good as it should be. Such "gray area" situations are harder to distinguish than an obviously shorted or open-circuited junction. The conduction characteristics shown in Fig. 7-1 are helpful in evaluating many devices if this behavior is understood and skillfully applied in practical situations.

Secondly, the pn junction behaves in somewhat the same way with reverse polarity applied to the junction, but the threshold of conduction then occurs at a higher voltage than for forward polarization. (A little thought will reveal that this must be so in order that the diode can function as a rectifier.) The reverse threshold voltage, more appropriately referred to as "breakdown voltage," is also called the avalanche or zener voltage. The term "breakdown" suggest an abrupt transition, and this does tend to be the case when the reverse voltage is raised sufficiently. Also "breakdown" infers destruction. Actually, the pn diode tends to conduct so heavily that, if allowed its current demand, destruction would, indeed take place. In practice we always make certain that sufficient limiting resistance is effectively in series with the diode and this provides complete protection. A pn diode with a current-limiting resistance in series can be repeatedly cycled in or out of its reverse-breakdown region, or can be indefinitely operated with this region with no damage.

In order to perform meaningful forward- and reverse-conduction tests, it is necessary to know the basic construction of the semiconductor device. Figure 7-2 shows the arrangement of the junctions in commonly encountered devices. For simplicity, the depletion layers are omitted.

The use of the word "threshold" merits some discussion. The onset of conduction can vary from a rather gradual change to a very abrupt one. For example, an inexpensive germanium diode or transistor would tend to approach the "hard conduction" region of forward conduction gradually compared to the sudden transistion into reverse conduction which occurs in a silicon zener diode. However, for the purposes of establishing go, no-go tests with an ohmmeter, most junction devices can be considered to be well represented by the generalized characteristics depicted in Fig. 7-1. For more refined semiconductor tests the actual leakage current in the below-threshold regions must be evaluated in terms of the tolerances allowed for specific types. Because of the extensive variations between thousands of types—the wide tolerances and the temperature and voltage-dependent characteristics—it would be of little help to list specific numerical values here. Instead, we will simply deal with tests that determine whether or not a device is satisfactory for the purpose at hand. Actually, when you become involved with such device evaluations, testing is not as indefinite as might first appear to be the case. For the most part, simple tests can identify the operating modes which characterize a good pn junction. These modes are pictorially and symbolically shown in Fig. 7-3 as a further illustration of the behavior depicted by the generalized curve in Fig. 7-1.

Using an Ohmmeter to Test Semiconductors

It is only natural that ohmmeter test techniques should be used for semiconductor devices. These meters are among the most commonly encountered test instruments and have certain features nicely suited to semiconductor testing. For example, most VOMs, multimeters, and electronic meters have internal voltage sources of 1.5 or several volts to operate the low ohms scales. This comfortably exceeds turn-on voltage in germanium and silicon junction devices, thereby enabling us to check one mode of pn behavior. At the same time, this internal voltage source is low enough so that the very low or negligible reverse conduction condition can

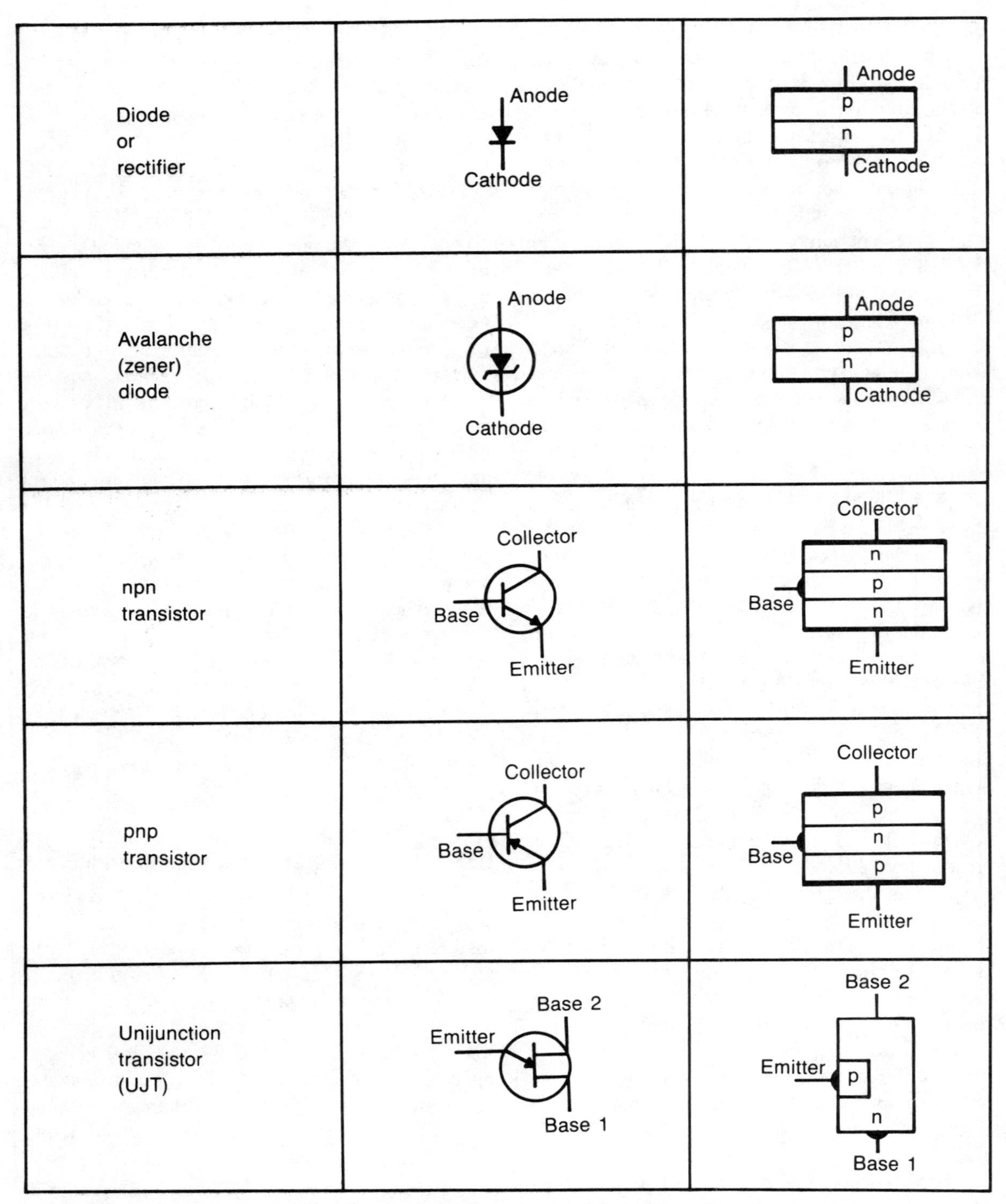

Fig. 7-2. Symbols and structural drawings of common semiconductor devices. (Part 1 of 3 parts.)

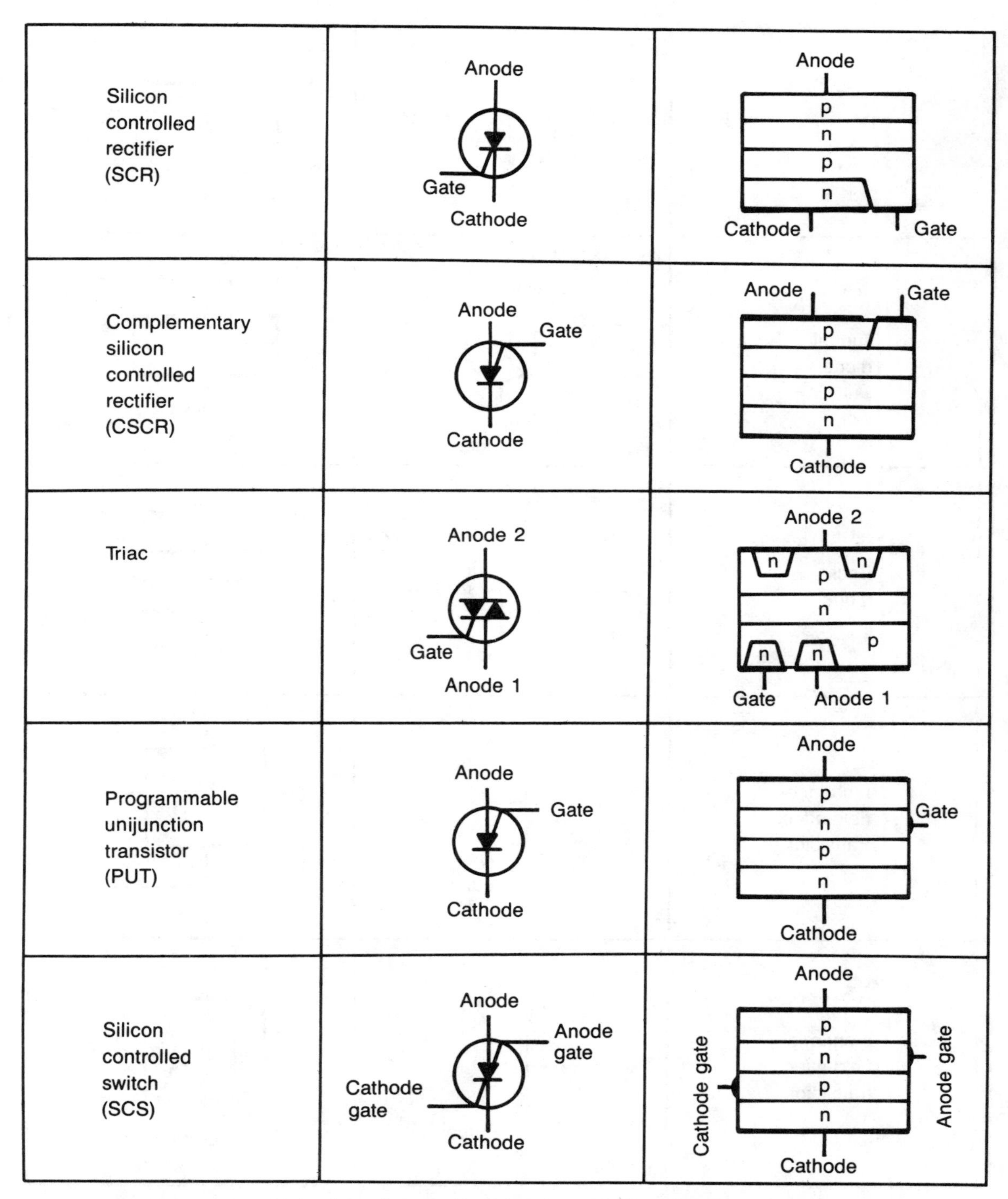

Fig. 7-2. Symbols and structural drawings of common semiconductor devices. (Part 2 of 3 parts.)

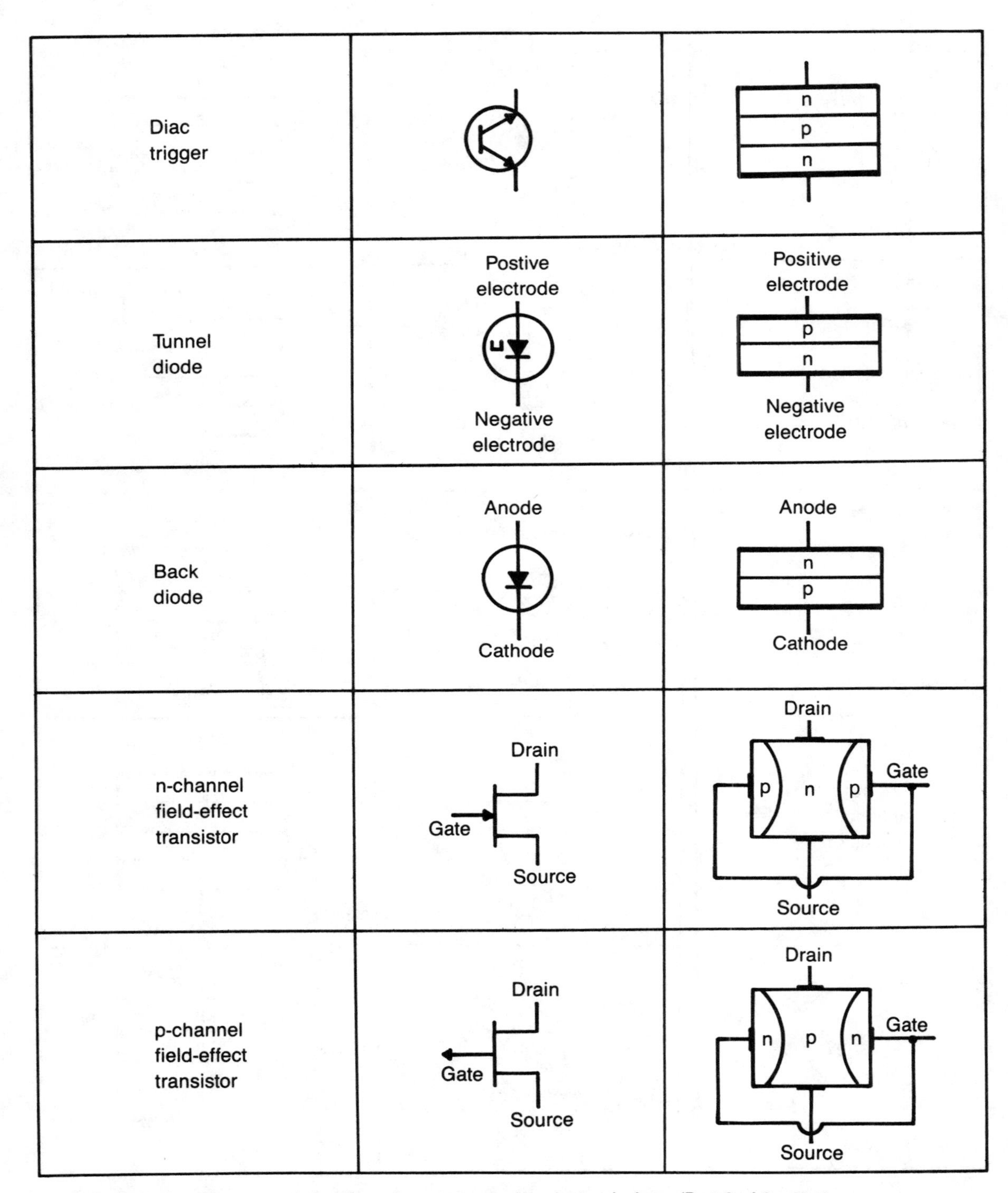

Fig. 7-2. Symbols and structural drawings of common semiconductor devices. (Part 3 of 3 parts.)

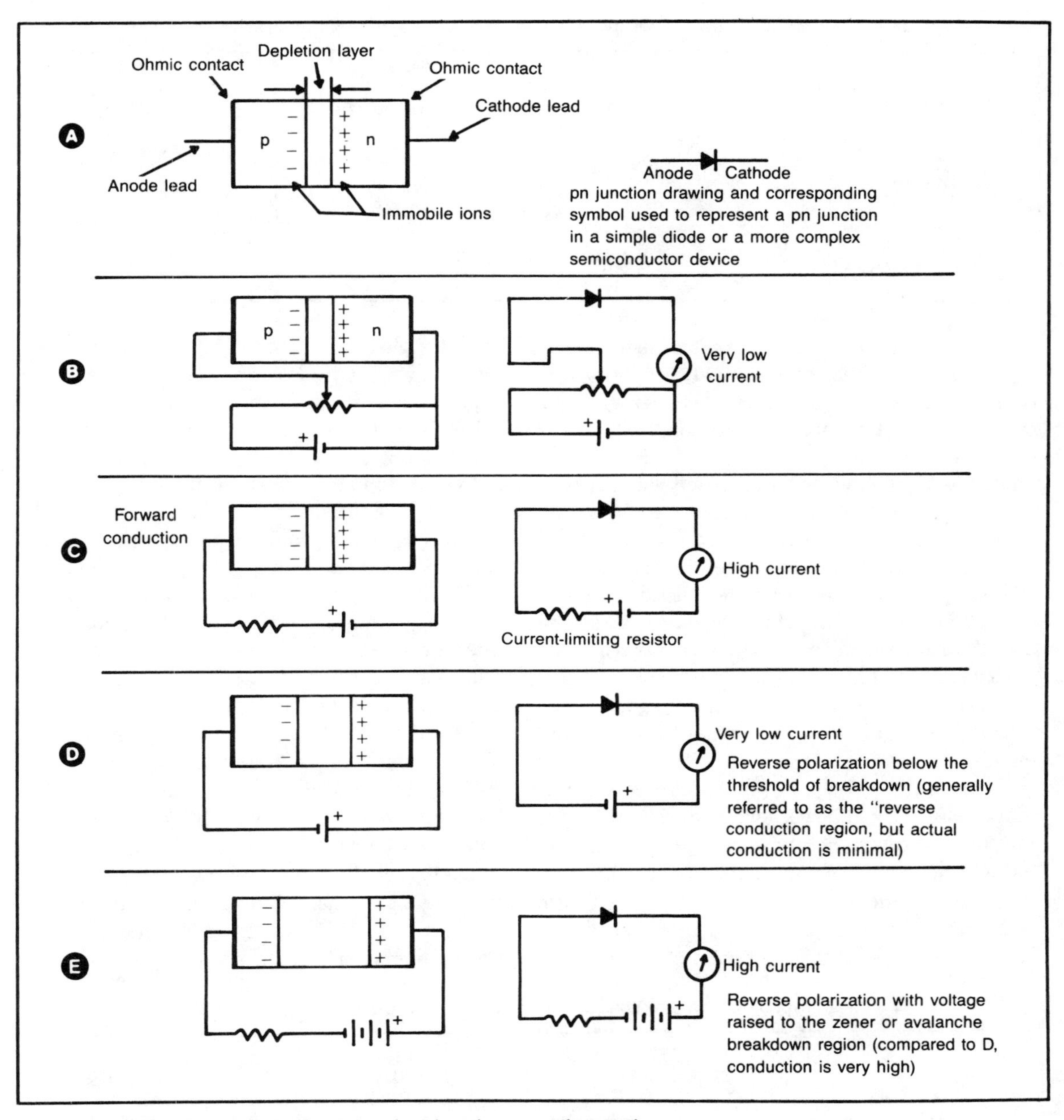

Fig. 7-3. Pictorial and symbolic drawings of pn junction operating modes.

be checked without driving the junction to the zener breakdown point. Thus, high-conduction (low resistance) on the R × 10 scale for the reverse-conduction test means a shorted or defective junction for most devices.

There are certain precautions we must observe when testing semiconductor devices with the ohmmeter. In particular, a VOM can deliver a rather high current on its R × 1 scale. In fact, many transistors were damaged by ohmmeter tests during the

Table 7-1. Typical Ohmmeter Output Characteristics of Six VOMs Selected at Random.

RANGE	OPEN-CIRCUIT VOLTAGE	SHORT-CIRCUIT CURRENT	MAXIMUM AVAILABLE POWER
R × 1	1.5 V	360 mA	140 mW
R × 10	1.5 V	36 mA	14 mW
R × 100	1.5 V	3.6 mA	1.4 mW
R × 1K	1.5 V	.36 mA	.14 mW
R × 10K	31.5 V	.80 mA	6.2 mW
R × 100K	31.5 V	80* μA	0.6 mW

first several years of the junction transistor. Later transistors, even the small-signal types, proved to be more rugged and less vulnerable to damage from the injection of high currents. Nonetheless, the ohmmeter remains a potentially destructive instrument and should not be blindly used without heed of its characteristics and a knowledge of the device to be tested. For example, an insensitive ohmmeter which has proved useful in general resistance and continuity testing can destroy the junctions of small transistors such as are typically found in the imported transistor radios. Also, the junctions in other devices such as tunnel diodes, junction FETs, and microwave transistors often are not rugged enough to withstand the currents supplied by some ohmmeters.

Tables 7-1 & 7-2 give you an idea of the currents, voltages, and power levels which ohmmeters can supply through their test prods. As you can see, there is much variation to be found in practical instruments. Some meters are relatively safe to use on all range scales, whereas others could easily destroy some solid-state elements. Although the direct-coupled output of an electronic meter is usually lower, there is the possibility of ac power-line leakage current. This is supposed to be minimal because of human safety considerations, but leakage due to capacitance and resistance is not always negligible, especially under unfavorable grounding conditions. An electronic meter which has an ordinary 2-prong line plug, rather than a 3-prong or polarized plug, should be checked for any 60-Hz voltage between the prods and from the prods to ground. Of course, an electronic meter which operates from its own self-contained batteries is safe in this respect. (If the meter is a combination ac-dc instrument, do not have the ac plug connected to the power mains, even though the panel switch is set at "internal battery").

Most of the forward-conduction tests applied to small devices, say, under 1/2-watt dissipation rating, can be satisfactorily carried out with the ohmmeter set at its R × 10 range. This avoids the much more potentially dangerous current that may be present on the R × 1 range in some meters. A large number of silicon junction devices tend to pro-

Table 7-2. Typical Ohmmeter Output Characteristics of Common Electronic Meters.

RANGE	OPEN-CIRCUIT VOLTAGE	SHORT-CIRCUIT CURRENT	MAXIMUM AVAILABLE POWER
R × 1	1.5 V	155 mA	58 mW
R × 10	1.5 V	16 mA	6 mW
R × 100	1.5 V	16 mA	6 mW
R × 1K	1.5 V	160 μA	60 μW
R × 10K	1.5 V	16 μA	6 μW
R × 100K	1.5 V	1.6 μA	0.6 μW
R × 1 Meg	1.5 V	160 nA	60 nW

duce readings in the central region of the ohmmeter scale when so tested. Germanium devices tend to show somewhat higher indications, that is, lower resistance.

There is another aspect to this precautionary consideration. In some cases, there has been evidence of cumulative damage to junction devices from minor electrical abuse. Even though you might not detect any change after applying a high test current, most sophisticated tests might reveal an increase in noise generation, degraded linear operation, or a decrease in the ratio of forward-to-reverse conduction. Such small amounts of damage are almost unnoticeable. In any event, it behooves the tester to use low, rather than high, dc currents and voltages whenever possible.

Wherever feasible, a test made on a device should be conducted under known conditions rather than on a cut-and-try basis. Thus, the polarity of the meter prods should be known prior to testing. It is well to keep in mind that the polarity of the test prods changes with different settings of the range switch on some meters. Of course, certain tests are bound to be exploratory in nature. For example, the unidentified leads of an older-type or foreign-produced transistor will necessarily involve blind tests. Even so, the polarity, current and voltage available at the prods of the ohmmeter should be known at all times. Otherwise, we may come up with a "strange" device with substantially zero resistance between its mysterious elements!

UNDERSTANDING TEST RESULTS

Generally, but not always, the condition of a junction is determined by its forward-to-reverse conduction ratio. In silicon devices this ratio tends to be extremely high, approaching infinity for many practical purposes. Thus, it will often be found that an approximate half-scale deflection will be obtained on the R × 10 range one way and no deflection can be detected on the highest range on the meter, say an R × 10,000 range, the other way. If the device is a high quality unit with passivated fabrication of the junction, a more sensitive ohmmeter with an R × 100,000 range might still reveal a negligibly small deflection. And this is in spite of the fact that the higher range scale is likely to involve 20 times the voltage of the lower ranges. It is almost as though we were dealing with a vacuum device! Ultimately, however, we could expect to detect a small leakage resistance if we resorted to yet higher resistance ranges, say R × 1 megohm on an electronic meter.

If a small silicon diode showed appreciable reverse conduction on the R × 10,000 range, it is true that the ratio of forward-to reverseconduction might still be fantastically high. Such a diode might, indeed, appear to perform as well in many applications as a new diode with negligible conduction on the R × 100,000 range of the ohmmeter. Before the advent of junction devices, older type semiconductor diodes such as selenium rectifiers had very much lower forward-to-reverse conduction ratios than that of our hypothetical "high-leakage" silicon diode. But this is not a valid criterion of comparison. The fact that the silicon diode tests poorly compared to new units of the same type is grounds for concern. It is always best to replace such devices, for it is difficult to guess what the degradation actually signifies. It could be the result of moisture or other contamination, or could be due to an impaired crystalline structure. Experience teaches that such junction characteristics often indicate a dying device.

The reference to forward-to-reverse conduction ratio should not lead to any confusion. It is true that the ohmmeter scale is calibrated in resistance units, that is, ohms. However, the meter movement is actuated by current, with low-resistance indications corresponding to high current; that is, high conduction. Conversely, high-resistance indications correspond to low current or low condition. Therefore, a large forward-to-reverse conduction ratio implies low resistance for forward conduction and very high resistance for reverse condition.

In practice, you will find a considerable variation in forward conduction, but most readings will tend to fall conveniently on the R × 10 scale. On the other hand, a tremendous variation will be found in the reverse conduction resistance. Junctions in inexpensive germanium devices may show reverse conduction resistances which are readable on the R × 1000 scale, whereas high quality silicon devices often show negligible reverse conduction when tested with an electronic meter on the R × 1 megohm scale. A point to keep in mind is that the exact value of resistance indicated by an ohmmeter cannot be taken too seriously. This is readily shown by reading the forward resistance of a diode on more than one range, say, the R × 1, R × 10, and R × 100 ranges. Three different values will be obtained, and this is to be expected because the diode has a nonlinear voltage-current relationship, at least for the currents supplied by ohmmeters. Not only will it be found that the forward resistance of a junction diode increases with the ohms range, but different readings for the same ohms range may be obtained from different meters. The reverse resistance tends to be less variable, but is often difficult to measure because of its high value. Germanium devices often are found to vary more than silicon devices in response to different measurement conditions. This is because of the greater sensitivity to temperature rise from current-induced heating. Uncertain as all this may appear, the high forward-to-reverse conduction ratio of junctions is a unique condition and it is generally not difficult to determine whether a semiconductor device is basically operational or defective.

Another way to look at the concept of forward-to-reverse conduction ratio is to recall that conduction bears a relationship to resistance. Thus, the measurement of the forward and reverse resistance of a good diode will yield a small fraction, say 300 over 300,000, where the numbers are the forward and reverse readings in ohms. The reciprocal of a small fraction is a large number, however, and represents the conversion from resistance ratio to conduction ratio. If the test evaluation were made with a meter calibrated in current units, the ability of the junction to pass current would probably be more naturally related to its conduction characteristics. In actual practice, the ohmmeter works rather well, since we learn to involuntarily respond to indications of good and defective junctions without much conscious analysis.

Procedure for Determining Activity and Leakage of a Transistor:

Objective of Test:To demonstrate a useful transistor test circuit utilizing a VOM and several simple components.

Equipment Required:

Transistor socket
VOM
Pushbutton switch (press to make contact)
330K, 1/2-watt 10 percent composition resistor
0.1μF 200-volt Mylar capacitor
A half dozen or more small-signal junction transistors, preferably comprising npn, pnp, silicon and germanium types.

Test Procedure: Connect the transistors as shown in Fig. 7-4. Although the circuit is quite primitive, it is well to make sure that inadvertent shorts will not be likely. Also, the leads should be no longer than necessary. Otherwise, undesirable effects may be produced from RFI and power-line pickup. The 0.1 μF capacitor should be connected directly between the base-emitter terminals of the transistor socket. Notice that the VOM is connected to test npn transistors. When testing pnp tran-

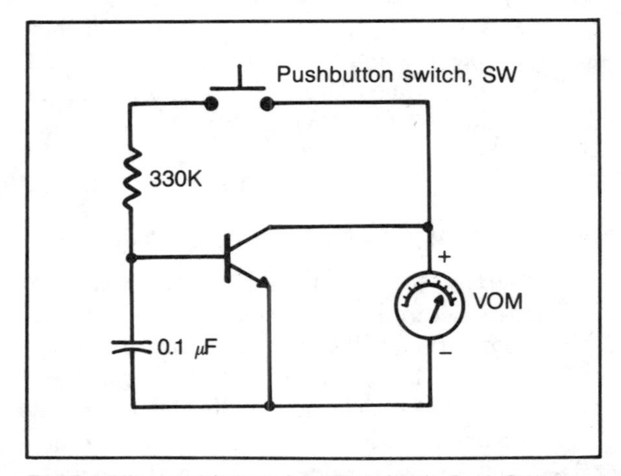

Fig. 7-4. Ultra-simple transistor test circuit. Push SW down to read the approximate dc beta on the R × 100 scale. With SW up, read the leakage on high range scale (generally R × 10,000 or higher). Quality silicon transistors will have extremely high leakage resistance and, if in good condition, may not cause a discernible meter deflection.

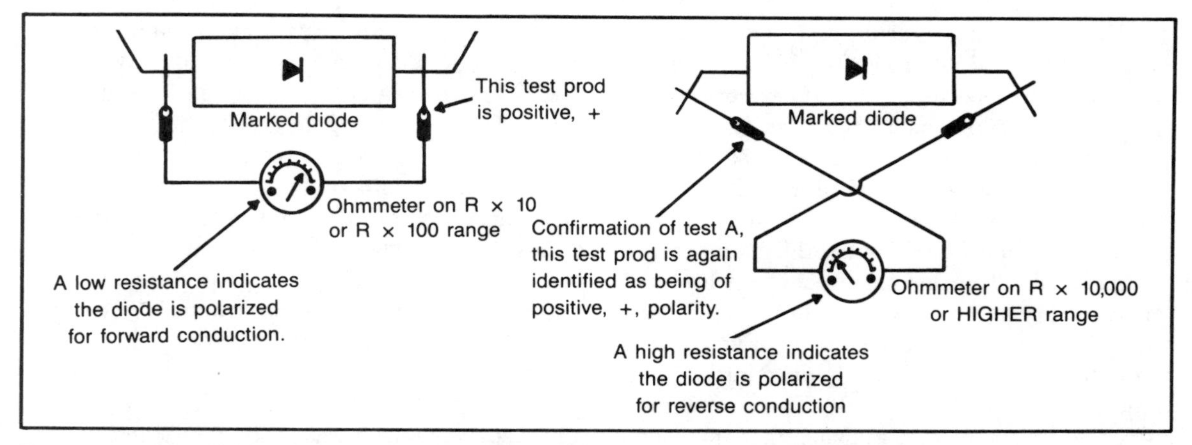

Fig. 7-5. A marked diode can be used to determine the polarity of an ohmmeter's test prods.

sistors, the meter polarity is reversed from that shown. The polarity of the ohmmeter test prods can be determined with another meter set to read volts. If the second meter is not available, you can determine the test prod polarity with the aid of a marked semiconductor diode. When the prods are connected to the diode terminals in such a way that conduction (low resistance) is indicated, the prod which is then contacting the cathode symbol on the diode is positive (see Fig. 7-5).

The tester can be built on a small chassis with pin jacks for the meter test prods. Also, a double-throw, double-pole switch can be used to conveniently change from npn transistors to pnp types. Despite its simplicity, this tester can be extremely useful in servicing transistor equipment.

In order to use this tester, a transistor is plugged into the socket. With the pushbutton switch, SW, in its up position, any reading obtained on the meter represents leakage. Specifically, such a reading indicates I_{ceo}, the collector-emitter current with the circuit open. Negligible, if any, meter deflection should be produced for good silicon transistors, even on the R × 100,000 range.

To determine whether the transistor is basically "alive," notice the deflection (away from infinity, towards zero resistance) on the R × 100 range when SW is depressed. The deflection will be approximately proportional to beta, the dc current gain of the transistor. The basic idea is to ascertain the existence or absence of "transistor action." Damaged transistors will produce no deflection or, judging from experience with a given type, one in which the conduction is abnormally low. If the previous leakage test indicated abnormally high leakage, the activity test tends to lose its significance and such a transistor should be considered defective.

Comments on Test Results: Obviously, the tester is used to determine the quality of a transistor. You learn from experimenting or experience with a given type how to assess the relative value of the transistor being tested. The important feature of the tester is its simplicity, since it is designed around a commonly available VOM and its capability of providing go, no-go information about transistors. The tester is intended for common small-signal transistors. These generally have dissipation ratings of less than 1 1/2 watts, with most being rated in the vicinity of several hundred milliwatts. Most of them are packaged in TO-5, TO-18, or similar cans, or are small epoxy types. The tester is not intended for "power" transistor checks but could be adapted for some of these with some experimentation. More base drive would have to be provided by means of a lower value base resistor, and the forward-gain indication would be read on the R × 10 or R × 1 range.

This tester can be calibrated to read the approximate value of beta by inserting current meters in the base and collector leads. Beta is then the ratio of the collector current to the base current. Although the ohmmeter deflection is actually the sum of collector and base currents, rather than collector current alone, this need not cause error if the beta calibration is carried out with current meters as suggested. The principal effect of the base current through the ohmmeter is to make beta calibration of the scale a bit nonlinear. Even this is negligibly small for commonly encountered beta values, say between 20 and 200. The arrangement and choice of components is such that the collector current will tend to be between a substantial fraction of a milliampere to about three milliamperes. This represents a good collector current range for testing the beta of most small-signal transistors.

A meter with a 1 1/2 volt cell on the R × 100 range and with a 20,000 ohms-per-volt sensitivity in its voltmeter modes is particularly well suited for a tester of this type. Meters with other voltages and sensitivities can be used if the base resistor is changed so that the operation as described above is approximately simulated.

The capacitor between base and emitter terminals of the transistor socket bypasses any rfi or electrical noise which may be picked up. Otherwise, unrealistic indications of leakage could be obtained. This can easily happen in the vicinity of a powerful radio station.

TRANSISTOR TESTS WITH A CURVE TRACER

The curve tracer is a specialized oscilloscope designed to show the dynamic operating characteristics of a wide variety of semiconductor devices. Because of its cost, and perhaps other factors such as bulk, it is not yet a common test instrument in servicing work. However, in laboratories where research, design, and development are being conducted, the curve tracer is often considered indispensable because it is the only test instrument that provides at a glance the complete story of the performance of a semiconductor device. Additionally, the tests provided by a curve tracer are extremely instructive. The insights gained from inspecting and evaluating a dynamic plot of device characteristics are bound to make test information from other instruments and approaches more meaningful. This is one of the basic reasons for investigating curve-tracer tests at this point. We will deal primarily with the interpretation of transistor curves, although the basic logic involved will naturally relate to tests with other devices.

We will not concern ourselves with the step-by-step setting of the knobs on the curve tracer. The display of the characteristics of a transistor are easy enough to obtain when you are aware of the basic idea of the instrument. The curve tracer is designed to produce a sweep of the collector-emitter voltage on its horizontal axis while the collector current is depicted on the vertical axis. The main feature of the curve tracer, however, is that the display of collector current vs collector voltage is deliberately split into a multiline pattern by means of a stepped variation in base current. Because of this "staircase" of base currents, the collector characteristics are seen as a family of curves with the base current as the variable parameter. The series of base currents is repetitive and the cycle of advancing increments of collector current is fast enough to appear stationary on the screen. Therefore, the overall pattern is an instantaneous picture of transistor performance which can cover the area of particular interest as well as behavior at other voltages and currents. How much of the dynamic range we want displayed is controllable by the panel selector knobs. Obviously, such an instrument must save the tedious time necessary to manually plot such curves where interest is centered not on go, no-go information, but rather on really useful engineering information.

It is not usually realized how much useful test information is contained in a single display of transistor dynamic characteristics on a curve tracer. As an example of the wealth of data available from such a test, consider the dynamic display shown in Fig. 7-6. Notice that the base-current increments have been selected to be 100 microamperes.

1. dc beta $= \frac{I_c}{I_b} = \frac{21 \text{ mA}}{200\ \mu\text{A}} = 105$ @ V_{ce}

$= 10$ volts and $I_c = 21$ mA

2. ac beta $= \frac{\Delta I_c}{\Delta I_b} = \frac{20.6 - 11.0 \text{ mA}}{200 - 100\ \mu\text{A}}$

$= \frac{9.6 \text{ mA}}{100\ \mu\text{A}} = 96$ @ V_{ce}

$= 8$ volts, $I_c = 15.8$ mA

3. Saturation voltage, $V_{ce(sat)}$ equals 0.4 volt when I_c is 15 mA.

4. Collector leakage current, I_{ceo} equals 1.0 mA when V_{ce} is 10 volts.

5. Collector cutoff current, I_{cbo} is approximately I_{ceo}, divided by dc beta for silicon transistors or 1 mA divided by 105 (9.5 microamperes) when V_{ce} is 10 volts.

6. The collector-emitter breakdown voltage is a function of the base drive: 18 volts when I_b is 100 microamperes.

7. Collector dynamic output resistance

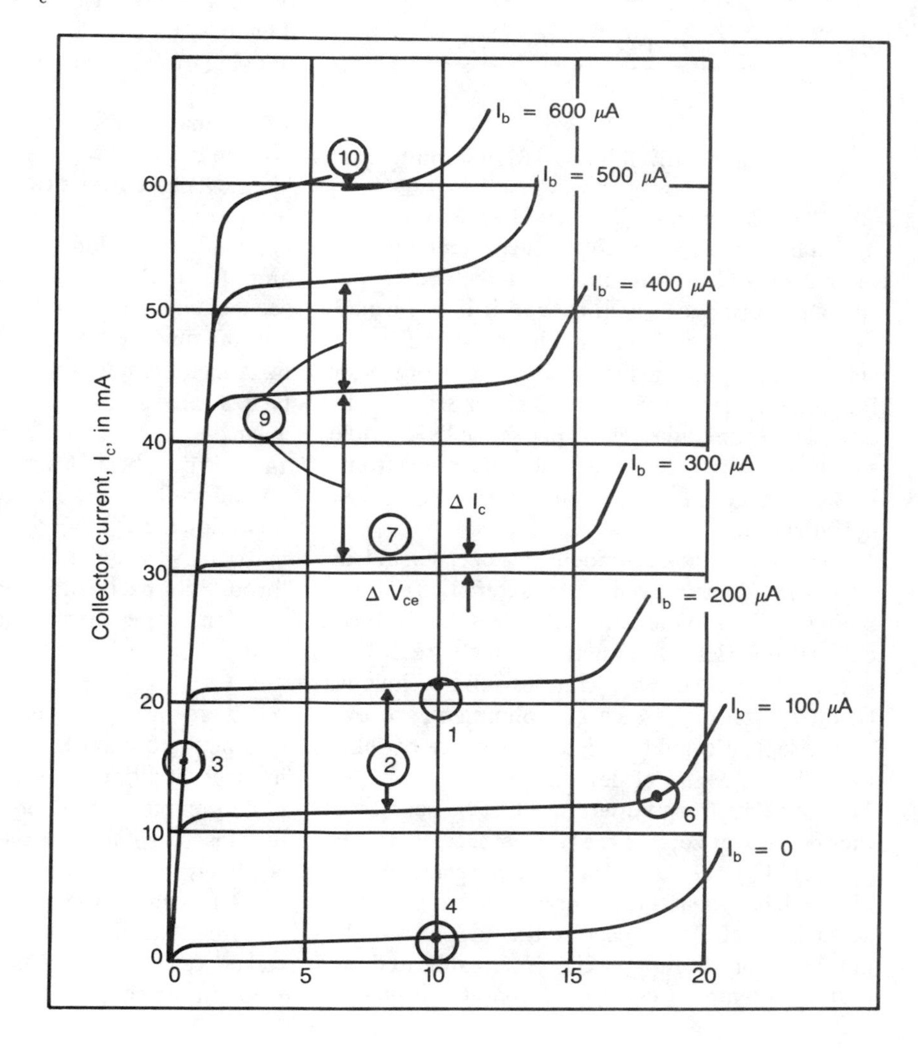

Fig. 7-6. Family of transistor collector current curves displayed on a curve tracer.

$$@\ V_{ce} = 7 \text{ volts},\ I_c = 30.5 \text{ mA} = \frac{\Delta E_c}{\Delta f_c}$$

$$= \frac{10 = 4 \text{ volts}}{31 = 30 \text{ mA}}$$

$$= \frac{1}{1 \times 100^{-3}} = 6000 \text{ ohms}$$

8. Dynamic output admittance, h_{oe} is the reciprocal of the dynamic output resistance. Notice that h_{oe} is one of the standard parameters found in transistor specification sheets.

$$h_{oe} = \frac{\Delta I_c}{\Delta E_c} = \frac{1 \times 10^{-3}}{6} = \frac{1}{6000}$$

$$= 0.000167 \text{ mho} = 167 \text{ micromhos}$$

9. The relative linearity of the transistor is determined from the uniformity of the spacing between the curves. This is an important feature of transistor performance, even though it is not usually expressed in general terms. The more uniform the spacing throughout a simulated range of operation, the less distortion that the transistor will develop as a Class A amplifier. Here, one quick look enables the designer to select the best transistors from a group which otherwise satisfy the basic type specifications.

10. One quick look also identifies occasional transistors which have strange characteristics in one or more members in the family of curves. Such defects can cause various malfunctions which are difficult to pin down with conventional servicing techniques. Ordinary static tests with an ohmmeter, or even with a sophisticated transistor tester, are not likely to reveal this kind of defect.

11. The effect of temperature can be seen by either heating or cooling the transistor while it is being tested by the curve tracer. The results of such tests will be essentially general in nature if the transistor is carefully exposed to the heat of a soldering iron, or more specific if a calibrated and regulated temperature cycling chamber is used. In the latter case, an extension cable is available as optional equipment by the major manufacturers of curve tracers.

12. Although not shown in the family of collector current curves in Fig. 7-6, the input characteristics of transistors can be readily evaluated by testing the base emitter section as a diode. The reverse breakdown voltage of the base-emitter junction is often omitted from most specification sheets. In some applications, this characteristic must be known if trouble is to be avoided. For example, if the transistor operates as a Class C amplifier, it is often best to avoid those transistors which have inordinately low base-emitter breakdown voltages. This may mean selecting from a number at hand, or going to a different brand or substitution of a different type.

13. In many circuits using pairs of transistors, it is often better to select two transistors with nearly identical characteristics than to look for "hot" performers in terms of beta or other parameters. Such applications include push-pull and differential amplifiers. Some curve tracers have a double-socket arrangement with a 2-position switch so that you can quickly make such a comparison. In other instances, photographs of the tracer display may be desirable so that the curve may be evaluated at leisure.

14. It is possible to learn more about less frequently considered characteristics such as second breakdown, negative resistance regions, and thermal runaway by actually putting the transistor through its paces in its equipment or breadboard position, rather than in the test socket of the curve tracer.

It seems that the curve tracer is not only destined to become increasingly commonplace, but in those situations where initial cost is a stumbling block, adapters are available to convert an ordinary oscilloscope into a curve tracer. Such units are relatively inexpensive and represent a sound investment for the well-equipped service shop.

Junction FETs can be tested in some commercial curve tracers by connecting a 1000-ohm resistor from the base terminal to ground. The base in-

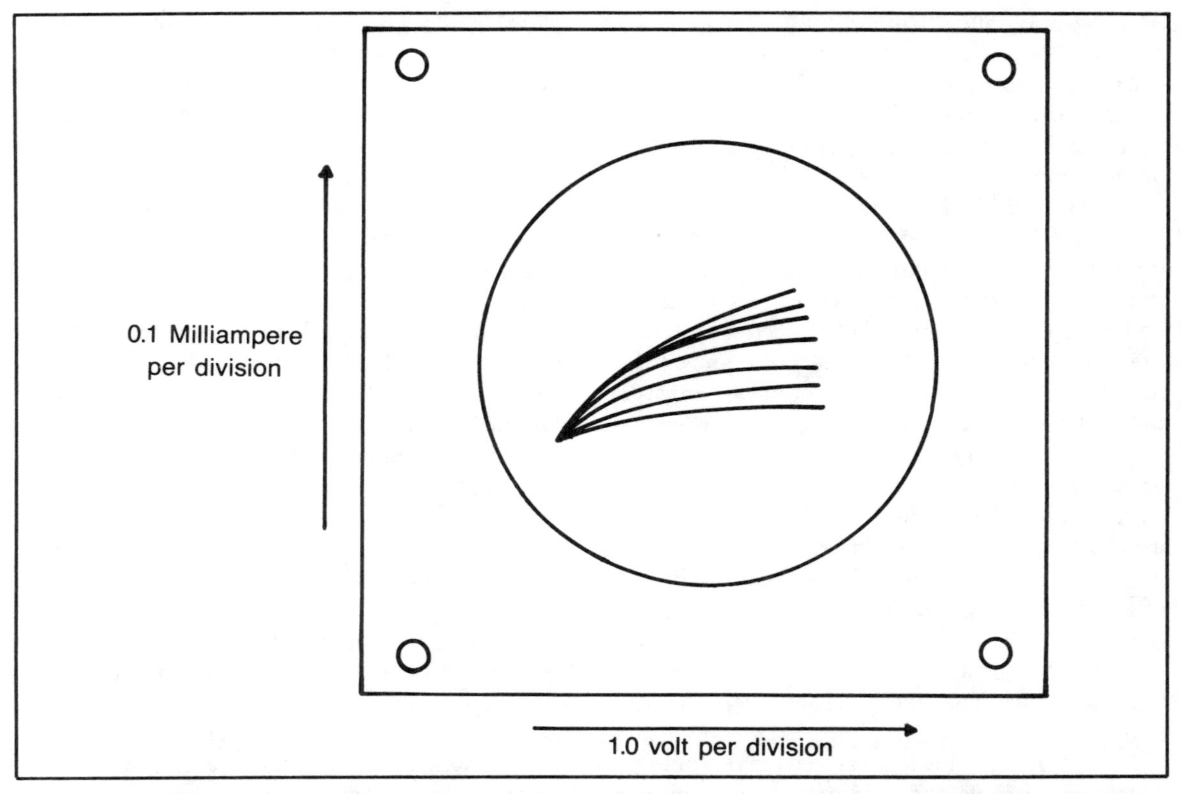

Fig. 7-7. FET characteristic curves displayed on a curve tracer. Top curve is for zero gate voltage; successive lower curves represent 0.5-volt steps.

cremental current selector is positioned at 0.5 milliampere. This produces an incremental IR drop of 0.5 volt, which is suitable for the gate circuit of the FET. The FET undergoing test is connected to the test terminals, or plugged into the test socket as if the drain were the collector of a transistor, the source were the emitter, and the gate were the base. The test should produce a family of curves resembling those of a bipolar transistor or a pentode tube (see Fig. 7-7).

Diode tests are similar to those used for transistors. The diode is connected between collector and emitter test terminals with no connection made to the base test terminal. The polarity of the diode is not important because forward and reverse characteristics can be displayed by alternating the type selector knob from npn to pnp. Of course, if you observe polarity when connecting the diode to the test terminals (anode to collector and the cathode to the emitter test terminal), the npn mode will display the forward-conduction characteristics and the pnp mode will display the reverse-conduction characteristics. The collector-voltage selector knob will have to be changed for the best display of the two diode characteristics. The diode test will not produce a display of a family of curves, since there are only two parameters involved, not three as in transistors and FETs. The diode test also provides an excellent evaluation of zener diodes when the reverse voltage is cranked up sufficiently to display the breakdown region.

POWER TRANSISTORS

Transistors having collector dissipation ratings of several watts and greater are very easy to test with a VOM. This is true for both silicon and germanium units. Two possible hindrances involved with smaller transistors generally do not apply to the go, no-go testing of power transistors. In the first place, there need be no hesitancy in the use of the ohmmeter, even on its R × 1 range. Indeed, it turns out that the R × 1 range is extremely useful for testing the basic diode action of the emitter-base junction and the collector-base junction. The junction of such power transistors is rugged enough to be immune to damage from the current output of the ohmmeter. Secondly, experience shows that there is a pronounced tendency for power transistor failure to be catastrophic. A test of a power transistor is likely to result in either clearcut acceptance or rejection, with relatively few units falling into a gray area of uncertainty. Defunct transistors will usually be found to have very definite shorts or to be completely open-circuited. A common indication of a defunct power transistor is virtually zero resistance across its collect-emitter terminals without regard to polarity.

For applications where the activity of the transistor is important, dc beta may be readily measured by injecting a measured current into the base-emitter diode and reading the resultant current which then flows in the appropriately biased collector-emitter circuit. Since beta may vary considerably with base drive, an attempt should be made to approximately simulate the base current at which the transistor normally operates. Somewhat more leeway is allowable for the collector-emitter voltage. Between 5 and 20 volts suffices for a wide variety of valid tests. The collector circuit should be fused in order to protect the current meter and it is desirable to use a power supply with a limited current capability. This objective can be approached by inserting a resistance in the collector circuit, but too high a value will detract from true beta measurements. A pushbutton switch or similar component should be used so the circuit can be turned on and off quickly in the event the transistor has a short. Simple power transistor tests are shown in Fig. 7-8.

The above static test can be extended to provide data for a set of collector current vs collector voltage curves with base current as the variable factor. This, of course, is tedious and probably justifiable only when the linearity of the transistor is an important consideration. Some power transistors are notoriously nonlinear elements. Circuit applications where this can be detrimental generally employ techniques such as feedback to overcome the bad effects which would otherwise occur from nonlinearity. In many power transistors applications, such as series-pass elements in regulated power supplies or switching elements in inverters or converters, linearity is of little or no consequence. Probably the best way to check linearity where it may be of importance, such as in hi-fi audio amplifiers (despite feedback), is to use a curve tracer. However, a special adapter is generally necessary to accommodate the current ranges of power transistors. The curve tracer also allows selection of approximately matched pairs for push-pull stages.

An important thing to be aware of when testing the output transistors in automobile radios is that the metal container, which is usually internally connected to the collector, is insulated from its heatsink, or chassis. The insulation is usually a thin mica spacer. (In some cases, more so in power supplies and inverters than in radios, the insulation is an electroplated film on the heatsink.) Very careful ohmmeter tests should be made to ascertain that the transistor is, indeed, insulated from its mounting. Before replacing the mica spacer, it should be coated with a silicone compound made for this purpose. Such a coating improves the thermal conductivity of the transistor-mounting interface. This can be quite important in auto radios because the power transistor is often a germanium type which operates under electrical and thermal conditions allowing little margin of safety. If the transistor has inadequate thermal contact, it will run hotter; this may lead to thermal runaway or other catastrophic failure. At best, the audio fidelity of the radio may be impaired at high volumes.

Germanium power transistors are pnp types. Npn types are usually made of silicon, but pnp types have become more popular. These are often found in association with npn types in push-pull or complementary-symmetry stages. The collector-emitter leakage resistance of silicon power transistors can be expected to be seen on the R × 100,000 or the R × 10,000 ohms range, whereas that of germanium power transistors often produces a considerable deflection on the R × 1000 ohms range.

Procedure for Measuring Leakage Current in Semiconductor Junctions

Objective of Test: To measure the residual current in reverse-biased junctions with common instruments.

Test Equipment Required:

VTVM with 13-megohm input resistance or equivalent solid-state instrument. Also, the meter should have a 1.5-volt dc voltage range. (These are commonplace meter characteristics).

9.0-volt transistor radio battery.

Rotary switch with six (or more) positions.

The following 5 percent 1/2-watt composition resistors: 110 megohm (can be made up from a 100-megohm and a 10-megohm resistor in series), 1.1 megohm, 100K, 10K, 1K.

A handful of transistors, diodes, SCRs, etc.

Test Procedure: Connect the components as shown in Fig. 7-9. The 110-megohm resistor is connected across the voltmeter terminals, whereas the other resistors can be switched. The purpose of the 110-megohm resistor is to convert the 1.5-volt range into a 150-nanoampere current meter. Thus, with the switch in position A, a full-scale reading corresponds to a current of 150 nanoamperes. The converted current scale is linear, so that a half-scale deflection of 0.75 volt corresponds to 75 nanoamperes, etc. When the switch is in position B, the full-scale deflection is 1.5 microamperes and the remaining positions provide sequentially higher current ranges by multiples of ten as indicated. The 9-volt battery is suitable for a wide variety of junc-

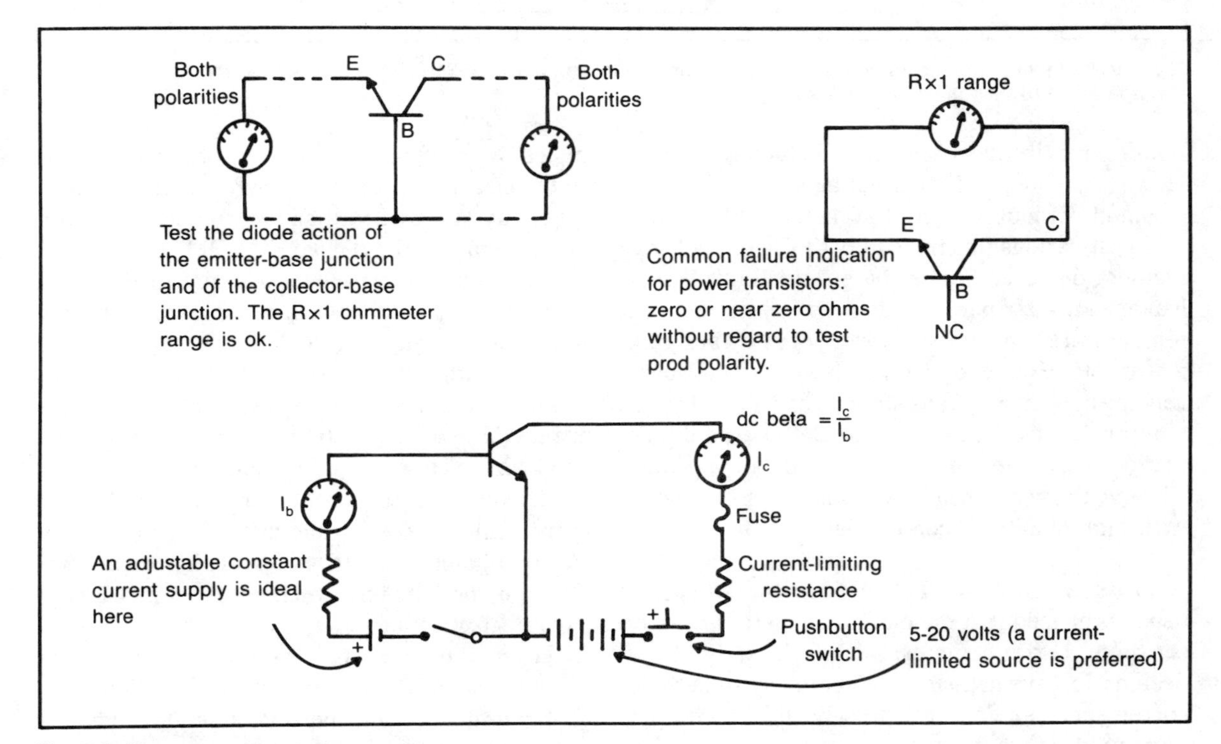

Fig. 7-8. Power transistor tests.

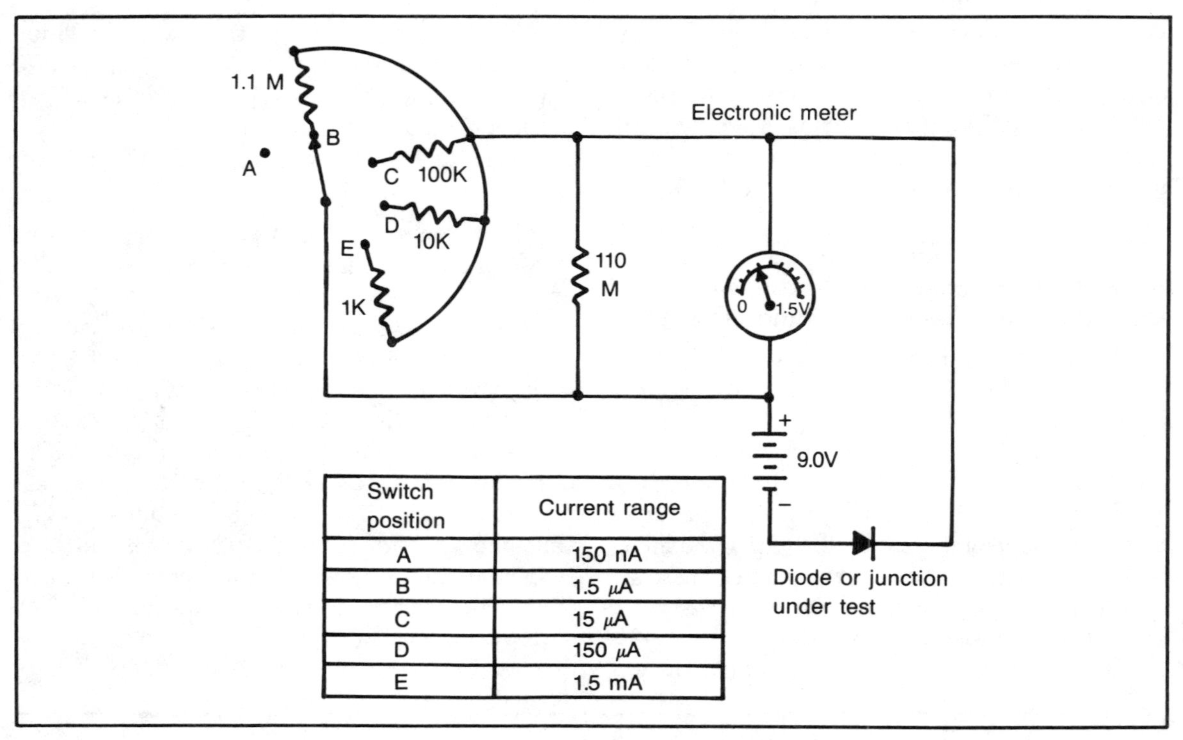

Switch position	Current range
A	150 nA
B	1.5 μA
C	15 μA
D	150 μA
E	1.5 mA

Fig. 7-9. Test setup for measuring the leakage current in junction devices. The circuit is designed for meters with a 1.5 V range and 110 megohms input resistance.

tions found in everyday semiconductor devices. However, if desired, this voltage source can be changed without altering the meter calibration.

Test various junctions in both silicon and germanium devices. Notice the substantially higher leakage in germanium devices. Examine the temperature sensitivity of the devices by bringing a heat source close to them. Test the collector-base leakage current in a transistor and also its collector-emitter leakage current. The latter should be greater than the former by the dc beta value. Observe the very small leakage currents associated with high quality silicon diodes.

Comments on Test Results: It is not usually realized that the electronic meter is a very sensitive current meter, although some later designs do have nanoampere and microampere dc current modes. This test procedure is somewhat more refined than the leakage measured on the high ranges of an ohmmeter because of the ease with which the actual current can be read. Also, greater sensitivity is provided than is available with even the high ohmmeter ranges on VOMs.

The leakage current is actually made up of several factors, but the two major ones are probably saturation current and currents through surface resistive paths. For most practical purposes, we combine these under the one designation—leakage current. The ideal junction is considered to have zero leakage current, and it should be pointed out that leakage current has been reduced by factors of thousands since the advent of commercially available junction devices. This has been accomplished by both material technology, primarily the general substitution of silicon for germanium, and better methods of fabrication, primarily the development of the process of "passivation."

Even the most inexpensive junction diode has vastly superior leakage characteristics when compared to the older "semiconductor" devices such as selenium or copper-oxide rectifiers. However, when the leakage current of a diode or other device is substantially higher than the usual worst-case value of another of the same type this should be cause for concern. Although a high-leakage junction may have a forward-to-reverse conduction ratio more than adequate for its circuit function, it is wise to replace such a device. High leakage is often the symptom of a developing failure. With the passage of time, you can safely wager that the leakage will increase, perhaps to the point of total failure. Also, inordinately high leakage is often accompanied by high noise generation. In transistors, high leakage can upset biasing and thereby degrade or destroy the operation of the circuit. If this doesn't occur at room temperature, the probability is often great that only a slightly elevated temperature is required to increase the leakage to an intolerable level.

If a higher voltage source than the 9.0-volt battery is used, the increased probability of avalanche breakdown should be kept in mind. This will be nondestructive even on the 1.5 milliampere range because of the high limiting resistance. On the other hand, due to the basic nature of electronic meters, a shorted junction will not burn out the meter movement.

This measurement technique can also be used to check leakage current during forward polarization when the voltage is below the turn-on point (mode B in Fig. 7-3). To do this, the voltage source should consist of a carbon-zinc cell and a mercury cell in series-opposition; that is, with either the positive or the negative battery terminals connected together.

The net voltage, somewhat greater than 200 millivolts, is suitable to measure forward leakage current before turn-on in silicon junction devices. Such information is useful for some circuit applications where it is desirable that negligible conduction takes place until the forward voltage is in the vicinity of about 600 millivolts. For example, a diode used as an amplitude clipper will provide a flatter topped (or bottomed) wave if forward leakage prior to the onset of conduction is relatively low. A check of germanium junctions by this method can also yield useful results, but tends to be of a more marginal nature because germanium junctions begin to conduct in the 200-millivolt range.

ZENER DIODES

Figure 7-10 shows four useful junction device tests where the reverse-voltage breakdown characteristic is an important feature of the circuit application, or in any event must be considered in both test procedure and design. Thus, in addition to zener diodes, it is useful to at least get a general idea of the reverse breakdown characteristics of other junction devices such as the emitter-base or collector-base junction of bipolar transistors, signal or rectifier diodes (even though normally forward-biased) and varactor and "tuning" diodes which are most useful in the reversed-biased condition, although sometimes driven to varying penetrations into the forward region as well.

The simple ohmmeter tests depicted in Fig. 7-10A can be useful, but must be interpreted with some caution. A near-zero ohmmeter reading for both polarizations on the R × 10 scale (or the R × 1 scale if the diode is a rectifier type or is known to be a rugged all-purpose type) is a reliable indication of a shorted junction. Similarly, an extremely high or infinite reading for both polarizations reveals an open-circuited diode. However, conduction for both polarizations must be interpreted carefully, especially for ranges higher than R × 10. For example, if the voltage across the ohmmeter prods is no longer 1 1/2 volts, but is higher than several volts, you must consider the possibility of zener or avalanche breakdown of the reverse-polarized junction. The internal batteries of many meters develop several tens of volts or even more on the moderate and especially the high ohms ranges. For the majority of meters, however, the R × 10 range is suitable for making a valid test. You should see the evidence of forward conduction with the prods connected one way (low resistance) and virtually no conduction (near-infinite resistance) with the prods connected the other way. It is presumed that the reverse breakdown voltage of

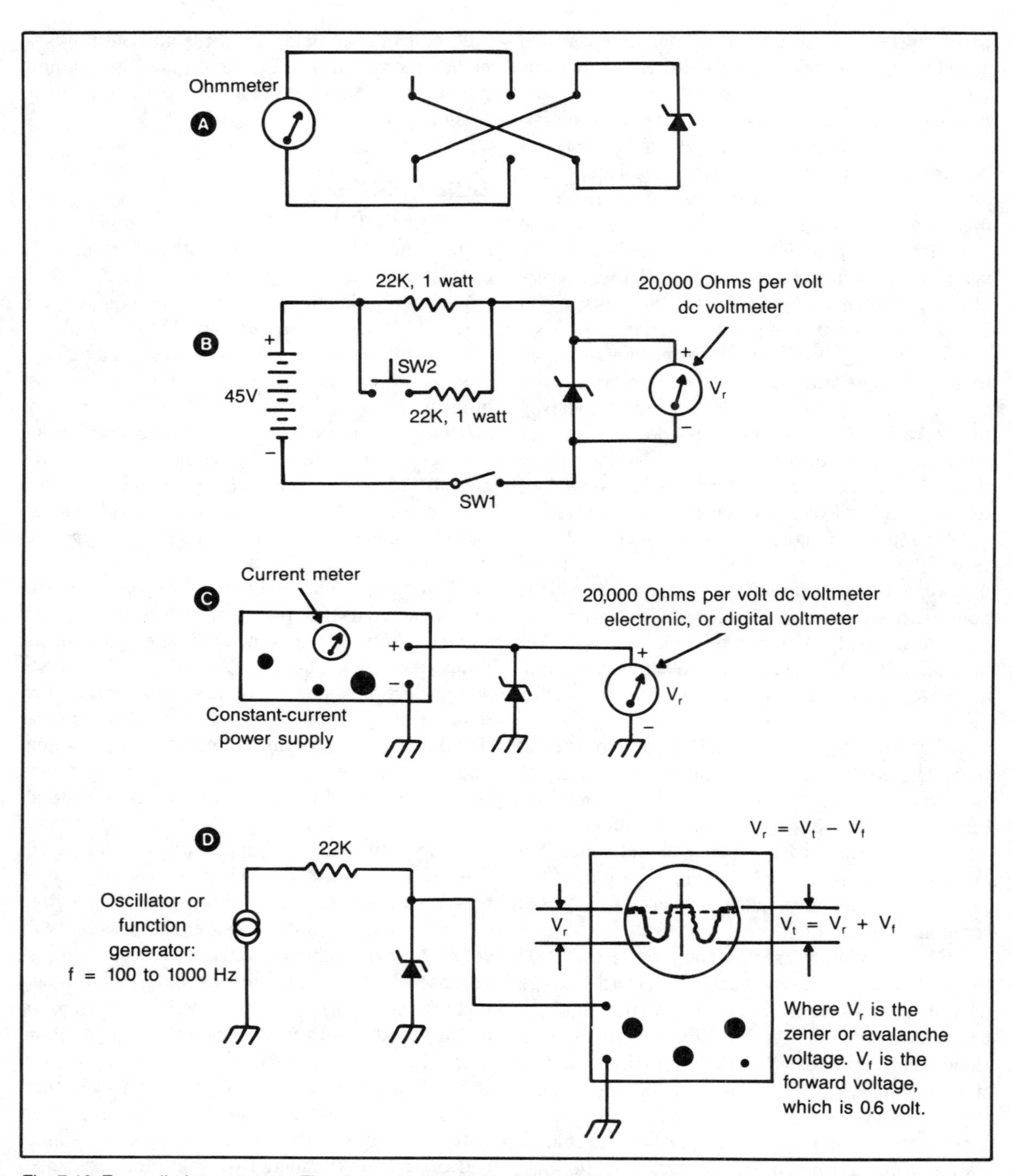

Fig. 7-10. Zener diode test setups. The ohmmeter test (A) is useful for detecting shorted and open diodes. The breakdown voltage test (B) is suitable for many situations. The breakdown voltage test at C is capable of better accuracy than test (B). For the wave-clipping method (D), a sine wave is OK, but a triangular-wave is preferred.

the diode is known and that the test is conducted as a go, no-go evaluation of the condition of the junction. More accurate information can be derived from the test procedures in Fig. 7-10B, C and D.

The simple test setup in Fig. 7-10B is suitable for testing a wide variety of diodes and other common semiconductor devices. You simply read the zener breakdown voltage from the voltmeter when switch SW1 is closed. In order to verify that the indicated voltage results from avalanche or zener breakdown and is not caused merely by the resistance of a defective junction or by insufficient battery voltage to reach breakdown, switch SW2 is momentarily closed. If the diode breakdown behavior is normal, very little change in voltage will be noticed. The usefulness of this test will be enhanced if it is ascertained that the battery is fresh, but only in cases when a battery is weak or a very tiny battery is used is there likely to be a big enough drop in the terminal voltage to distort the results. If, when switch SW2 is closed, voltage V_r drops more than several percent in zener diodes, it should be cause for concern. In the case of other devices, a greater tolerance can be allowed because the breakdown characteristic probably has a rounded "knee."

The test procedure shown in Fig. 7-10C makes use of a constant-current source, and is capable of greater accuracy than you would normally expect from the text procedure in Fig. 7-10B. Here, a digital voltmeter can be used to specify the observed V_r with a known current. With this method, more meaningful results can be obtained in terms of the values listed on the manufacturer's specification sheet. Also, you can get a good idea of how the diode operates in the actual circuit. (Notice that there is no danger of damage to the power supply if the diode is shorted because constant-current supplies limit the output current right down to a zero resistance load.) In order to avoid misleading results from this test, make sure that the compliance voltage of the supply exceeds V_r, the reverse-breakdown voltage of the diode. The compliance voltage is the highest voltage the supply can provide and yet maintain a constant load current. Although the supply can deliver an even higher output voltage at no load, this does not mean that the diode under test is in danger of receiving greater than the intended current. Rather, the higher than compliance voltage is accompanied by a near-zero current capability. Thus, there is no danger to either the supply or the diode if the desired supply current is first adjusted to flow into a short circuit. The inability of the supply to develop a high enough compliance voltage will simply prevent the diode from being operated in its zener or avalanche mode and it will behave as an open circuit.

The combination of a good constant-current supply, a modern sharp-knee breakdown diode, and a digital voltmeter can result in refined test data. For example, the variation of breakdown voltage as a function of temperature can be investigated by placing the diode in a temperature chamber.

The oscilloscope display of junction voltage breakdown shown in Fig. 7-10D is also useful and can sometimes reveal certain dynamic discrepancies not readily detected with meters. Frequencies between 100 and 1000 Hz are suitable for this test. If a function generator is available, the triangular wave is actually to be preferred to the sine wave because the relative abruptness of the transition from nonconduction to conduction is more readily evaluated. In any event, notice that the peak-to peak excursion of the clipped wave represents the sum of forward and reverse breakdown voltages. Therefore, the reverse-breakdown voltage is the peak-to-peak voltage V_t less the forward voltage. For practical purposes, the reverse-breakdown voltage, V_r, equals $V_t - 0.6$ volt (for silicon devices). If the device is germanium, use the relationship: V_r equals V_t – 0.2 volt. If a dc-responding scope is used, V_r may be read directly in relation to the base line, but this should not be attempted from the display of an ac oscilloscope because of the difficulty of estimating the "true" base line.

When using this method, it must be ascertained that the voltage available from the ac source has sufficient amplitude to reach the reverse-breakdown level of the diode. If this is not the case, the complete shape of the half cycle representing reverse polarization will be displayed. If such a display is also accompanied by a lack of clipping for the alternate half cycle, the test reveals an open-circuited diode.

An inspection of a half cycle of the reverse-breakdown region expanded to fill the screen can reveal useful information, especially on a dc scope. The clipped region should be essentially horizontal and should begin and end abruptly if "good" breakdown behavior is an important factor. If an ac scope is used, this evaluation is best made at 1000 Hz, or even higher to avoid the waveshape distortion caused by the lack of dc response. A stepup transformer can be used in conjunction with the ac source in order to reach junctions having higher breakdown voltages.

EFFECT OF THE JUNCTION BEHAVIOR CHARACTERISTIC IN BOARD TESTS

The fact that most semiconductor devices tend to be non-conductive at low voltages allows us to use a simplified test procedure for circuit boards containing various mixtures of semiconductor devices and passive elements. It is often with considerable reluctance that we decide to remove individual components in order to electrically isolate a component to be tested from the rest of the circuitry, or to free a component on the board from shunting effects which might mask the test results. Such a procedure may endanger the printed-circuit connections and risk thermal damage to semiconductor devices. Additionally, removing and replacing parts on a PC board can be quite time consuming. Certain manufacturers make ohmmeters which have very low voltages at the test prods. If such a meter has, say 50 millivolts instead of the commonly encountered 1.5 volt, is obvious that we can make many tests without disconnecting any components. We are thinking now of the components associated with the semiconductor devices, not the semiconductor devices themselves.

Suppose that Fig. 7-11 represents a portion of a PC board where we want to test resistor R1 with a low-voltage ohmmeter as described above. The prods of the meter could connect to circuit points A and B without regard to polarity, because diode CR1 would have an extremely high resistance in both forward and backward directions. Moreover, diode CR2 would be effectively out of the circuit for the same reason. Similarly, resistor R2 could be measured across circuit points C and D. The low test voltage would neither drive diode CR1 or the base-emitter junction of transistor Q1 into appreciable conduction. And, unless R2 was extremely high in resistance, a substantially accurate measurement could be made. Silicon devices can be very effectively "uncoupled" from the associated circuitry by this test approach.

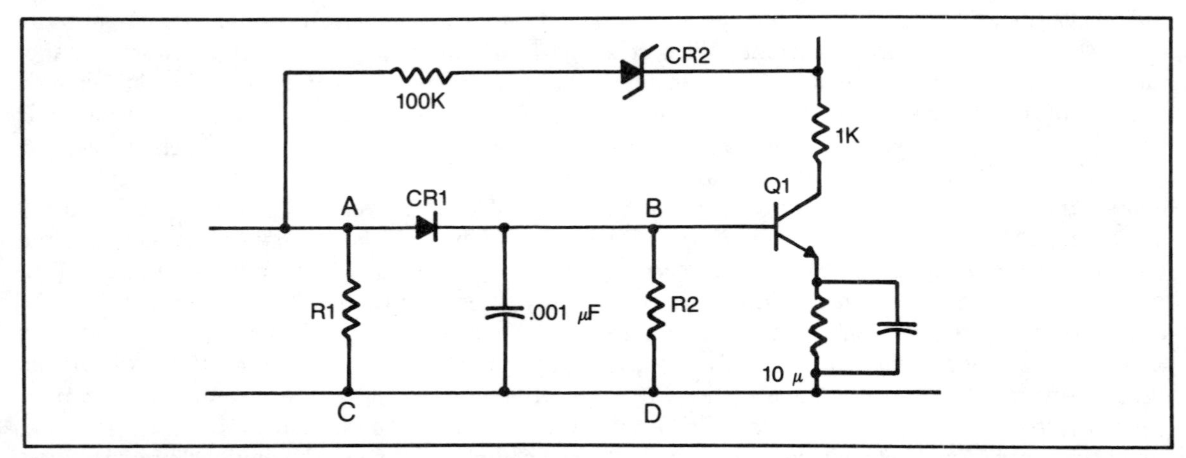

Fig. 7-11. With PC boards, some components can be tested with a low-voltage ohmmeter due to the use of polarity isolation provided by pn junctions.

Germanium devices sometimes conduct appreciably at relatively low voltages, thus depriving us of the isolation we had with the silicon devices. For example, if transistor Q1 were a germanium unit, it might be a good idea to experiment with the polarity of the test prods while measuring R2. The highest resistance reading would then tend to be the most accurate. In using this test technique we must exercise good judgment and cross-checks in order to avoid false conclusions due to defective semiconductor devices. Shorted junctions, at least, are very easy to detect with either a low voltage or a more conventional ohmmeter, for such a defect generally is not likely to be masked by associated components.

Often, a conventional meter can be employed in a similar way, provided that due consideration is given to polarity. For example, resistor R1 could be measured by contacting the negative test prod at circuit point A and the positive prod to point B. Under this condition, diode CR1 and diode CR2 would be reverse-biased at a low voltage, thereby uncoupling the rest of the circuitry from resistor R1. However, the conventional meter would not, in general, be useful for measuring R2 in the circuit. With the prods connected one way, the base-emitter junction of transistor Q1 would conduct, while the other would make diode CR1 conduct. Resistor R2 would be shunted in both cases.

Some semiconductor devices do not have the voltage-gap characteristic that is common to conventional junctions found in junction diodes, junction transistors, the gate circuits of junction FETs and SCRs, etc. With devices such as point-contact diodes, tunnel diodes, back diodes, the emitter circuit of unijunction transistors, and certain other devices, the "uncoupling" test technique may not be so easy to implement. If other circuit considerations permit, some of these devices will behave as a reasonably high resistance if the test prods of a low-voltage ohmmeter are connected in the correct polarity. Tunnel and back diodes, however, should be removed from the circuit by physically disconnecting one lead. It is not likely that reliable results are possible by attempting electrical isolation of these devices.

IN-CIRCUIT TRANSISTOR TESTS

A parallel situation to the one just discussed involves testing semiconductors without removing them from the circuits. Of course, there would be no particular reason for being interested in testing transistors in a circuit where transistor sockets are used. Sockets provide a convenient and rapid means of testing and servicing, but for various reasons the sockets which may have been used during breadboard and development stages generally do not appear in the finalized equipment. Cost is one factor. Also, many designers feel that the sockets subtract from the overall reliability that might otherwise be attained. This is because it is not easy to provide positive contact when fractions of a volt or very low currents are involved. Such contacts tend to accumulate an extremely thin corrosive film. Then, too, the strength of the mechanical grip provided by sockets varies greatly and it is not uncommon for transistors to jump out of their sockets under some conditions of vibration or shock. Other reasons can be cited, too. The alternative to the socket is the soldered-in-place transistor, a technique which is subject to its own unique shortcomings.

The principal drawback of the soldered-in-place technique and the one which complicates test procedures is the difficulty encountered in removing the transistor. Special skills or special tools are needed to remove soldered-in transistors if thermal and mechanical damage is not to be inflicted on the PC wiring or on the transistor itself. Even so, it is no fun—and poor economics—to unsolder transistor after transistor only to find them basically OK. For some strange reason, it invariably happens that the defective transistor is the last one to be checked in the equipment. It's easy to say that you should first establish which transistors are likely causes of the existing trouble, but often the situation is confused by the circuit complexity and the interdependence between stages and sections of the circuitry. In-circuit transistor tests can provide a good solution to these test dilemmas. Most in-circuit testing has so far dealt with bipolar transistors, but the basic logic of the various techniques often apply to diodes and semiconductor devices.

An ohmmeter will reveal some defects in transistors, diodes, and other junction devices. It happens that a shorted junction generally has such a low resistance that its effect is not likely to be masked by the shunting effects of associated components. It will usually be found that such damaged junctions have resistance of several ohms regardless of the polarity of the ohmmeter test prods. And it is not uncommon to measure substantially zero resistance in both directions. Thus, one class of semiconductor defects may be detected with fair reliability without removing either the devices or associated circuit components. There are some exceptions, however. For example, the gate junction of an SCR is sometimes directly fed from the secondary winding of a pulse transformer. In such a case, the low dc resistance of the winding could mask the resistance test of the gate junction. Even here, there is a good chance that the even lower resistance of a shorted gate junction will show up if the ohmmeter deflection is carefully observed. Another situation where test results can be misleading involves the use of back-to-back diodes, or equivalent. Although we cannot expect to determine the forward-to-reverse conduction ratio with any reasonable approach to accuracy during in-circuit testing, the one-to-one ratio of back-to-back diodes seemingly suggests one or more defective junctions. Of course, this is not the case; we simply obtain forward conduction readings for both test

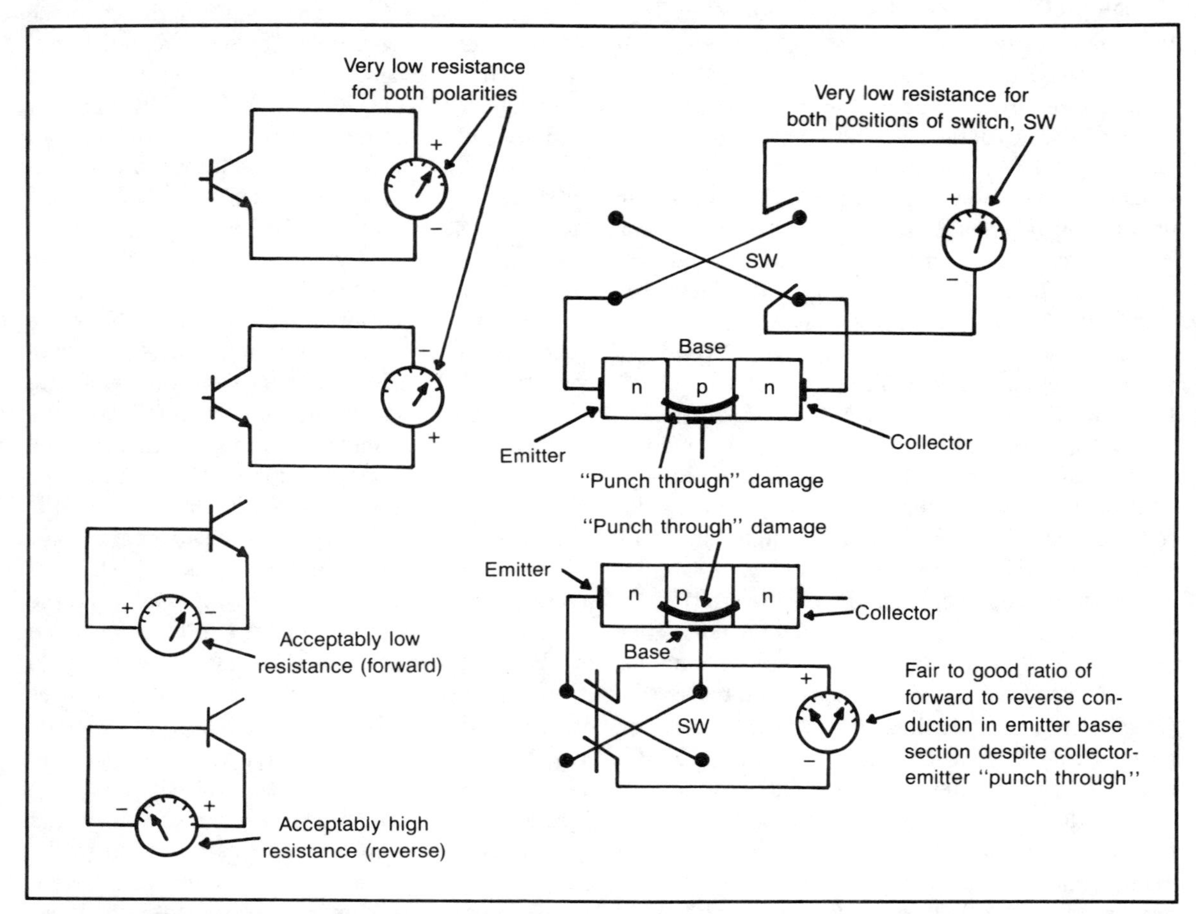

Fig. 7-12. Schematic and pictorial representations of strange but common transistor defects. Switch SW accomplished the same thing as reversing the test prods of the ohmmeter.

polarities. (Diodes connected in the fashion are often encountered as circuit-protection techniques.)

A very common type of junction failure occurring in transistors is a short from collector to emitter. This actually involves both transistor junctions, but often in a way which appears strange from the testing viewpoint. Although the ohmmeter might reveal substantially zero resistance between collector and emitter in both test polarities, a check of the emitter-base junction may show, surprisingly, a good forward-to-reverse conduction ratio (see Fig. 7-12). (For the moment, let us suppose that the transistor has been removed from its circuit so that there will not be any possibility of complicating factors.) Such test results seemingly violate some of our most sacred laws of the electric circuit. It would seem that a bipolar short of the collector-emitter electrodes existed.

The answer to this dilemma is not simple and it is doubtful that any useful knowledge can be gained about testing by bogging ourselves down with the physics of transistor action. Suffice it to say that this type of transistor damage is probably caused by a phenomenon known as a "punch-through," where the collector region reaches across the base region and merges with the emitter region. It can do this within a restricted area which need not involve the entire base region. Hence, the emitter-base junction may be left in an essentially operative condition. Because it is so commonly encountered, time will be saved during in-circuit testing if the collector-emitter electrodes of transistors are tested first. (Power transistors, especially, are susceptible to this kind of damage.) Defective emitter-base junctions are less likely to be found in most equipment. When encountered, it will be found that collector-emitter measurement will also be abnormal.

In-Circuit Transistor Testers

Techniques involved in the in-circuit testing of transistors and other junction devices with a VOM serve a purpose. Skill in the use of the simple and commonly available ohmmeter is commendable and, in some practical situations, necessary. However, it should be realized that ohmmeter evaluations of in-circuit semiconductor devices are quite primitive; the best that can be hoped for is a fairly reliable go, no-go result. This is of value and is not to be ignored. However, we must consider more sophisticated ways of testing transistors.

For this purpose there are many in-circuit testers. Some are being used with ever greater frequency in developmental laboratories and on production lines. The cost of these instruments is quickly amortized by the worth of the service they can provide. For example, it often is not good enough to prove by ohmmeter tests that the junctions of a transistor are basically OK. Perhaps the transistor current gain is too low to properly perform its intended function. It would be much more revealing to be able to determine the relative activity of the transistor. As we know, some transistors are "hot," while others are almost dead in comparison. In-circuit testers can not only indicate the relative activity of a transistor, but can provide a readout of actual dc or ac beta as well. Under favorable conditions, which are not too difficult to find, fairly good accuracy can be attained.

There are several types of in-circuit transistor testers. One type forces the transistor undergoing test to assume the role of an oscillator. This type reveals information about the quality of the transistor—whether there is sufficient "transistor action" for oscillation to occur. By introducing different amounts of attenuation in the feedback network, it is possible to describe the transistor as "poor," "average," or "high-gain." However, most of these instruments simply indicate whether the transistor is reasonably active or not. Thus, you might produce oscillation for all transistors having betas in excess of, say ten. The oscillations are rectified and used to actuate a dc meter with a "good-bad" scale. Often the oscillator circuit is such that approximately the same meter deflection is produced by any transistor that will oscillate. Figure 7-13 shows the basic circuit of an oscillator-type tester.

A second type of in-circuit tester operates the transistor as a low-frequency amplifier and provides a readout of ac beta. One way of accomplishing this

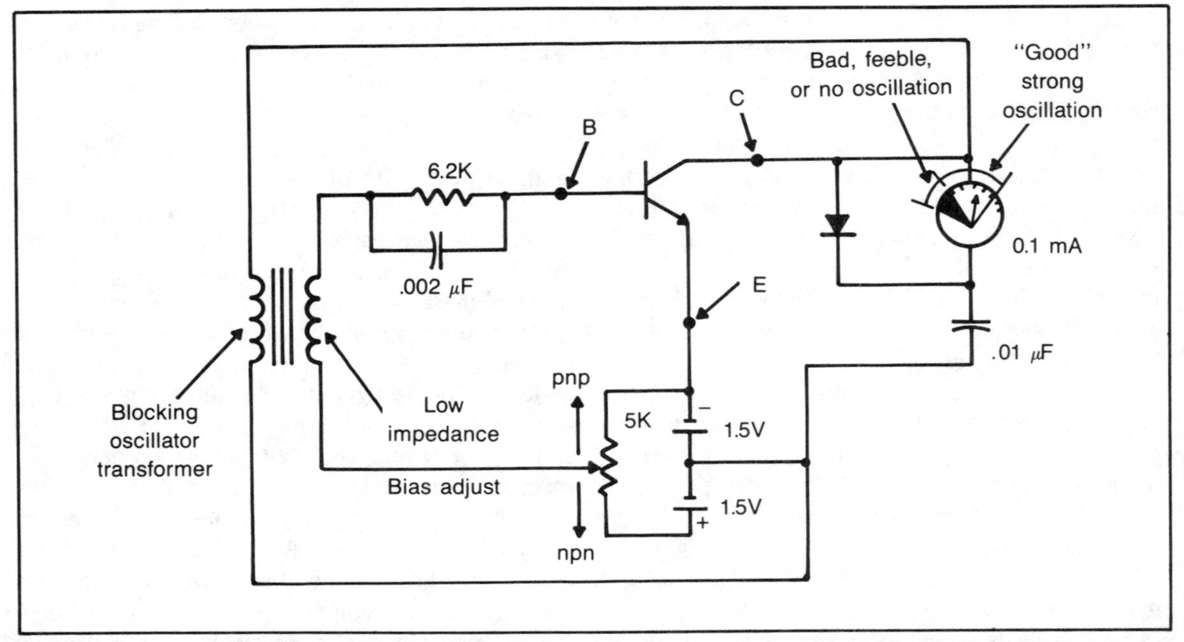

Fig. 7-13. Basic circuit of an oscillator-type in circuit transistor tester. The bias adjust enables the forward bias to be set to suitable values for both npn and pnp transistors.

is shown in Fig. 7-14. Q1 represents the transistor being tested. Its biasing network and other associated components are not shown because it is assumed that they are of negligible consequences. Such an assumption is based on the fact that the secondary winding of transformer T1 presents such a low impedance that the shunting effect of circuitry associated with the transistor is not a significant factor.

Connections are made to the base, emitter, and collector of Q1 as indicated. The output amplitude of the 1-kHz multivibrator is adjusted so that the current meter produces a full-scale deflection with switch SW in its calibrate position. This usually corresponds to 1 milliampere—a good collector current value for most small signal transistors. (The inertia of the meter movement averages the unidirectional pulses so that a steady reading is presented.) Notice the mechanism involved in making the transistor operational in the absence of conventional dc biasing. During the positive excursion of the squarewave voltage developed across terminal 1 with respect to terminal 2 of transformer T1, the base-emitter section of the npn transistor is forward biased. Simultaneously, the collector-emitter section is supplied with a reverse voltage. Thus, the transistor operates as a Class B common-emitter amplifier. The same is true when switch SW is placed in its beta position, the essential difference is that the current meter is now connected to respond to the average value of base current. The scale of the current meter is calibrated in units of current gain; that is, beta. It is seen that the smaller the base current, the higher the beta.

Both of the in-circuit transistor testers described are designed to present a much lower impedance to the input of the transistor under test than the combined impedance of all components associated with it. With many of these units, the test can be made if the input impedance seen at the base of the transistor is at least 150 ohms at 1000 Hz. Some testers can accommodate lower impedances, on the order of 50 ohms. In any event, there are bound to be circuit situations where an in-circuit tester will not work. Several of these are shown in Fig. 7-15. Notice in Fig. 7-15A that the

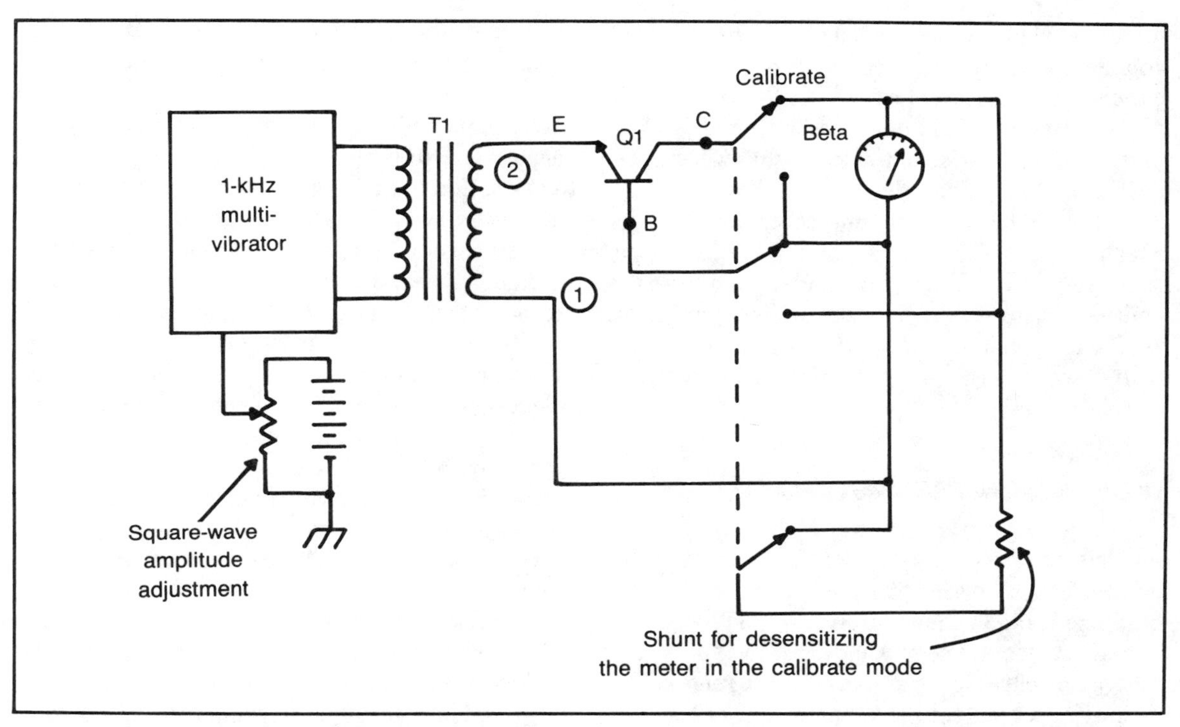

Fig. 7-14. Basic circuit of an amplifier-type in-circuit transistor tester.

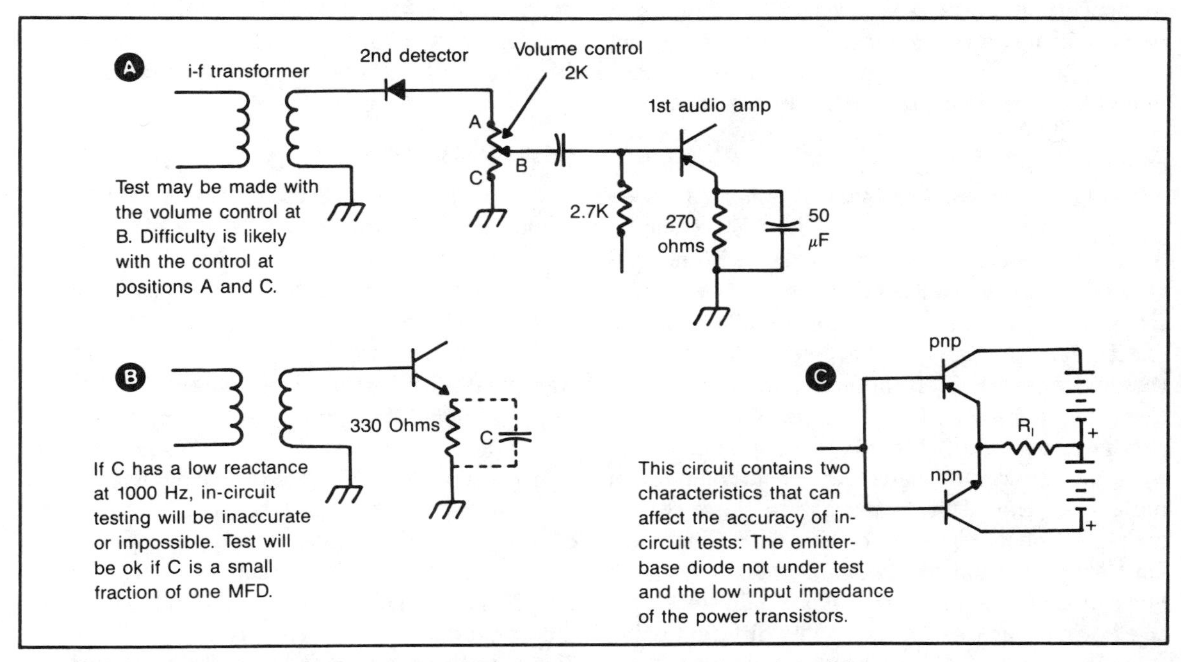

Fig. 7-15. In circuits such as those shown, in-circuit transistor tests may not be reliable.

input impedance to the transistor will be low if the volume control is positioned at A or C, that is, at either of its extreme. However, at B, in the middle of its range, the minimum loading effect will be presented to the base of the transistor. Therefore, it is very likely that the transistor can be successfully tested if the volume control is set at or near its mid-position, B. In Fig. 7-15B, the transistor cannot be tested if the capacitor across the emitter resistor is a large electrolytic type. In Fig. 7-15C, the input shunting effect of one transistor with respect to the other tends to make in-circuit testing difficult.

UNIJUNCTION TRANSISTORS

Unijunction transistors are tested by measuring the reverse current which flows through the emitter/base-two section with no connection to the base-one lead, as shown in Fig. 7-16. The battery voltage can be between 9 and 22 1/2 volts. A good unit generally shows a reverse current in the vicinity of several tens of nanoamperes or less. Appreciably higher leakage currents should be cause for concern in most types. Additional test information can be gained by connecting an ohmmeter, set on its R × 100 range, to the emitter/base-one section of the unijunction transistor. Forward conduction should be indicated by a reading near the center of the scale. When the meter prods are reversed, the meter should show an essentially open circuit (infinite ohms) reading. A final check should be made between bases one and two with no connection to the emitter lead. Most units will show from several kilohms to perhaps ten kilohms. The exact value is not important in determining whether the device is "alive" or not. This interbase resistance should be the same when the ohmmeter test prods are reversed, since no junction is involved.

Another way to evaluate the basic condition of unijunction transistors is to make up the very simple testers shown in Figs. 7-16D and E. These circuits are relaxation oscillators in which a sawtooth wave or a low-frequency audio tone indicates an active device. The tone will vary with different units, since the frequency depends on the intrinsic standoff ratio of the transistor. This parameter can be considered somewhat similar to the ionization voltage of a gaseous diode. Gaseous diodes such as neon lamps with different ionization voltages will also produce different pulse repetition rates due to the variation in time required for the charging capacitor to develop a firing voltage. In the unijunction transistor relaxation oscillator, the firing voltage is a nearly constant fraction of the supply voltage rather than simply a fixed level such as, say, 70 volts as in a neon bulb. Therefore, an advantage of the unijunction oscillator is that the generated frequency is substantially independent of the supply voltage. Because of tolerances and differences in types, two unijunction transistors may generate quite different tones. However, the production of a tone is itself a fairly good indication that the unijunction transistor is OK for many practical applications.

Complementary unijunction transistors are similar to the type discussed here, except that the polarity of the dc supply must be reversed. Thus, the complementary type is similar to pnp bipolar transistors and the hitherto available types are similar to npn transistors. The circuit in Fig. 7-16F operates in similar fashion to the circuit in Fig. 7-16D, except that the sawtooth wave is inverted.

TUNNEL DIODES

Although the tunnel diode is a junction device, its characteristics differ considerably from those of the junctions encountered in a wide variety of other semiconductors. Although other devices or circuits have overall characteristics similar to the tunnel diode, the tunnel diode is unique in that its negative-resistance behavior is achieved within a single junction. Although negative-resistance is a characteristic of such devices as SCRs, unijunction transistors, and certain microwave diodes, the tunnel diode probably involves the most adverse combination of circumstances, making it both difficult to test and vulnerable to damage. Its power requirements, in the several hundred millivolt and often fractional milliampere area, make ordinary ohmmeter tests hazardous. The negative-resistance feature is more easily destroyed than is the rectification characteristic of conventional junction diodes.

In ordinary devices, the power supply need not have a lower dc output resistance than is required by Ohm's Law considerations of the voltage and current demanded by the device. The tunnel diode, however, will not even operate with respect to a

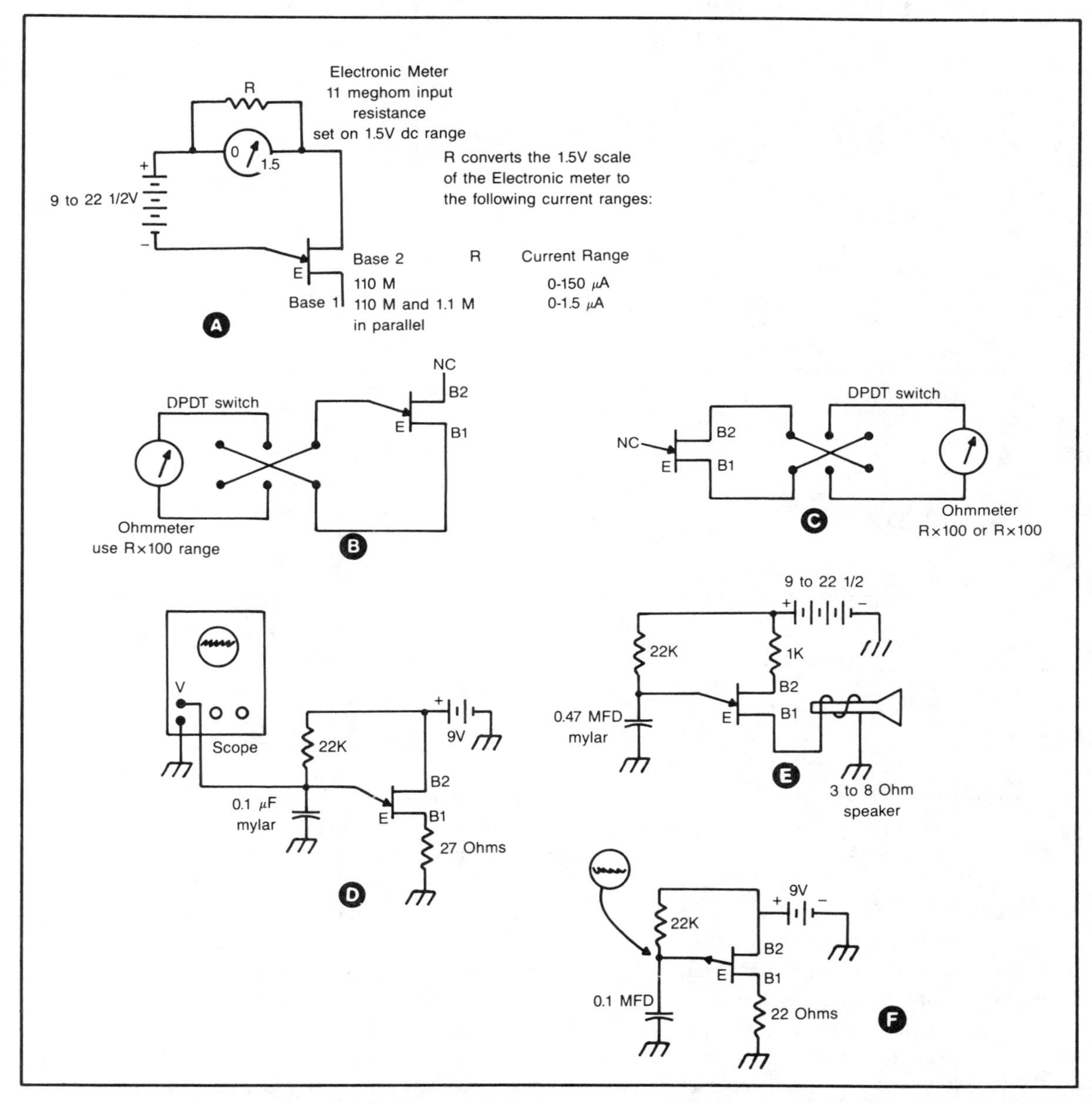

Fig. 7-16. Unijunction transistor tests. Circuit A measures the leakage current of the emitter junction; B tests the diode behavior of the emitter junction; C measures the conductivity of the interbase region; D is an activity test as a relaxation oscillator; E is an activity test as a tone-producing relaxation oscillator. F is a complementary unijunction transistor circuit. Notice the dc supply polarity.

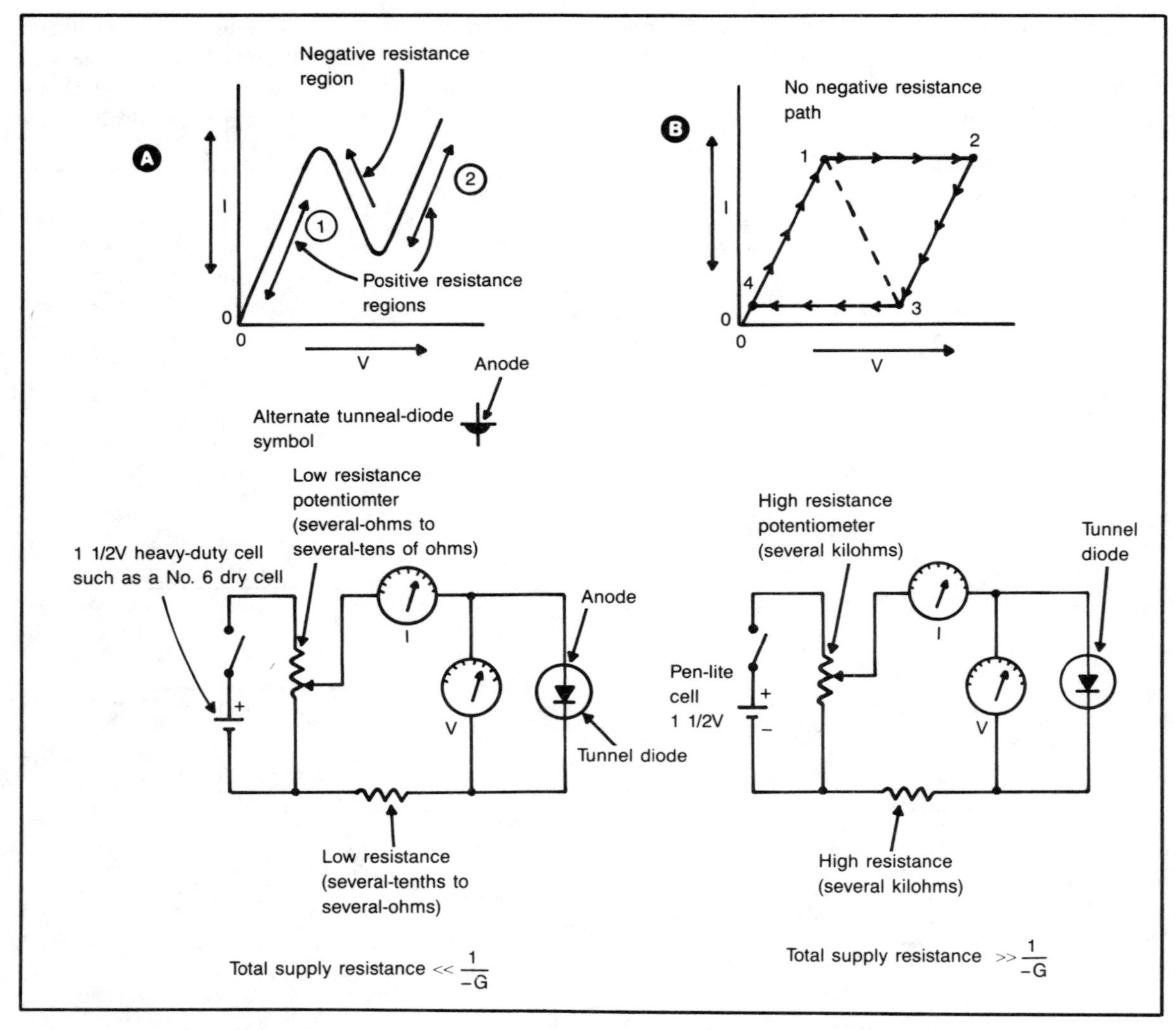

Fig. 7-17. The two operating modes of the tunnel diode: Circuit A has a stable negative-resistance characteristic and B is a switching circuit. An abrupt jump takes place from point 1 to point 2 and again from point 3 to point 4.

fixed point within its negative-resistance region unless the power supply is shunted by a very low resistance, or has a current supply capability are in excess of that actually required by the tunnel diode. We can comply with such a requirement, but testing then becomes awkward in terms of readily available equipment. If we construct a special power supply, we must be prepared to provide a smooth variation from zero through the several hundred millivolt region without transients or appreciable ripple. Batteries can be used, but various difficulties are encountered here, too. Unless the batteries are of fairly high capacity and are fresh, trouble is likely to be experienced with voltage and current stability.

Perhaps we can figure out how to get around these problems by examining the two tunnel diode modes of operation showing Fig. 7-17. In Fig. 7-17A, a "brute-force" dc supply is used in which the tunnel diode sees a resistance much less than its negative resistance. (Specification sheets usually list the reciprocal of negative resistance; that is, negative conductance, designated by $-G$. It follows, therefore, that negative resistance is $1/-G$.) A classic "N" type current vs voltage curve results from such a measurement technique. This method actually could be used for test purposes if we were careful not to exceed the current capability of the tunnel diode in its second positive resistance region. Also, consideration would have to be given to the current meter because its internal resistance would tend to defeat the low resistance feature of the supply. If such a test were conducted successfully and the N plot were derived, the tunnel diode would have to be considered good. For all but the most specialized of applications, the mere existence of the negative-resistance region is an indication of a good device.

In Fig. 7-17B, the dc supply is made to appear as a high resistance to the tunnel diode. We might suppose that this would make no difference, because it would seem possible to achieve the same combination of voltages and currents to operate the tunnel diode. Apparently, the tunnel diode has its own peculiar reactions to the situation; notice that the negative resistance region is missing from the characteristic curve. Beginning at zero, the plot is "normal" until point 1 is reached. A tiny increase in applied voltage then produces an abrupt jump to point 2 instead of following the dotted negative resistance line. If the applied voltage is decreased, the plot is again "normal" until point 3 is reached. A slight reduction of voltage then produces the abrupt jump to point 4. From point 4, the entire cycle can be repeated. Thus, the effect of the higher power supply resistance makes the tunnel diode behave as a voltage-sensitive switch.

The question now arises concerning the usefulness of the switching cycle for testing. Although we no longer see the negative-resistance characteristic, we may rest assured that it is present. Otherwise, the switching would not occur. Moreover, the plot is not so very different after all; we can easily visualize the missing segment as a line connecting points 1 and 3 in the Fig. 7-17B plot. And best of all, the tunnel diode is now protected by the current-limiting action of the higher series resistance! From a practical viewpoint, an additional advantage is the relaxed demand on the supply. Instead of the physical resistance R, the higher resistance can be incorporated and in the supply itself via a higher resistance potentiometer and a smaller cell. Finally, a more conventional current meter will serve the purpose; its internal resistance can now be considered an asset.

Further consideration of these matters lead to a dynamic, rather than a static test technique. This is easily accomplished because an oscilloscope display can be produced without regard to linearity or the actual readout quantity. A display which reveals the switching behavior in response to changing terminal voltage tells us that the tunnel diode is operating. A simple setup for testing tunnel diodes in this way is shown in Fig. 7-18. There is virtually no danger to the diode involved in this test technique.

JUNCTION FIELD-EFFECT TRANSISTOR

Like bipolar npn and pnp transistors, junction FETs are available in two types. The n-channel most JFET most closely resembles the electron tube in the operating potentials required. Like the tube, the n-channel JFET control element is generally at zero or at a negative voltage with respect to its commonly grounded electrode, and, like the tube, the electrode which usually connects to the load impedance is biased positively. The n-channel JFET is also similar to an npn transistor in the power supply polarity required for operation. Unlike the ordinary or bipolar transistor, the control element (the gate) of the JFET does not consume bias current. The p-channel JFET is similar to the n-channel type, except that the dc polarity of the electrodes is reversed. We are reminded here of a pnp transistor. (In electron tubes, there is no counterpart of the p-channel JFET.)

The gate of a JFET is a reversed-biased junction. The source and drain of the JFET are at opposite ends of the so-called channel, which is semiconductor material but does not involve any junctions. Therefore, current can flow in both direc-

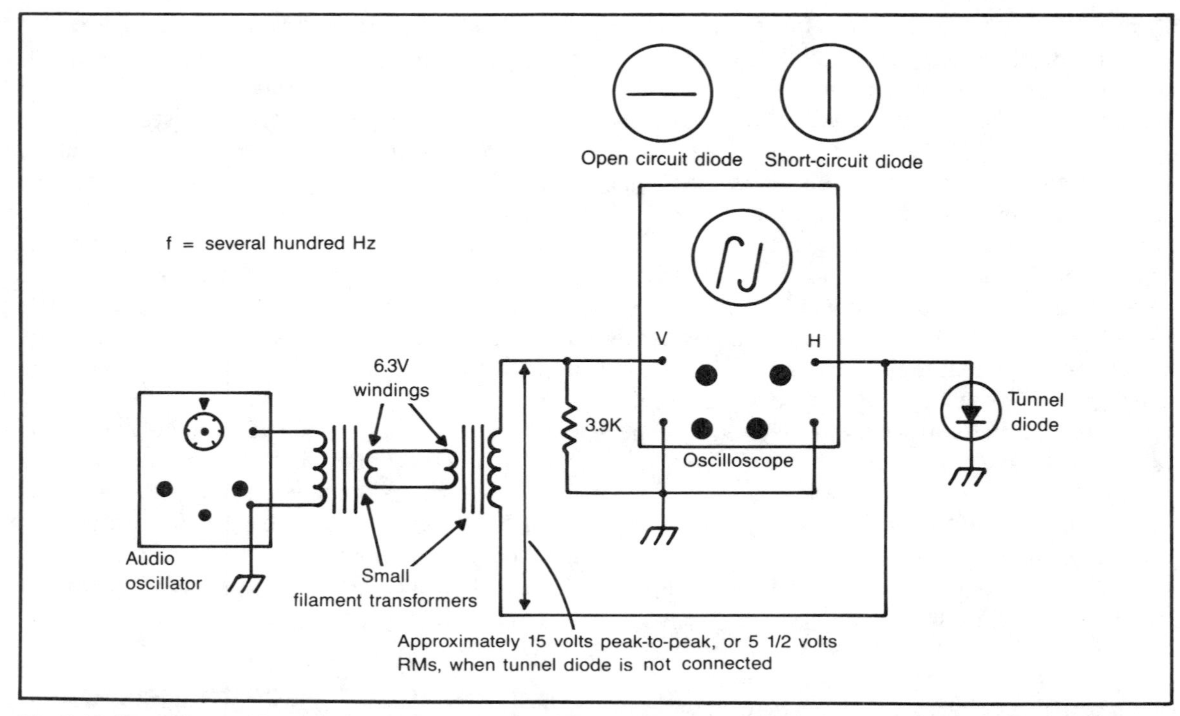

Fig. 7-18. Dynamic test circuit for a tunnel diode. On some oscilloscopes, the slope of the two lines will be opposite to that shown. This does not signify negative resistance and does not alter the basic test evaluation.

tions through the channel. The source-drain channel arrangement could be perfectly symmetrical, and indeed was in many early units. The source and drain leads can be transposed without adverse effects in many applications. In operation, a variation in channel resistance is produced by variations in the voltage applied to the gate.

Although the great preponderance of JFET defects involves destruction or degradation of the gate junction. It is well to test the channel resistance in both directions. (It is not very likely that an open or shorted-channel will be found.) When checking the channel with "normal" polarity, the gate should be connected to the source. (See Fig. 7-19.) Otherwise, it will respond to static electrical charges and make the channel resistance check unreliable. The channel resistance will generally be from several to many kilohms. Its exact value is not often useful as a test parameter. If the ohmmeter leads are switched to reverse the polarity across the channel, the gate should first be connected to the drain so that the gate junction will not be forward biased during this test. As previously mentioned, the channel resistance should be substantially the same for both directions of current flow.

The test of the gate junction is generally the more useful of the simple tests. The gate junction of JFET's is not designated to carry current, although the nature of the device can result in a junction of appreciable area. It is advisable to use the R × 100 ohms range to check the forward conduction of the gate-source circuit. No connection is made to the drain. Most JFETs will give a mid-scale or lower-resistance reading on this test. Here, you simply determine that the gate junction is neither shorted nor open. That being the case, the next test measures the reverse resistance of the gate-source junction. Since these devices are fabricated according to advanced silicon technology, the reverse resistance should be extremely high. Preferably, it should not even move the meter on the 10,000 × 1 range. And if the circuit application requires the

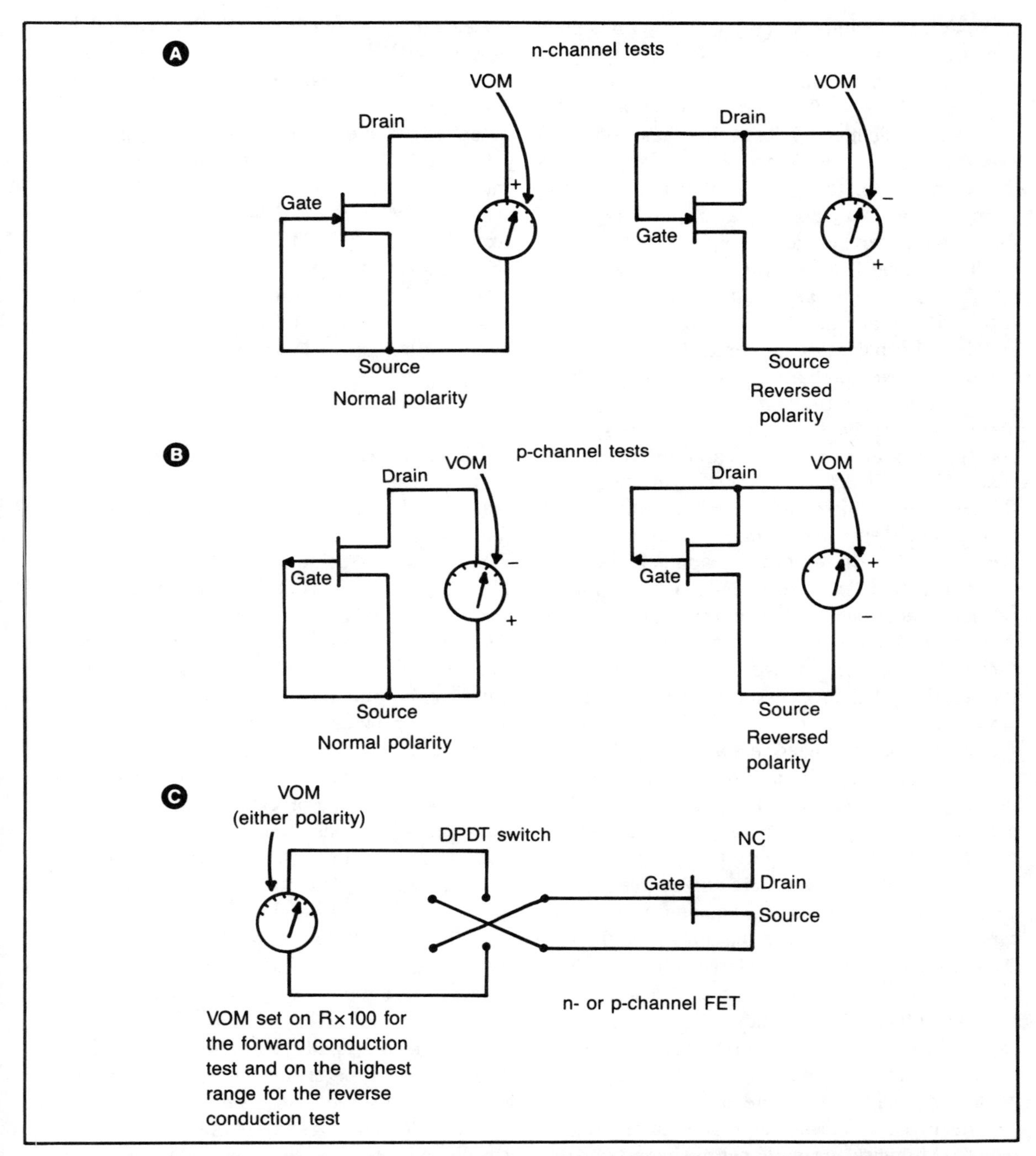

Fig. 7-19. Ohmmeter tests of junction FETs. Circuit A is a resistance test of an actual n-channel type. Resistance values are usually not of much significance, although a blown gate junction can result in an abnormally low reading. However, the resistances measured with the two ohmmeter polarities should be about the same. Circuit B illustrates the same test for p-channel units. The gate-junction test is shown in circuit C. This is generally a much more revealing test. A high forward-to-reverse conduction ratio is desirable.

very high input impedance possible with a good condition JFET, the reverse assistance of the gate-source junction should not cause a discernible indication on the R × 100,000 range, whether a VOM or an electronic meter is used. In other words, we look for an extremely high forward-to-reverse conduction ratio even though the conduction in the forward direction may not appear as good as that of the base-emitter junction of bipolar transistors.

The typical JFET characteristic curves shown in Fig. 7-20 remind us of those for bipolar transistors. Two major differences, however, affect test and design considerations. First, the control junction, being reversed-biased, consumes virtually no current. Secondly, the voltage breakdown, which occurs at excessive drain-source potentials, is not due to zener or avalanche characteristics in the channel between drain and source connections. The channel conduction characteristic remains essentially unchanged even when potential differences greatly exceed those normally applied. Also, the channel itself usually recovers quickly from the heating effect of higher than normal currents. This is why an ohmmeter test of the channel rarely shows any damage between drain and source. Rather, the breakdown associated with excessive drain-source voltage occurs in the gate junction. It should be clearly realized that a higher voltage applied between the drain and source also increases the reverse voltage across the gate junction. This, perhaps, becomes more clear when it is recalled that virtually all circuit applications have some kind of a return path between gate and source. In fact, a curve such as that for V_{gs} equals 0 in Fig. 7-20 can be duplicated for the situation where the gate is connected directly to the source. The breakdown is a drain-gate phenomenon, but need not be destructive if there is sufficient resistance in the drain or gate circuit to limit the breakdown current.

From Fig. 7-20, we can see that the reverse voltage applied by an ohmmeter during a test of the channel must not closely approach the breakdown region. Otherwise, we might get the impression that the FET is defective, because of the inordinately low resistance.

For a more reliable test, the use of a curve tracer is recommended. A 1000-ohm resistor should be connected across the gate-source electrodes of the FET. Then, if the curve tracer is set to produce 0.5 milliampere base-current increments, we will actually have a 0.5-volt increments applied to the gate of the JFET. (Notice that the polarity of the increments must be opposite to that which we would select for a transistor with the same collector polarity as the FET drain polarity.) Make sure that the source for FET connects to the emitter test terminal, the drain of the FET connects to the collector test terminal, and the gate of the FET connects to the base test terminal.

SCRs

Some characteristics of the silicon controlled rectifier can be easily and quickly evaluated, whereas others require time, consideration, and often, sophisticated instruments and test procedures. The reason for such a broad statement is that the firing characteristics of SCRs tend to be extremely variable. A given type, for example, might require a gate voltage anywhere from a fraction of one volt to several volts and the required gate current can vary over a three-to-one or greater range. Since we shall concern ourselves with simple test procedures, the focus will be on the basic quality of the device, rather than on its ability to perform under various conditions of anode or gate rise time, gate pulse duration, and wide temperature variations. Nor shall we investigate such matters as turn-on or turn-off time. First, what can be accomplished with a VOM?

The best VOM for making simple tests on SCRs is the very one which we must be very careful using to test most other semiconductor devices! Here, good use can be made of a VOM which can provide several hundred milliamperes on the R × 1 range. Other meters can be used for some of the basic short-open tests, but a meter which does not deliver at least several tens of milliamperes of short-circuit current may not provide the required "holding current" for one of the basic tests. In any event, one of the most reliable indications of a defective SCR is a zero or near-zero ohm measure-

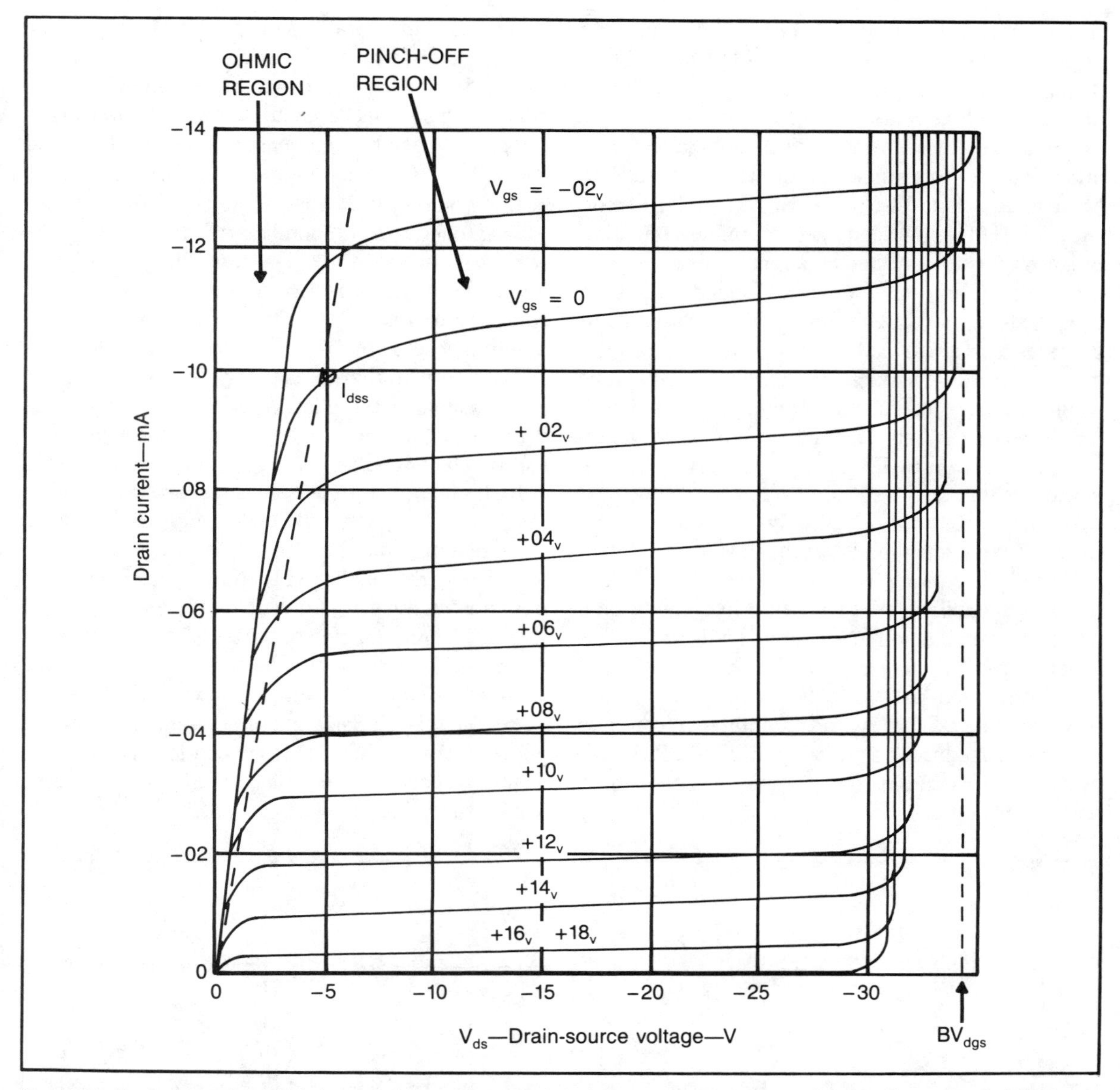

Fig. 7-20. Typical junction FET characteristics. Characteristics shown are for p-channel units. For n-channel FETs, reverse the sign of all polarities shown.

ment between anode and cathode. Usually, this low-resistance reading is independent of polarity. The gate is left open when making this basic test. (See Fig. 7-21A.)

Fortunately, many SCRs, whatever the initial failure characteristic, quickly stop working. This is particularly true when the SCR is utilized to control appreciable power levels such as is the case in inverters, converters, power supplies, cycloconverters, and "crow-bars." On the other hand, if the circuit function of the SCR is more on the order of a logic or signal-processing element, failure tends to be less catastrophic and an anode-cathode short is less likely to be found. Thus, small

SCRs, those controlling tens or perhaps hundreds of milliamperes of current are not so likely to display the short symptom as are devices carrying ampere level currents.

If an SCR of any size does not have a low anode-cathode resistance on the R × 1 scale, the same check should be made for both polarities on the R × 1000 ohms scale. A discernible deflection is reason for concern and an appreciable deflection indicates a defective device. Indeed, at room temperature, good SCRs, even with many tens of amperes of average load current capability, may not show perceptible leakage on the R × 10,000 ohms range when so tested. For many applications, the latter test is probably safer to use as a go, no-go criterion. If an SCR survives both the anode-cathode short test and the leakage test as well, it is probably—but not necessarily—a good unit. We have not yet determined that the device is not open circuited, nor have we ascertained basic SCR action—the ability to be triggered into conduction and to remain latched in that state as longer as sufficient anode-holding current is available.

Rather than makes tests on the gate-cathode section of the SCR, it is more profitable to continue the test procedures as shown in Fig. 7-21B. When switch SW is closed, the resulting gate signal should turn on the SCR which will be indicated by a low resistance reading of the VOM on its R × 1 range. Notice that only a momentary gate signal can be delivered even if switch SW were to remain closed. As soon as capacitor C attains full charge, the gate is essentially open circuited insofar as dc is concerned. Thus, a sustained low resistance reading on the VOM shows that the SCR has probably latched into its conductive state. Next, momentarily break the circuit at X. This should allow the SCR to return to its blocked or non-conductive state. Then the test can be repeated. The purpose of the resistor across capacitor C is to provide a discharge path for the capacitor after a test is made. Otherwise, repetitive tests would not be successful since a charged capacitor will not carry a current pulse to the gate. The resistor value is high enough so that it does not provide turn-on current to the gate, but low enough to discharge the capacitor in a small fraction of a second.

Suppose that the SCR did not turn on. This does not yet mean that it is defective, because some units do require in excess of the 1.5 volts available from VOMs. Repeat the described test procedure with the circuit arrangement shown in Fig. 7-21C. Here, a 1 1/2-volt flashlight or penlight cell has been added to the gate circuit. (In exceptionally stubborn cases, two cells can be tried.) If the SCR still cannot be turned on, it must be considered defective.

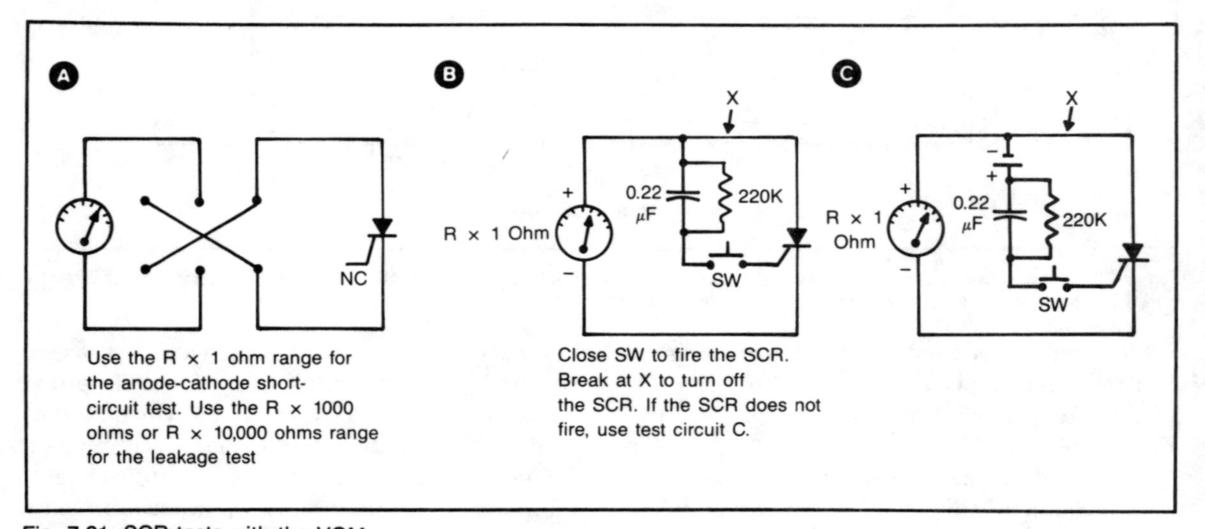

A
NC
Use the R × 1 ohm range for the anode-cathode short-circuit test. Use the R × 1000 ohms or R × 10,000 ohms range for the leakage test
B
X
+
0.22 μF
220K
R × 1 Ohm
–
SW
Close SW to fire the SCR. Break at X to turn off the SCR. If the SCR does not fire, use test circuit C.
C
X
+
–
0.22 μF
220K
R × 1 Ohm
SW

Fig. 7-21. SCR tests with the VOM.

Of course, the clincher in a test such as this is to demonstrate that a known good unit will perform in the test setup.

In many cases it may be desirable to replace SCRs with a slow turn-on characteristic with a responsive one. At the same time, it must be remembered that the performance of such a unit may still be quite satisfactory once it is triggered. In the circuit in which it operates, adequate trigger signal may be available. Moreover, there are applications where a sensitive SCR might not be desirable because of its susceptibility to noise signals and transients.

SCRs become increasingly difficult to fire as the temperature is lowered. It has been assumed that the SCR was at room temperature (about 21 degrees C) for the VOM tests. Sometimes an inordinately insensitive unit can be provoked to turn on if its temperature is raised 10 or 20 degrees. This is best done with a hot air source. If such a test causes a unit to fire when it didn't at room temperature, it would appear that sufficient gate voltage and/or current was not applied to the gate at room temperature.

A deceptive situation can occur if you are not aware that stud-mounted SCRs have the anode connected to the stud, or, in "reverse polarity" construction, the cathode connected to the stud. Generally, the symbol of the SCR is printed on the package in such a way that the type of construction is made clear. Once this is determined, the tests are performed in exactly the same way for both types. If you were unaware of the two packaging arrangements, the test procedure (in Fig. 7-21A) may indicate the possibility of a good device, but no SCR action (firing and latching) will be obtainable from the test procedures in Fig. 7-21B or C.

Procedure for Determining the Firing Characteristics of an SCR

Objective of Test: To demonstrate a simple method for observing the firing characteristics of an SCR under dynamic operating conditions.

Equipment Required for Test:

Oscilloscope (preferably dc)
Two 6.3-volt filament transformers
Variac
1N91 germanium rectifier diode
100-ohm 2-watt 10 percent composition resistor
Single pole, double-throw switch
No. 47 pilot lamp

Test Procedure: Connect the components as shown in Fig. 7-22. Notice the phasing indications on the secondary windings of the filament transformers, T1 and T2. If the phasing is not as indicated, the SCR cannot be fired because positive gate voltage will not occur when the anode is positive with respect to the cathode. If the phasing is correct, zero ac voltage will exist across switch terminals 1 and 2 when the variac is turned up to apply line voltage to the primary of transformer T2. If the phasing is not correct, this same test will result in 12.6 volts across switch terminals 1 and 2. This phasing test should be made with no SCR in the circuit.

The evaluation of the SCR firing characteristics begins by slowly turning the Variac up from its zero position. Continue it until the lamp just glows. (The lamp will burn brighter if the Variac is momentarily advanced beyond this point, but we are interested in the threshold of conduction.) Gate characteristics can now be determined. With switch SW in its No. 1 position, the oscilloscope shows the peak voltage which must be applied to the ate with respect to the cathode in order to fire the SCR. With the switch in its No. 2 position, the oscilloscope shows the peak voltage dropped across the 100-ohm resistor in series with the gate. If this peak voltage is divided by 100 ohms, you get the corresponding peak current required by the gate for conduction of the SCR.

Comments on Test Results: The germanium diode provides a convenience for this test, but it is not essential. If it is not used, the SCR gate will receive the alternative negative half cycles and these will be displayed on the scope. When the germanium diode is used, the peak voltage appearing on the oscilloscope screen is actually greater by about 200 millivolts than that which is effectively

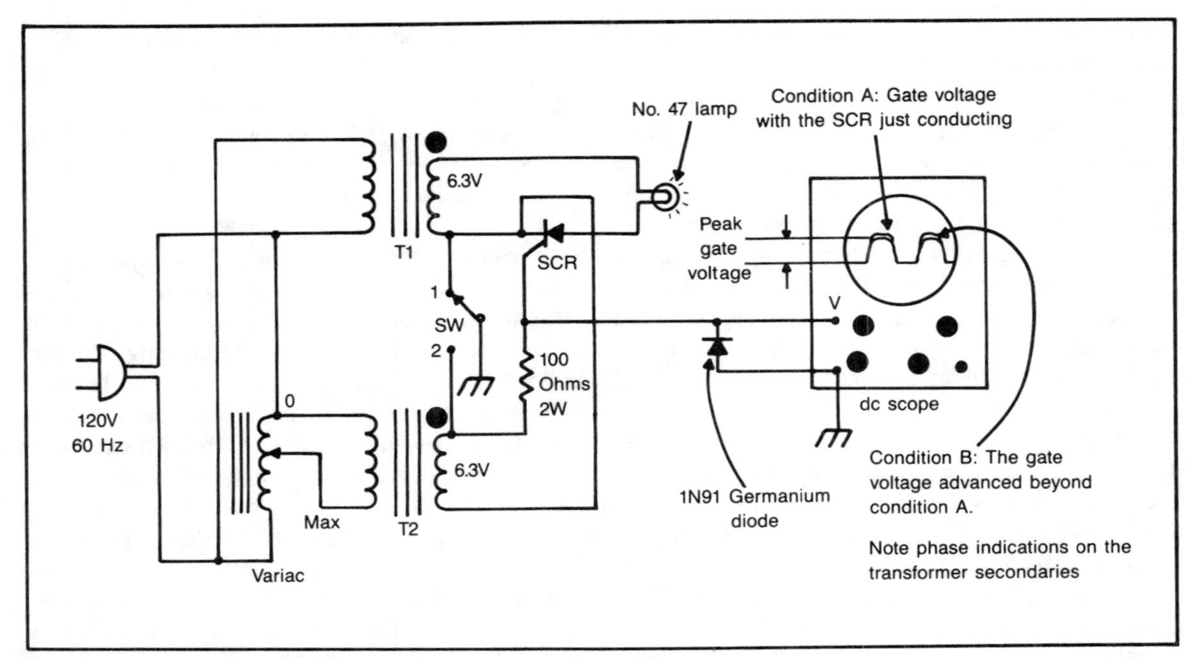

Fig. 7-22. Dynamic test of SCR firing characteristics.

present between gate and cathode of the SCR or across the 100-ohm resistor. This is of negligible importance with the voltages and tolerances involved in this test procedure as a substitute for the 1N91 germanium type.

This test procedure is a very useful one because the voltage and current needed to fire identical type SCRs varies greatly. Where SCRs are used in pairs, units with similar firing characteristics can be selected with this setup. Additionally, this test procedure provides a quick go, no-go assessment of whether basic SCR action is present or not. For this purpose, the oscilloscope may not be considered necessary. A grossly defective SCR will either allow the lamp to glow without gate voltage (with the Variac at its zero position), or it will be found that the lamp will not be turned on even with the Variac advanced to supply maximum gate voltage. The most common symptom of "blown" SCRs—an anode-cathode short—is a higher than normal lamp brillance. When a good SCR is conducting, the lamp receives current only on alternate half cycles and its brilliance is accordingly less than when the full cycle is applied by a shorted SCR. An additional reason for limited lamp current in good SCRs is pointed out below.

In this setup, a phase displacement between the gate and anode voltages can be as great as 90 degrees, despite the fact that no phase shift networks are present. The phase or timing displacement occurs because the anode voltage may have to reach a relatively high amplitude before the gate voltage is high enough to produce firing. However, this is of little consequence for the purposes of the test. We are interested in the magnitude of the gate voltage needed to fire the SCR, not in its time of occurrence. The effect of the phase displacement is to make the lamp burn with less brilliance. This is why the lamp brilliance increases if the Variac is advanced beyond condition A, in Fig. 7-22, corresponding to the threshold of conduction, because then the SCR is fired earlier in the anode voltage cycle.

TRIACS

The firing characteristics of a wide variety of triacs can be tested with the simple setup shown

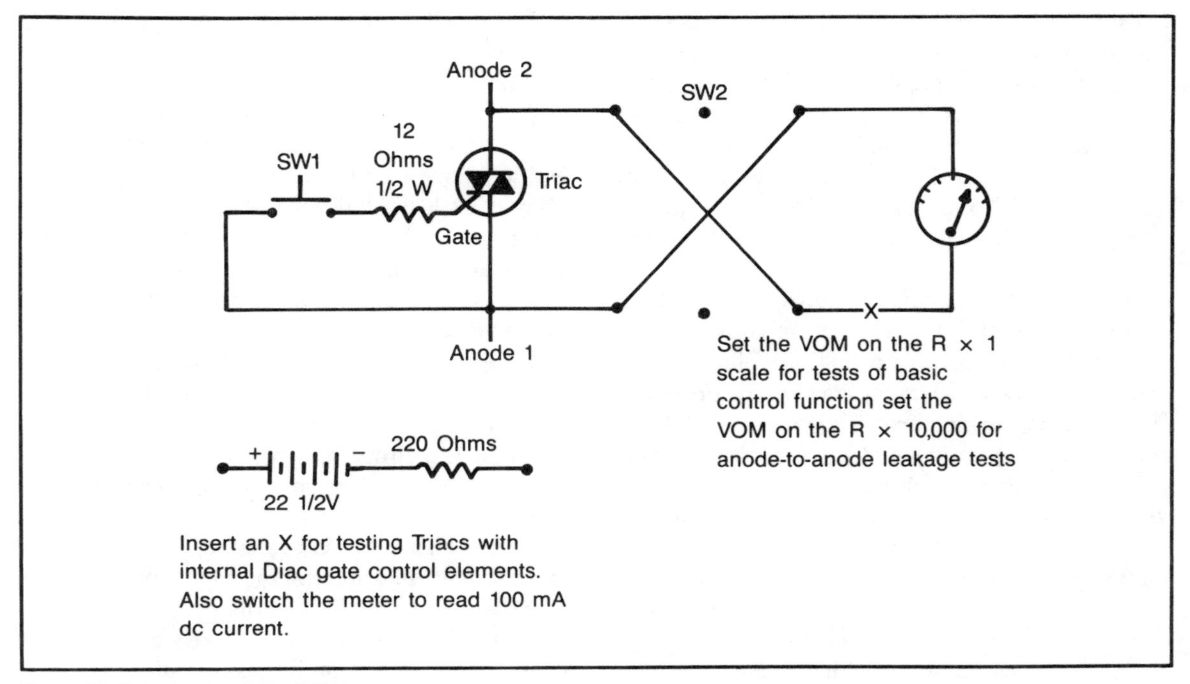
Anode 2
SW2
12
Ohms
1/2 W
SW1
Triac
Gate
Anode 1
X
Set the VOM on the R × 1 scale for tests of basic control function set the VOM on the R × 10,000 for anode-to-anode leakage tests
220 Ohms
+
−
22 1/2V
Insert an X for testing Triacs with internal Diac gate control elements. Also switch the meter to read 100 mA dc current.

Fig. 7-23. Triac tests with a VOM.

in Fig. 7-23. However, this does not include more specialized types such as those with integrated or internally connected diacs. (All types of triacs can be tested for anode-to-anode shorts with the arrangement in Fig. 7-23.)

Make certain of the correct identity of the two anode terminals by consulting the specification sheet. Notice that the gate voltage is applied with respect to anode No. 1. The VOM should have at least a 100-milliampere short-circuit current capability on the R × 1 range if the test technique is to be nearly universal. However, several tens of milliamperes is sufficient for testing many smaller capacity triacs.

The test begins with SW1 open. It should be proved that the triac passes no current for either position of SW2. That is, with the VOM on its R × 1 range, infinite resistance should be measured for both positions of SW2. Next, momentarily close SW1. This should fire one side of the triac and produce a deflection in the vicinity of half scale on the VOM. Next, place SW2 in its alternate position. This accomplishes two things: It reverses the anode 1-anode 2 polarity, and it extinguishes conduction in the triac. Having gotten this far, we know the triac will behave as a simple SCR, which is a half-wave control device. Momentarily close SW1. This should trigger conduction through the triac again and a near mid-scale deflection should be again seen on the VOM. To determine if the correct full-wave control characteristic exists again reverse the position of SW2. In a good triac, conduction will be extinguished, as indicated by an infinity reading of the VOM.

Anode-to-anode leakage can be checked with the VOM on its R × 10,000 range. Good triacs tend to produce no discernible deflection. This test should be carried out for both positions of SW2. (No gate control is involved here, so do not close SW1.) A deflection beyond what might be termed negligible is caused for concern, even if the triac has passed its full-wave control testas outlined above.

For triacs with self-contained diac gate-firing elements, replace the 12-ohm resistor with one of 220 ohms. Switch the VOM to its 100 milliampere dc current range and insert another 220-ohm resistor in series with it and a 22 1/2-volt battery. The polarity of the battery is not important since SW2 reverses polarity. The test for full-wave control characteristic is performed as in the previous case. The on condition will be indicated by deflection in the vicinity of full scale on the 100 ma dc current range. To preserve the life of the battery, do not leave the triac in its on state longer than necessary.

The simple test circuit shown in Fig. 7-24 is useful in determining the basic control action of a wide variety of 120 watt or higher voltage triacs. One of three possible conditions will be observed: If the triac is working properly, you will be able to control the lamp brilliancy from off or from a very pale glow to its normal line voltage brilliancy, or very nearly so. A defective triac will still provide continuous control of lamp brillancy, but will perform as a half-wave rather than a full-wave element. Behaving in this way, it will essentially operate as an ordinary SCR and will, therefore, be useful for some purposes; however, the maximum lamp brilliancy will be considerably reduced. A triac defective in this way cannot perform its intended purposes, and must be replaced by a new unit. The third possible test result is no control at all. Here, the lamp remains either fully on or off, or, in some types of damage, at an intermediate brightness. Such "dead" triacs have been permanently damaged and cannot serve any useful control function.

The half-wave type of defect can be confirmed in the following way: Substitute an ordinary silicon rectifier diode for the triac (any such diode with rating of at least 200 volts and 500 milliamperes will be suitable). The diode polarity is of no consequence and the gate connection is ignored. If half the triac is working, the lamp brilliancy will be the same with the diode in its place. In other words, if the lamp glows as brightly with the diode in place of the triac, the triac under test is capable of working only as a half-wave rectifier.

In the test circuit in Fig. 7-24, the back-to-back arrangement of filament transformers provides isolation from the ac power line. If a one-to-one ratio isolation transformer is available, this, of course, may be used. A .02 microfarad capacitor connected from the junction of R3 and the neon bulb may result in smoother control in then early off condition, but it is not essential for the purpose of the test procedures. Some neon bulbs are not satisfactory because of its extremely high ionization voltage, or because of polarity effects. A suitable neon bulb may be selected with the aid of a 90-volt battery, a 9-volt battery and a 100K resistor. The batteries are connected in series opposing so as to produce

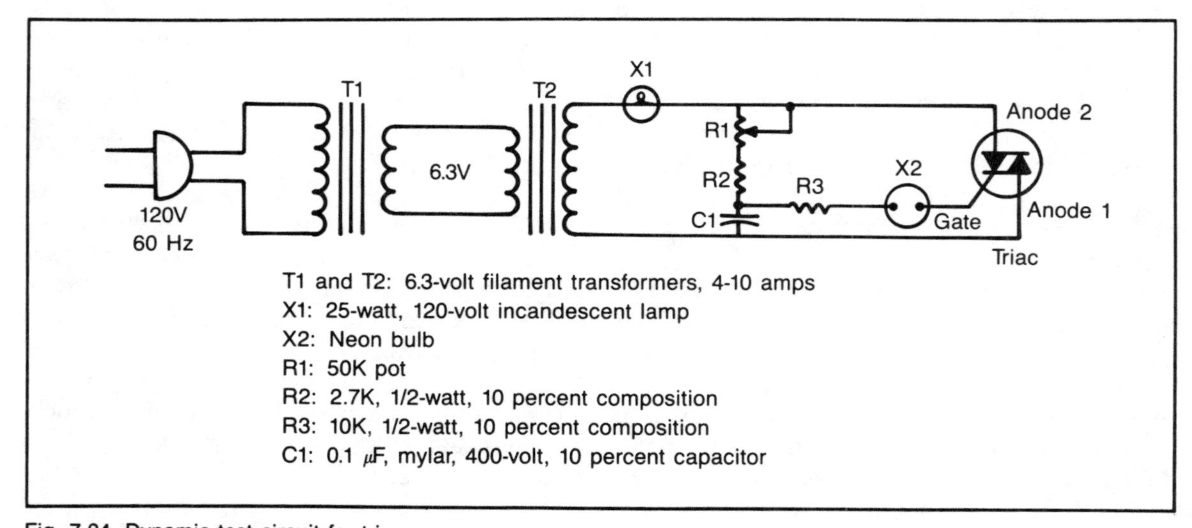

Fig. 7-24. Dynamic test circuit for triacs.

82 volts. If a neon bulb will glow when connected both ways in this setup, it will be suitable for the triac tester.

QUESTIONS

7-1. A simple crystal radio set is constructed with an old-style point-constant diode. Later, it is decided to upgrade the performance of the set by replacing the older diode with a high quality modern junction diode. The results prove disappointing, for it is found that the only reception attainable is from a very powerful local station. What has been overlooked in this quest for improvement?

7-2. A battery charger using a half-wave rectifier circuit with a large germanium diode tends to run hot. It does not appear practical to attach a larger heatsink to the rectifier diode. So it is suggested that perhaps a silicon rectifier connected in parallel with the germanium diode will divert enough of the current to enable the germanium unit to run cooler. Because of the much higher temperature rating of the silicon rectifier, and because it also has a higher wattage rating than the germanium unit, it is decided to try the experiment with no heatsink for the silicon rectifier. What can be immediately predicted?

7-3. An FET is being tested to determine how much signal swing its drain-source circuit can handle before encountering the reverse-breakdown region which appears on both FET and bipolar characteristic curves. With the gate circuit open, it is found that no breakdown appears when quite high voltage is applied through a limiting resistance to the drain-source terminals. The same test procedure applied to a bipolar transistor with its base circuit open causes the expected breakdown. Since the curves of these two devices are so similar, why are there no results from the FET test?

7-4. An attempt is made to measure the transconductance of bipolar transistors, since this parameter is much used as a figure of merit for other active devices such as the vacuum tube and the FET. A voltage-regulated power supply is used and other precautions are observed in order to make the test mathematically valid. However, transconductance in the neighborhood of several hundred thousand micromhos and greater are obtained! This appears unreasonable when compared with values to several thousand to perhaps several tens of thousands of micromhos commonly realized for tubes and FETS. What might be the trouble?

7-5. An ohmmeter is being used to test a pnp transistor with physically unidentifiable leads. How can the leads be identified?

7-6. The ohmmeter function of a VOM is set for its highest range, R × 10,000. The negative test prod is clipped to the base lead of an npn silicon small-signal type transistor rated for several hundred milliwatts collector dissipation. The positive prod is touched to the collector lead, and there is no noticeable leakage. (It is known that the base-collector diode is not open because a previous opposite polarity test showed good forward conduction.) The positive prod is next touched to the emitter lead; this results in approximately a half-scale deflection. Apparently, the base-emitter junction of the transistor has quite high leakage. Should the transistor be discarded, or should the apparent defect by confirmed in some other way?

7-7. Four silicon rectifier diodes are arranged in a bridge circuit and ecapsulated in epoxy. The markings designating the input and output terminals as well as the polarity of the output terminals have been obliterated. How can an ohmmeter be used to identify the terminals.

7-8. Do the tests described above also indicate that all four diodes are good?

7-9. Replacement transistors for a direct-coupled oscilloscope amplifier are being evaluated in a transistor tester. Many of the required transistors are available so it is decided to select high beta replacements. The beta spread appears to be about four to one (not uncommon). Only those transistors which fall in approximately the upper ten percent of this spread are chosen. Explain why such an approach may not be a wise one.

7-10. An SCR with a load current rating of several tons of amperes is being tested with a VOM for basic SCR action. The VOM is on the R × 100 ohms range and is properly connected with respect to polarity to the cathode and anode of the SCR. A gate signal is applied from a 3-volt source consisting of two series-connected flashlight cells and a 1000-ohm current-limiting resistor. When the ate is made positive with respect to the cathode, the VOM indication changes from infinity to a reading of about 1500 ohms. Because of the low current supplied by the ohmmeter test prods, such a reading would suggest that the SCR is basically operational. However, the SCR is regulated to the scrap heap because it is later observed that the VOM again indicates infinity when the gate signal is removed. (The SCR apparently does not display characteristic "latching" behavior.) Explain why the evaluation of the SCR might have been a overly hasty one. What simple test could be made to confirm the apparent defect?

ANSWERS

7-1. Although the junction diode may excel in such specifications as forward-to-reverse conduction ratio, linearity, and high-frequency characteristics, its weakness for this application is its voltage gap, approximately 200 millivolts for germanium and 600 millivolts for silicon. Thus, no station can be heard unless the signal is strong enough to develop the voltage needed in the circuit to reach the conductance point of the junction. Conversely, the older point-contact diode starts conducting at substantially any forward-voltage exceeding zero. (Some of the older point-contact diodes were later "improved" by making them junction devices, but with no change in the designation!)

7-2. The immediate failure of the experiment can be predicted, but not because the silicon rectifier might fail to stand up thermally. Actually, the silicon rectifier would not work in the circuit at all! Its required forward voltage of about 600 millivolts would not be allowed by the germanium rectifier, which develops a voltage drop in the vicinity of a couple hundred millivolts.

7-3. The reverse-breakdown region which appears on the collector current vs collector voltage curves of bipolar transistors represents an avalanche condition in the base-collector junction of the output circuit. In the FET, the output circuit is a silicon region uninterrupted by a junction. Therefore, there is no evidence of change in the conductive state when the test is carred out as stated. However, if the gate is provided with an external path to the source, a sufficient difference in potential between the drain and source will ultimately cause reverse-breakdown to occur in the gate junction. This condition is responsible for the abrupt voltage saturation displayed in FET drain current vs drain voltage curves.

7-4. Although the fact is not widely specified, transistors do, indeed, have transconductance ratings in the range mentioned. At first, the concept of transconductance was shunned as a meaningful indicator of transistor performance. However, some of the major manufacturers now show transconductance curves in addition to hybrid data for some bipolar transistors. The transconductance of a transistor is the ratio of a small change in collector current to the change in base-emitter voltage causing it. The collector voltage must remain constant for this definition to be valid. A statement of the average values of base current and collector current should accompany transconductance test data.

7-5. First, you must know the polarity of the ohmmeter test prods. The idea is to locate the base-lead first. The base connection of a transistor is unique in that once forward-conduction is established with respect to one of the other two leads, the ohmmeter polarity must be reversed in order to obtain an indication of forward conduction with respect to the remaining lead. Such tests are generally best carried out on the R × 10 ohms range for ordinary small transistors. (For large power transistors, the R × 1 range can be used without danger of damage to the transistor.) Once the base lead has been identified, the other leads can be quickly determined as follows: With the negative-polarized test prod of the ohmmeter connected to the base lead, the positive prod is connected to first one and then the other of the remaining two leads. The contact which results in an indication of forward conduction on the meter identifies that lead as the emitter. Therefore, the lead that produces no meter deflection is the collector. The test should be confirmed by connecting the positive-polarized test prod to the base lead.

7-6. On the R × 10,000 ohms range, many VOMs make use on internal batteries of 15 volts or higher. However, the reverse breakdown voltage of the base-emitter junction in many small signal transistors is less than this value. Therefore, the apparent base-emitter short could be a normal indication of a zener or avalanche condition. In other words, the base-emitter portion of the transistor behaves as a zener diode under this test condition. If the junction breakdown voltage in this case were in the vicinity of 7 or 8 volts, the ohmmeter would indicate somewhere in the central portion of its scale. However, the actual ohms indication would not be accurate and should be ignored. What should be done is to repeat the test on a lower ohms range. Usually, the R × 1000 range uses a 1 1/2-volt battery. If there is no detectable deflection on a lower range, you can safely assume that the "leakage" seen on the R × 10,000 ohms range does not actually exist. In all probability the transistor is OK. However, if you want to evaluate the base-emitter junction for reverse leakage, an electronic meter is best bet. There are sensitive VOMs with R × 100,000 ohm and R × 1 megohm ranges at 1 1/2 volts, but these are not as popular as the less expensive varieties, which must employ high internal voltages on the higher resistance ranges to overcome the inherent lack of sensitivity in the meter movement. All 1000, 5000, 10,000, and 20,000 ohms per volt VOMs fall into this category.

7-7. The R × 10,000 range of a 20,000 ohms per volt instrument can be used to good advantage here. The pair of terminals which produce no meter reading regardless of test prod polarity are the ac input terminals. The other pair of terminals will be polarity sensitive. With the prods one way, there will be a high reading. With the prods the other way, the reading will be low. When the prods are connected so that the reading is high, the terminal contacting the negative test prod will be the output terminal which is positive during normal operation of the rectifying bridge.

7-8. No. One or two of the diodes could be open. However, once the input and output terminals have been identified the cathode and anode from each of the four diodes can be contacted at the appropriate terminals to test for basic diode action. An open diode will be revealed by a lack of forward conduction. These tests are most easily performed if the diode bridge arrangement is sketched on paper as soon as the input and output terminals and the polarity of the output terminals have been determined.

7-9. Although much depends on how biasing is applied,and feedback is used in the amplifier circuit, extra high beta transistors tend to upset the bias operating points of the stages. This tendency is particularly evident in dc-coupled amplifiers because the high-bet transistor will also tend to have a higher collector-to-emitter leakage current. This leakage current appears as bias to the subsequent stage. Often overlooked is the fact that two transistors with identical collector-base leakage currents (I_{cbo}) will not necessarily have the same collector leakage currents when used in the common-emitter configuration. In this circuit, the transistor with the higher beta will have a proportional higher collector leakage current (I_{ceo}). If, because of higher beta than the circuit design can accommodate, the biasing of one or more stages is shifted appreciably, the amplifier will be unusually susceptible to overloading, distortion, and other malfunctions. Also, using such "hot" transistors can result in oscillation, or interfere with the flat frequency response.

7-10. This mistake made during the above described test procedure is not an uncommon one. In order for an SCR to turn on and latch, a minimum "holding" current must be available from the source supply the cathode-anode circuit. When the VOM is on its R × 100 ohms range, sufficient current is often not available for this purpose. When this test is conducted with the VOM on the R × 1 range, the same SCR which failed to remain on in the previous test when the gate signal was removed will very likely continue to conduct (assuming, of course, that the SCR is otherwise OK).

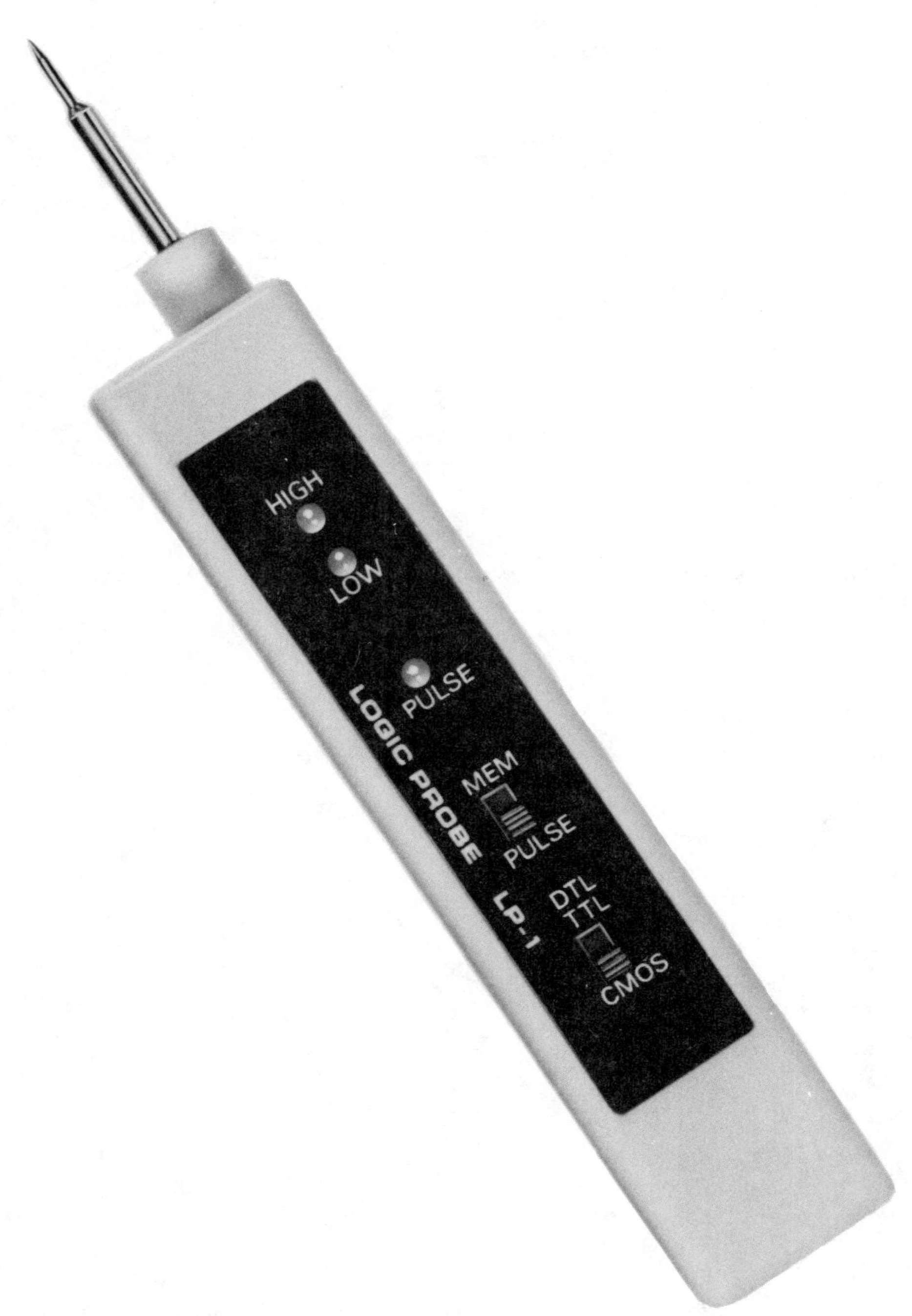

In this commercial Logic Probe, the PULSE mode produces flashes of light from the PULSE LED on rising or falling logic transitions. The MEMORY mode produces a steady light from the PULSE LED from glitches or transients, thereby revealing their presence. Courtesy of Global Specialties Corporation.

Chapter 8

Rf Test Techniques

In the material thus far covered on test procedures, discussions regarding frequency range or limitation have been purposely held to a minimum. There are several reasons for this. First, it has been assumed that the reader is at least somewhat knowledgeable concerning the effects of frequency. Secondly, it is misleading at times to say that a certain test technique is good only up to a certain frequency. More care in construction and implementation can extend such an alleged "limit" many times. Components with more ideal characteristics appear almost by the day. It is not difficult to find statements in older books which say that the transistor was destined to be confined to the audio frequency spectrum, that SCRs were developed only for use at 60 Hz, that direct readout oscilloscopes and frequency counters were "inherently" limited to the several MHz range, etc. Indeed, it is within the author's recall that prevailing technical opinion held the FCC's requirements for citizens-band operation as impractical and unattainable at such high frequencies (27 MHz).

It is true that at some point after we pass the audio-frequency range, the generalized approach to test procedures can be best implemented by certain instruments, techniques, and approaches. But, again no rigid frequency line will be drawn. For example, link coupling, an "rf" technique, can be used right down to the low audio frequencies. Another example is the technique of heterodyning (frequency translation). Although useful in the realm of megahertz, we recall that in the synchrodyne radio receiver,the "i-f" frequency goes right down to dc (zero frequency!). Just as we could think of "extending" many power-frequency and audio-frequency techniques into higher frequency regions, we should give thought to the possibility of "extending" rf techniques into the lower frequency regions. All this raises the question, where do the radio frequencies begin? Because of reasons already cited for staying away from strict lines of demarcation, we will not try to specify where each frequency region begins or ends. Rather, the prevailing attitude will be that if you are resourceful enough to use a dip meter to determine audio-frequency resonance, the objective of this chapter will have been precisely served.

USING THE DIGITAL COUNTER IN RF TEST PROCEDURES

Traditional ways of using electronic devices and test instruments are sometimes based too much on limitations which prevailed at an earlier state of the art. The digital counter was originally restricted to relatively low frequencies. When its range was first extended into the rf spectrum, the costs and complexities of the device made it a status symbol of well-funded laboratories. As technology advanced, direct-readout instruments with frequency capabilities of several tens of megahertz and higher became available. Moreover, the cost of these counters has steadily declined to the point that we can at least say that the digital counter is found in more and more service shops. With the advent of ICs, the digital counter is today a standard instrument in every well-equipped amateur radio station. This being the case it is appropriate to investigate methods of using the digital counter at radio frequencies.

In Figure 8-1 a sampling of the rf voltage is picked up by a link and cable arrangement rather than by direct connection. An immediate advantage provided by this rf technique is the isolation of the counter from dc potentials. If the length of the coaxial line is close to or in excess of a quarter wavelength of the frequency being measured, a resistance equal to the characteristic impedance of the line should be connected across the input terminals of the counter. In practice, a 68 to 100-ohm composition resistor will serve this purpose. A noninductive load terminates the line to avoid a resonant rise of rf voltage as a protective measure. Where it is known that the rf source is at a very low power level, a resonance situation could be used advantageously. When sampling a frequency in this way, you should always begin with the link too far away from the source to pick up appreciable rf energy. Then gradually move it closer until the readout appears on the counter.

Two additional methods of sampling rf energy for frequency measurement are shown in B and C in Fig. 8-1. In Fig. 8-1B the rf choke tends to suppress disturbances from the 60-Hz power line and from fluorescent lighting fixtures. The radiation field of the rf source around the whip antenna can result in a simple and clean method of monitoring frequency, if no other strong rf sources are in nearby operation. In Fig. 8-1C a resonant circuit serves as an "antenna" and provides considerable attenuation of unwanted signals. Often it is only necessary to ascertain that the LC circuit is near resonance at the desired frequency. This scheme is often comparable in performance with the whip antenna setup in Fig. 8-1B because of the physical separation that can be maintained between the pickup device and the rf source.

When the power level of the rf source is known to exceed the level of several watts a protective technique should be used such as shown in Fig. 8-1D, even though similar protection may already be built into the counter. Regardless of the instrument protection provided, we should be ever alert to avoid conditions that could result in inordinately high rf inputs to the instrument. This is particularly important when dealing with transmitters.

SIMPLE RESONANCE INDICATORS

Often, it is not necessary to determine the approximate frequency of a source of radio frequency energy, but it is desired to tune for the maximum rf power level at a particular point in a system. In a Class C rf amplifier, the dc current meter in the plate or output circuit provides the means of resonating the output tank circuit. Resonance is indicated by a pronounced dip in the dc current. See Fig. 8-2A. Amplifier stages used for moderate and high power levels must not be operated for any appreciable time in the nonresonant condition. In fact, it is often wise to begin the resonating procedure with greatly reduced dc power. Then, more refined resonating adjustments can be made at progressively higher power levels. This precaution applies to tube stages with operating power input levels of a hundred watts or more. For transistor stages of several watts or more, the same precau-

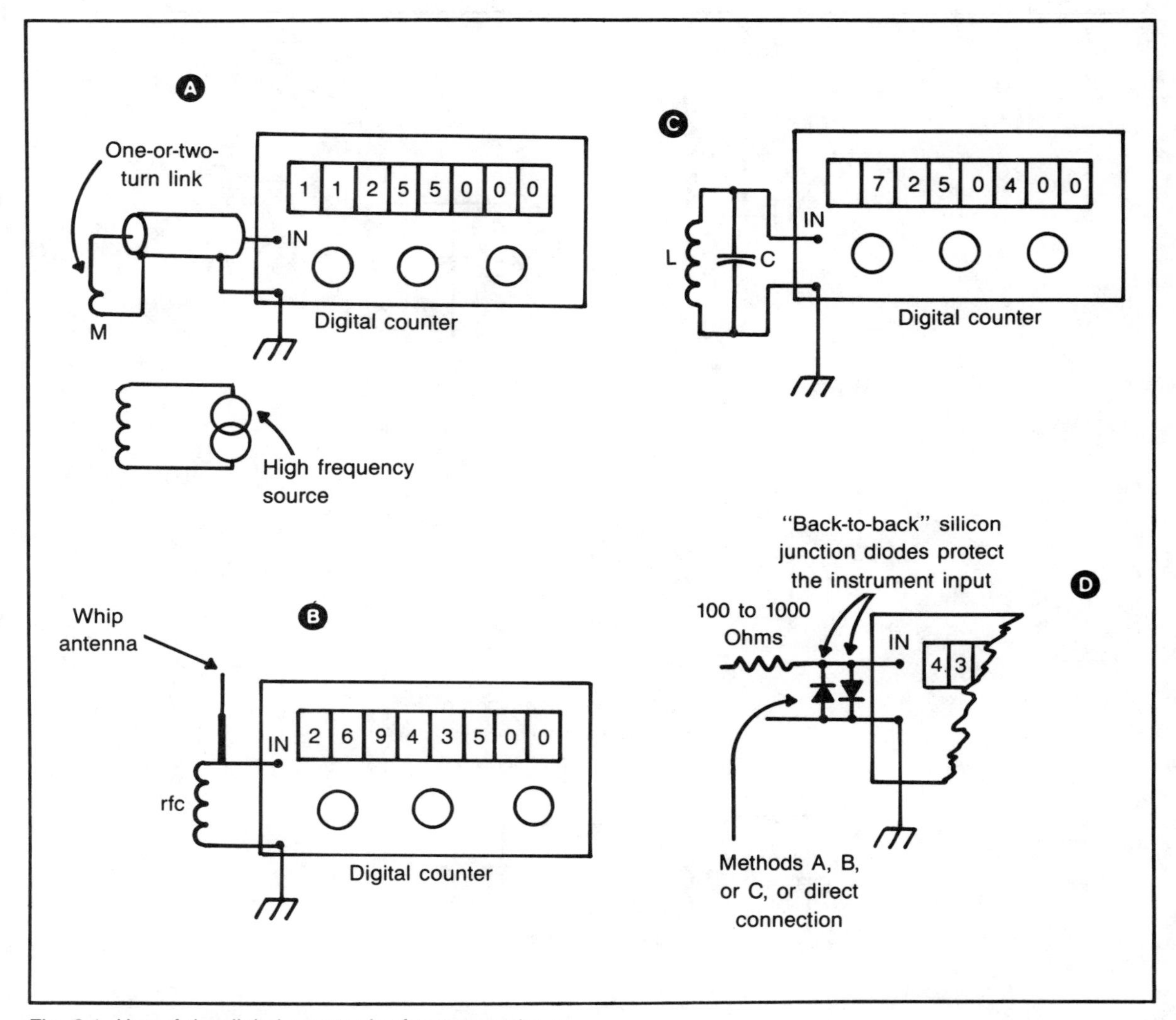

Fig. 8-1. Use of the digital counter in rf test procedures.

tion should be observed, although not necessarily for the same reasons. If it is not convenient to reduce dc power, the resonating procedure can be carried out by observing momentary deflections of the dc plate or output meter as the tank circuit is tuned. Some judgment is necessary concerning the safety of this technique in individual cases. For tube stages with more than several hundred watts input, and for transistor stages with more than several tens of watts input, this procedure may subject the tube or transistor to damaging conditions.

The link and lamp rf indicator shown in Fig. 8-2B has at least a psychological advantage over the technique of monitoring the dc amplifier current. When we see the lamp glow, we know it is due to the rf energy present. At any fixed coupling between the link and the test circuit, the intensity of the lamp filament increases with the rf power level in the tank. Therefore, it is very convenient to resonate the tank by watching the lamp. It also is very important that the coupling be maintained as loose as possible, not only to protect the lamp

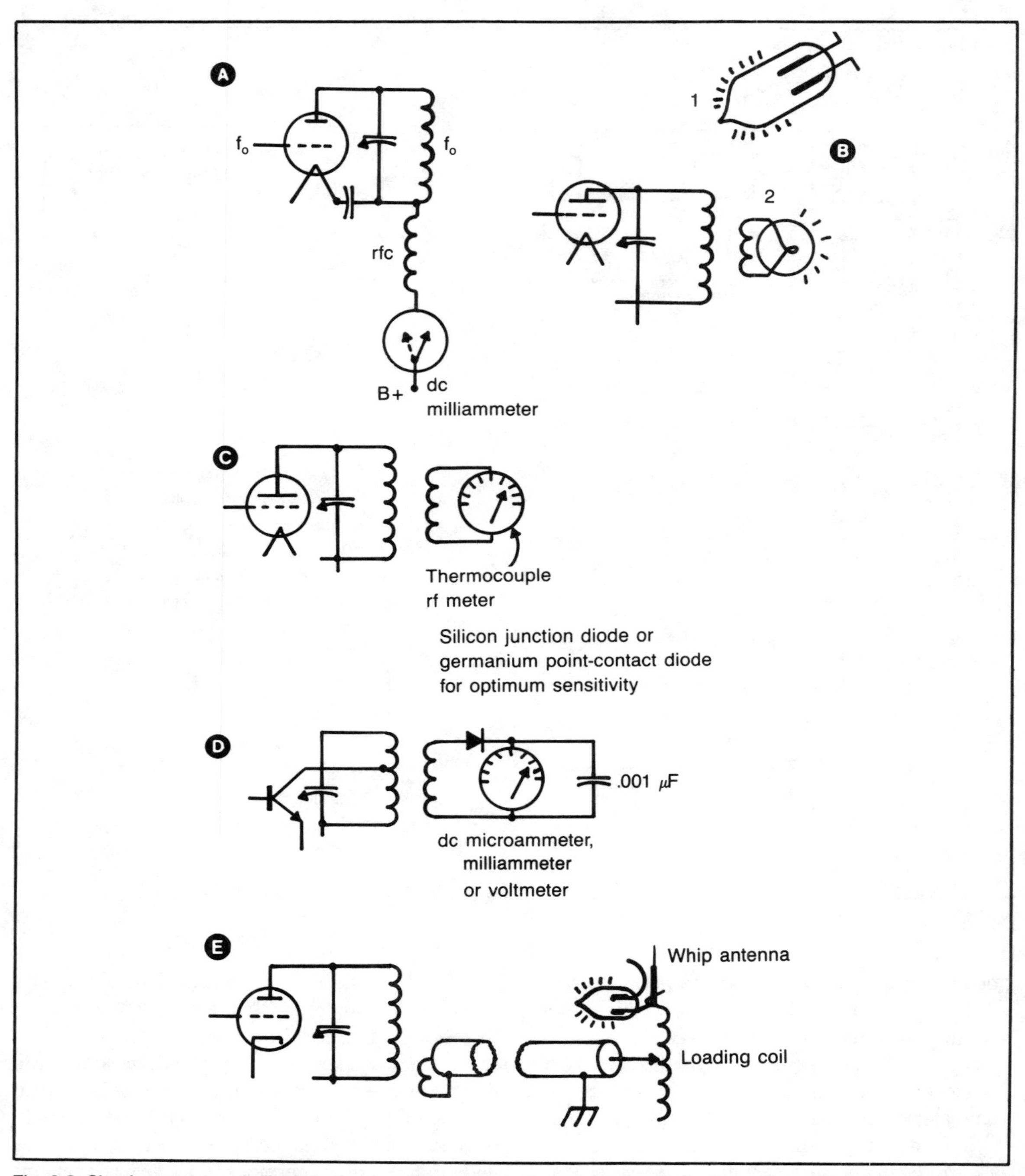

Fig. 8-2. Simple resonance indicators. A—dc milliammeter in the output circuit of a Class C amplifier. A pronounced dip occurs when LC resonates at f_0. B—A neon bulb in the electric or radiation field (1), and a loop and filamentary bulb in the electromagnetic field (2). C—Loop and thermocouple meter. D—Loop and dc meter for low-power rf levels. E—Resonating a loaded whip antenna with a neon bulb.

from burnout, but to avoid detuning of the tank circuit. At the final resonating adjustment, the subdued glow of the lamp filament should scarcely be visible, and only at exact resonance. Other things being equal, the more sensitive the lamp, the more accurate will be the resonating procedure because less loading will be imposed on the tank circuit. On the other hand, there may be occasions when this technique will be used to simulate an rf load, such as that of an antenna. Such a "dummy load" is very useful for determining the performance of a transmitter without radiating a signal. Too much significance should not be taken from the glow of a bulb in such a simple arrangement, however.

Also shown in Fig. 8-2B is a neon bulb. For tube transmitters with power inputs in excess of several tens of watts, the bulb readily glows in the electric and radiation field of the tank circuit. Therefore, it provides a means of resonating the tank. However, the glow of the neon bulb, especially when produced by rf excitation, is a very coarse indication. It is not always so easy to determine maximum intensity as it is with a filamentary lamp. However, for test purposes, the neon indicator often proves useful. Sometimes one terminal of the neon bulb must be attached to a "hot point" of the tank circuit if the power or voltage level is marginal. Caution: Do not explore tube stages with the neon bulb while the power is on unless the neon bulb is at the end of a nonconducting stick or rod at least one foot in length. In general, this tuneup technique should be used only with lower power transmitters, those with less than, say, 500 volts of available dc.

The indicator in Fig. 8-2D is a refinement of the lamp-type observation technique. Its greater sensitivity and the presence of a meter readout make it particularly well suited for alignment of transistor rf stages. It is also useful in neutralizing tube rf stages, and is an excellent test device for revealing the presence of parasitic oscillations because of its continuous response.

A simple means of resonating a loading coil associated with a whip antenna is shown in Fig. 8-2E. Surprisingly, this method is often useful with transmitters in the several-watt range. This is because of the rf voltage stepup in the high Q loading coil. A tiny neon lamp, with the lead attached as shown, will not exert an appreciable detuning influence.

The wavemeter is an ultrasimple device. In its most primitive form, it consists of an inductor and a variable capacitor. The inductor is a coil wound and arranged so that it can be electromagnetically coupled to a tank circuit; that is, placed close to the inductor of a tank circuit. How close is "close"? This depends considerably on the power level of the oscillations in the tank circuit, and often varies from near-physical contact to an appreciable fraction of a foot or more. If the axes of the two coils are aligned, the distance can be greater; in other positions the coil will have to be closer. Wavemeters and test situations are shown in Fig. 8-3. The basic idea is to extract some energy from the tank circuit; maximum energy is picked up when the wavemeter is tuned to the frequency of the oscillations in the tank circuit. There are ways of knowing when this condition exists. With the most primitive wavemeter shown in Fig. 8-3A, you must be aware of a unique disturbance occurring in the electronics associated with the tank circuit. In an rf amplifier, the plate (or collector or drain) dc current meter will undergo a sharp increase. Conversely, grid current (or base or gate) current will sharply decrease. In an rf oscillator, the dc currents will behave similarly. Additionally, oscillation can abruptly cease if excessive energy is taken from the circuit. It should be realized that the wavemeter works because it dissipates the extracted energy in its own losses. During a wavemeter test, the wavemeter is actually coupled to the tank circuit and becomes part of it.

This brings up a basic precaution in the use of the wavemeter, and that is to always try to take as little energy as possible from the circuit under test. This means using the minimum coupling which will give a reliable indication. This will mean greater accuracy, because only then is the frequency "pulling" effect at its minimum. You might suppose that this would be an important precaution in oscillator circuits, but not in amplifiers. This is not the case, however. To be sure, you are not likely to disturb the frequency at the output of an amplifier

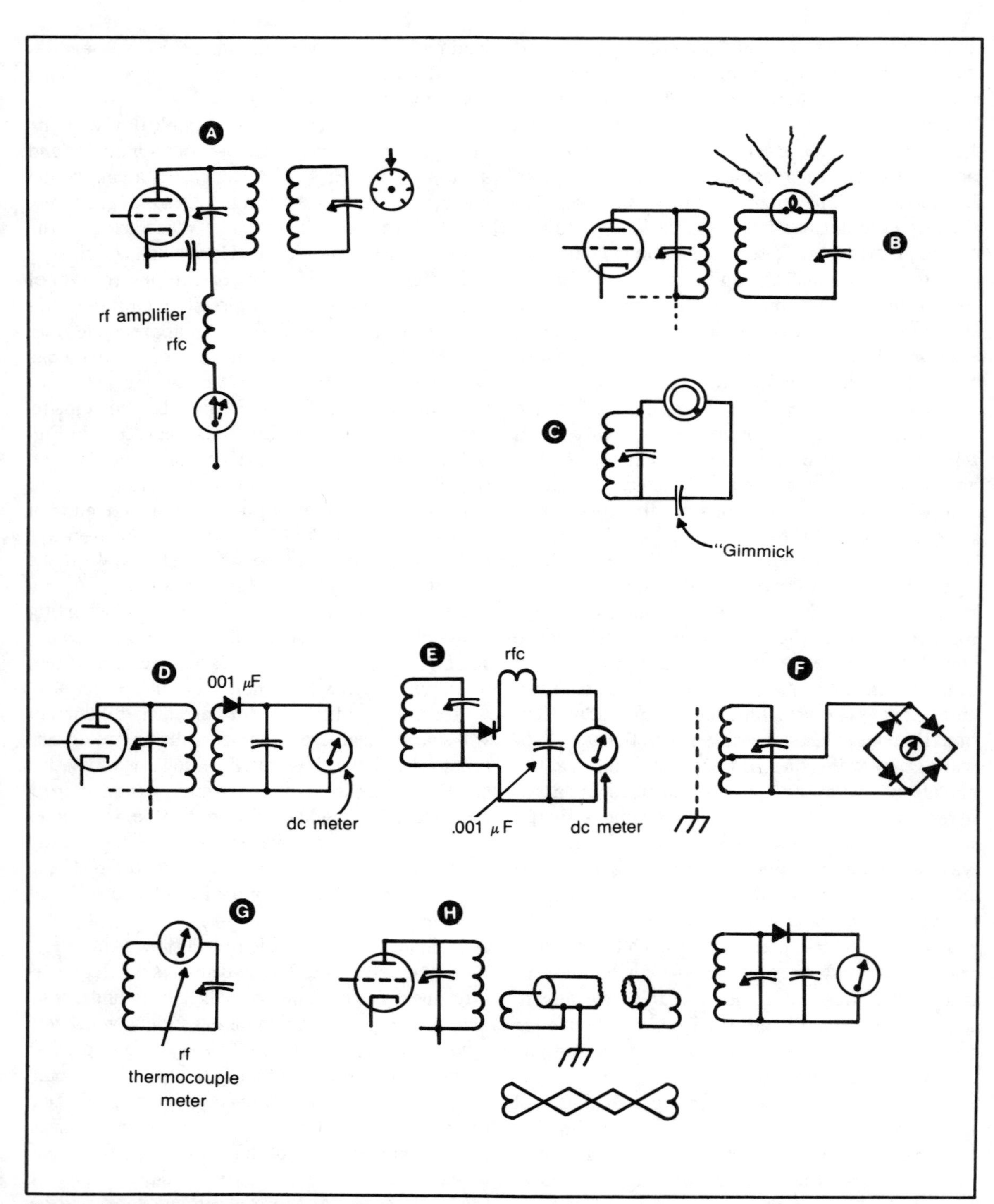

Fig. 8-3. Absorption wavemeter measurement techniques.

(especially a second stage or a frequency multiplier). However, the pulling effect also involves the tuning of the wavemeter itself. Even though the calibrations on the dial of the wavemeter may be used primarily for a rough frequency measurement, there is no point in using sloppy test procedures. For example, if you are concerned whether or not a frequency multiplier is delivering the intended fourth harmonic, extreme care would be necessary in using the wavemeter to be certain that the energy extracted from the multiplier tank circuit was not at the fifth or sixth harmonic, for example.

In Fig. 8-3A, the condition of wavemeter resonance must be observed in the rf source itself. If, as is often the case, the rf source is a Class C oscillator, amplifier, or frequency multiplier, the wavemeter resonance will be indicated by a pronounced increase of dc current supplied to the plate or output circuit of the stage. The wavemeter technique shown in Fig. 8-3B utilizes a filamentary bulb to indicate resonance directly in the wavemeter itself. The bulb may range in size from a "rice grain" type to a large 120-volt multiwatt light bulb. For most purposes associated with transmitter tests, various common flashlight bulbs have long proved suitable.

The scheme shown in Fig. 8-3C has been used for neutralizing "straight-through" amplifier stages, and for tracking down parasitic oscillations. This wavemeter is not always easy to use because of the tendency for the neon bulb to be ionized in the electric or radiation field of the rf source. Although the indicator does not require much rf power, it needs about 70 volts and, therefore, may not work at all if the rf source is in the "flea-power" category. The wavemeters shown in Fig. 8-3D, E and F utilize a dc meter which indicates the rectified output. These wavemeters are suitable for a wide range of power levels and provide better calibration accuracy than can ordinarily be expected from lamp indicators. Dc microammeters, milliammeters, and voltmeters make good indicators. Maximum sensitivity results from the use of germanium point-contact diodes, but for most purposes the silicon junction diode is adequate and is less susceptible to damage from overload. The tapped inductor scheme in Fig. 8-3E and F are recommended when maximum calibration accuracy is important, because the resonant tanks are less loaded by the indicator circuits in such arrangements. Additionally, the wavemeter setup in Fig. 8-3E makes use of a Faraday shield to reduce capacitive effects on tuning.

The simple arrangement in Fig. 8-3G can be called the "classic" wavemeter. It is well suited to the demonstration of resonance and the absorption of rf energy. The thermocouple meter probably degrades the Q of the LC circuit less than do many lamps. However, in the rough and tumble procedures of wavemeter tests, this type of meter tends to be vulnerable to burnout. Also, the relatively sluggish response of some of these meters can be a less than desirable feature.

The arrangement shown in Fig. 8-3H enables wavemeter measurements to be made remotely from the rf source. At least one and preferably both of the links should be so loosely coupled that the meter cannot exceed a full-scale deflection. If the rf stage shown operates at a high power level, then the six or ten inches of possible separation between the link and the tank coil provide safety from high dc voltage. Links generally consist of between one and three turns of heavy wire with link diameter ranging from 1 to 3 inches. Fewer turns and smaller diameters are better for frequencies exceeding several tens of megahertz.

It is important that the resonant frequency of the links and all associated circuitry be much higher than the frequencies involved in the test. At frequencies up to several megahertz this condition tends to exist more or less automatically and the loop is essentially noncritical. The length of the coaxial or twisted line separating the links should be considerably less than one quarter wavelength at the test frequency. Otherwise, line resonances may interfere with the desired "lumped circuit" resonance. Again, at frequencies on the order of a few megahertz, this usually is no problem. If long lines must be employed, 100-ohm resistors connected in series with the sending-end link and in parallel with the receiving-end link will tend to make line resonance effects negligible.

Using a Dip Meter

The above situation, where energy is absorbed from the circuit under test, is reversed with a dip meter because it is the source of rf energy and a passive LC circuit becomes the test object. This means that oscillators, rf amplifiers and other active circuits associated with LC tanks must be turned off before testing begins. When the coil of the dip meter is coupled to the LC tank being tested, either the LC tank or the dip meter is tuned, depending upon what we wish to accomplish. In any event, when the resonant frequency of the LC tank is the same as the frequency generated by the dip meter, the dc meter on this instrument will, as suggested by its name, undergo a pronounced dip. The calibrated tuning dial on the dip meter then indicates the resonant frequency of the coupled tank circuit. This test situation is illustrated in Fig. 8-4.

When the dip meter is being tuned, always proceed from lower to higher frequencies. Initially, fairly close coupling may be necessary so that you can get a rough idea of the frequency where the dip will occur. However, before the frequency is read from the calibrated dial, the dip meter should be physically backed away from the LC tank or oriented with respect to it so that a dip is just perceptible. This will ensure minimum disturbance of both the LC tank and the oscillator frequency in the dip meter, thus contributing to optimum accuracy. This sounds simple enough, but some practice is needed to develop the required skill. A dip meter usually does not maintain a constant deflection over a given frequency range. On the one hand, dips may be masked by this undesired motion of the meter pointer; on the other hand, various circuit conditions sometimes produce effects which can be, at least initially, mistaken for a low Q dip.

A dip produced by a low-Q resonant circuit will tend to be broad, and, other things being equal, will require closer coupling to produce a dip of appreciable depth. High Q tanks usually produce very deep dips, but these are the ones which are often passed by because such dips are so sharp. Either they occur while the eyes blink or the attention is very momentarily diverted. Also, if the dip meter is tuned too rapidly, the inertia of the meter, though small, can obscure a sharp dip. An extreme case of this kind is encountered in the rough measurement of unknown quartz crystals which are connected to a loop which, in turn, is coupled to the dip meter. The dip is very difficult to see despite the deep dip which results.

Perhaps the most common use of the dip meter involves the design of LC circuits. Because of the influences of many factors difficult to accurately account for in formulas, it is not always easy to wind a coil which will perform as specified. Whether calculations are used or you rely more on past experiences, an LC combination can be made to resonate or to have a certain frequency range with sufficient accuracy for most applications by skillful use of the dip meter. With the dip meter, the frequency can be raised by increasing the turn spacing, by removing turns, by tuning with a smaller capacitance, or by inserting an appropriate shorted turn or a nonferromagnetic slug. The opposite in each case will lower the frequency.

Figure 8-4B depicts a technique to measure the resonant frequency of a tank from a remote location. This is useful in tight situations where you can't get the dip meter close enough. Here, as in virtually all situations, the coupling should be as loose as possible at both the tank and the coil of the dip meter. If at all feasible, the link associated with the tank undergoing test should be physically positioned towards the "cold" (ground end) of the tank coil. The same reasoning applies to the dip meter coil. However, with the oscillator employed in most dip meters, both ends of the coil are "hot." This indicates that the zero or low rf potential region of the coil is roughly at its midposition.

A link of several turns and of relatively large diameter will usually provide enough inductive coupling at a sufficient physical separation so that detuning effects from capacitance are negligible. At higher frequencies, however, the self-resonance of such a link may cause it to behave as a tank circuit and confusing indications of the dip meter will be the result. Obviously, there are various options

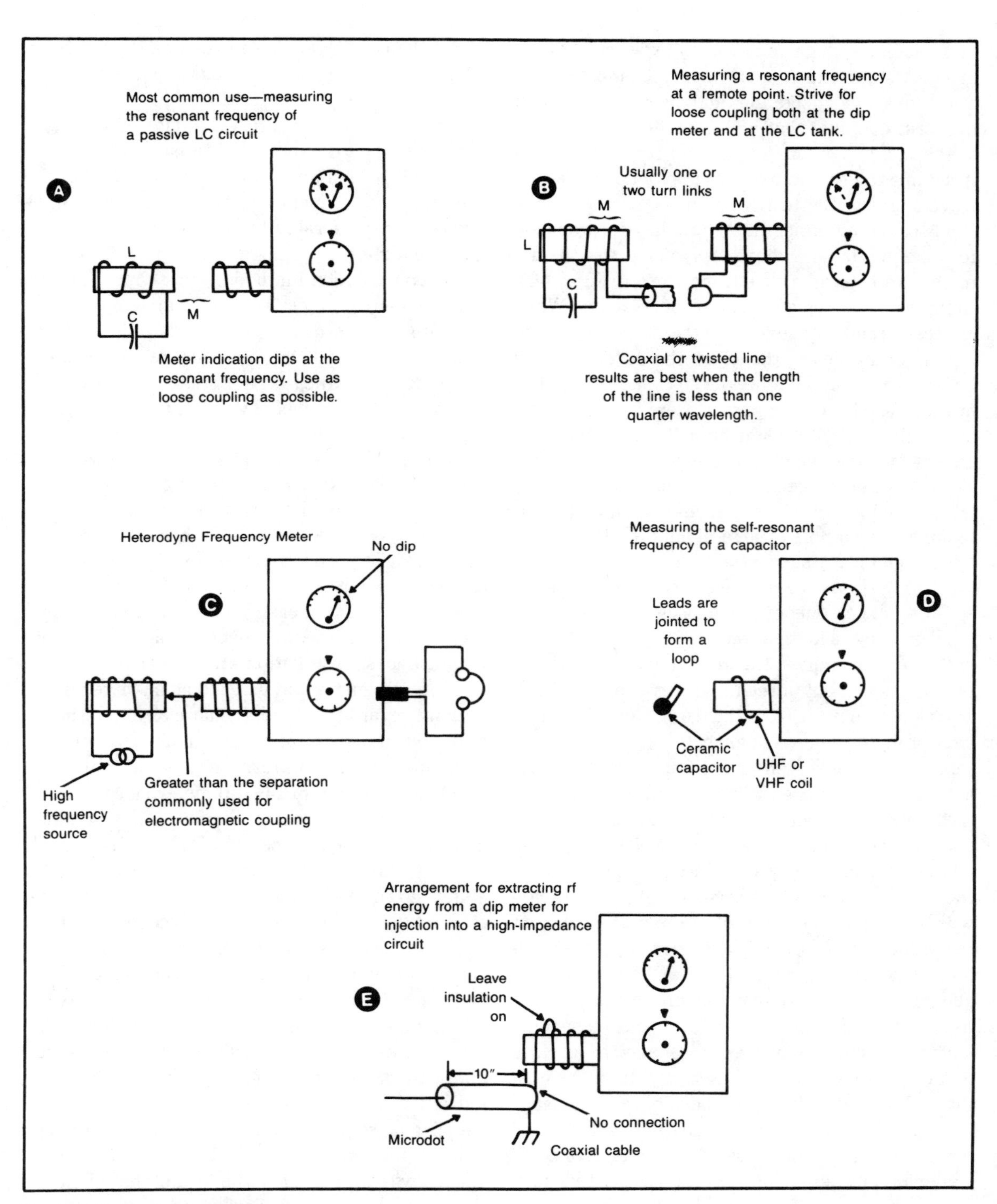

Fig. 8-4. Dip meter applications.

and trade-offs which you learn by experience to use in a given test situation. However, the most important is to always make the final measurement with the least coupling that will give a discernible dip of the meter. Of course, you can run into trouble if the line length too closely approaches a quarter wavelength at the test frequency.

Most commercial dip meters have a provision for plugging in headphones so you can use the dip meter as a heterodyne frequency meter (Fig. 8-4C). In this case, the tested circuit must be active. When the test circuit frequency and the frequency of the dip meter are nearly the same, the difference will be heard as an audio tone, the so-called "beat" frequency. The idea is to approach zero beat as closely as possible; if you could actually hear zero frequency (dc), the two frequencies would be identical. For the purposes generally served by the dip meter, and in light of the accuracies and stabilities attainable, a beat note of several hundred Hz is often close enough for such measurements.

In Fig. 8-4D the dip meter is used to test the self-resonant frequency of a capacitor and its leads. Until you have acquired some experience with VHF and UHF techniques, this test may appear trivial, perhaps novel, but unrelated to anything practical. Such a notion is far from the truth. Consider a service problem where a blown bypass capacitor is replaced in the screen grid circuit of a 220-MHz amplifier in a mobile transmitter. Replacement of the original .002 μF ceramic capacitor somehow does not work out and the stage oscillates. Surely, a larger capacitor, say a .005 μF unit should solve the problem. But the trouble is still present even with a .01 μF capacitor. Part of the problem is that the inductance of the capacitor leads prevents effective bypassing, and part of the dilemma centers about the fact that the manufacturer did not merely use short leads. Rather, he used leads of a specific length so as to produce self-resonance at or close to the operating frequency. But why? It will be seen that although the dip-meter check of self-resonance sees the capacitor-lead combination as a parallel tuned tank, it becomes a series-tuned tank once installed in the equipment. And it turns out that such a series resonant circuit can provide better bypass action than can a capacitor even of a larger size in which the lead inductance has not been advantageously used.

In the event that series-resonant bypassing is not used, such dip-meter tests are still valuable, because it is then important to make sure that the self-resonant frequency is at least several times higher, and preferably by a ratio of several tens, than the highest operating frequency involved. Due to the construction methods, some capacitors are quite inductive regardless of lead length. Such capacitors can prove troublesome at the Citizens Band or even the lower amateur frequencies. Ceramic disc and mica capacitors are usually quite suitable for high-frequency work. But even so, the dip-meter evaluation is often priceless in avoiding trouble both in design and in testing. Two facts should be kept in mind regarding this situation. First, all capacitors have a self-resonant frequency. Second, a series resonant circuit becomes inductive at frequencies above resonance, where its bypass capability is impaired or nullified. The parallel resonance frequency measured by the dip meter and the series resonance frequency which occurs in a circuit bypass installation are virtually the same.

The dip meter can be used as an rf generator without regard to its "dip" characteristic. Energy may be coupled out of the dip meter by inductive coupling, and this, in turn, may be accomplished by placing the coil of the dip meter close to a coil in the receiving circuitry. Similarly, link coupling and coaxial cable may be used. These two methods of implementing inductive coupling are essentially similar to the applications shown in Fig. 8-4A and B. The only difference is that the receiving circuitry generally is not resonated. Constant rf output power can be maintained by varying the coupling to preserve a constant deflection of the meter across the tuning range. When the receiving circuit has a high impedance, say, above 1000 ohms, and certainly above 10,000 ohms, the rf energy may be efficiently extracted from the dip meter by the capacitive-coupling technique shown in Fig. 8-4E. Other things being equal, the coupling will be greater if the insulated wire is looped at either end of the coil rather than around the center. The less the coupling, however, the more accurate the frequency calibrations of the dial. A longer length of Microdot cable may be used than shown, but this will be accompanied by higher attenuation of the rf output voltage. It is not desirable to run any appreciable length, such as a test lead, with no shielding. At lower than rf frequencies, open test leads may be OK. Because of radiation and capacitive coupling, such open wires may couple rf energy into circuits where all types of disturbances can be created.

Procedure for Dip Meter Calibration

Objective of Test: To use a simple means of calibrating and checking the dial indications of a dip meter.

Equipment Required for Test:
Dip meter
Communciations receiver

Test Procedure: Arrange the equipment as shown in Fig. 8-5. The basic idea is to reduce the receiver's capability to pick up stations, but to enable it to receive the signal from the dip meter. The "attenuated" antenna shown in the illustration should serve this purpose. In the event difficulty is encountered, do not provide any antenna for the dip meter, since this will tend to make the calibrating procedure inaccurate. Rather, either close the distance between the dip meter and the receiver or expose more inner conductor in the

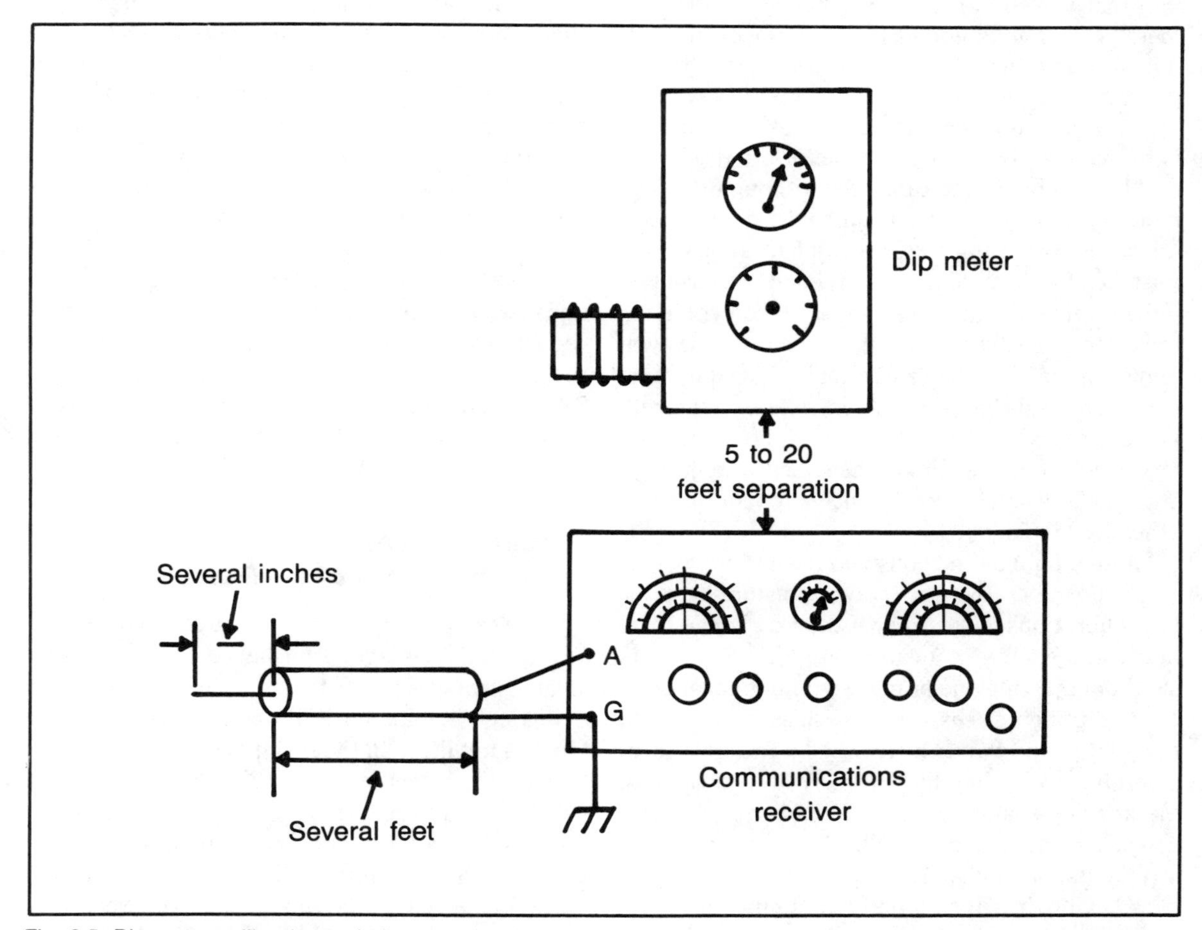

Fig. 8-5. Dip meter calibration technique.

coaxial cable receiver antenna. We assume that the receiver is known to provide reasonably accurate frequency indications. If not, this will have to be attended to before proceeding. For the purpose at hand, the overtones of a 100-kHz crystal oscillator and/or checking with known stations such as WWV will be suitable.

Allow both the dip meter and the receiver a warmup time of at least 20 minutes even if they are both solid-state. Calibration is accomplished by listening to the signal from the dip meter. This should be done with the beat-frequency oscillator in the receiver turned on. With the receiver tuned as close to zero beat as possible, the dial indicates the frequency of the dip meter signal. Make certain that the coil of the dip meter is not close to metallic or conducting masses or surfaces. Indeed, it is desirable that all objects be removed from the proximity of this coil.

A basic problem with the calibration technique is to make sure that a harmonic of the dip meter oscillator is not being tuned in. Although the harmonic output of a good instrument tends to be low, it is often surprising how easily a harmonic can be mistaken for the fundamental frequency. Of course, if the received frequency is obviously not within the range covered by the dip meter coil in use, you know you've picked up a harmonic. Also, any time it is suspected that a harmonic is being received, you can quickly attempt to locate the fundamental near one half or one third of the received frequency. The fundamental is always much stronger than any other harmonic.

Depending on the construction and provisions of the dip meter, there are ways to restore the present calibrations to accuracy if they are found to be in error. This might mean rotating the dial on its shaft or adjusting one or more trimmer capacitors. If the dip meter uses a tube oscillator, the condition of the tube should be checked if the dial is considerably in error. In commercial dip meters, making any change in the coil windings should be a last resort, if used at all. For home-made instruments, such changes may be necessary.

Usually, it will be found that a number of stations can be received, although weakly, with the antenna shown. If the exact frequency of any of these stations is known, it can be heterodyned with the dip meter frequency. When this is done, the beat-frequency oscillator in the receiver is not needed. Again, you should approach the condition of zero-beat frequency to determine the dip meter frequency. In some cases, it may be more expedient to keep a record of dial marking deviations than to try to make the dip meter dial track accurately over an entire tuning range. This is particularly true with an inexpensive instrument.

Comments on Test Results: The dip meter is not ordinarily considered a precision instrument and the calibration technique described is sufficiently accurate for most practical purposes. On the other hand, things can happen to introduce unacceptable error in the oscillation frequency of the instrument. Rough handling of the coils and physical abuse of the dip meter itself often cause frequency changes, particularly noticeable on the higher frequency ranges. The aging of components can impose another variable factor,and weak tubes can affect both amplitude and frequency.

This test procedure makes use of a communications receiver, but an even better calibration procedure would substitute a digital counter for the receiver. The whip antenna setup shown in Fig. 8-1B would then be more appropriate, and the dip meter would have to be situated a foot or two away.

Measuring Frequencies Beyond the Range of the Digital Counter

The setup in Fig. 8-6 is a useful and economical way to derive more performance from a digital counter than was designed into it. Interestingly, the technique depends on a dip meter characteristic which is not listed in the manufacturer's literature namely its harmonic output. The basic idea is to zero-beat a known harmonic of the dip meter with a received signal. Under this condition, the readout on the digital counter multiplied by the order of the harmonic yields the frequency of the received signal. Because of the approximate calibration of

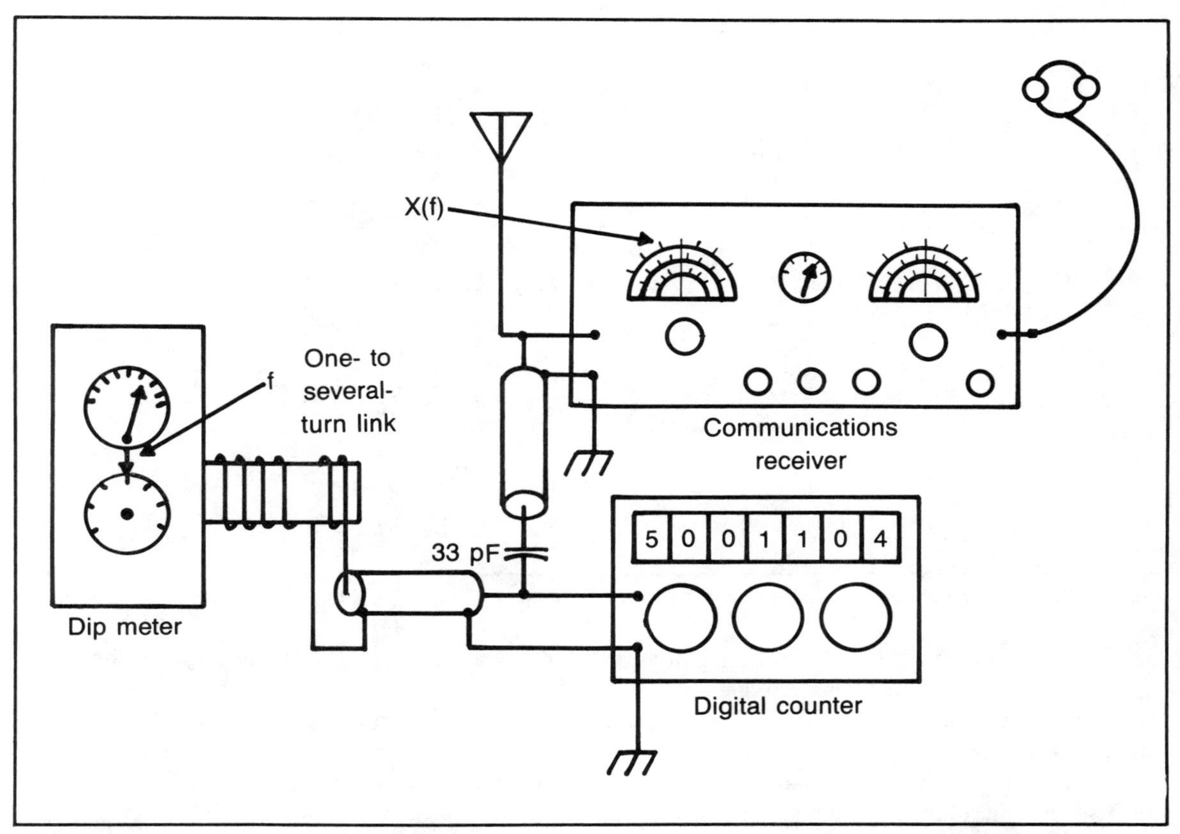

Fig. 8-6. Use of the dip meter as a scaler for extending the range of a digital counter. "X" is the order of the harmonic from the dip meter which zero beats with the received station.

the dip meter tuning dial, and the generally better calibration of receiver dials, there need be no trouble in identifying the order of the harmonic. For example, if the dip meter dial and plug-in coil inform us that a fundamental frequency of about 5 MHz is being generated, this can be readily confirmed by the readout on the digital counter, which may indicate, say, 5,001,104 Hz. From the receiver dial, we see that zero beat corresponds to about 25 MHz. This at once establishes the fact that heterodyning is occurring between the received signal and the fifth harmonic of the dip meter fundamental frequency. (No other harmonic order times 5 MHz is close to 25 MHz.) Therefore, the received frequency is $5 \times 5{,}001{,}104$ Hz, or 25.00552 MHz.

Notice that the receiver dial can also be calibrated by this means. If the receiver dial had read 24 MHz or 26 MHz, we would know that it would have to be adjusted to read as close as possible to the actual received frequency; that is, 25 MHz. Also, this technique can be used to measure the frequency of the received signal if this is desired.

Determining Capacitance and Inductance

A dip meter is helpful in determining unknown capacitance and inductance. Figure 8-7 shows how this is done. Since the procedure is similar for capacitance or inductance, a single dip meter is shown in the drawing. The basic idea is to use only enough coupling to produce a noticeable dip. The frequency corresponding to this dip is read from the dial of the dip meter. Then, as is appropriate to the situa-

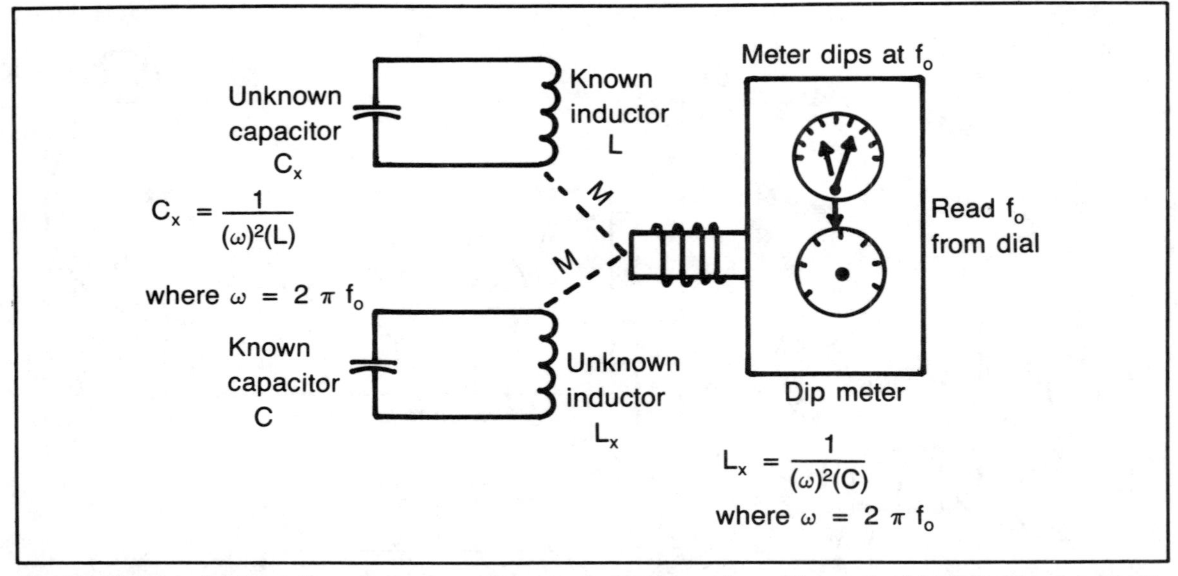

Fig. 8-7. Method of determining unknown capacitance or inductance values with a meter (single-frequency method).

tion, either the formula for G_x or the formula for L_x is used. In these formulas, frequency is expressed in Hz, inductance in henrys, and capacitance in farads. However, more appropriate to the practical world of rf techniques are the MHz, the pF, and the microhenrys. If these units are used, the indicated formulas are modified as follows:

$$C = \frac{25{,}330}{f^2 L}$$

$$L = \frac{25{,}330}{f^2 C}$$

When determining unknown capacity, you should at least be mindful of the effect of distributed capacitance in the known inductor. This error is inherent in the test technique and it is a question of whether or not it can be dismissed as being negligible. The known inductor should have winding and constructional techniques which result in relatively low distributed capacitance. A coil with a "universal" type winding configuration and a ferrite core is the best in this respect. Such windings are found in i-f transformers and certain rf chokes. A single layer coil with spaced turns on a low-dielectric material is probably more common and usually is suitable for determining capacitance from about 25 pF and up. Lower values can be measured, but extra care is necessary.

It is a good idea to record the self-resonant frequency of the known inductor. This should be at least three, and preferably more, times the frequency resulting from the test. If this ratio is close to three, the computed value of the unknown capacity will be about 10 percent too high. If this ratio is four or higher, the error resulting from distributed capacitance will be 6 percent or less, and it will not, in the light of other factors, prove profitable to strive for greater accuracy.

The above described test conditions are not difficult to attain. But it is not recommended that this procedure be extended to the situation where the ratio of the self-resonant frequency to the test frequency is less than three even with an attempted correction for the effect of distributed capacitance. In many instances, the self-resonant frequency is not in the frequency region up to, say, four times the test frequency. If the search is carried out patiently and carefully, this is a favorable indication in that you can assume that self-resonance occurs at a higher frequency. However, high-frequency self-resonance is sometimes quite difficult to find because the Q of the self-resonant circuit tends to be low. Sometimes you can be guided to the vicinity of self-resonance by connecting a small capacitance, generally several pF to several tens of pF, across the inductor. Obviously, some judgment may have to be exercised here, and this is best cultivated from experience.

The measurement of inductance involves a very similar test procedure. Of course, it must be realized that we are talking about rf windings only—inductors up to several hundred microhenrys. The known capacitor can be a Mylar or mica type with 5 percent tolerance. A value of 100 pF is convenient and is large enough to swamp the error introduced by the distributed capacitance of the inductors commonly used in high-frequency work. For larger inductors, the known capacitor can be made larger in the interest of accuracy.

Notice that the known reactor in this case is of a much "purer" form than was the case with the known inductor in the previous test procedure. Thus, it is possible to ignore the effects of self-inductance in the known capacitor if some initial thought is given to the situation. Self-inductance in a 100-pF Mylar capacitor will have a negligible effect throughout the several MHz range. Mica and ceramic capacitors are superior if the test takes place in the frequency range of several tens of MHz.

Other things being equal, both the capacitance and the inductance measurements should be made at relatively low rf frequencies; the range between 2 MHz and 6 or 7 MHz is suitable. The known inductor and the known capacitor should be chosen to be sufficiently large, if such a selection is feasible. Of course, the frequency of self-resonance will fall where it will. Measurements can be made at higher frequencies, but the accumulation of the effects from stray parameters tends to detract from the reliability and accuracy.

Alternate test procedures for measuring capacitance and inductance with the dip meter are shown in Fig. 8-8. In order to measure capacitance, you need not know the value of the associated inductance in the resonant circuit. This can, in certain cases, prove advantageous over the previously described method. For a reasonable degree of accuracy, the distributed capacitance of the inductor must be small compared to the known capacitance, C_s. For most rf coils used in high-frequency work, this situation will exist with the use of a 100-pF mica capacitor for C_s. In any event, the self-resonant frequency of the inductor should be greater than three times the resonant frequency resulting from connection of C_s.

In Fig. 8-8B, the two-frequency method is used to measure the value of an unknown inductor. Here, the value of one of the inductances must be known, but the value of the resonating capacitor need not be known. In order to swamp out errors from distributed capacitance,the resonating capacitor should be at least 100 pF and perhaps as large as 1000 pF. A mica or ceramic type is suitable and a Mylar type can be used if the resonances are not beyond the several MHz range. Since the exact value of the capacitor does not enter into the computations, any tolerance is acceptable. Notice that the solution for the unknown inductance, L_x, is carried out in two steps.

Although the dip meter test techniques here have been conducted in the several MHz region, the same approaches can be extended through VHF and UHF. However, it then becomes exceedingly important that a good quality dip meter is available and that you develop a good feel for the art and science of high-frequency work. Very close attention must be given to lead length, proximity effects, and component characteristics.

USING AN OSCILLOSCOPE TO MONITOR RF

Using the cathode ray tube itself, even oscilloscopes in which the vertical response is approximately flat only to several hundred Hz can be used for extended response to tens of MHz. The frequency response limitation is the vertical amplifier. Thus, in order to view rf, it is necessary to apply it directly to the CRT's vertical plates.

There are two possible objections or obstacles to overcome with this technique. First, the rf must be at a high voltage level in order to deflect the beam. Secondly, although it depends on the frequency of the rf and the maximum horizontal sweep rate, it is not likely in most cases that individual rf cycles will be seen. These objections are valid, but the technique can still be useful. With regard to the voltage level of the rf, this is not always an obstacle when performing tests on transmitters, for many tens and hundreds of volts of rf are often available. And since this kind of rf monitoring is used principally to compare relative levels and for viewing various modulation envelopes, it is not necessary that individual rf cycles be visible.

Figure 8-9 shows two ways of applying rf to the

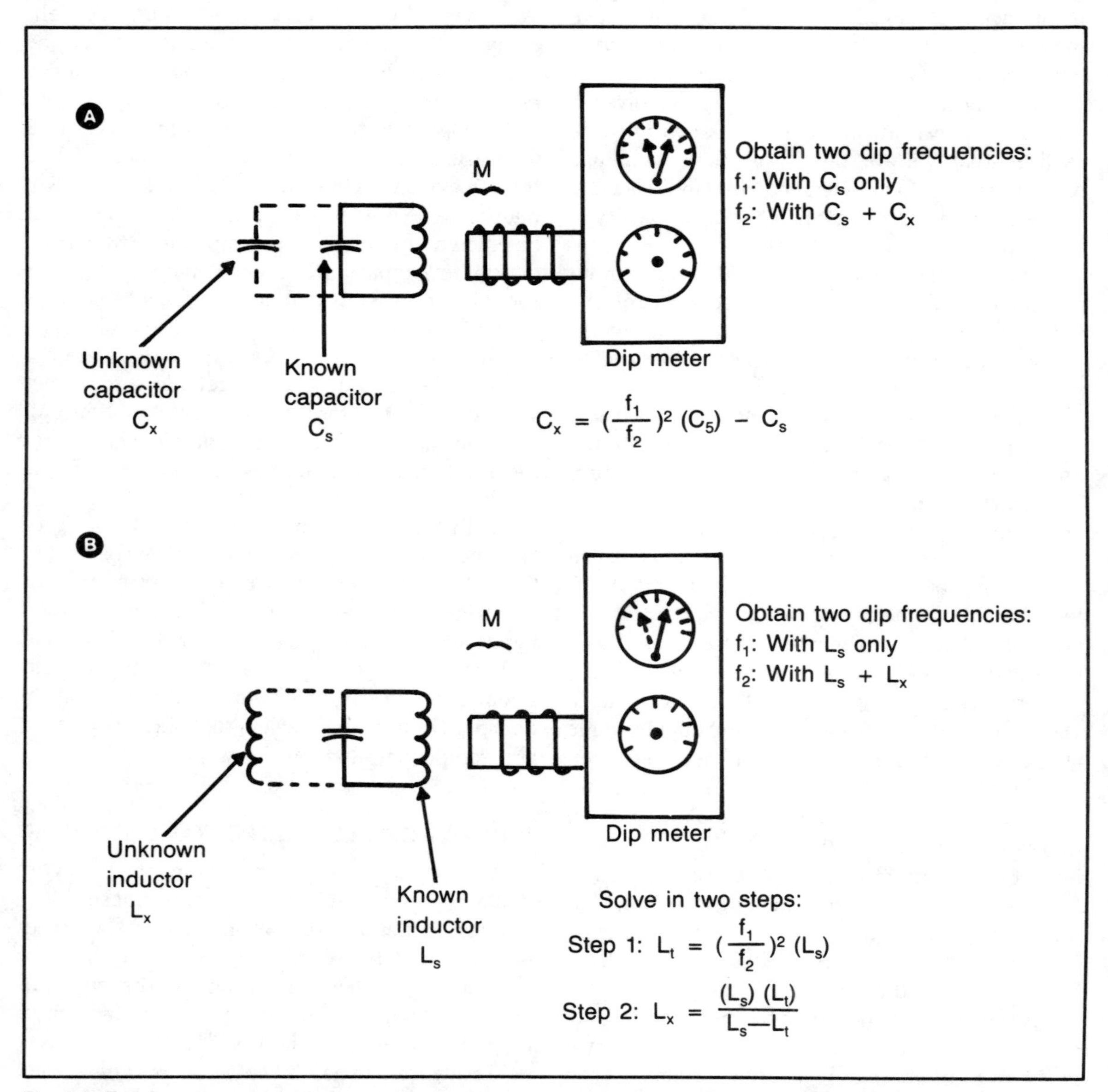

Fig. 8-8. Two-frequency method of determining unknown capacitance or inductance values with a dip meter.

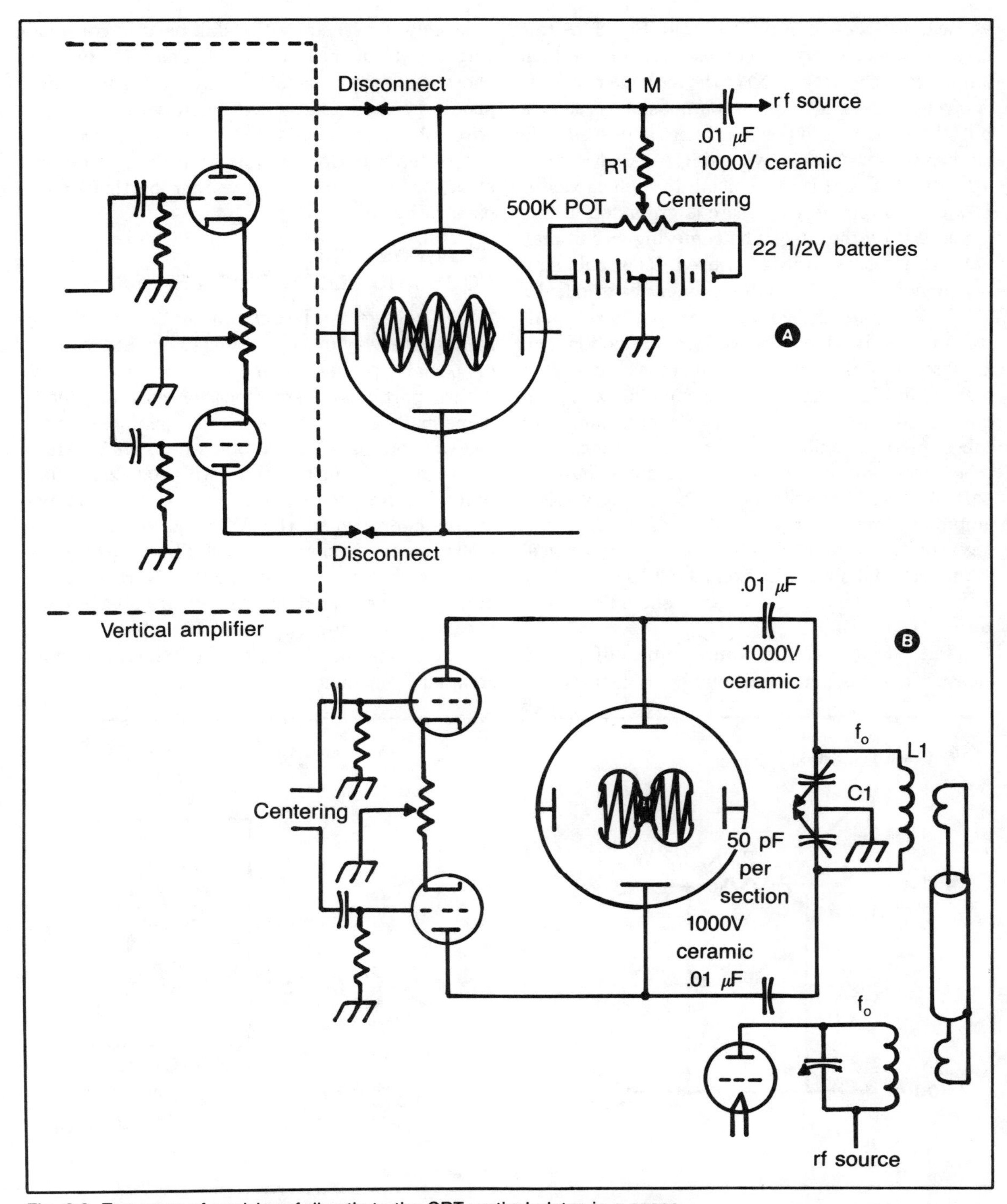

Fig. 8-9. Two ways of applying rf directly to the CRT vertical plates in a scope.

vertical plates of an oscilloscope. In Fig. 8-9A this is accomplished by disconnecting the vertical amplifier at the socket (X) of the cathode ray tube. A push-pull vertical amplifier output stage is shown, but the modification is made in an essentially similar way on scopes with single-ended vertical amplifiers. The rf is introduced through capacitor C1. The bottom vertical plate is grounded, if this is not already the case. The centering circuit may not be necessary. If it is not used, R1 should simply connect to ground in the manner of a grid leak.

In Fig. 8-9B the voltage stepup of a resonant transformer is used. The vertical deflection sensitivity to rf is appreciably increased over the method in Fig. 8-9A. Also, the cumulative capacitance associated with the vertical plates (amplifier tubes, plate capacitance, wiring capacitance, and other stray capacitances) is used advantageously as part of overall capacitance C1. Since the vertical amplifier remains in the circuit, no external centering provision is needed. When in use, the vertical input terminal should be connected to ground by a short wire and the vertical gain should be set at zero.

The upper rf response limit of these measurements depends on a number of factors, but is largely governed by the skill used in constructing high-frequency circuits. Leads must be kept short, components should be physically small, parts should be mounted to avoid long wire runs, and wires should be kept away from possible sources of interference. Generally, it is wise to confine construction and wiring to the rear portion of the oscilloscope.

RF PROBES FOR VOMS AND ELECTRONIC METERS

Rf voltage can be read in either a relative or an absolute way on common instruments with aid of special rf probes. Such probes are made for use with a particular instrument, or may be simple home-made affairs. A very simple arrangement for use with ordinary 20,000 ohms per volt VOMs is shown in Fig. 8-10. This setup will produce a deflection of approximately one volt RMS of rf per microampere when the VOM is set at its 50 microampere dc current range. This type of probe is responsive to the peak value of the rf wave. Its high-frequency capability can exceed 50 MHz and it can be used down to about 150 kHz.

The rf probe shown in Fig. 8-11 converts a commonly encountered electronic meter with the

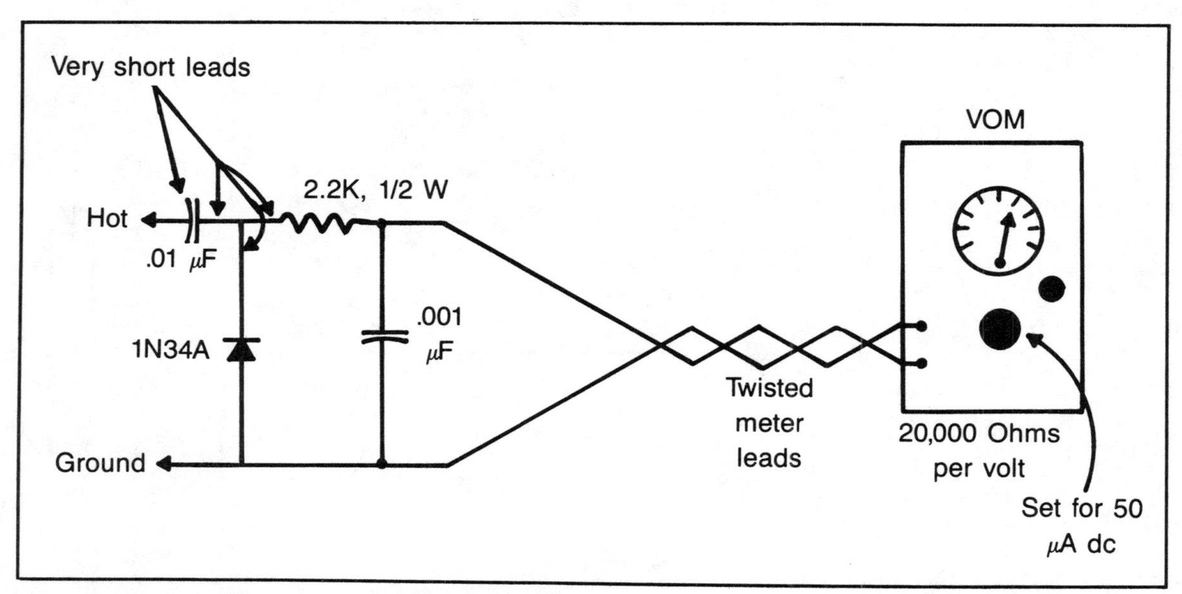

Fig. 8-10. Simple rf probe circuit designed for use with a VOM.

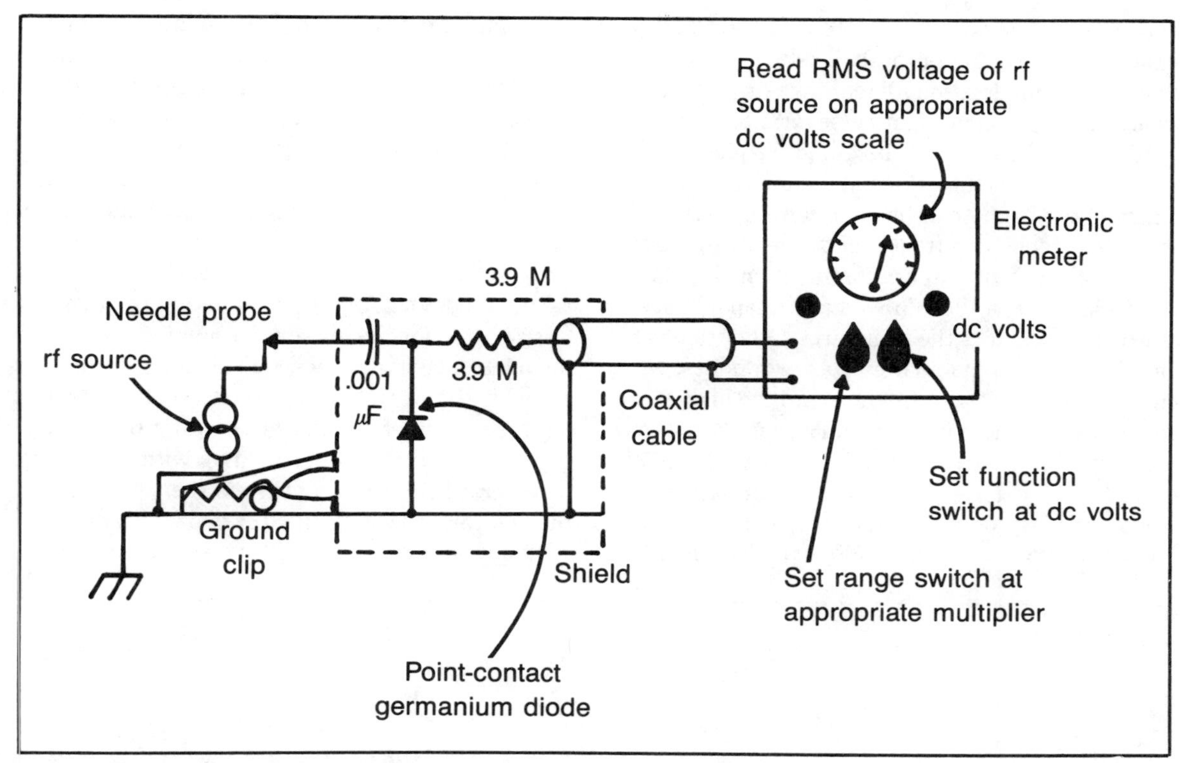

Fig. 8-11. An rf probe circuit for an electronic meter.

more or less universal 10-megohm (without the one-megohm isolation probe commonly used) input impedance into an excellent rf voltmeter. The 3.9-megohm resistor serves a triple purpose here. First, it functions as a voltage-divider arm to convert the peak dc voltage to an RMS voltage. Thus, you can read the RMS voltage of the rf source on any appropriate dc volts scale. Secondly, it operates in conjuction with the capacitance of the coaxial cable to filter the rectified dc voltage, thereby keeping rf out of the electronic meter. By virtue of the high value, the resistor tends to linearize the dc output of the probe. Because of the sensitivity of many electronic meters, such a probe can be very useful in investigating the performance of low-level rf circuits, such as is encountered in receivers. Also, because of its extremely broadband characteristic, it is well suited for tracking down the source of parasitic oscillations and for neutralizing rf amplifier stages.

It should be realized that the input capacitance of rf probes is most certainly not the value of the input capacitor. Because this capacitor remains charged to the peak value of the rf wave, its capacitive reactance has little effect. Rather, the input capacitance of typical rf probes is on the order of several picofarads at most; therefore, such probes generally cause negligible detuning of resonant circuits. The input capacitance is primarily a function of the diode and the hardware configuration of the probe.

DEMODULATING PROBE

The probe shown in Fig. 8-12A is similar to the rf probes discussed for use with electronic meters and VOMs. However, the time constant is such that the input capacitor cannot retain a charge for many rf cycles. This enables the dc level at the cathode of the diode to freely follow the positive excursion of the rf envelope, that is, the modulating wave. Such a "demodulator probe" is, in essence, an amplitude modulation detector. For the values shown, the rf carrier can be several tens of MHz and the modulation can be audio frequencies. By varying component values and construction techniques, the probe can be used with a wide range of rf and modulating frequencies. The 220K resistor does not serve any calibrating function; it is the series arm of an rf filter in which the capacitance of the coaxial cable is the shunt arm. Other things being equal, the demodulator probe works best when the ratio of the carrier frequency to the highest frequency component in the modulating signal is very high. Notice that the frequency capability of the oscilloscope only has to accommodate the modulating signal. The input capacitance of this probe is very low, perhaps a few picofarads. However, care must be taken that the input resistance does not load down high-impedance circuits. Up to 10 MHz, the input resistance goes down rapidly so that at 100 MHz it is only about 5000 ohms.

Although the arrangement shown in Fig. 8-12B would not ordinarily be referred to as a probe, it serves a function similar to the above demodulator probe and provides extra convenience in some situations. The "probe" can be useful over a wide range of both carrier and modulating frequencies. Carrier frequencies of several hundred MHz and modulating signals in the video range are commonly tested with this circuit. The pickup link can be two or three turns of hookup wire with a diameter of two or three inches when dealing with carriers up to several MHz. For carriers in the tens of MHz, the link can consist of a turn or two with a diameter of about an inch. A single turn and a diameter of

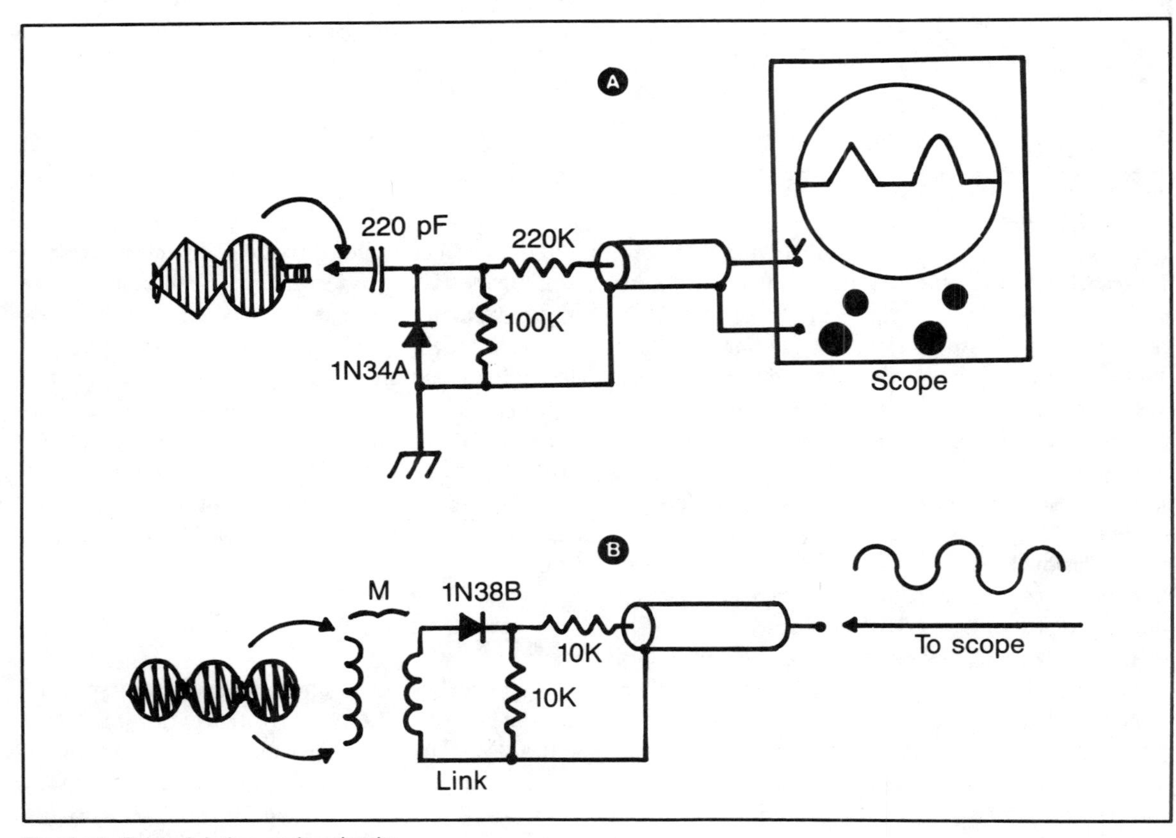

Fig. 8-12. Demodulating probe circuits.

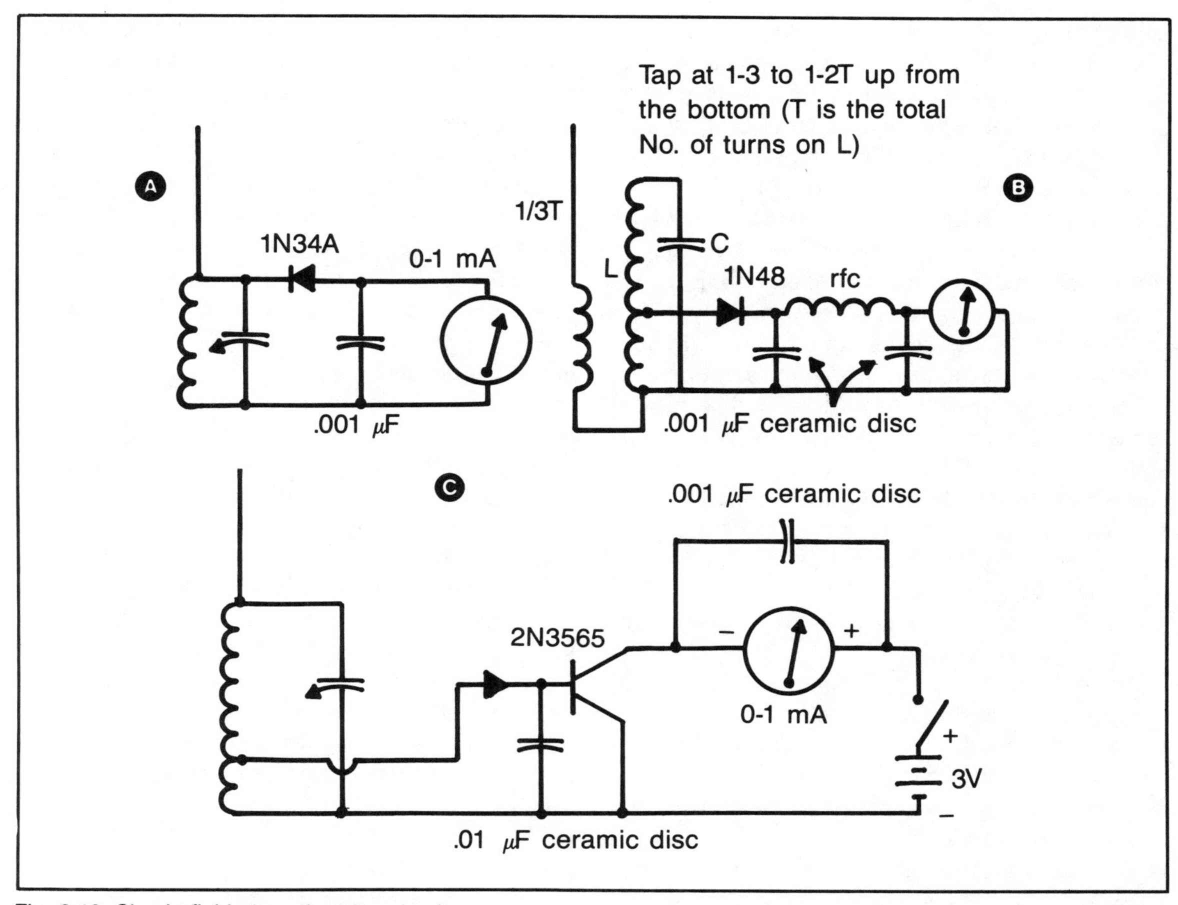

Fig. 8-13. Simple field-strength meter circuits.

a half inch should suffice for carrier frequencies in the hundreds of MHz. Such information is very general and is intended to serve as a reasonable starting point. The physical layout and impedance conditions of each test situation must be considered if the most favorable results are to be obtained.

The demodulators in Fig. 8-12 have an approximate square-law response at low signal levels and become more linear with increasing signal amplitude. The modulation information is generally derived in a relative way. What calibration may be needed is usually performed to meet the test situation at hand. These demodulating devices can be extremely helpful in tuning up and debugging an AM transmitter or a CB station. Very little energy need be taken from the stage under investigation, and a comparison of the recovered modulating signal waveform with the original modulating signal waveform can give you an abundance of information regarding drive levels, linearity, bias, and other circuit and operating conditions.

FIELD-STRENGTH METER

The field-strength meter is similar in some ways to the absorption wavemeter and test procedures using the two devices do, indeed, have some overlap. It can be used in more ways than a wavemeter and one of its principal functions, as might be gathered from its name, is the evaluation of radiation from an antenna. Several field-strength meters are shown in Fig. 8-13. A primitive but useful circuit is shown in Fig. 8-13A. Figure 8-13B is a somewhat more elaborate circuit with several operational advantages. Because the whip antenna is coupled to the resonant tank through a primary winding, this circuit has better calibration possibilities than the simpler circuit in Fig. 8-13A. The inductive coupling, if not too tight, tends to make the tuning more independent of the length of the dipole and its proximity to other objects. On the other hand, calibration is not always a vital factor in the use of the field strength meter; often, it is only necessary to merely tune for maximum response. This circuit also incorporates a low-impedance tap on the inductor, thereby relieving some of the loading imposed on the resonant circuit by the diode. The higher Q resulting from this technique leads to sharper tuning, again a calibration advantage where this is desired. The use of a more sensitive meter allows the instrument to be physically situated farther away from the antenna or other rf source.

The field-strength meter shown in Fig. 8-13C is even more sensitive and it eliminates the need to use such a sensitive and possibly delicate meter as is the case with the simple diode circuits. The residual reading (if descernible) from transistor leakage can be nulled by a mechanical adjustment of the meter movement. Notice that the forward base bias for the transistor is derived from the rectified rf.

Basically, field-strength meters are radio receivers. As such, it does not tax the imagination to think of many features that could be incorporated. Actually, for professional work the field strength meter assumes the form of a very elaborate receiver from which we can derive accurate frequency indications and data leading to the measurement of absolute field strength in microvolts per meter. However, for most test purposes, the simpler varieties of this device are entirely adequate. Indeed, the portability and the capability of field-strength meters to yield quick results are features that are quite useful for the general run of rf work.

When using a field-strength meter to get the best performance from an antenna, the transmitter feeding it, or the combination of the two, several things must be kept in mind if the tests are to be meaningful. The field strength meter should be located at least four or five wavelengths from the antenna. Thus, for tests involving CB transmitters, the distance should be on the order of 50 yards or 150 feet. If this rule is not observed, the induction field surrounding the antenna can actuate the field strength meter. We are interested in the radiation field while peaking up the performance of the antenna system. (If the purpose of our test is to coax maximum efficiency from the transmitter, this rule, can be ignored. However, the adjustments made in the transmitter should not include tuning the final tank or its output coupling circuit, because such adjustments involve the antenna system.)

The field-strength meter antenna should be polarized the same as the transmitting antenna; i.e., if the transmitting antenna is vertical, the antenna of the field strength meter should be likewise. It is also desirable that the field strength meter be at the same height as the transmitting antenna. The field strength meter antenna should only be long enough to produce the desire deflection of the meter. In general, it is best to avoid the possible situation in UHF work where the antenna of the field-strength meter is dimensionally close to the elements of the transmitting antenna. Under such conditions the field strength meter antenna may act as an additional coupled element to the transmitting antenna rather than an as a passive probe of the electric field. This can lead to confusing results.

Generally, the use of the field-strength meter is a two-man job. Therefore, some type of voice communication is desirable so that the effect of transmitter adjustments is known instantly. A walkie-talkie system serves this purpose very well. Or a simple telephone system can be devised from two sets of electromagnetic headphones interconnected by twisted line and energized by one or two flashlight cells. An alternative, particularly after a transmitter-antenna system has been successfully placed in operation, is to locate the field-strength meter at a permanent weather-shielded position and connect it by coaxial cable to another series-connected meter at the transmitter location. Then you can ascertain whether or not rf is being radiated at an acceptable level. This technique is usually more practical at the higher frequencies for the average ham installation.

Frequency calibration of the field-strength meter is not a must for the tests involved in radiation efficiency or in attaining the desired radiation. However, once the field-strength meter has been tuned to the radiated frequency, it is important to make sure that this adjustment is not disturbed during subsequent phases of the test procedure. Also, it should be realized that the amplitude response of most field strength meters, such as those shown in Fig. 8-13, is not linear. Ordinarily, this is of little consequence because the test procedures are based on the relative response with respect to adjustments made in the transmitter and on the transmitting antenna. (A bypassed emitter resistance can be used in Fig. 8-13C to produce negative feedback at dc. Such feedback can make the meter response an approximately linear function of the input rf voltage.)

A field-strength meter can be calibrated quite readily with a dip meter. See Fig. 8-14. Minimum spacing or coupling between the two instruments should be used during the calibrating procedure. The dip meter is best employed as a "distant" source of rf radiation; electromagnetic coupling between the coil of the dip meter and the coil of the field-strength meter should be avoided. With the dip-meter coil several inches to several feet from the fully extended antenna of the field-strength meter, vary the frequency of the dip meter until the field strength meter peaks. This can be done by tuning from one end of the range of the field-strength meter to the other end. Do not look for indications on the dip meter. An rf generator with a foot or so of antenna can be used in place of the dip meter. Make certain that the response is not a harmonic of the rf source. This can be quickly checked by tuning the source to the vicinity of double and to triple the frequency which produced the first response on the field-strength meter. If this test results in an even greater response, then the field-strength meter is actually tuned to a harmonic of the dial indication on the rf source. This test need only be made once for each band calibrated.

The field-strength meter is an excellent device

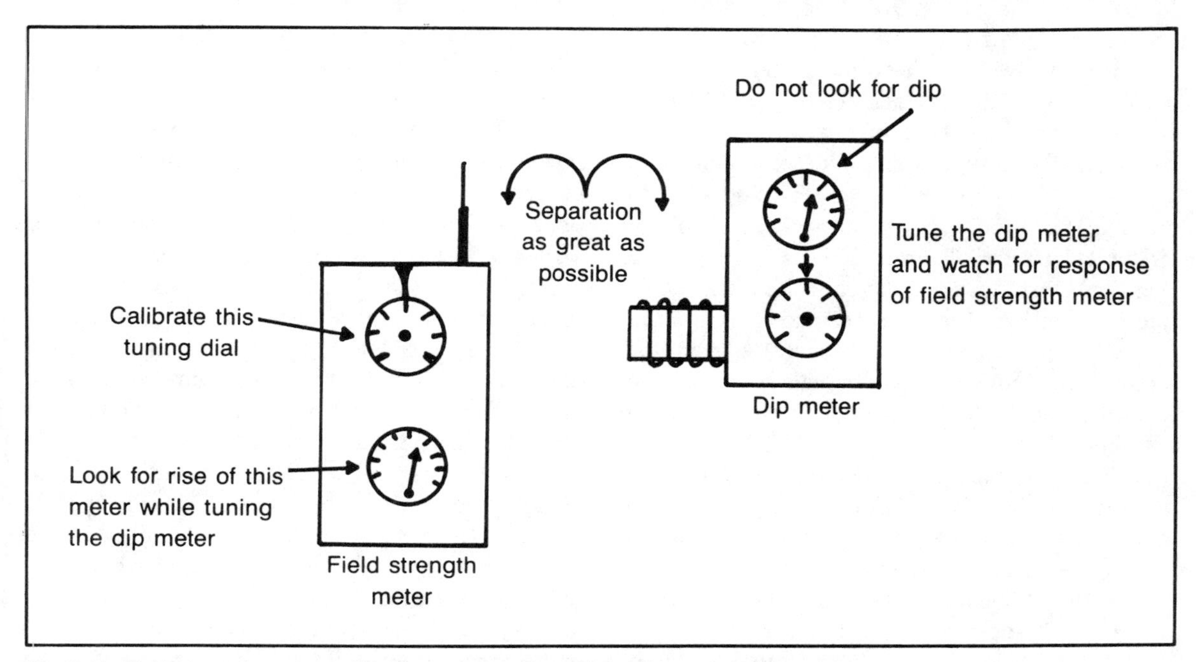

Fig. 8-14. Field-strength meter calibration setup using a dip meter as an rf generator.

for discovering harmonics radiated from a transmitter. Sometimes it happens that harmonic generation is acceptably low in the transmitter and that the antenna coupling circuit further attenuates harmonic energy to the point where it is difficult to believe reports of harmonic interference. Often the harmonic in such instances is radiated by a nonlinear conductor in the induction field or in the strong radiation field of the antenna or transmitter. Such nonlinear elements can be plumbing connections, wire splices, the junction of two dissimilar metallic objects, or an overloaded electronic device in the rf field. With the aid of the field strength meter and with a little exploratory work, you can track down the source of such "secondary" transmissions. Such tests are greatly simplified if the transmitter itself does not contribute significantly to the overall harmonic energy being radiated from the station.

FREQUENCY MARKER

The frequency marker is a harmonic-rich oscillator which can be used to calibrate receivers, transmitters, and other radio frequency apparatus. Probably, the simplest frequency marker is an oscillator using a 100-kHz quartz crystal such as is shown in Fig. 8-15. Two main features of this oscillator circuit are the lack of a resonant tank circuit and the vernier tuning capacitor, C1. The reason for the absence of a tank circuit is not simply because the Pierce oscillator will function without one; more importantly, since there is no tank circuit harmonics will not be attenuated. (Of course there are other ways of accomplishing this objective both with and without resonant circuits, but this is by far the simplest approach.) A useful objective is to obtain harmonic signals at 100-kHz intervals through the 14 MHz amateur band, and hopefully, higher.

Since the harmonics become progressively weaker as their multiple of the 100-kHz fundamental becomes greater, the limit to which they can be reliably detected is a function of many variables, such as the harmonic characteristics of the marker, the receiver sensitivity, the noise situation, etc. However, assuming that the harmonics can be heard, it is first necessary to know that they are, indeed, spaced at 100-kHz intervals. This is the reason for the variable capacitor, C1. When the output of the frequency marker is compared with an accurately known frequency, the crystal can be made to oscillate at exactly 100 kHz with C1. Once standardized in this manner, the frequency marker can be relied upon to produce 100-kHz spaced signals throughout a wide portion of the rf spectrum. The setup shown in Fig. 8-15 uses signals from station WWV, WWVH, or a broadcast station with a carrier frequency which is a whole-number multiple of 100 kHz. The beat-frequency oscillator in the communications receiver is not used. Rather, the heterodyne is developed between a harmonic from the frequency marker and the carrier frequency of the received station. The basic idea is to adjust C1 in the frequency marker until this heterodyne is at zero beat. For tests requiring the highest accuracy attainable, headphones may not be the best way to identify the zero-beat condition, but it is surprising how proficient you can become with a little practice. By carefully listening to the background noise from the receiver, zero beat can be closely enough attained for most practical purposes. (Zero beat means that the particular harmonic from the frequency marker and the reference frequency are the same.)

It is not necessary to use a communications receiver if you choose to reference the frequency marker to a broadcast station rather than such stations as WWV or WWVH. Since the receiver serves only the intermediate function of amplifying the two signals, an inexpensive broadcast band receiver is just as accurate as a high quality communications set. Make sure the carrier frequency of the selected station is a whole-number multiple of 100 kHz. Thereafter, the procedure is much the same as with the communications receiver. Simply adjust C1 in the frequency marker until the pitch of the audio beat note becomes essentially zero. If possible, the final determination should be made during an interval when no modulation is being applied to the carrier.

In both of the above test procedures, the frequency marker, even if it is solid-state, should first

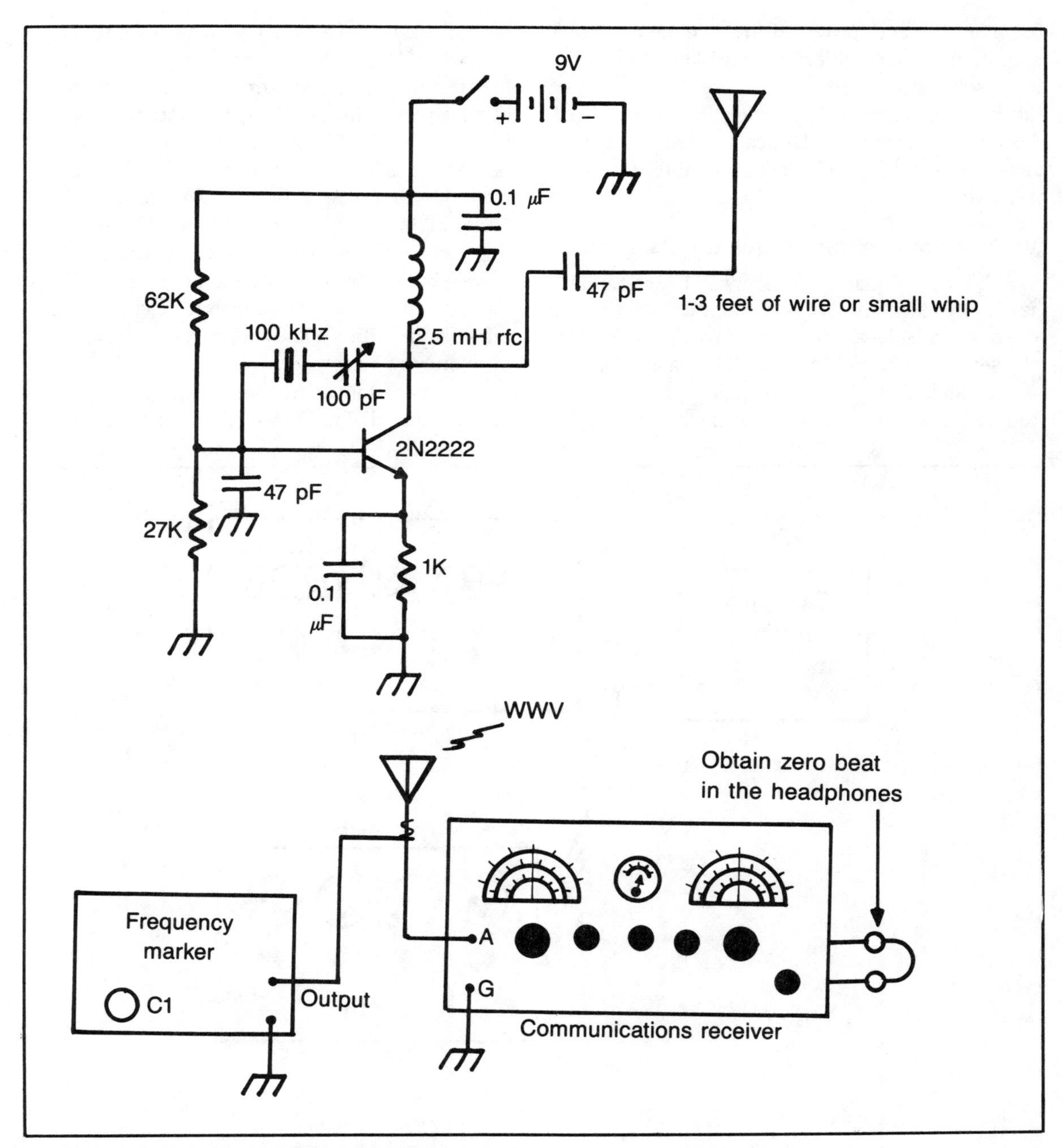

Fig. 8-15. One method of standardizing a frequency marker.

be allowed a half-hour warm up period. Best results are generally obtained when the frequency marker is operating continually, but this may not be convenient for all situations. In any event, some consideration should be given to the operating environment so that the 100 kHz quartz crystal operates at a constant temperature. A voltage-regulated power supply is also desirable, although the convenience provided by ordinary battery operation is undeniable. Frequency markers made with solid-state elements, rather than vacuum tubes, can be expected to retain adjustment longer. Also, electrical and mechanical ruggedness always pays dividends in precision and reliability for such test instruments.

VFO or Transmitter Frequency Calibration

After a frequency marker has been standardized, the position of a transmitter carrier frequency can be localized in the rf spectrum by using the setup shown in Fig. 8-16. The basic idea here is to establish known 100-kHz points on the receiver tuning dial in relation to a known station. Suppose the dial position of a station with a known frequency of 3.450 MHz has been logged. It must follow that the first frequency-marker harmonic encountered at a frequency higher than this station is the 35th harmonic of 100 kHz, or 3.500 MHz. The next higher one is 3.600 MHz, etc. Going the other way from the reference station, we can mark off 3.300 MHz, 3.200 MHz, 3.100 MHz, etc. Not only does this technique serve to calibrate the dial of the receiver, but the frequency generated by the VFO in the transmitter can be adjusted or can be determined to lie within a known 100-kHz interval of the frequency spectrum.

The need for the frequency marker stems from the fact that dial calibrations on receivers and on

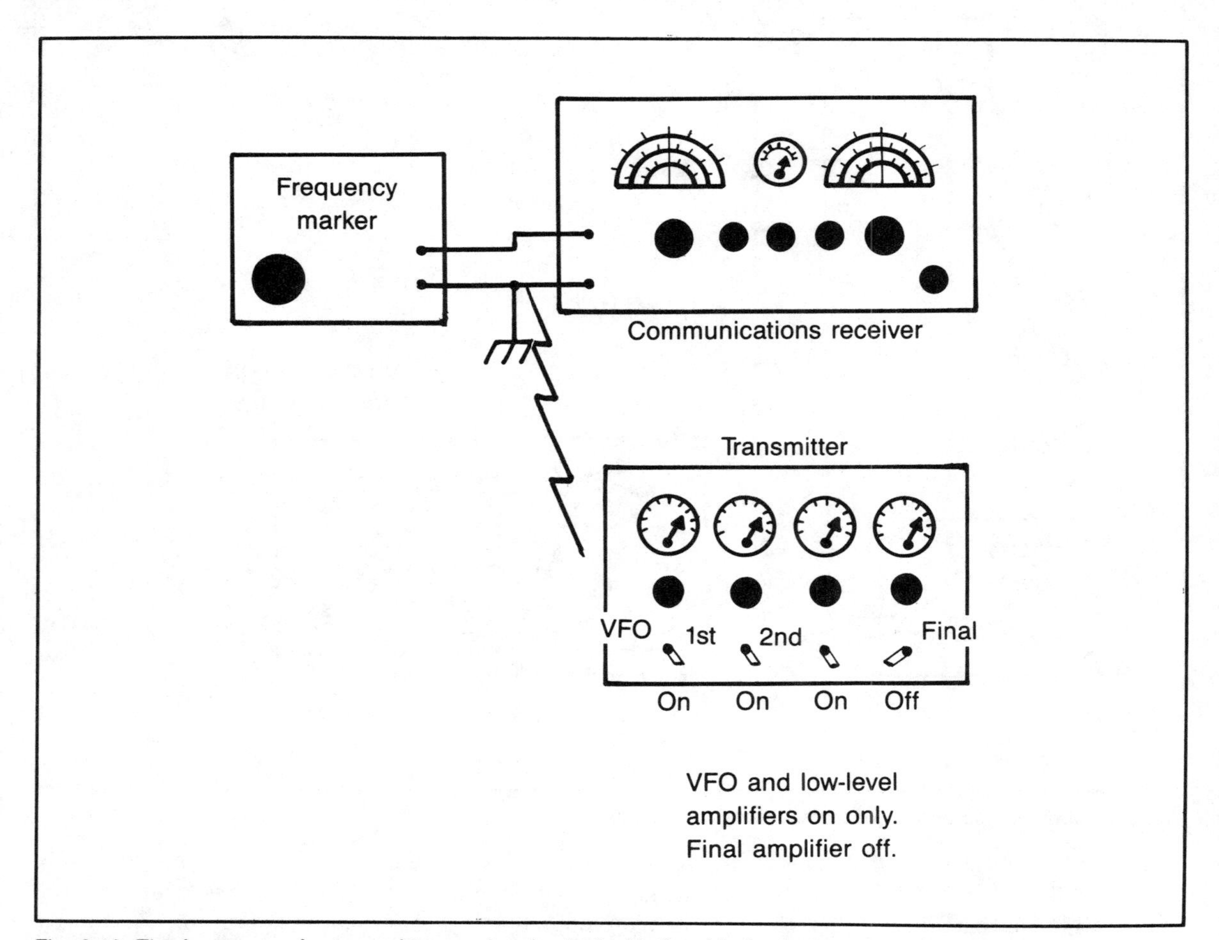

Fig. 8-16. The frequency of a transmitter can be checked with the aid of a frequency marker.

VFOs may not be trustworthy for a variety of electrical and mechanical reasons. Moreover, when the 100-kHz harmonic intervals are also accompanied by 50-khz and 10-kHz harmonics from a more elaborate frequency marker, the accuracy of tuning mechanisms definitely exceeds that which can be ordinarily obtained from the dial indications. Frequency markers capable of supplying 1-kHz harmonics have been used for high resolution calibration in the rf spectrum. However, the 100-kHz signals from a simple oscillator suffice for many calibration needs.

In Fig. 8-16, it is desirable that the harmonic signals and the transmitter signal be about the same strength. To help bring this about, the antenna should be disconnected from the receiver and the frequency marker should be connected directly to the receiver antenna terminals. Unless the transmitter has a very low power output, its final amplifier should not be operating. If for the stated reason, or because of other circumstances, the final amplifier must be on, the transmitting antenna should be disconnected in order to prevent interference with other stations during the test procedures. Connecting the frequency marker directly to the antenna terminal of the receiver can introduce a small frequency error in some cases, but generally this will be negligible in comparison with the overall possible deviation from accuracy. If good results can be obtained by means of "gimmick" coupling, so much the better. The relative signal strength of the harmonics of the transmitter signal will vary widely with the circumstances of the particular situation and some experimental work will have to be done to find the best techniques for a given setup.

Identifying Harmonics

The identification of harmonics is sometimes difficult, particularly at higher multiples. Figure 8-17 shows the general harmonic spectrum for several waveshapes of 100-kHz or other fundamental frequency. Notice that for the perfect square wave, the even-order harmonics are missing. In practice, it would be found that even-order harmonics, though present because of an imperfect square wave, would be relatively weak compared to the odd-order harmonics on either side. This fact has been used for identification purposes in more sophisticated tests with marker signals. Since the triangular wave is not much different from a sine wave, it is not rich in harmonic energy and is not very useful for marker test procedures. (The negative amplitudes imply phase reversal.) The sawtooth harmonic spectrum provides a useful array of gradually decaying odd and even harmonics. Obviously, the sawtooth is well suited to the needs of frequency comparison based on marker signals. The Pierce oscillator circuit shown in Fig. 8-15 is capable of approaching such a waveform with an active crystal.

Another difficulty in identifying marker harmonics is the confusion caused by unmodulated carriers, whistles, birdies, heterodynes, etc. This can be overcome by skimping on the filter components in the dc power supply to the frequency marker. This will result in some 120-Hz modulation of all the harmonics, and will tend to make them identifiable in the presence or proximity of various interferences.

SWEEP-MARKER TEST PROCEDURES

A frequency-modulated rf signal and accurate marker frequencies provide an effective and thorough means of determining the frequency response of all kinds of passive and active circuits, devices, and systems. The frequency-modulated rf generator is known as a sweep generator because of the way in which it is used in the test setup. The basic idea of this procedure is shown in Fig. 8-18, where you can see that the horizontal channel of the oscilloscope is driven by a voltage from the sweep generator. This voltage is, or is a sample of, the very voltage which causes the frequency modulation. The horizontal movement of the spot on the oscilloscope screen is, therefore, synchronized with the changing rf frequency applied to the circuit undergoing a response test (in this case, a parallel resonant LC tank). Several features of this technique may not be immediately apparent.

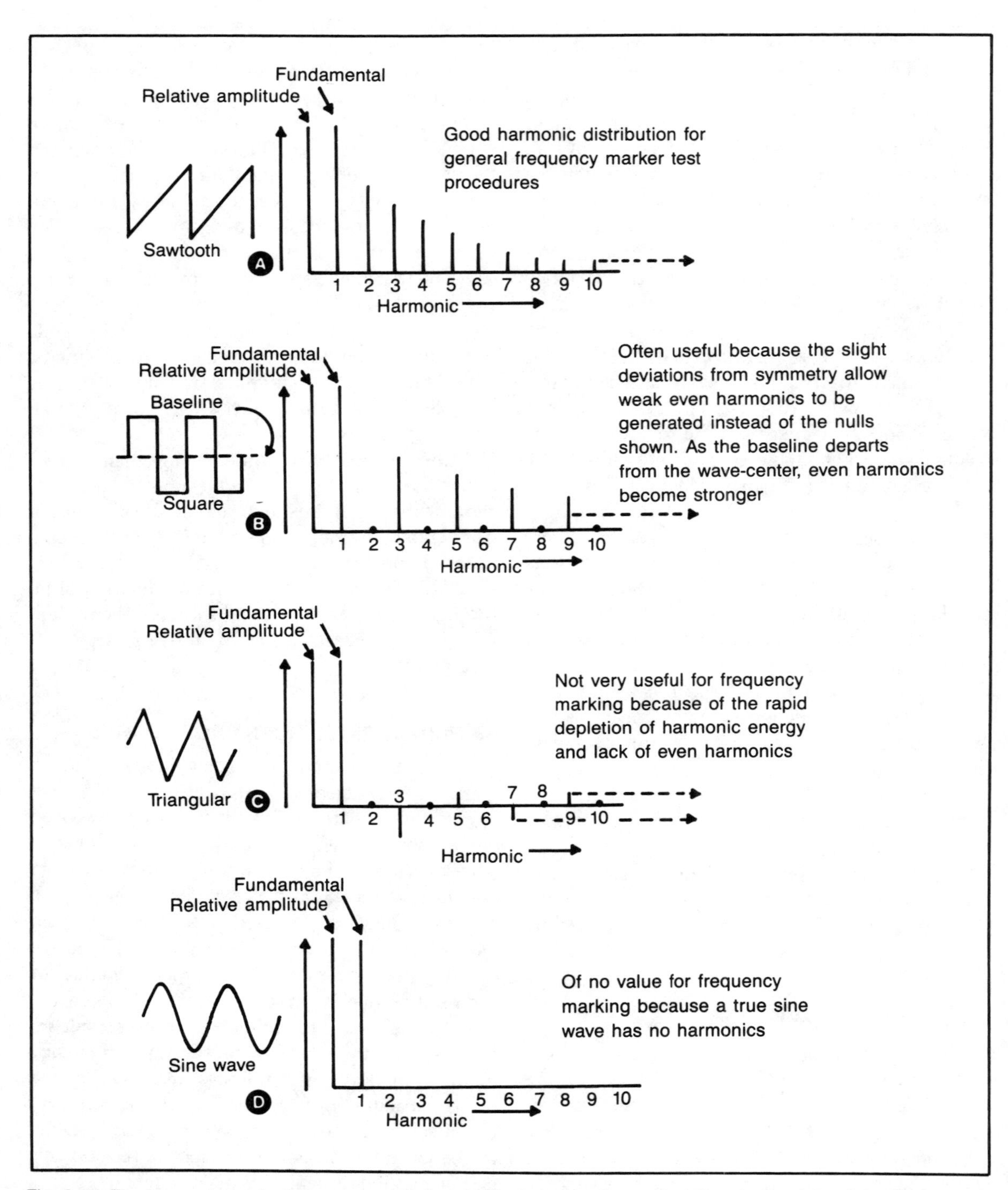

Fig. 8-17. The harmonic content of various waveform is illustrated by the vertical lines to the right of the fundamental frequency.

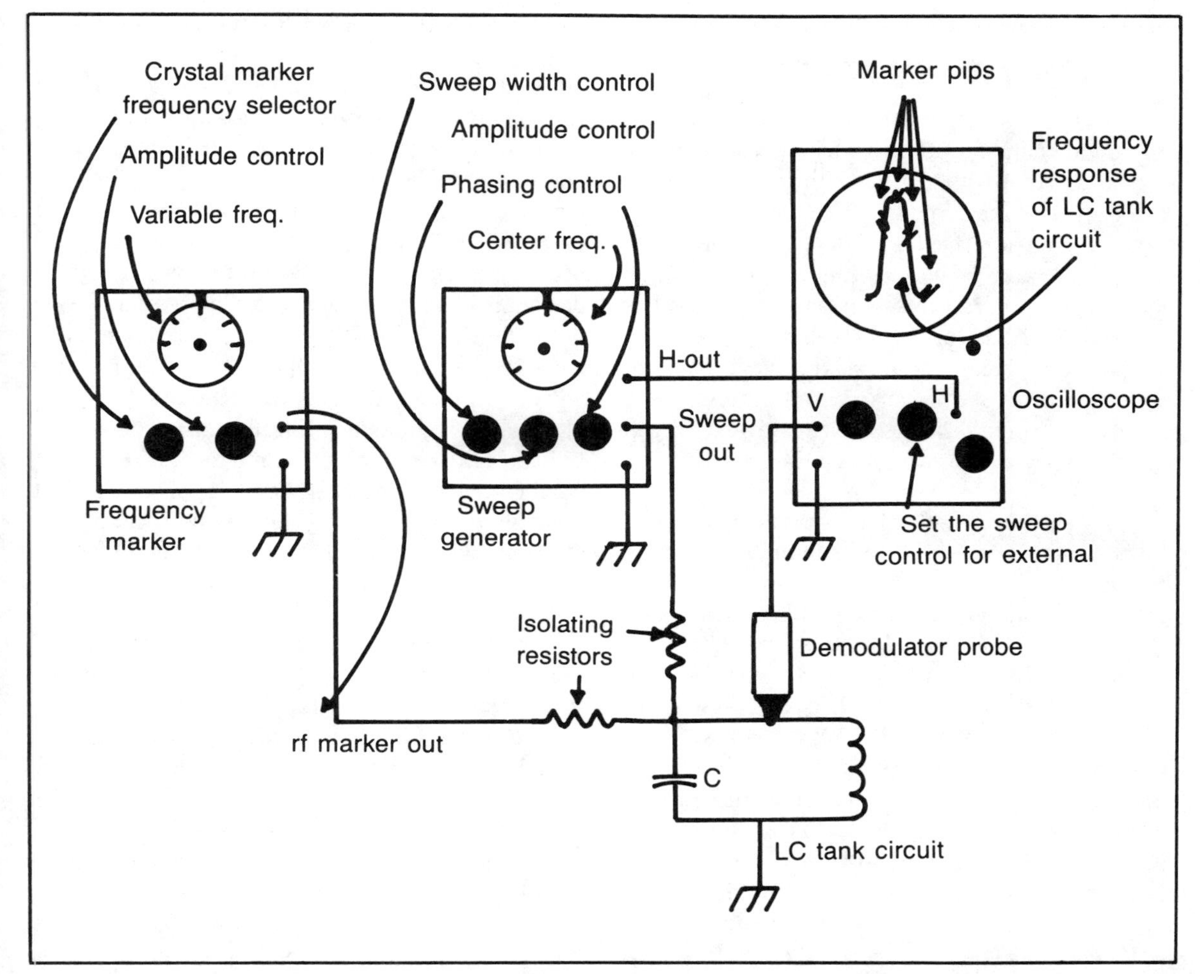

Fig. 8-18. Sweep-marker test equipment setup. The frequency response curve is caused by the frequency-swept rf signal from the sweep generator. The marker pips, which identify specific frequencies on the response curve, come from the frequency marker.

They are as follows:

1. The oscilloscope can be a narrow-band type regardless of the range of swept frequencies involved. This is because the swept rf sampled from the circuitry being tested is demodulated. Only the slowly varying sweep frequency is applied to the vertical channel of the scope. In Fig. 8-18 this is accomplished by a demodulator probe. When checking the alignment of TV sets, the rectification or demodulation is often accomplished by the video detector. Notice that the scope need not be a dc type because the voltage processed by the demodulator probe or other rectifier is at the sweep frequency under the actual test conditions. This is often 60 Hz.

2. In most instances, no calibration of either the vertical or horizontal channels of the oscilloscope is necessary. Horizontal calibration is not required because the marker pips accurately define important points on the response curve. Vertical calibration is not required because you can evaluate important frequencies in terms of their

relative distances up from the base line or down from the peak of the response curve.

3. The sweep voltage presented to the horizontal channel of the oscilloscope need not be a linear sawtooth such as is commonly required in oscilloscope sweep circuits. Since the same voltage wave is used for the frequency deviation in the sweep generator as well as the horizontal deflection of the beam in the scope, the display will be "linear" regardless of the waveshape of the sweep voltage. This being the case, the most often encountered sweep voltage is a sample of 60 Hz, which by one of several methods modulates the center frequency in the sweep generator.

4. The 60-Hz deflection voltage impressed on the horizontal plates of the scope must be exactly synchronized with the 60-Hz rf sweep voltage. Because of inadvertent phase shifts in both the sweep generator and oscilloscope circuits, this condition is not likely to occur without special attention. The effect of phase shift between these 60-Hz signals is a Lissajous-type display. Such a display may not be immediately recognizable as such, but its obviously objectionable feature is the presence of two response images. This can be confusing, particularly when there is small separation in the two displays. Fortunately, the solution to this problem is simple. The sweep generator has a control to adjust the relative phase between the 60-Hz voltage which sweeps the rf and the 60-Hz voltage which sweeps the electron beam along the horizontal axis of the scope. It is an easy matter to merge the double display into a single response curve.

Some oscilloscopes have a 60-Hz phasing control. In such cases, this control will serve the same purpose. When using such a scope, the 60-Hz horizontal sweep voltage is derived within the scope, rather than from the sweep generator. This is permissible because both instruments derive the ac sweep voltage from the same source—the 60-Hz power line.

5. Because the rf signal is alternately swept to a higher frequency, then returned, there is a natural tendency for two displays to appear in a mirror-image relationship to one another. This problem is different from the one discussed above. Even if it were practical to cause these two mirror images to come together on the scope screen, it would be impossible to truly merge the two because the high-frequency response portion of one curve is, timewise, the low-frequency response portion of the other. Fortunately, this is taken care of by the designer of the sweep generator. The return sweep of the 60-Hz wave which frequency modulates the rf center frequency is blanked out so that the scope will discontinue the display during that interval. An alternative technique is to kill the rf oscillator for the interval over which a mirror-image display would otherwise occur. If the 60-Hz phasing adjustment is not set correctly and if the blanking circuitry within the sweep generator is also defective, four response curves could appear on the scope!

From the previous discussion, it should be evident that the user must be skilled in operating a good instrument in order to produce usable response curves. Figure 8-19 depicts several common troubles which tend to plague such test procedures. Figure 8-19A shows the typical double display due to a phase difference between the 60-Hz voltage sweeping the rf frequency and the 60-Hz scope sweep voltage. As already explained, the two curves can be merged into one by adjusting either the phase control on the sweep generator or the 60-Hz phase control on the scope when it is so equipped.

The display in Fig. 8-19B initially appears to be OK, but what have we here? By turning the variable-frequency dial on the frequency marker, it is discovered that marker frequency A is a higher frequency than marker frequency B. Although we might be prepared to temporarily indulge in reversed-direction reading, this is not the solution. The remedy is quite simple. Just reverse the ac line plug of the oscilloscope.

The curve in Fig. 8-19C can occur with active circuits such as i-f amplifiers. The flat top is caused by saturation (overdrive) of the circuit or some defect which prevents the circuit from amplifying the

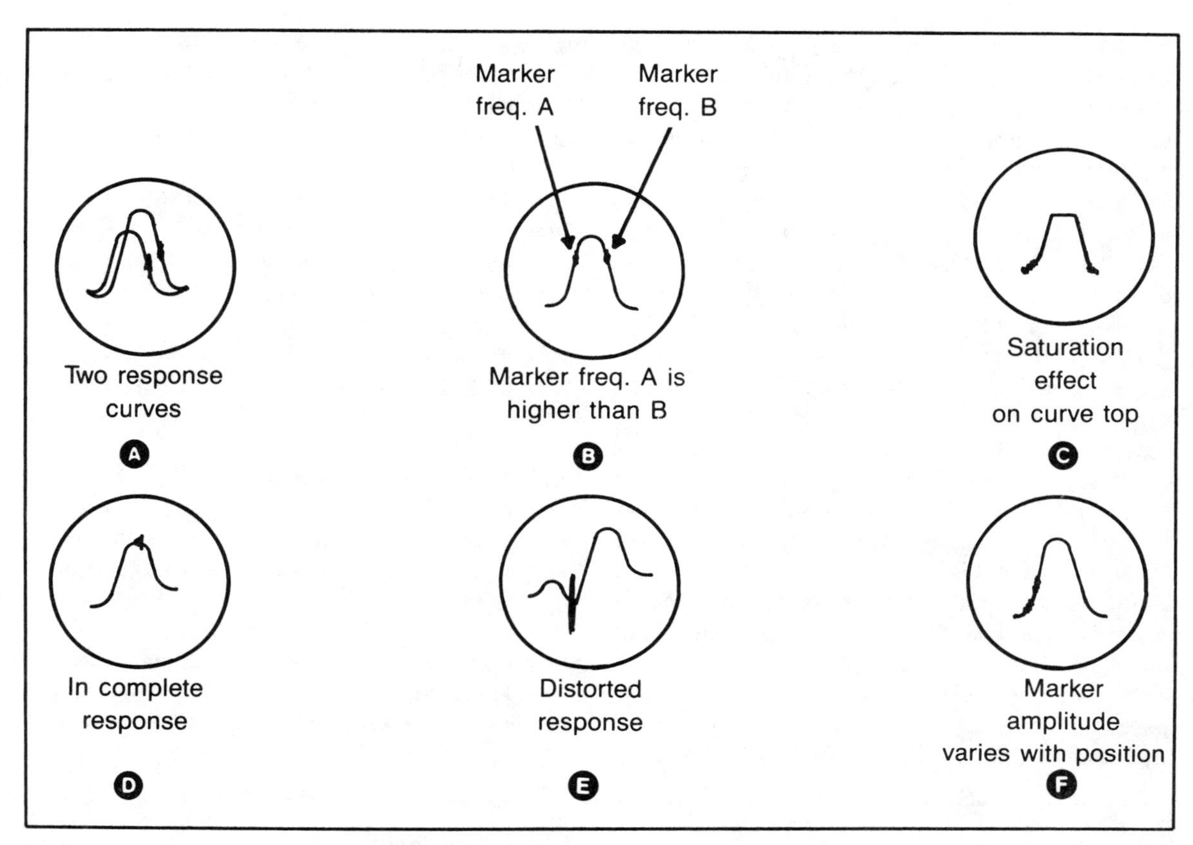

Fig. 8-19. Common frequency response curve troubles.

entire input signal. Sometimes judgment is necessary to ascertain whether such a response is the natural response of the active circuit or is the result of overloading. If the top portion of the display changes when the level of the swept signal is reduced, overloading is indicated.

Another precaution which must be observed with active circuits is the effect of automatic gain control. In order to obtain a true sweep frequency response of the active circuit, the automatic gain control circuitry must be disabled. For example, when evaluating i-f channels in TV receivers, the grid bias (or other control voltage) in the i-f amplifiers must be supplied from a fixed source so that constant gain will prevail regardless of output signal level. Many sweep generators provide a manually adjustable bias for this purpose.

The display shown in Fig. 8-19D is incomplete because the sweep does not cover a sufficient frequency range. This condition is corrected by the sweep width control on the generator panel. (Such an incomplete display invariably appears before a clean, completely swept response curve is attained.) In the event that an extra wide range of frequencies must be swept, which taxes the deviation ability of the sweep generator, both the center frequency and the width of the sweep range may have to be adjusted for the best compromise. Or, if a compromise is not acceptable, the lower and upper frequency portions of the response can be observed separately by shifting the center frequency.

The troubles depicted in Fig. 8-19E and F are corrected by a modification of the test setup. The troubles in these displays stem from the fact that both the sweep frequency and the marker signals go through the circuit undergoing test together

(Fig. 8-18). In Fig. 8-18E, the true response of the circuit is greatly distorted by the effect of an overly strong marker signal. Because of the tendency toward interaction between the response curve and the marker pip, the amplitude of the marker pip should never be much greater than needed to make it visible on the response curve. Otherwise it can "bite" into the curve and can even become the predominant feature of the display. In Fig. 8-19F, a barely discernible marker pip is seen at the base line, while the one farther up on the skirt of the response curve borders on being too big. This is a natural consequence of the setup in Fig. 8-18, and you must exercise skill and judgment in such situations.

The difficulties depicted by the displays in Fig. 8-19E and F can be avoided with the setup shown in Fig. 8-20. Here, another functional block, the "marker adder," is added to the procedure. The essential difference is that now the marker pips are added to the response after the swept rf signal has emerged from the circuit undergoing test. Therefore, the markers are independently superimposed upon the display and there is no interaction. Although it might still be desirable to keep the marker pips within certain amplitude limits, all marker pips will be the same amplitude regardless of their location on the response curve. Also, the marker will not distort the shape of the response when this setup is used.

In addition to the sweep generator function, some instruments also contain a built-in frequency marker, marker adder, and dc bias voltage supply to replace the bias ordinarily provided by automatic gain circuitry in tuned amplifiers.

The marker pip displayed on the response curve in Fig. 8-21A is not as useful as a sharp pip because it affects much more of the response curve than is necessary. The marker pip is actually the result of heterodyne action between the marker frequency and the swept rf. As such, sum and difference beat frequencies are generated, which begin at zero frequencies and display respectable amplitudes through the low radio frequency region, at least. The useful portion of the beat signal is a narrow group of very low frequencies clustering around zero frequency or zero beat. These low frequencies develop the actual interference recognized as the pip. It is obviously desirable to get rid of the

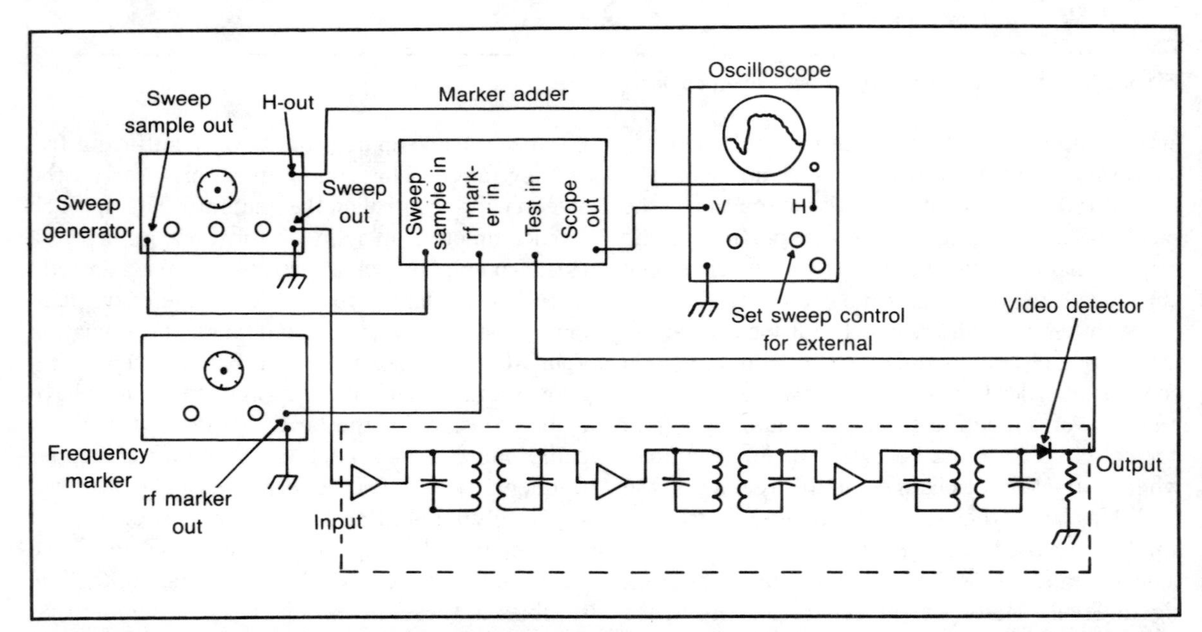

Fig. 8-20. Improved sweep-marker test procedure. Because of the marker adder, the markers are introduced independent of the circuitry being tested.

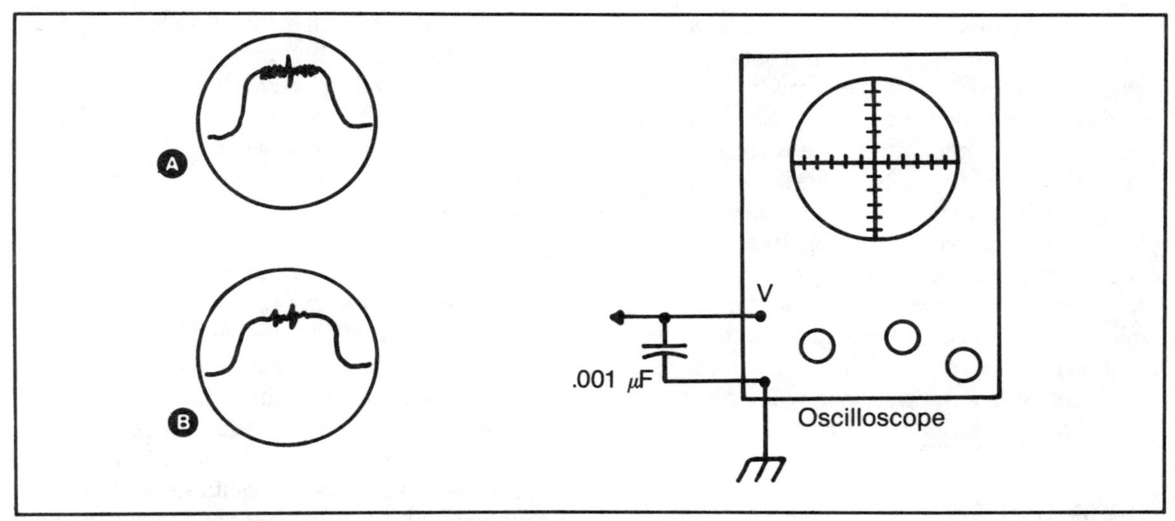

Fig. 8-21. The capacitor across the scope terminals will sharpen the marker pip.

higher beat frequencies, but to retain the lower ones. This is readily accomplished by connecting a small capacitor, usually in the vicinity of .001 μF across the vertical input terminals of the oscilloscope. This simple addition is particularly beneficial when the setup in Fig. 8-18 is employed, that is, when no marker adder is involved.

The setup shown in Fig. 8-22 has also been used. Here, the sweep generator linearly sweeps the rf signal. The internal sweep circuit of the scope

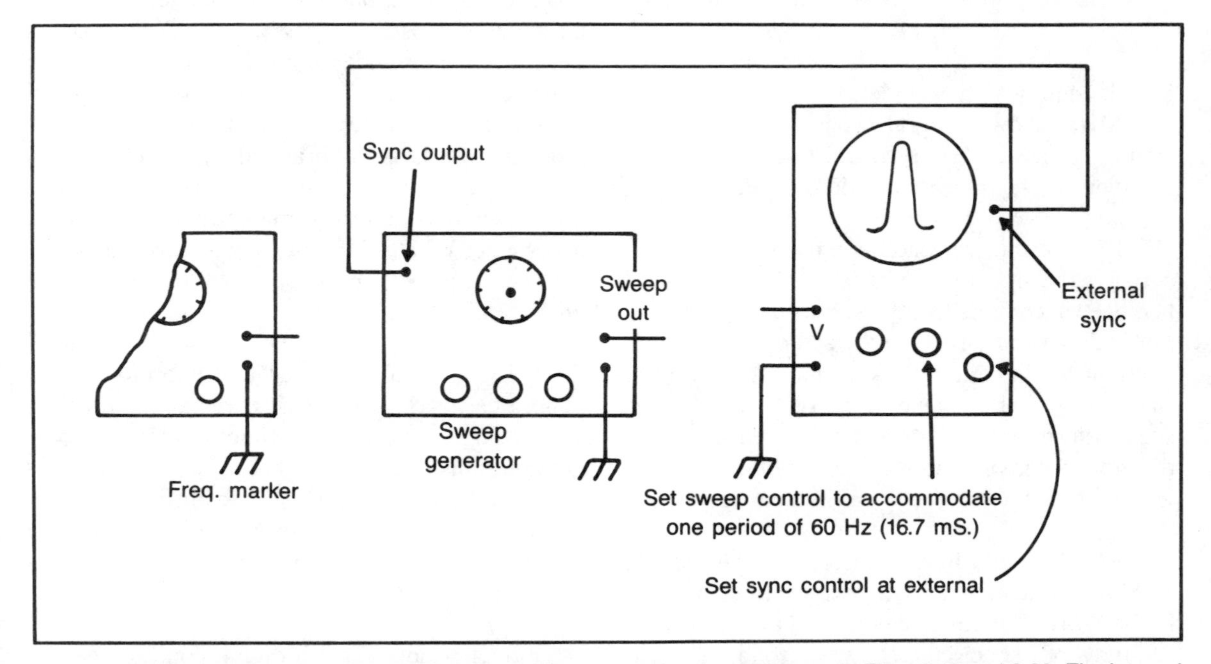

Fig. 8-22. An alternate method of developing the oscilloscope sweep for the setups in Figs. 8-18 and 8-20. The internal sweep of the scope is used by synchronizing it with a sync signal from the sweep generator.

is employed to also provide a time-linear sweep of the electron beam across the horizontal axis of the screen. Synchronization between the generator sweep and the scope sweep is provided by a pulse from the sweep generator. The technique shown in Fig. 8-22 provides a 60-Hz sweep rate. However, other frequencies are used, too. Actually, in this application, it is probably desirable to get away from 60 Hz. Because a sweep generator of this type tends to be more expensive to manufacture than one which does not use the linear sweep technique, it is not often used. However, in research and development laboratories, this type of test instrument often produces useful results.

QUESTIONS

8-1. The coil of an absorption wavemeter is wound in the opposite direction to that of the coil in an active resonant circuit which is to be checked for the presence of rf energy and also for the approximate frequency. How should the wavemeter be positioned in relation to the tank circuit being tested?

8-2. It has been found that a dip meter with a hairpin loop, which provides a range upward from 150 MHz, shows several pronounced dips when coupled to a 3.5-MHz resonant tank. What is the meaning and possible implication of these readings?

8-3. A dip meter had been used to tune some series resonant trap circuits in a TV set. After this had been accomplished, the serviceman put the dip meter down and turned on the set, eager to see the fruits of his labors. To his dismay, a severe herringbone pattern appeared on the screen. Since such patterns are indicative of interference reaching the video detector, it appears that his work has been in vain. What could cause a problem such as this?

8-4. Difficulty is encountered in the use of a 100-kHz frequency marker because the harmonics in the 7-MHz amateur phone band are very weak. A study of the circuit shows it to be a Pierce oscillator designed around an FET. A little experimentation shows that the 100-kHz output can be increased appreciably by replacing the load resistance in the drain circuit with a 100-kHz resonant tank. However, it is then found that the 7-MHz harmonics have virtually disappeared. What similar technique can be used to actually increase the 7-MHz harmonics?

8-5. With a sensitive calibrated wavemeter, it is determined that the tuning range of a VFO (a self-excited oscillator) should be lowered a very small amount. Accordingly, a small fixed capacitor is connected across the tank circuit. This very nearly accomplishes the objective and it is believed that after a copper shield can is placed over the tank inductor, the additional stray capacitance will drop the frequency range very close to the desired limits. A subsequent listening test on a communications receiver indicates that the frequency range is even higher than it was at the start. What can be responsible for the unexpected shift in the tuning range?

8-6It is desired to align a 60-MHz i-f ampilfier in a radar system, but only a signal generator and a narrow-band oscilloscope are available. The signal generator can provide a constant rf output from 30 MHz through 100 MHz and has conventional features such as rf amplitude control, internal audio modulation, a large calibrated dial, etc. The – 3 dB point in the frequency response of the scope is about 200 kHz and its deflection plates are considered inaccessible. What relatively inexpensive accessory can be acquired or constructed to use in the alignment?

8-7. A crystal-stabilized multivibrator circuit generates nearly symmetrical rectangular waves with sharp rise and fall times. Odd harmonic generation is satisfactory, but at high frequencies the even-harmonic markers are difficult to find amidst the background noise of a receiver being calibrated. What can be done to improve this situation?

8-8. It is noted that a digital counter readout varies between 59, 60, and 61 Hz when sampling the power-line frequency. Because of this, a technician is hesitant to use the instrument for rf measurements where much greater accuracy and stability are required. (It is assumed that the alternator at the power plant is driven at constant speed and is not responsible for this variation.) Does this behavior mean that the readout will be unstable at 6 MHz?

8-9. A vertical antenna used with a 20-meter transmitter is approximately three meters in length. At the base of this antenna is a loading coil which makes the overall antenna installation accept maximum rf power from the transmitter just as a physically longer antenna would. A field-strength meter located about 15 meters (five antenna lengths) away is used to tune the transmitter for maximum output. Then it is noticed that the tuning required for maximum output as indicated by an rf thermocouple ammeter at the base of the antenna is quite different from the tuning required for a maximum indication on the field-strength meter. Why are the readings different?

8-10. A push-pull power amplifier has a parallel-resonant tank circuit. When a wave-meter is coupled to this tank circuit, it is found that appreciable second-harmonic power is available. A check with a dip-meter when the amplifier is turned off shows the resonance to be proper for the intended output frequency. What is a likely cause of the inordinately high second-harmonic output?

ANSWERS

8-1. The difference in the way the two coils are wound is of no importance. The principle of energy transfer is the same as that which occurs in a 2-winding transformer. However, the objective is to tune the wavemeter to resonance while taking only enough rf energy from the circuit to produce a usable indication of the lamp or meter. This means loose coupling by adequate physical separation or by angular orientation.

8-2. The VHF and UHF resonant indications are probably provided by "linear" circuits composed of lead lengths and hardware such as the plates of the tuning capacitor. Such conductors form crude but often effective transmission line resonators. The implication is that parasitic oscillations can exist at those frequencies when the tank circuit is energized. Indeed, this is an excellent way of tracking down known parasitics, or determining whether or not certain harmonics can be enhanced in the tank circuit.

8-3. It would be well for the TV serviceman to make sure that he turned off the dip meter! Since the dip meter is essentially an oscillator, it functions as a flea-power transmitter, whether by intent or otherwise!

8-4. The parallel resonant circuit increased the 100-kHz output but attenuated harmonic energy at the same time, since this is the nature of a tank circuit. However, if more dc voltage could be applied to the drain of the FET, and a high impedance could still be maintained for 100 kHz and its harmonics, we could expect improved harmonic operation. This can often be accomplished by replacing the load resistor with an rf choke. Since rf chokes are not ideal because of their distributed parameters, it would probably be a good idea to experiment with different types of rf chokes. A 10 mH ferrite-core choke would be a reasonable start for such an investigation.

8-5. The copper can behaves a a shorted turn and, therefore, decreases the inductance of the VFO tank coil to which it is inadvertently coupled. This, in turn, increases the resonant frequency of the tank circuit. Such an effect can be substantial if there are few turns on the tank inductor and/or it is compact, and if the shield separation from the coil is not sufficient. In this example, a quick check with and without the shield in place would show the effect of the shield on the tuning.

8-6. A demodulator probe can be purchased for little cost, or one can be made in a short time. Such an accessory allows the circuit to be tuned for maximum rf amplification while observing the detected modulated signal on the scope.

8-7. The rectangular waves should be made into unipolar pulses by simply connecting a diode across the output of the multivibrator in order to clip or rectify the wave. Normally, the diode will reduce odd-harmonic amplitudes a small amount, but this should be accompanied by a relatively large increase in the strength of even harmonics.

8-8. The instability in the readout at the power-line frequency is due to the plus-or-minus one-count variation which is characteristic of digital instruments. If the power-line frequency is measured with several decimals of resolution, the fluctuation would again be seen in the last digit. Thus, a reading of 60.0056 Hz might fluctuate between 60.0055, 60.0056, and 60.0057 Hz. By the same reasoning, the measurement of a 6-MHz signal would be very accurate, with only the last digit subject to instability, providing that the decimal in the readout were positioned thus: 6,000,000.0 Hz. Even 6000 kHz would provide sufficient resolution for many purposes, with the variation again in the last digit. Readouts would vary between 5999, 6000, and 6001 kHz. Essentially the same situation would prevail if the readout were 60.00 MHz, for this would result in the following readouts, 59.99, 60.00, and 60.01 MHz. However, if the readout were displayed as 60 MHz, then the readings could be expected to fluctuate from 59 to 60 to 61 MHz, exactly as with the power-line measurement first discussed. Additionally; the digital counter is limited in accuracy by the bounds of a small fraction of 1 percent due to overall accumulation of drifts in its time base and to other factors. However, these tend to vary slowly, and are not related to the type of inaccuracy discussed.

8-9. The field strength meter is well within the induction field of the antenna because it is less than one wavelength away! (The distance of five antenna lengths has no significance, even though the antenna can be made to behave electrically as though it were one quarter, one half, or one wavelength long.) The field-strength meter should be moved to a distance of about five wavelengths of the transmitted frequency, which in this case is a distance of approximately 100 meters. Under this test condition, the tuning for the two indications should not vary greatly if the standing-wave ratio of the antenna feed line is low and there are no obstructions or objects too close to either the transmitting antenna or the field-strength meter. When the rf ammeter and the field-strength meter readings show approximately the same tuning adjustments, the transmitter should finally be tuned for maximum output on the field strength meter.

8-10. Much depends upon the interpretation of "appreciable" second-harmonic inasmuch as at least some is bound to be available despite the attenuation provided by the resonant tank circuit. However, the suggestive evidence is that the tubes or transistors are grossly mismatched; this could be due to the fact that one of the power devices was damaged or was delivering impaired performance. For, otherwise, push-pull rf amplifiers tend to cancel even-order harmonic energy in their output circuits.

Similar evaluations can be made in single-ended power amplifiers using pi-section output filters. Connect an appropriately-rated 50 ohm resistor in series with a single-turn loop across the output of the filter. The loop provides coupling to the dip-meter or to the wave-meter. The dip-meter will then indicate the approximate cut-off frequency of the filter. And, with the amplifier turned on, the wave-meter can be used to evaluate the attenuation of the filter for second, third, and higher harmonics. Although such harmonic measurements are qualitative, they are useful in a relative way. For example, it is often found that the filter components may be "tweeked" to provide better second-harmonic discrimination perhaps at the expense of more-than-adequate attenuation of the third harmonic.

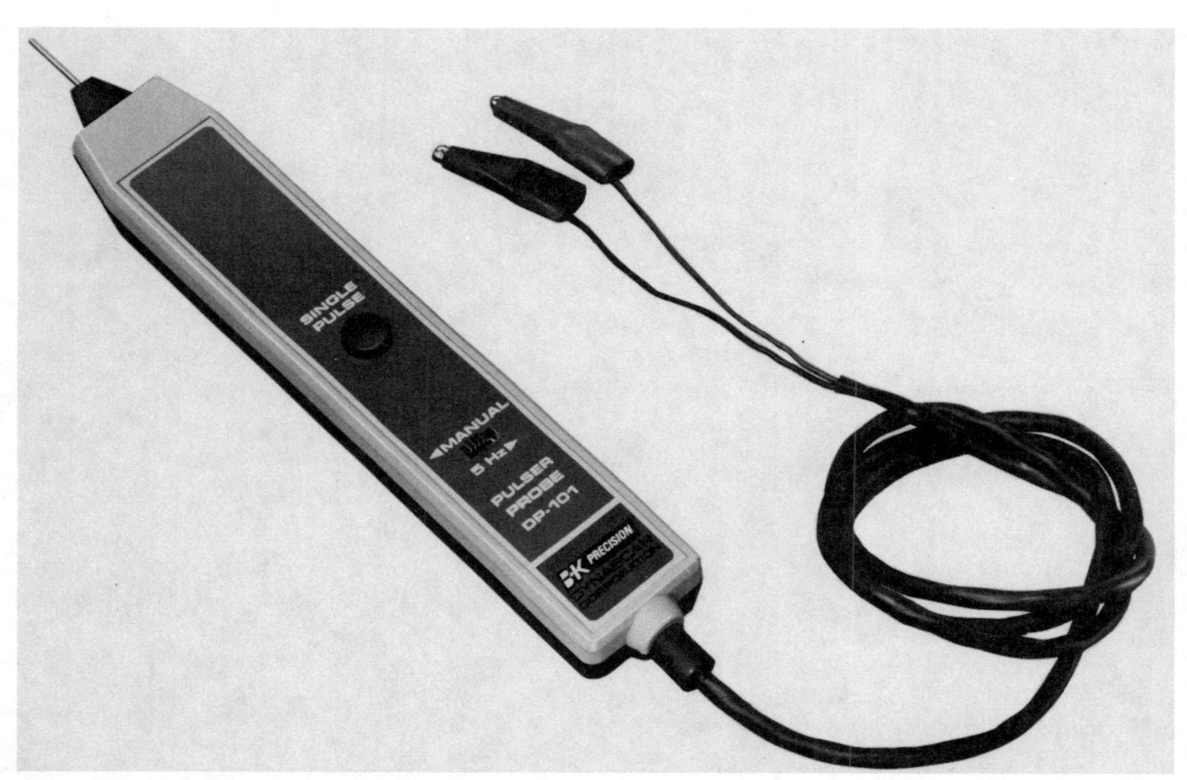

With this Digital-Logic Pulser, a single pulse or a pulse-train is available for injection at a selected circuit-node. The pulses are very narrow, but have sufficient power to over-ride the prevailing logic-state of the digital device. Operating power is obtained from the power supply of the circuitry undergoing test. Courtesy of the Dynascan Corporation.

Chapter 9

Digital Multimeters and Digital Circuit Testing

Insofar as measurement and test techniques are concerned, the digital multimeter readily classifies as an electronic meter. Indeed, in the numerous techniques illustrated and discussed in the previous chapters, it is generally optional to use any of the measuring instruments utilizing some variety of signal processing ahead of the actual indicating device. Included are the VTVM, its solid-state equivalents using FETs, bipolar transistors, or ICs, or the digital multimeters. A common feature of all these types is that they have input impedances of ten megohms or better. (The very first digital meters were electromechanical in nature and often were faulted by the limitation or annoyance of relatively low input impedance. Although they made use of certain electronic techniques, their usefulness was in the same category as the first heavier-than-air flying machine—an invitation for subsequent designers to do better.)

Many refinements have been incorporated in digital multimeters over the years and their use has proliferated enormously. At the same time, their price has decreased to the extent that they have become one of the commonly-available measuring instruments. In one sense, the digital multimeter is just an electronic instrument with a digital, rather than an analog readout (as with "conventional" electronic-meters, FET voltmeters, etc.) In another sense, there are differences beyond the method in which data is communicated to the eye. These differences are sufficiently profound, and are numerous enough to merit a special discussion of these measuring instruments.

READING DATA DIGITALLY

The manifold technical features of digital multimeters are hardly arguable. Such performance parameters as accuracy, precision, speedy readout, and a wide variety of peripheral refinements, such as auto-ranging, polarity accommodation, RMS capability on non-sinusoidal waves, portability, line-frequency rejection, etc. are there for all to see. Best of all, the readout is truly an *objective* one—all beholders perceive the same indication. This contrasts to the more subjective situation pertaining to analog-indicating meters where different viewers can be expected to report at least slightly different readings. These differences are due to both visual and psychological perception factors. Agitating the situation even more is the fact that the analog meter may not provide identical indications in horizontal and vertical orientations. All this is obvious enough. Why then should there be a very appreciable residue of user-objection to testing and measuring with such a technically-superior instrument as the DMM?

In varying degrees we all possess goodly amounts of resistance to change; it tends to be more comfortable to adhere to the tried and proven ways of yesteryear. Our reflexes have long been acclimated to needle-pointing meters and there may appear to be something "unnatural" in reading data from digital displays. One would think that the computer age would have overwhelmingly overcome aversion to digital readouts but such has not yet been accomplished. Those who object to digital readouts on test and measuring instruments are quite content to let the computer buffs have their digital display. At the same time, they complain that they lose the "feel of the road" when conducting test and measurements with DMMs.

More specifically, they say that during such adjustments as peaking a resonant circuit, or tweaking a potentiometer to optimize a voltage, the DMM does not provide a satisfying indication that the task has been properly done. What is lacking is a continuity of visual information that the final adjustment is better than previous ones. (A somewhat related complaint is heard from car owners with digital speedometers. While viewing an indication of, say 55 mph, they feel the reading would be more meaningful if they could simultaneously see where 50 mph and 60 mph is.)

Not all instrument manufacturers fly in the face of such "irrational" grievances, however. Several of the more progressive ones acknowledge that feelings have much to do with the way humans perceive the world and have taken some unique steps to make DMMs more "friendly" with the user. One manufacturer includes a small analog meter alongside the digital display. During measurement, one can observe trends and be comfortably alerted to optimizations, nulls, and peak readings by watching the analog meter. Finally, fine adjustments can be carried out with the aid of the digital readout; in the end, one has the accuracy of digital instrumentation, having been assisted by the analog indications. Here, then, is the best of both worlds.

Other approaches to such a compromise make use of a linear bar-graph, or a circular readout format in which an electro-optic "pointer" provides similar analog data to the needle of an analog meter. Indications are digitally determined and are therefore more accurate than those shown by an analog meter. Note the bar-graph in the frontispiece meter of Chapter 7.

Another technique involves audio tones for low-ohm and continuity tests. Whatever the psychological explanations of these auxiliary functions, wise marketing policy does not put down the allegations of those who use these instruments. For it is well known that many DMM users habitually connect analog meters to test junctions "to make sure that DMM readings are in the right ballpark." Such is the present situation; at some time in the future, analog meters may be seen only in museums but that day has not yet arrived.

DIGITAL ACCURACY

Before investigating the nature of digital multimeters, it would be well to dispel the commonly-held notion that digital instruments are *per-se* accurate measuring-devices because of their numerical readouts. What is true is that there is no optical parallax—you see the same indication regardless of the viewing angle; also, the displayed indication is an objective fact—there is no ground for subjective interpretation of the reading. But, beyond these factors, the designer has a formidable task in producing an accurate measuring instrument. This task can be very rewarding in the sense that the actual usable accuracy of a digital instrument can be made tens, hundreds, or thousands of times better than that readily provided by the best analog types. This obviously requires a very dedicated effort and is not found in most consumer goods featuring digital readouts. Thus, scales, blood-pressure monitors, thermometers, odometers, and other measuring devices are not necessarily accurate or reliable, their digital indicators notwithstanding!

It stands to reason that the accuracy of the digital multimeter can be no better than its weakest link. Admittedly, the digital display can be considered its strongest link. Virtually all digital displays yield perfect accuracy plus-or-minus *one* count; that is, the least-significant digit is always subject to this "built-in" uncertainty. Obviously, the greater the number of digits in a digital display, the less significant is this basic uncertainty. It is paradoxical that the overall accuracy of the digital instrument is, for the most part, dependent upon *analog* circuitry that precedes the display.

One thing that immediately comes to mind is the range network. In analog instruments, it pays but little dividends to go overboard in specifying the tolerance, temperature coefficient, and stability of the resistors in the range network. In a quality digital instrument, the situation is otherwise—whatever precision can be designed into the range network can be advantageously utilized in the performance of the meter.

The same philosophy prevails in other signal-processing functions. For example, it would be a penny-wise and dollar-foolish procedures to use simple rectifying-circuits for the ac measurement-mode. Junction rectifying-diodes have an energy gap of about 0.6 volt to begin with, and are far from being linear devices thereafter. However, when these silicon diodes are appropriately associated with a high-gain op-amp., very high linearity maybe attained virtually down to zero volts. Such a circuit is known as a precision ac-to-dc converter. It "idealizes" the rectification characteristics of diodes.

Similarly, the digital multimeter (DMM) for the first time, makes possible an everyday test-instrument capable of high accuracy measurements of the RMS value of any waveform. But in order to get this feature, it must be ascertained that the proper analog processing circuitry has been designed into the meter; otherwise accuracy may be restricted to sine waves and the digital display will "lie" for other waveforms!

THE BASIC CONCEPT OF THE DIGITAL MULTIMETER

As already pointed out, the DMM is basically an electronic meter with a digital, rather than an analog readout display. Consider the FET electronic meter shown in Fig. 9-1. If digital multimeters were not already an accomplished fact of technical life, how would we go about "inventing" it, using the FET instrument as supporting circuitry? We would remove the analog meter and substitute a functional block called an "*analog-to-digital converter?*"; this would be followed by another functional block known as "*display logic*", and finally by the digital display itself. The nomenclature of the functional blocks is self-explanatory and their use and arrangement is obvious enough. Such an arrangement is shown in the block-diagram of Fig. 9-2. Although simplistic in concept, it nevertheless represents the basic approach to DMMs.

Actual DMMs involve a great deal more complexity than is evidenced by this simple depiction. This is primarily because the analog-to-digital converter often comprises several signal processing sections and great care is necessary to attain precise performance over wide ranges of voltage, frequency, and temperature. Then, too, many sophisticated metering techniques must be accommodated, such as auto-racing, polarity sensing, decimal point manipulation, protection, and automatic zeroing. Moreover, there are at least a half-dozen analog-to-digital conversion methods. This leads to trade-offs in precision, cost, stability, speed, and other performance parameters. Many stages of inter-related circuitry are needed in a practical instrument. As in many other electronic applications, circuit complexity is greatly simplified in actual implementation via the use of monolithic ICs. As the art has developed, *dedicated* chips have emerged. As with hand-held calculators, these chips have given the DMM the appearance of trivial simplicity to casual inspection.

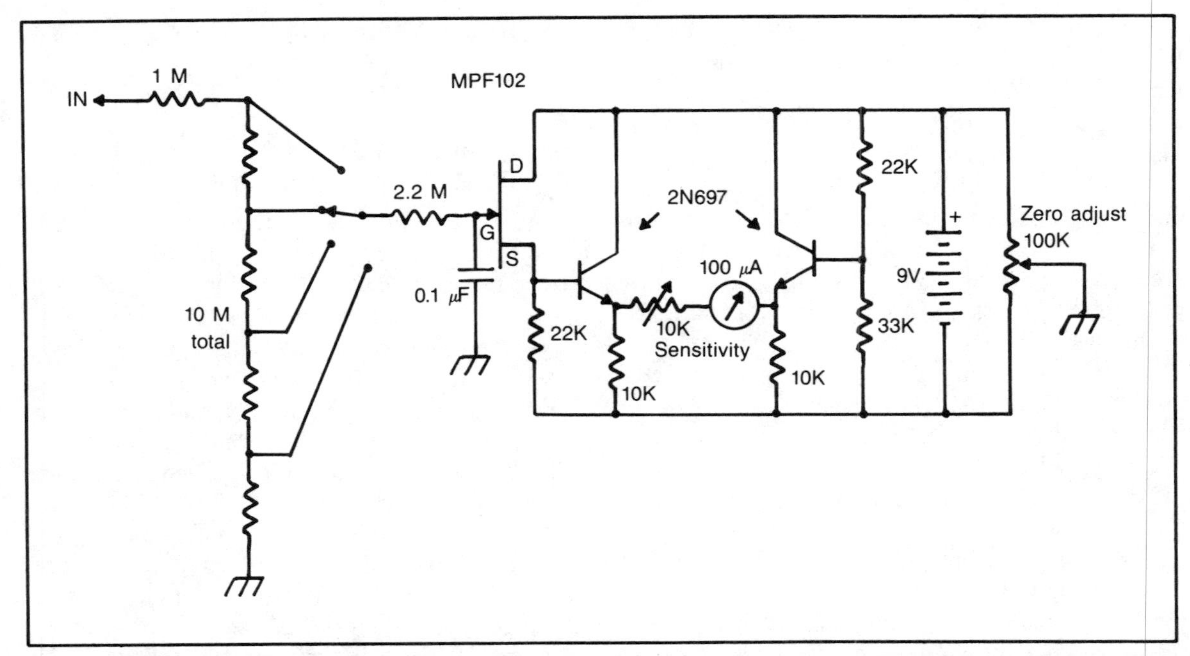

Fig. 9-1. The FET Voltmeter—an electronic meter with analog readout. An elemental digital meter can be devised by replacing the 100 μA D'Arsonval meter with a analog-to-digital converter, display logic, and a digital display.

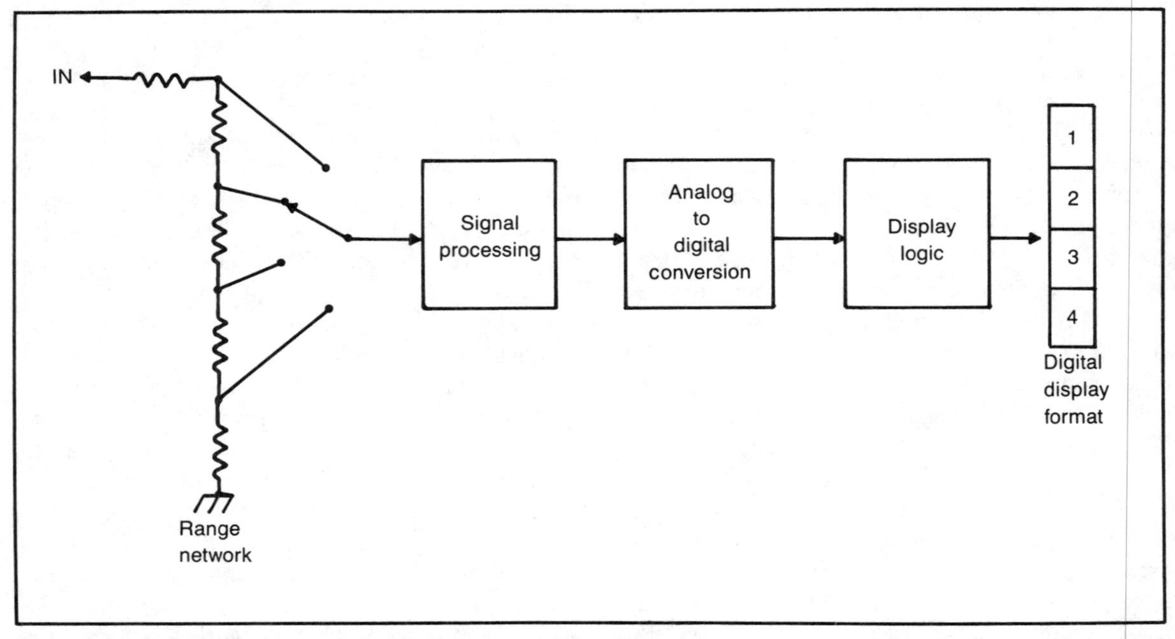

Fig. 9-2. Block diagram of the basic digital multimeter (DMM). The range network and the circuitry for signal processing could conceivably be similar to the FET instrument of Fig. 9-1. Instead of an analog readout meter, appropriate circuitry is needed for actuating a digital display format.

Digital readouts have crept into consumer products such as scales and thermometers. Profits have been made on the misconception that this display format is intrinsically accurate. In DMMs, the digital readout, itself, confers no accuracy beyond that due to unambiguous perception of the numerals, and the elimination of parallax. The accuracy is thereafter dependent upon the resistive divider network, the analog-to-digital converter, and upon such matters as temperature coefficients, internal power supply stability, immunity to fields, transients, and line-frequency harmonics, and stable voltage-references. It is interesting, however, that an "economy-class" DMM can easily surpass the accuracy readily attainable from quality analog meters. Even though the last digit of a DMM always has an uncertainity of plus or minus one count, the three or four decimal places indicated by small hand-held DMMs provide measurement resolution never expected from analog meters.

Characteristics and Features of Digital Meters

The idea presented of retrofiting an analog to digital converter to an FET electronic meter serves the purpose of suggesting what a DMM is all about. Such an implementation would, indeed, demonstrate the *feasability* of instrumentation with digital readout. As might be expected, actual DMMs are complex electronic systems and involve many performance features and trade-offs. While the DMM user need not have intimate engineering knowledge of his instrument, a generalized comprehension of operational basics cannot but help result in better measurement and test results. Even these sophisticated meters have their limitations. For example, an RMS measuring DMM may or may not provide accurate measurements for non-sinusoidal waves. Again, when ac is superimposed on a dc level, it is necessary to make separate ac and dc measurements with some instruments. With all DMMs it is necessary to know the frequency capability of the instrument before attempting high-frequency measurements. And, the resolution of the DMM should comfortably exceed the intended accuracy of the particular measurement. We are reminded of an automobile driver; those with some understanding of the goings on beneath the hood develop better intuitional approaches to the ever-changing impositions of road, traffic, and weather.

One of the important design differences in DMMs is in the way in which the analog input level is converted or transformed into digital data suitable for driving the digital logic section, which in turn, causes the display to indicate in decimal notation. A number of circuit techniques and mixes thereof have been implemented. Mathematical or theoretical superiorities of a given method cannot always be relied upon to be manifest in the actual marketed instrument. Much depends upon the manufacturer's engineering capability and dedication to the use of high quality components. Five methods for bringing about the all-important analog to digital transformation of the measured signal level are as follows:

Voltage to Frequency Integrating
Single Ramp
Dual-Slope Integrating
Successive Approximation Potentiometric
Continuous Balance Potentiometric

The operating principles of these conversion techniques will be described shortly. In the meantime it would be instructive to list the major performance parameters of DMMs. This will drive home the considerable variation in measurement capabilities of these instruments. Because their physical appearance tends to be dominated by the digital display, users sometimes are led astray by notions of assumed accuracy. The performance parameters and basic features of DMMs are as follows:

Type and Format of the Digital Display

(a) Incandescent lamps...superceded by LEDs.
(b) Nixie tubes......used on earlier models
(c) Fluorescent display tubes.....large bright numerals
(d) Mechanical digits......obsolete
(e) LED.........presently popular

(f) Liquid crystal display.......very easy on batteries in portable units

(g) Combined digital display and analog meter or readout

The more digits, the better the resolution. However, all readouts have inherent uncertainity of plus or minus one with respect to the least-significant digit. Also, in most DMMs the left, or most significant digit, is known as a half-digit because it can only display zero or one. All of the other digits can display numerals from zero to nine. Because of this, a DMM with four digits is called a three and one-half digit instrument. Similarly, four and one-half digits designates an instrument actually having five in-line digits.

Number and Range of Functions

(a) dc volts....polarity indication?

(b) dc current

(c) Resistance in ohms.....audible output for continuity checks?

(d) Conductance in ohms or siemens

(e) ac volts

(f) ac current

Note: The ac measuring capability must be approached with care. Obviously, one is interested in the frequency range claimed for the instrument. Often, so-called RMS capability pertains only to sine waves. An instrument with true RMS capability can accommodate non-sinusoidal waves as well. But even some of these require separate dc and ac measurements when measuring ac superimposed on a dc level. The very best of the "true" RMS instruments yield an accurate readout for all waveshapes, including those associated with a dc level.

(g) dB and/or dBm.....some DMMs make automatic correction for different impedances.

(h) Diode and transistor forward conductance measurements

Auxiliary Features

(a) Automatic polarity accommodation on dc.

(b) Range extension probes, such as a 30kV probe for monitoring CRT voltages.

(c) Portability, size, and convenience.....NiCad battery and charger? Optional ac operation.

(d) Autozeroing.

(e) Autoranging....some models dispense with range switch entirely.

(f) Illuminated function indicator.

(g) Four-terminal leads for Kelvin resistance determinations.

(h) Peak-hold for monitoring transients.

(i) "Beeper" for audible testing of logic circuits, and for continuity.

(j) rf probe.

(k) Clamp-on current probe.

(l) Direct temperature readout from factory-calibrated thermocouple.

(m) Input circuit floated from chassis ground.

(n) Fusing and protection provisions—overvoltage alarm.

(o) Microprocessor-assisted function blocks.

(p) Variable pitch tone to provide "analog feel" when peaking, nulling, or comparing.

(q) rfi shielding.

(r) Selectable mode to convert instrument into a logic probe.

(s) Adjustable or selectable constant-current source for testing diodes and pn junctions.

(t) IEEE 408 interface—General Purpose Instrument bus (GPIB).

(u) Frequency-measuring capability.

(v) Immunity against interference from power line and its harmonics.

(w) Capacitance measuring mode.

(x) Range lock—holds selected range.

(y) Tilt stand, carrying case, and other amenities.

(z) Factory warranty—it is unwise to plan on self-servicing.

Considering the nature of the digital display, the ranges and functions, together with the auxiliary features and options, it becomes obvious that knowledge is power when it comes to selecting and using DMMs. The need-to-know often surprises those whose first impulse is to grab a meter and conduct a test or measurement. Even the "smart" microprocessor-assisted instruments do not dispense with the human brain. All DMMs are only too willing to produce a digital readout. If the user knows something about the particular instrument and understands his test or measurement technique, then, and only then, is it possible to attain measurement accuracy far exceeding the capability of analog instruments.

The Electromechanical Digital Voltmeter

The electromechanical digital voltmeter is shown in Fig. (9-3). Although this type has long been replaced by solid state versions, it is interesting and instructive to contemplate this early approach to digital instrumentation. This is particularly true inasmuch as one type of solid-state all-electronic instrument employs the same basic principle of operation—servo-nulling of a sampled reference voltage against a sample of the measured voltage. Surprisingly, the electromechanical type was capable of measurement accuracies in vicinity of 0.05%. Its main drawback was its slowness of response. The mechanical readout comprised four wheels, each bearing numerals from 0 to 9. This readout device was mechanically linked to the servo motor, as was also a precision potentiometer. The output of the comparator governs the rotational direction of the motor and the overall action is to hunt for a null, i.e., zero voltage at the output of the comparator inputs are equal. Inasmuch as the motor is deprived of operating voltage at null, the digit wheels are stationary and can be read.

The ac filter is needed to attenuate picked-up 60 Hz power line interference. This has also been a problem with modern all-electronic digital meters. These continue to use filters, but more sophisticated techniques have been developed. In some instances, the filtering is accomplished by active, or digitized filters instead of, or in conjunc-

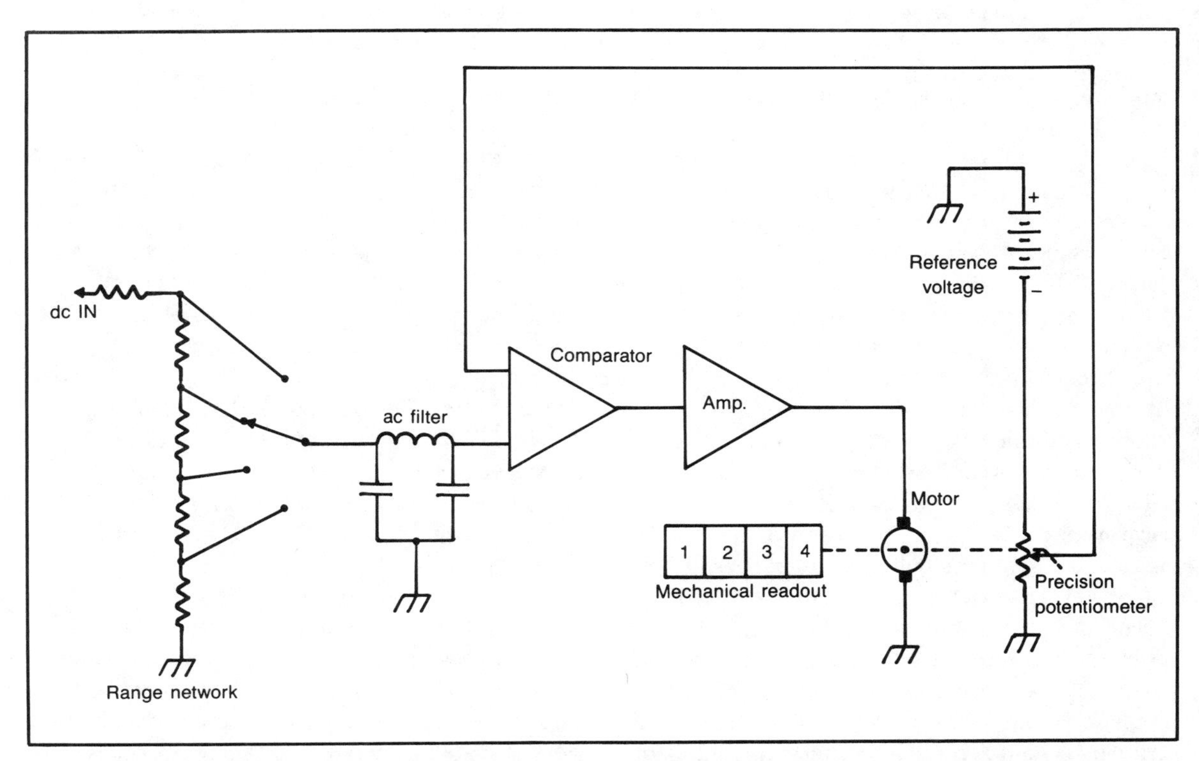

Fig. 9-3. An electromechanical digital voltmeter. This scheme, basically a servo-nulling technique, has industrial applications, but is now little used for instrumentation. It is instructive, however, having been the predecessor of certain all-electronic instruments now popular.

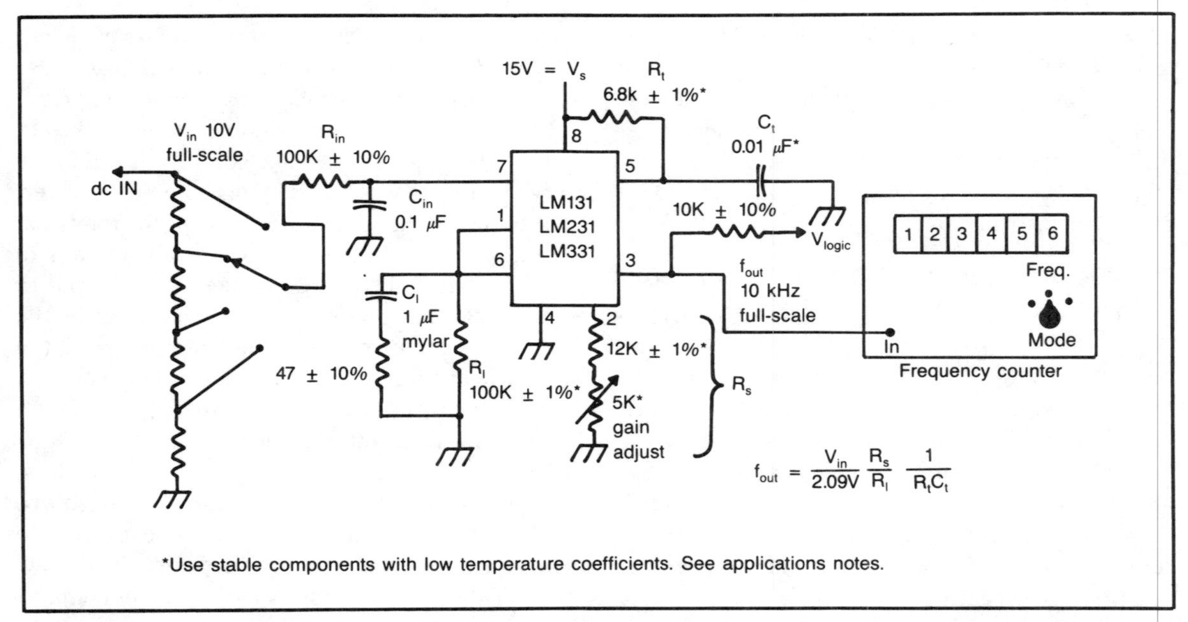

Fig. 9-4. An improvised digital voltmeter. The voltage-to-frequency converter enables the frequency counter to measure dc voltage-level.

tion with passive filters. The appropriate selection of clock frequencies often confers very high immunity against power line interference frequencies. All in all, much was learned from the electromechanical instruments; even though electronics engineers like to replace mechanical functions with electronic techniques, it should not be thought that these earlier instruments were crude implementations—some were in the same league as fine watches (of the mechanical variety, of course.)

From Fig. 9-3, it can be appreciated that stability of the reference voltage and the linearity of the precision potentiometer were largely involved in the accuracy of the instrument. The comparator and the amplifier were often discrete circuits, but hybrid modules and other packaged circuitry were also used. Usually, these instruments served as dc voltmeters only. By the time manufacturers began to provide other measurement options, the solid-state (except for the nixie readouts) digital meter became competitive. These newer instruments could provide almost instantaneous readout and thereby superceded the electromechanical types.

An Improvised Digital Voltmeter

Improvision often leads to a keener insight into the nature of complex techniques. With regard to digital meters, it is quite easy to demonstrate the basics of one way in which measurement instruments can be made to provide a digital readout of an analog quantity. It fortuitously happens that semiconductor firms have been marketing inexpensive ICs that can, with high accuracy and linearity, deliver an output pulse-rate directly proportional to a dc voltage-level impressed at the input of the IC. It follows that a frequency-counter connected to the output of such an IC will provide a digital readout which can be related to the actual dc voltage undergoing measurement. This implementation is shown in Fig. (9-4).

An arrangement such as this is not intended to simulate the performance of actual commercial DMMs. Many peripheral circuits and sophisticated techniques would be needed to accomplish this. Nonetheless, a little experimentation could produce very statisfactory results for some purposes. This is because, both, the IC and the digital counter are, in themselves, capable of very accurate operation.

If DMMs were not presently on the market, such an improvision would very likely merit serious consideration if demonstrated to an engineering or marketing executive. Its performance would certainly suggest the feasability of digital voltmeters.

Inasmuch as digital counters have become fairly commonplace measuring instruments, it is entirely possible that the ingenious experimenter can put this improvision to practical use. The idea, in itself, is not novel, but the task of designing and constructing an analog-to-frequency converter was formidable with discrete components. Moreover, voltage-to-frequency conversion is actually used in some DMMs. And, it is not far-fetched to suppose that an enterprising maker of digital counters might increase the versatility of his instrument by adding a voltmeter measuring mode via a technique such as this. (The converse situation is seen in certain DMMs that provide the user with frequency-measuring capability)

THE RAMP-TYPE DIGITAL METER

As mentioned, the basic design difference between a number of digital meters is in the technique for the conversion of analog to digital information. There are several approaches for accomplishing this transformation via decade or binary counters. The basic idea is always to produce an accumulated count proportional to the signal level undergoing measurement. One such method is implemented in the ramp-type digital meter. The block diagram of a simple digital voltmeter using this technique is shown in Fig. 9-5. Here, it will be seen that the "front end" resembles those used in the more primitive digital meters alluded to earlier. It should be understood that, although only dc voltages can be measured by the arrangement depicted in Fig. 9-5, ac levels can be readily measured by using a rectifier, or by the use of a more accurate ac to dc converter designed around an op-amp. And, with other auxiliary circuitries, other electrical parameters can be measured. For example, resistance measurements could derive from the dc potential drop produced across the unknown resistance by a pre-determined current.

The operation of this digital meter is best explained by referring to Fig. 9-6. Here it is seen that a voltage ramp commences at time t1. This ramp, which must be very linear if good accuracy is to be attained, rises until its amplitude is equal to that of the sampled voltage level undergoing measurement. The equality is sensed by the comparator at its input terminals. This event brings us to time t2.

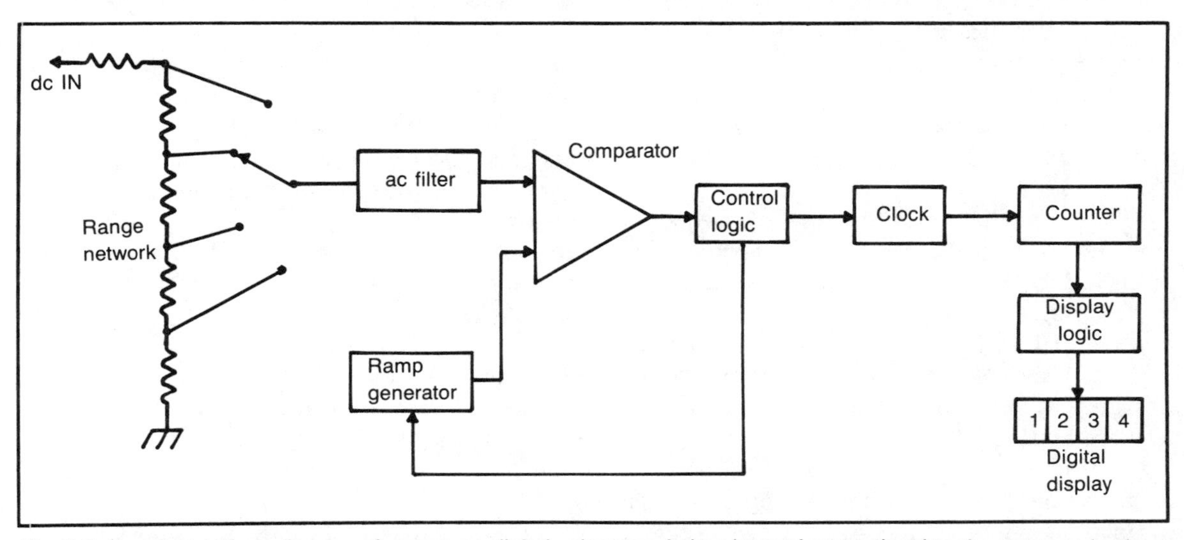

Fig. 9-5. Functional block diagram of ramp-type digital voltmeter. A time interval proportional to the measured voltage is sensed as a number of counts stored in a counter. The digital display thereby driven is calibrated in volts.

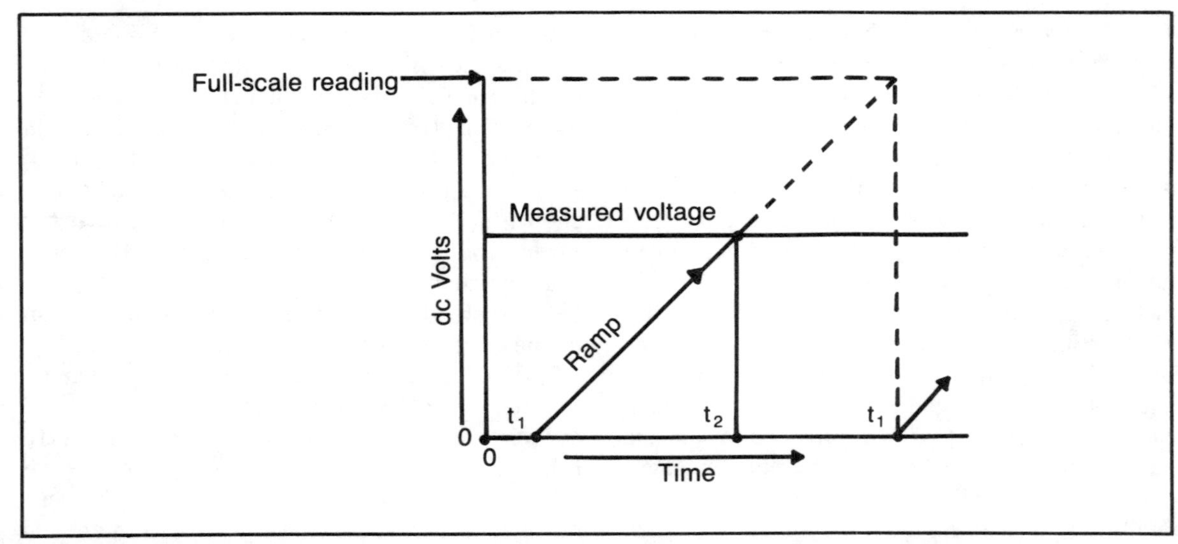

Fig. 9-6. Cycle of measurement events in ramp digital-voltmeter. At time t_1; ramp voltage starts (Counter has been set at zero). Clock pulses are gated into counter. At time t_2; comparator senses equality of ramp and measured voltage and stops clock pulses to counter. The counter is read and then reset. The sequence then starts again at t_1.

The important thing about the rise of the voltage-ramp is that during its excursion, clock pulses have been accumulating in the counter. The counter, of course, started from zero count at time t_1.

At time t_2, clock pulses are prevented from further increasing the accumulated count. The accumulated count is read out of the counter and displayed in decimal notation on the digital display. After being displayed for a short time, the readout, the counter and the ramp are reset to their zero or start conditions and a new measurement cycle commences with t_1.

Those familiar with the use of frequency counters will perceive that the counter in this instrument is caused to operate as an internal timer, i.e., it measures an elapsed period or interval. The time interval, of course, is $t_1 - t_2$. Because the ramp is linear, this interval is proportional to the measured voltage. For example, a meter could be designed so that 1000-counts corresponds to one volt. Obviously, the clock must have good frequency stability in this technique. The repetition or sampling rate of this measurement cycle can be controlled by other circuitry.

THE VOLTAGE-TO-FREQUENCY CONVERTER INSTRUMENT

The block diagram of the digital meter shown in Fig. 9-7 is essentially the inverse of the ramp-type of Fig. 9-5. Whereas the counter in the ramp instrument monitors a signal-proportional time interval, the counter in this meter monitors a signal-proportional frequency. In the arrangement of Fig. 9-7, the time base allows the gate to deliver signal pulses to the counter for a predetermined time. In this way, the accumulated count represents the amplitude level of the measured voltage. Thus, the higher the measured voltage, the higher is the frequency of the pulses generated by the voltage-to-frequency converter. Accuracy depends upon the linearity of this relationship, as well as the precision of the time-base.

As with other digital meters, precautions are necessary to prevent interference from the 60-Hz power line. Although an input filter is shown in the block diagram, it is often possible to gain additional immunity to such interference by appropriately relating the counting interval to the power-line frequency. The basic idea involves the fact that positive and negative noise produce opposite effects

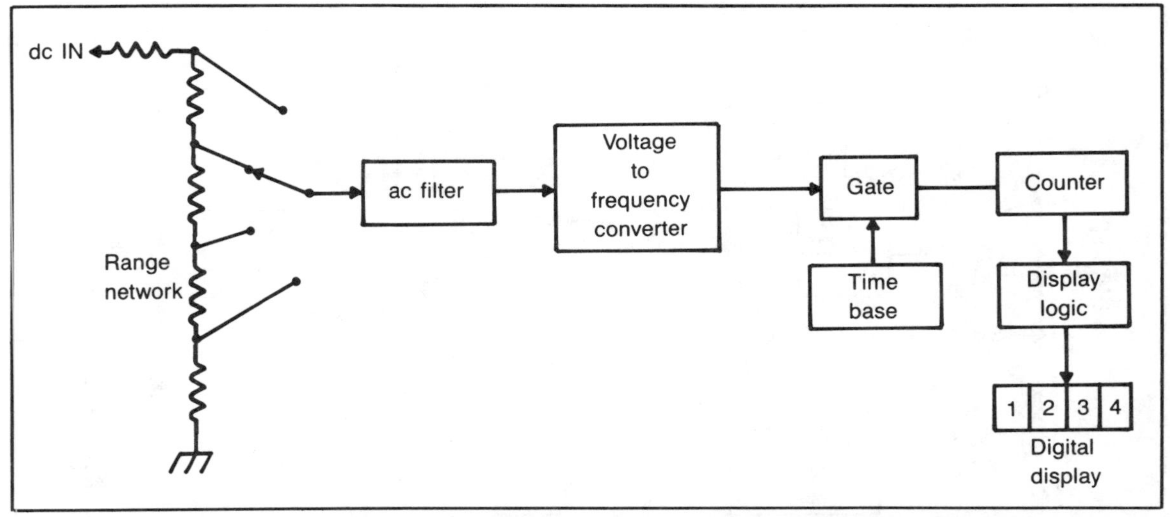

Fig. 9-7. Block diagram of digital voltmeter using a voltage-to-frequency converter. The time-base provides pre-determined durations for accumulation of counts in the counter. Thus, the basic parameter measured is frequency. This contrasts to the action in the ramp-type meter where the basic parameter measured is period.

on the pulse rate generated by the voltage to frequency converter. At certain counting intervals, the averaging effect of such noise can be optimized so that near-cancellation occurs and the number of pulses delivered by the voltage-to-frequency converter is very nearly what it would be without noise.

For those who happen to have digital counters, but no digital multimeter, there are now available inexpensive, but very precise monolithic voltage-to-frequency converters, such as the LM131, and LM231, and LM331 family. These can be used in conjunction with a digital counting instrument to convert it to a digital voltmeter. Such as implementation will not be as useful nor as convenient as a "real" DMM but will allow far greater accuracy for certain measurements than is forthcoming from analog instruments. This arrangement, for example, may well suffice for experiments with solar cells, thermocouples, and exotic batteries. In any event, it demonstrates the basic principle of the voltage-to-frequency converter.

THE DUAL-SLOPE INTEGRATING DIGITAL METER

Dual-slope integration is a very popular technique for accomplishing analog-to-digital conversion is DMMs. A functional block-diagram of such an instrument is shown in Fig. 9-8. The basic idea here is to charge a capacitor from the measured voltage-level for a predetermined time, t_1. Next, the capacitor is *discharged* by reverse-polarity from an internal voltage-reference. The time, t_2, needed to discharge the capacitor to zero voltage is monitored as a number of time-base pulses accumulated in a counter. The total of these pulses is a linear function of the voltage undergoing measurement; this voltage is then indicated by the digital display. The timing diagram of Fig. 9-9 shows the linear nature of both charge and discharge cycles. This lenearity is brought about by the association of an op-amp with the capacitor, C, in a circuit appropriately known as an integrator. (If the op-amp were not part of such an integrator circuit, the capacitor charge and discharge curves would be exponential in shape and the scheme used in this DMM would not be feasable.)

When the measurement cycle begins, the control logic opens switch SW2 and sets switch SW1 to its "A" position so that the integrator capacitor can be charged from a sample of the voltage being measured. At the same time, the AND gate is ena-

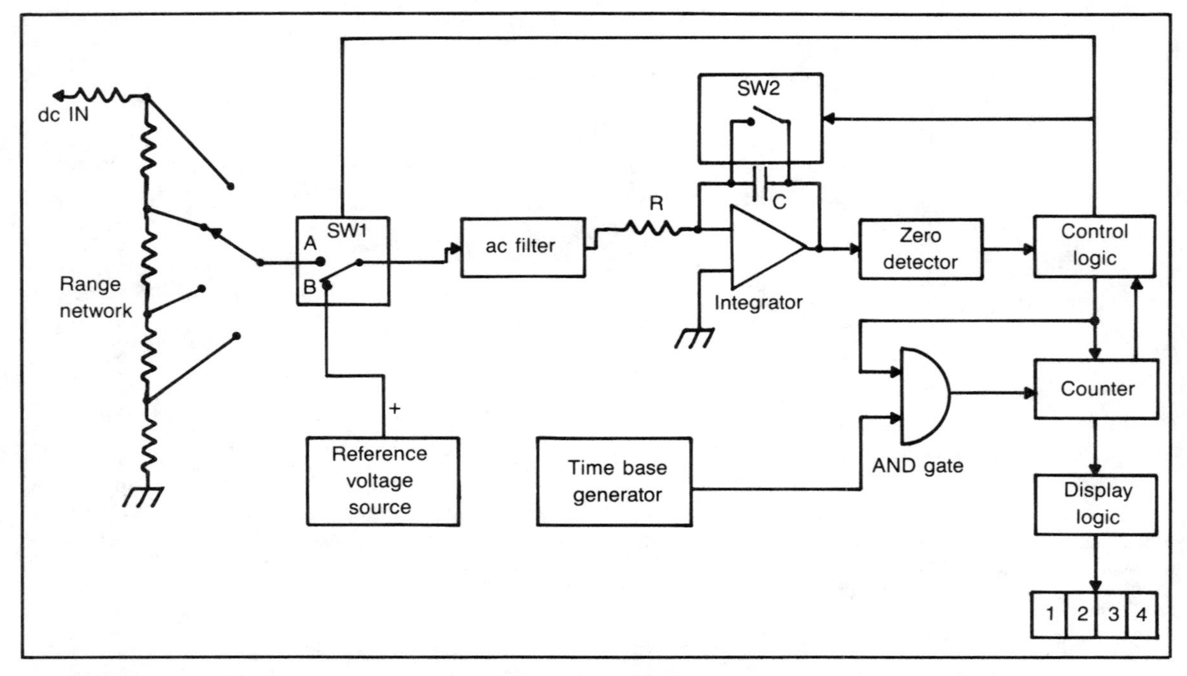

Fig. 9-8. Block diagram of digital voltmeter using dual-slope integration. The analog-to-digital conversion technique in this digital voltmeter is unique in that actual measurement does not take place in real time. What is actually measured is the discharge time of a capacitor, C, that has been charged by the voltage being measured.

bled; this allows the counter to accumulate counts from the time-base generator for a pre-determined time, t_1.

When the count reaches its predetermined limit, corresponding to t1, the control logic resets the counter and sets switch SW1 to its "B" position, thereby initiating the discharge cycle of the integrator source. Time, t_2, required to discharge C is directly proportional to the voltage developed across its terminals during the time of its charge cycle, t_1. Thus, time t_2 represents the magnitude of the voltage being measured by the meter.

As the reset counter accumulates counts from the time base generator, the capacitor is discharging. When finally discharged, the zero detector, actually a voltage comparator, causes the control logic to disable the ANd gate and to close switch SW2, holding the capacitor voltage to zero. These events define the end of time t_2 and the accumulated count represents the measured voltage.

This measuring technique has some interesting aspects. As might be expected, accuracy will be affected by changes in voltage and current offsets in the integrator circuit. The stability, of the voltage reference source is obviously involved in the ac-

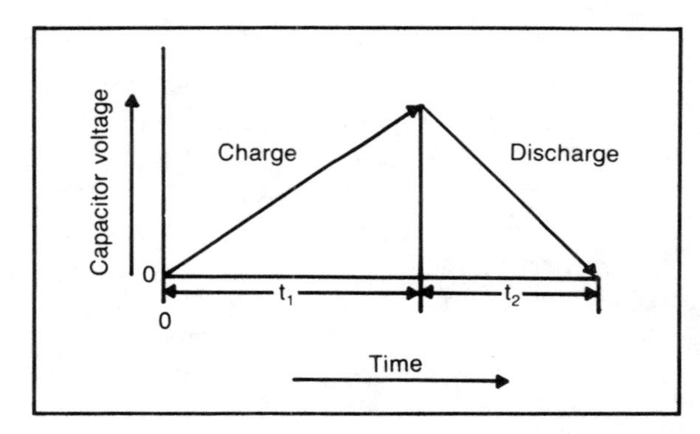

Fig. 9-9. Integrator capacitor-voltage during one measurement cycle. The voltage being measured is allowed to charge the capacitor for a predetermined time, t_1. Then, t_2, the time required to completely discharge the capacitor from a reverse-polarity reference voltage, is measured. Finally, t_2 is displayed as the indication of the measured voltage.

curacy. Surprisingly, the tolerances of R and C, the long-term aging of time-base frequency, and imperfect action of the zero detector all exert minimal, if any, effects on accuracy because their influences in positive and negative directions tend to cancel during a measurement cycle.

In the described sequence of events during a measurement cycle, it will be observed that the counter serves two distinct purposes. During the charge cycle, the counter determines the duration of time t_1. This time is fixed, being designed into the instrument. But, during the discharge cycle, the counter measures the elapsed time in terms of its accumulated count of pulses from the time-base generator. (As pointed out, the counter is reset when switched from one purpose to the other.)

Because of the highly-linear nature of the charge and discharge cycles, their rates have very simple relationships. Charging takes place at a volts-per-second rate given by E/RC where E is the sample of the measured voltage provided by the range network. Similarly, the discharge rate is given by V/RC, where V is the internal reference voltage. These relationships enable us to see why the time-constant, RC, is not primarily involved in the measurement accuracy. A change in the time constant will, to be sure, affect the voltage developed in capacitor C at the end of charge time, t_1. But t_2 will be unaffected—the discharge of a higher voltage in the capacitor will simply take place at a proportionally higher rate. The inverse argument holds for the discharge of a lower voltage in the capacitor. (this fortuitous situation does not prevail in the R and C elements used to form the ramp in the ramp-type DMM.)

Putting everything together mathematically, we have, $t_2 = (E \times t_1)/V$. Inasmuch as t_t and V are fixed, t_2 is directly proportional to the measured voltage, E. A natural question would pertain to the effect of variations in the value of voltage E during the measurement cycle. It happens that this type of instrument averages such variations, i.e., the readout in volts would be the average of the measured voltage excursions during the measurement cycle. This is not necessarily the case with DMM's employing other analog-to-digital techniques. Although most measurements are made for voltage levels that are steady at least during a measuring cycle, the averaging characteristic has favorable implications for noise immunity.

In block diagram of Fig. 9-8, only a negative voltage can be accommodated for measurement. In an actual instrument, means are provided for measuring either polarity, and often automatically. This can be accomplished by reversing the polarity of the reference voltage, or by inserting a precision inverter in the path of the measured dc voltage.

THE SUCCESSIVE-APPROXIMATION POTENTIOMETRIC METER

Potentiometric techniques for digital meters are, loosely-speaking, the electronic simulation of the early electro-mechanical instruments, such as alluded to with respect to Fig. 9-3. The successive-approximation potentiometric meter is interesting in that its comparator participates in a linear servo-action even though feedback is initiated with digital data. The block diagram of such an arrangement is shown in Fig. 9-10. When making a measurement, the control logic produces a reading which is too high. This reading is decremented downward in a one-digit-at-a-time sequence until the least significant digit is less than what is required for a null at the output of the comparator. Then, the least significant digit is incremented upward, finally "homing in" on the correct readout and halting the servo action.

This technique is capable of very accurate measurement, being primarily dependent upon the stability and integrity of the reference voltage. It is often found in 4 1/2-and 5 1/2-digit multimeters. The lack of an integrating circuit, such as those involved in the capacitor charge and discharge cycles of ramp and dual-slope meters is a mixed blessing. The integrating circuits also behave as filters, thereby imparting considerable protection from noise. In particular, these integrating circuits can be made to have time-constants which discriminates against interference from 60 Hz and its harmonics. In potentiometric meters, special filtering circuits (not shown in the simplified block diagram Fig.

9-10) are needed. But, because measurement and filtering processes are now independent, considerable design flexibility is possible. For example, panel-selectable filter operation may be incorporated.

Noise interference and filtering are important considerations in digital meters. Later, in the discussion of RMS measurements, it is pointed out that it is sometimes necessary to make two measurements, one in the ac mode, the other in the dc mode, and finally compute the RMS value. Some digital meters may not be suitable for measuring a small dc component or dc level in the presence of a high-amplitude ac waveform. A digital meter with panel-selectable filtering may offer advantages here.

It should not be construed that the successive-approximation potentiometric meter is necessarily slow. Sophisticated control logic can dramatically reduce the number of exploratory steps required to reach the correct indication. As with the other types of digital meters, the block diagram of Fig. 9-10 provides a "bare-bones" look at the operational scheme. To measure ac requires an ac-dc converter ahead of the range network. Similarly, the ohms-function requires a constant-current source so that dc voltage can be monitored across the unknown resistance.

THE CONTINUOUS BALANCE POTENTIOMETRIC METER

The continuous balance potentiometric meter makes direct comparisons in real time as does the successive-approximation potentiometric meter. However, the feedback path from the control logic operates on the reference-voltage source. A step-by-step sequencing takes place until the comparator output attains a steady null. Under this condition, the fluctuating digits of the readout come to a halt. The logic is ever ready to follow changes in the measured signal. This scheme tends to be susceptible to disturbance from noise. Also, the exploratory sequencing of the feedback is less amenable to speed-up techniques than is the case in the successive-approximation potentiometric meter. The net result tends to be a relatively slow measurement cycle. Filtering techniques are

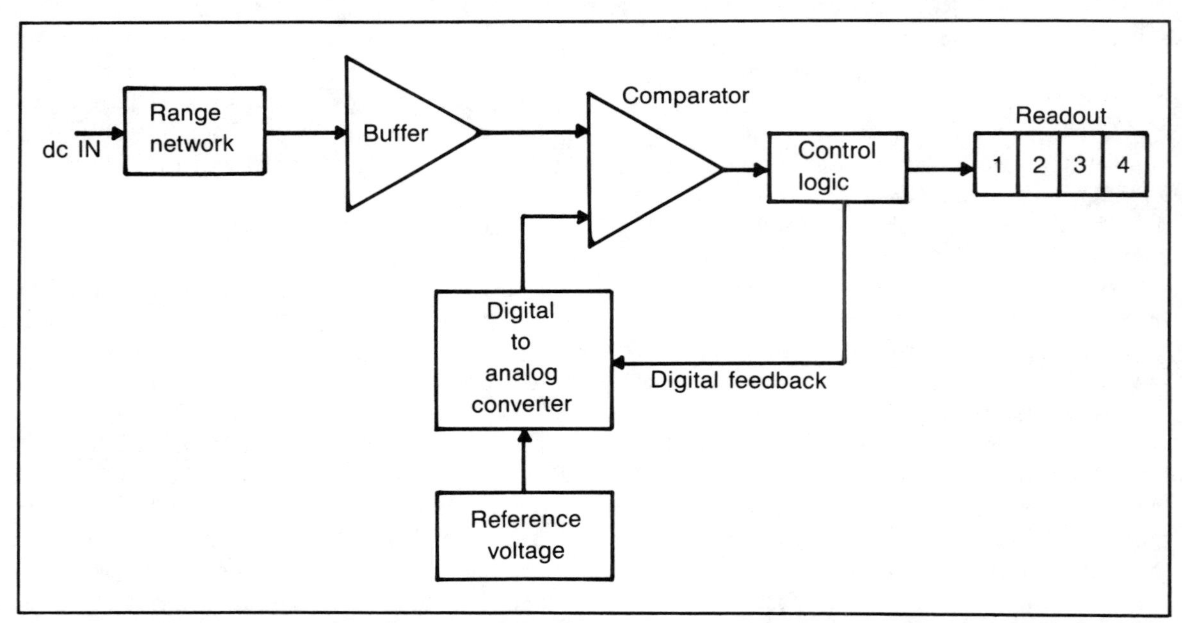

Fig. 9-10. Block diagram of the successive approximation potentiometric meter. The digital feedback loop allows incremental approaches to a null condition at the output of the comparator.

especially important in the design of this type of digital meter because of its natural tendency to follow noise pulses. The various other types generally seem to have cost, accuracy, or other performance advantages over this design approach, but it is not unlikely that its popularity will again increase as new devices become available. For example, digital filters could conceivably add noise immunity to this meter.

OTHER MEASUREMENT FUNCTIONS

A common denominator of all the various types of digital multimeters is that they are basically dc voltmeters. This is true even though the voltage undergoing measurement may be internally converted into a time interval, a frequency, or some other quantity. As a consequence, the most straight forward approach to the measurement of resistance, current, capacitance, temperature, frequency, or any other quantity is to first *convert such a quantity into a voltage* and impress this representative voltage at the input of the instrument, or at some appropriate junction of the range network. If such a derived voltage were truly proportional to the quantity undergoing measurement, the overall accuracy could be just as good as that of the DMM when used as a dc voltmeter.

The ohmmeter function of a DMM is brought about by passing a known current through the resistance undergoing measurement and measuring the potential drop developed across it. It would be awkward to attempt this from an unregulated voltage supply if for no other reason that the current is not readily determined. Some scheme using a regulated voltage source and a precision series-resistance might be devised but would be discovered to be impractical for a wide range of resistance measurements. Moreover, it would be found that dangerously-high voltages would be needed. The solution is to use a constant current source. Such a source has the immediate advantage that the current can be predetermined very precisely. Moreover, a constant current source is electrically equivalent to a high voltage source with a very high series resistance. (Indeed, the high series resistance converts a high voltage source into a constant current source.) The significance of this is that a constant current circuit using appropriate feedback and operating from a low-voltage source actually simulates the current "regulation" of the high-voltage, high resistance circuit, but without its danger to the operator and to sensitive components and devices, such as mosfets.

Figure 9-11 shows this concept and several constant current circuits. These are basic circuits. With good engineering implementation and some additional sophistications, it is possible to attain extremely tight consistency of the delivered current even though the dc operating source is only moderately well voltage-regulated. As an example, suppose that such a constant current source delivers 2 milliamperes. Then, a 1000 ohm resistance undergoing measurement will provide 2.000 volts to the DMM. Ohm's law works in our favor here with regard to linearity—a five-hundred ohm resistance will, accordingly, produce a 1.000 volt readout. Of course, the calibration unit for the readout is now ohms. This linear relationship enables the basic DMM to be used "as is." Insofar as it is concerned, it is still measuring dc volts in the manner in which it was designed to do so.

It can be seen that a very wide range of resistance measurements can be handled by this technique. It is only necessary to change the value of constant current for the different resistance ranges. Very high resistances, for example, need only micro, or nano-amperes.

Very High Resistance

With modern low-leakage op-amps and precise current sources for exceedingly low currents, it is feasible to measure the very high resistances involved in the insulating material of circuit boards, the dielectric leakage in small capacitors, and the PN junctions of semiconductors. Although such measurements have long been accomplished in the laboratory, considerable skill, time, and preparation have generally been required. The modern DMM intended for the mundane world of testing

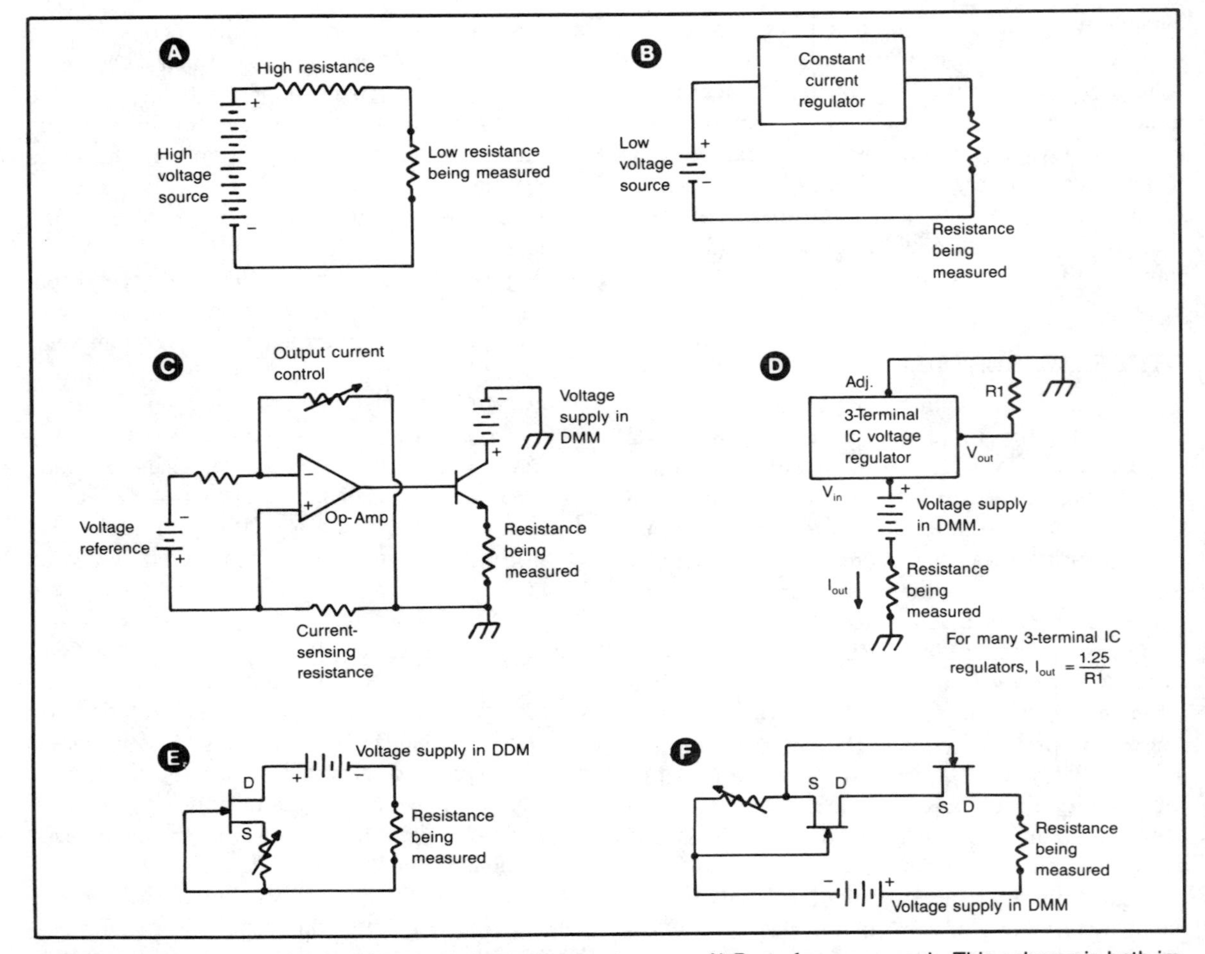

Fig. 9-11. Basic concepts for the ohmmeter constant-current source. A) Brute-force approach. This scheme is both impractical and dangerous. B) The use of a constant-current regulator allows operation from low voltage. C) A constant-current source comprised of an op-amp and a transistor. D) Three-terminal IC voltage-regulators can be used as constant-current sources. E) A junction FET provides constant current. No voltage reference is needed. F) Tighter current regulation is provided by two-FET circuit.

and everyday measuring can provide this capability with relative ease.

When measuring very high resistances, it has been found to be more convenient to display the result in reciprocal-ohms, that is, in units of *conductance*. The basic unit of conductance is the *mho* or *siemen*. (These designate the same unit; in late years, the siemen has become more popular and appears to be used more and more in the technical literature.) In any event, conductance and resistance are reciprocally related to one another. Thus ten ohms is equivalent to one-tenth siemen, one megohm is a micro siemen, etc. An interesting thing about such a reciprocal relationship is that its "natural" non-linearity can be straightened out by plotting resistance and conductance on logarithmic scales. This is demonstrated in Fig. 9-12. Another advantage conferred by the log-log plot is the easy accommodation of wide ranges in the two quantities. As will be seen in the smaller graph of Fig. 9-12, a reciprocal relationship plotted in the ordinary way is very restricted in the useful range which can be depicted. Over a tenfold range, the curve becomes cramped at both extremes. (This, incidentally, is the nature of the cramping on the ohmmeter scales of VOMs)

DMMs which function this way for high-resistance measurements are made by *Fluke*. On such a meter, the operator might read 10.00 nanosiemens and this would correspond to 100 megohms. As conductance indications become more commonplace, it will be less compelling to translate conductivity back into those "old familiar" resistance units. We will tend to intuitively evaluate the "goodness" of a one nano-siemen insulation path for a specific purpose. As far as the internal design of such a DMM, the reciprocal function can be manipulated very accurately, just as in a hand-held calculator.

High and Low Ohm Modes

Somewhat confusing at first is the "high" and "low" ohm selectable options on some DMMs. Advertising literature often describes these options to be provisions for high- and low-power ohms testing.

Notwithstanding these confusing references, such instruments possess an inordinately useful feature—the ability to test resistors, capacitors, and inductors in a circuit board without turning on the pn junctions in semiconductor devices. Thus, in one DMM, the low-ohm option results in only 0.2 volt across the test prods. This is sufficiently low to prevent forward biasing of silicon diodes, transistors, and the devices within monolithic ICs. Unless such a low voltage is used, it is often necessary to remove components from the PC board in order to subject them to a meaningful test. Not only does such procedure invoke a good deal of effort, but the heat from the soldering iron or tool, or leakage of power-line current through them, sometimes damages sensitive devices which were originally good.

A question may be raised here with regard to germanium devices. Admittedly, 0.2 volts is marginal—under some conditions a germanium device may conduct sufficiently with this voltage to

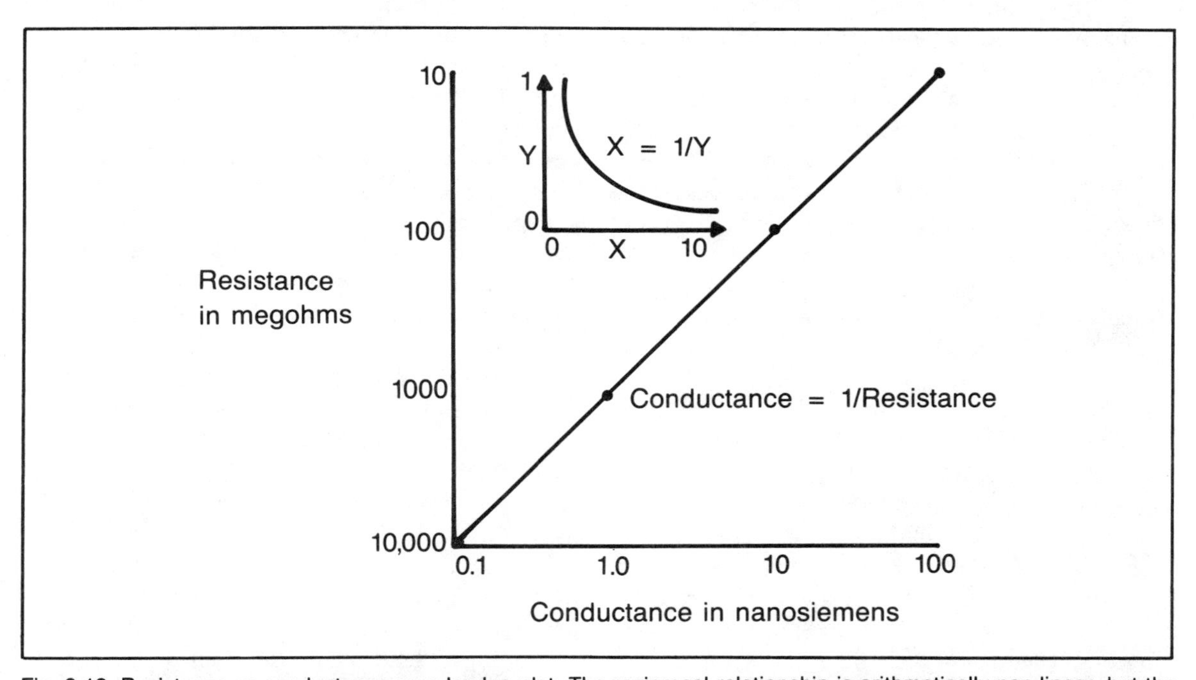

Fig. 9-12. Resistance vs conductance on a log-log plot. The reciprocal relationship is arithmetically non-linear, but the use of logarithmic scales straightens out the curve. This makes it more convenient to transform from one quantity to the other. The small plot shows the shortcomings of the reciprocal relationship when graphed to arithmetical scales.

becloud PC board measurements. This is particularly so if the germanium device is at a high ambient, or at a greater-than ambient temperature. In many instances, however, the germanium device will sufficiently simulate an open circuit. Of course, the best answer is that not many germanium devices are encountered these days. It would be well to keep in mind that germanium point-contact diodes, such as the IN34 and IN56, begin to conduct in the forward direction at practically zero volts.

In the "high" mode, resistance measurements are made with 2.0 volts across the test prods. This is sufficient to bias pn junctions into their forward region. Because of the constant current feature, such tests are excellent for comparing devices. For example, in differential amplifiers and similar circuits, it is possible to select closely matched transistors via such resistance tests. Of course, one might also be interested in matching other transistor parameters, such as beta, and reverse-current leakage.

Although the 2.0 volt level suffices for checking and measuring forward conductance in PN junctions, it is low enough to preclude the possibility of projecting semiconductor devices into their Zener or avalanche regions. Tunnel and back-diodes behave differently from most other junction devices. The constant current feature of the ohmmeter function enhances the meaningfulness of comparative tests, however. One can evaluate these devices by comparing their resistance measurements with those of devices known to be good.

The Current-Measuring Modes of DMMs

Digital multimeters with current-measuring modes do not make use of the "physical" shunts commonly used with VOMs. Such shunts are bulky, expensive, and in higher ampere ranges may be an objectional source of heat dissipation. And, in many cases, they are sources of measurement error. This is because their resistance, though low, is not low enough to prevent disturbance of the circuit in which the current is being measured. Such disturbance is not of importance in VOM measurements because it tends to be small compared to the summation of all the inaccuracies and uncertainties inherent in analog instruments. The opposite is true in the DMM where the digital readout would clearly indicate the reduced current in the circuit because of the resistance in the shunt. Although one could argue that this source of measurement error would not negate the features of the DMM, it turns out that it is relatively easy to provide an electronic shunt characterized by low-cost and high precision.

The basis for the electronic shunt is the operational amplifier in a very simple circuit arrangement. This is shown in Fig. 9-13. Note that a shunt resistance is actually used, but it is lower in resistance than the "equivalent" shunt in a VOM. (Shunts intended for the same, or nearly the same, current range can be considered equivalent). The reason, of course, that the shunt in the electronic arrangement can be a relatively low resistance is that the op-amp provides amplification of the voltage drop developed across it. The point to remember is that this amplification level can be made extremely precise and stable—this is readily accomplished by the two resistors in the feedback network (R2 and R3) of the op-amp. Because of the high open-circuit gain of the op-amp, the gain "thrown away" in the feedback network actually stabalizes the realizable gain with feedback. For the intended purpose of the electronic shunt, the realizable gain is ample to produce the required digital readout of the measured current.

The two diodes, D1 and D2, are for protection. For example, they would absorb the punishment if the current was much higher than anticipated, or if the probes inadvertently got across the voltage source. During normal measurement procedures, the voltage drop across the shunt resistor is too low to forward-bias these diodes—they remain effectively out of the circuit.

For measuring ac current, the electronic shunt would be followed by an ac to dc converter. Thus, for either dc or ac measurement of current, there is minimal disturbance to the circuit under measurement.

Some DMMs feature the added refinement of

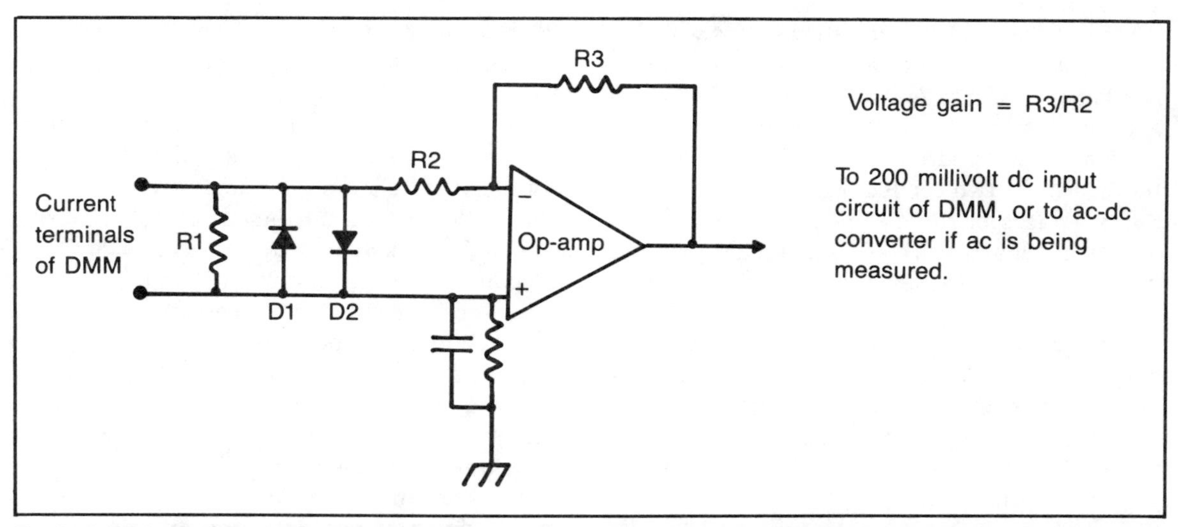

Fig. 9-13. Basic arrangement for current-measuring mode of DMMs. The voltage drop developed across shunt resistance R1 is boosted by the gain-stabilized op-amp. The scheme is, in essence, an electronic shunt.

a clamp-on current probe in which the measured circuit need not be physically interrupted. Here again, the stabilized gain of op-amps is used to process the induced or monitored voltage so that it is suitable for presentation to the dc voltage-measuring function of the DMM. Inductive-coupled current probes can only pick up ac. Current probes utilizing Hall-effect devices can sample dc. Both types of current probes clamp on to the current-carrying conductor making it unnecessary to sever any connections.

EXTENDING THE CURRENT RANGES OF THE DMM

Many still perceive the DMM as a measuring instrument suitable for the voltages and currents that actuate typical small-signal devices and ICs, for logic levels, for general communications and audio work, and the like. However, by means of suitable add-ons and probes, the DMM is able to greatly extend its measuring ranges. High voltage probes enable the accurate measurement of the 30 kV used with color TV tubes in the same manner as has long been done with analog instruments. Lesser known is the availability of current-sampling devices which permit measurement of hundreds of amperes in motor circuits and in other industrial applications.

There are two types of extended current sensors. For measuring dc or ac, a Hall-effect device is used. For ac measurements, one principle employed is electromagnetic induction, i.e., the device is actually a current-transformer in which the current-carrying conductor constitutes the "primary" winding of the transformer. These current-sensing devices are made with split yokes so they may be clamped over the conductor. Thus, there is no need to sever the connection as is necessary in conventional current measurements. Figure 9-14 shows such a clamp-on "ammeter." It delivers its output to the 200 millivolt range of any DMM. (It is to be noted that most DMMs have a 200 millivolt range by virtue of the 2 1/2, 3 1/2, 4 1/2, etc. readout format which is so popular.)

The clamp-on ammeter of Fig. 9-14 has a switch-selectable option of two current ranges—0-20 amperes, and 0-200 amperes. It is generally used in 50 Hz and 60 Hz power systems, but will probably work well at 400 Hz also. When using this device, be sure to clamp on to a single current-carrying conductor. If more than one are embraced, the sensed current will be the algebraic sum of currents, taking into account their phases.

The most practical manifestation of this would be the result of embracing the two wires involved in feeding an ac circuit or load. This result would be an indication on the DMM of zero current even though tens or hundreds of amperes were actually flowing in the two conductors. Also, be sure that the pole-faces of the clamp-on yoke are free of foreign material which could alter the magnetic circuit and thereby cause an erroneous sampling of the current.

The clamp-on ammeter is not an afterthought of the instrument designer, but represents the most practical way of measuring tens and hundreds of amperes—it would not be feasable to endow the DMM itself with this capability. Most meters have 1, 2, or 10 ampere ranges.

PRELUDE TO DISCUSSIONS OF RMS MEASUREMENTS

In the ensuing discussions on the measurement of RMS values of voltage and current, there are several guideposts for facilitating navigation in somewhat turbulent waters. It should help resolve confusion by keeping in mind the following points:

□ There are two basic instrumentation techniques for measuring RMS values: One makes sure of input circuitry which actually *responds* to the RMS value of a waveform. The other technique, appropriately called "average-responding" senses the average value of a measured wave, but the readout is calibrated to provide the corresponding RMS value. But, this readout calibration or scaling factor is valid only for sine waves.

□ Whether the instrument is RMS or average-responding, it may employ either ac or dc (direct) coupling. Ac coupling generally implies there is a capacitor somewhere in the path of the measured signal. With ac coupling, the meter is not responsive to any dc component or dc level which may be involved in the measured signal. Even if the manufacturer claims "true" RMS capability for the measuring instrument, it does not indicate true RMS value when the measured ac waveform contains a dc component or is accompanied by a dc level.

□ RMS value is the common link among thousands of waveforms inasmuch as it is the measure of the very basic phenomenon of heating power. No matter how divergent waveforms may be, if they have the same RMS values, their ability to produce heat energy in a given resistance is the same. One of the sure signs of progress in digital multimeters is their ever-improving capability of measuring RMS values of voltage and current over extended frequency and dynamic ranges.

□ Ideally, the best instrument would be a

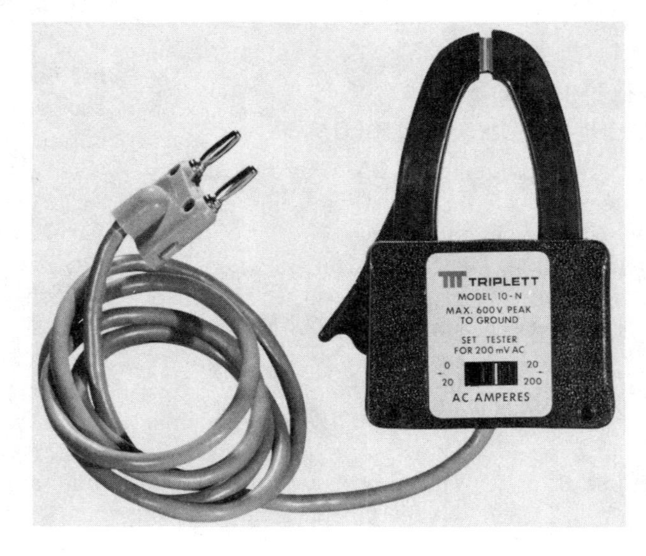

Fig. 9-14. A clamp-on ammeter for extending the ac range of current measurement on the DMM. Courtesy of the Triplett Corporation.

WAVEFORM	CORRECTION FACTOR MULTIPLY READING BY:
Sine	1.000
Full-wave rectified sine	1.000
Half-wave rectified sine	1.414
50% Duty-cycle symmetrical square wave	0.900
Rectified 50% duty cycle square wave	1.272
Rectangular pulse (X, Y)	$\frac{0.9}{D}$ where D = X/Y
Triangular	1.038

Table 9-1. Extension of the Use of Average-Responding Meters for Non-Sinusoidal Waves. Average-Responding Meters Indicate RMS Value Correctly Only for Sine Waves. The RMS Value of Other Waveforms May Be Obtained by Multiplying the Readout by an Appropriate Correction Factor.

"true" RMS type with dc coupling. However, such a meter tends to be expensive. There are performance trade-offs too. Accordingly, it may be more realistic to use one of the other types even if it requires the application of correction factors, or involves the need of more than a single measurement, or makes it necessary to perform mathematical computations.

EXTENDING THE RMS MEASURING CAPABILITY OF AVERAGE-RESPONDING METERS

It is to be emphasized that the manufacturers of electronic meters acknowledge the "built-in" susceptability to error of averaging meters when measuring non-sinusoidal waves. This has little to do with the quality of the meter, but is predicted on mathematical principles. It happens that cost considerations favor the averaging meter; it, moreover provides acceptable accuracy if its waveform limitations are recognized. Not only is this meter quite accurate for sine waves, but correction factors are readily employed to remedy the error when reading RMS values of triangular waves, square waves, and other waveforms. Because the averaging meter can thus be used to yield RMS values of at least three of the frequently-encountered waveforms, many users feel this type of meter can suffice for their anticipated measurement requirements. See Table 9-1. For best results, however, the factory-specified bandwidth of the averaging meter should be ten or more times the repetition rate of the square-wave, triangular-wave or other nonsinusoid which will be converted by computation to "true" RMS value. (This same stipulation apples to a true-RMS meter, with the exception of course, that no correction factor need be applied to the indicated RMS value.)

Some insight into the divergent readings displayed by averaging meters subjected to nonsinusoidal waveforms is shown in Fig. 9-15. From what has been said, it is obvious that this divergence has nothing to do with the digital readout of the meter; indeed, VOMs and FET meters are subject to exactly the same error if they are average-responding types, as is often the case.

AC RMS MEASUREMENTS

A major difference among DMMs involves their ac measurement capabilities. This is a very important aspect in the evaluation of these instruments, yet is often confused by the practice of certain makers to claim RMS measurement capability for their DMMs. Such a claim may be a half-truth in the sense that the RMS value will be displayed only for a sine wave. And generally the sine wave has to be very pure in order that the basic digital accuracy of the instrument can be realized. Such capability is useful. It is, however, a limited one considering the proliferation of nonsinusoidal waveshapes in modern technology. For example, ac measurement involving switching power supplies, motor currents, and thyristor phase-control would be in serious error if the wrong

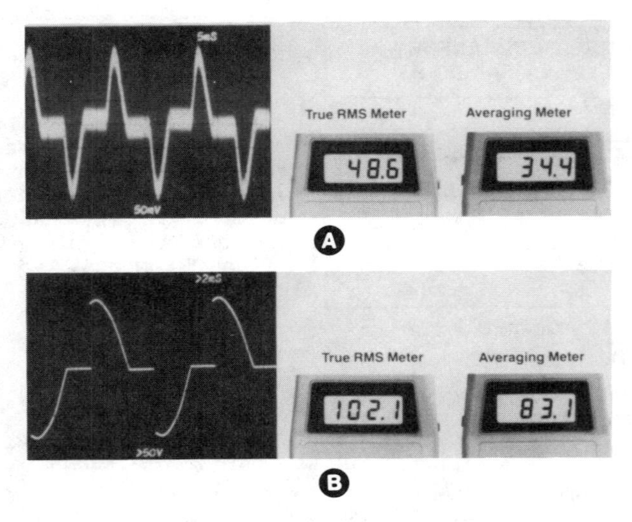

Fig. 9-15. Examples of errors from measuring non-sinusoidal waveforms with averaging metes. A) Motor current. B) Load voltage in Triac Circuit. In both cases, the correct reading is provided by the true-RMS meter. Courtesy of the John Fluke Mfg. Co., Inc.

type of DMM were used. In a sense, this situation is old hat, having been inherited from techniques used in VOMs, VTVMs, and FET meters. Somehow, however, the digital display of the DMM suggests measurement accuracy that may not actually be attainable.

Some manufacturers of DMMs have circumvented this performance restriction by designing "true" RMS capability into their instruments. This, indeed, is the proper approach to ac measurements. For, with "true" RMS capability, one obtains the measurement of the effective heating ability of any waveform. Square waves, triangular waves, SCR waveforms, and all others are accurately indicated in their RMS values. But even here, one must be aware that some "true" RMS instruments require two measurements if the ac wave is superimposed upon a dc level, or if the waveform has a dc component. See Fig. 9-16. One such measurement must be made in the dc mode; the other is then made in the "true" RMS ac mode. Finally, the operator must make the computation: RMS value = $\sqrt{(\text{dc value})^2 + (\text{ac value})^2}$

Obviously, it is much more convenient to use a DMM capable of measuring "true" RMS whether or not a dc level is involved. Such instruments are more expensive than the others, but it is probable that their price will decline to the extent that they will become commonplace. In the meantime, you must know whether your "true" RMS meter has an ac- or dc-coupled input circuit. In the former, and most usual case, dc levels in the measured waveform are rejected.

The less-versatile DMM's accomplish their ac measurement capability by first rectifying the incoming ac wave. Thus, such instruments respond to the average value of the wave, even though their display indicates the RMS value. This is easy enough to do for sine waves because of the known relationships between average and RMS levels. (RMS = 1.11 × average). Unfortunately, this ratio varies widely with non-sinusoidal waveforms and the instrument then displays an erroneous RMS reading. Some DMM's use an ac to dc converter circuit instead of the simpler rectifying formats. This technique takes care of the inherent nonlinearity in junction rectifier diodes, but does not overcome the limitation of being average-responding. Accordingly, accuracy is gained for sine wave measurements, but not for other waveshapes. Moreover, most average-responding meters also have ac-coupled inputs, thereby using unresponsive to any dc level in the waveform.

The Nature of RMS-Responsive Instruments

Examples of the two common types of RMS converters used in digital multimeters are shown in Figs. 9-17 and 9-18. In Fig. 9-17 matched thermopile units are used in a balancing circuit in such

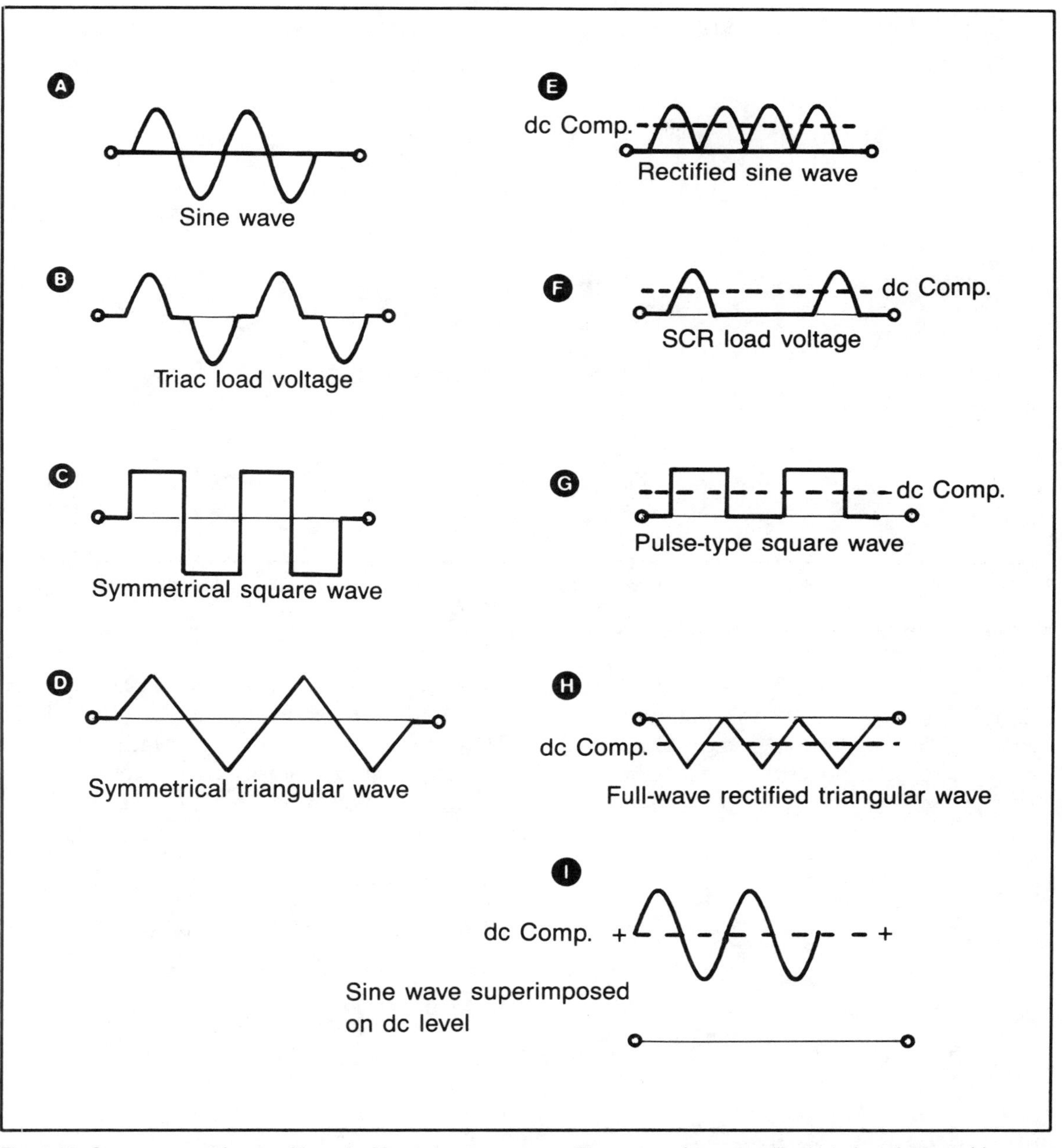

Fig. 9-16. Common waveforms with and without dc components. These waveforms are illustrated as they would appear on a dc oscilloscope. Waveforms (A), (B), (C), and (D) have no dc component. They may be accurately measured on a "true" RMS digital multimeter with ac-coupled or dc-coupled input. (Waveform (A) may be accurately measured on an average-responding meter with either ac or dc coupling.) Waveforms (E), (F), (G), (H), and (I) can be accurately measured on a "true" RMS instrument set at its ac measurement mode if the input circuit is dc coupled. If, the "true" RMS meter is ac coupled, an additional measurement must be made in the dc mode so that the actual RMS value can be computed. (Average-responding meters can provide RMS measurements for some of the waveforms by using the correction factors of Table (9-1).

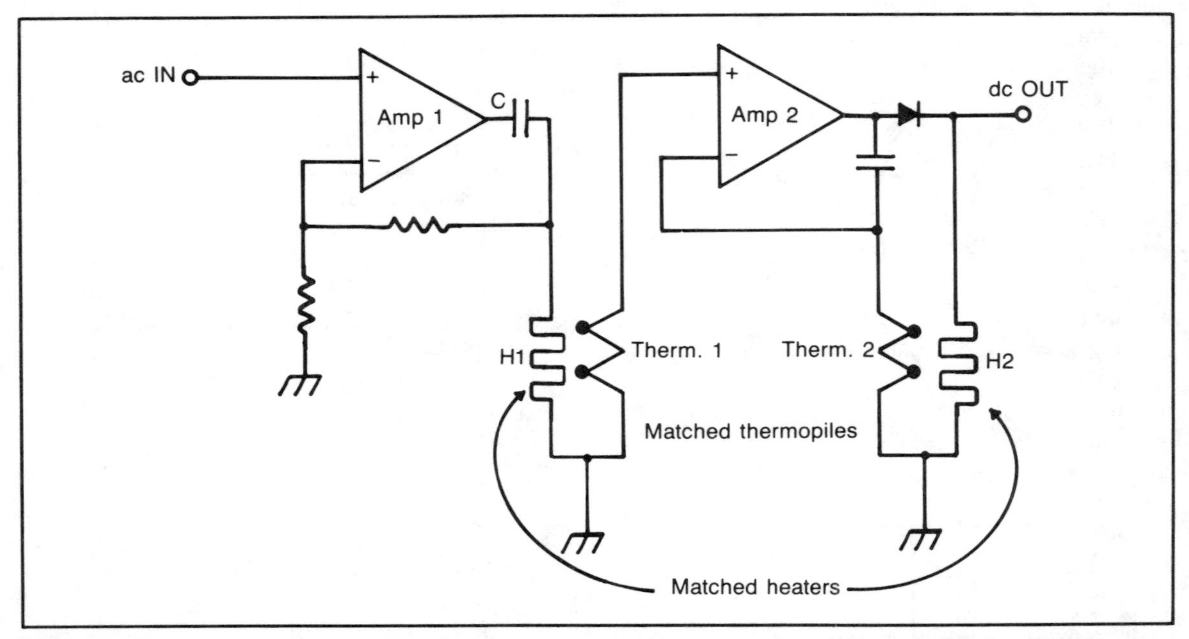

Fig. 9-17. RMS converter using thermopiles in a balancing circuit. Because of the presence of capacitor C, this arrangement is ac-coupled. It will, therefore, deliver a dc voltage output which is proportional to the ac RMS voltage developed across H1, but will not be responsive to dc components if present in the input signal.

a way that the dc voltage delivered from the system is proportional to the RMS value of the input ac waveform. This is a direct implementation from the very nature of RMS value which is based upon the heating capability of an ac waveform. Stated another way, any ac waveform has an electrical heating power corresponding to a certain dc level. Thus, the dc output of this arrangement represents the heat-producing power of the waveform being measured. Note that input-amplifier-1 delivers its

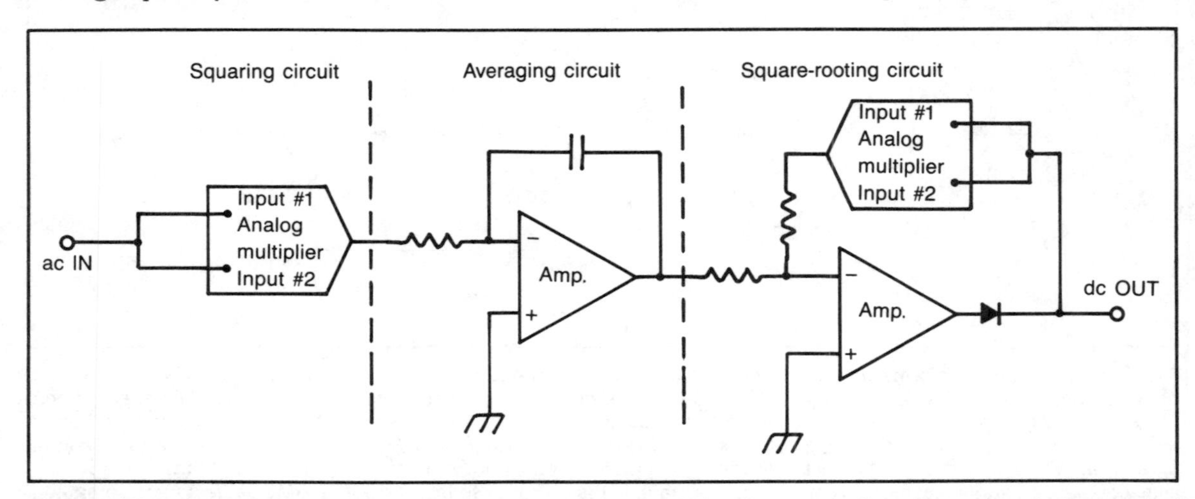

Fig. 9-18. RMS converter based upon analog operational technique. This scheme actually performs the calculation based upon the principle that the RMS value of a function is the square-root of the average-squared value of the function.

output into the heating element of thermopile #1. The small dc voltage thereby generated drives amplifier-2 which feeds into the heating element of thermopile #2. The dc voltage developed in thermopile #2 drives amplifier-2 in the opposite direction from the drive received from thermopile #1. This produces a state of balance in which the two inputs to amplifier-2 have the same magnitude. Under this condition, the dc RMS voltage impressed across the heating element of thermopile #2 is equal to the ac RMS voltage applied to the heating element of thermopile #1. Of course, the designer must take into account other factors such as the voltage gain of amplifier #1 and discrepancies due to non-ideal characteristics of the op-amps. However, this approach is capable of excellent accuracy.

A couple of disadvantages of the thermocouple or thermopile technique are not too hard to identify. The method is inherently expensive. This stems from the need for very closely-matched thermopile units, and also from the demands made on the input amplifier which must provide a rather high power-output with minimal sacrifice of bandwidth. Another difficulty has to do with thermal response time. One to several seconds is often required for the thermopile units to stabilize and in some cases this can exceed 10 seconds. This, at once puts a low-frequency limit on the measured ac waveform. For, unless the thermal time-constant is much greater than the period of the measured wave, the output will tend to represent the average rather than the RMS value.

The circuit scheme of Fig. 9-17 is one of two ways in which thermopile units, (or other temperature dependent devices, such as matched PN junctions) are commonly implemented. In the other approach, the heater element of the device corresponding to therm. 2 in Fig. 9-17 is operated from a fixed current source. Both thermopile voltages are applied to the inputs of an op-amp, but the output of the op-amp now actuates circuitry which varies the gain of the signal input amplifier. Again, a balanced condition results with equal-magnitude voltages delivered by the thermopile units. Thus, the gain of the input amplifier represents equal temperature rise in the two heating elements and this information is communicated to the readout. This method of RMS conversion is usually ac-coupled, whereas the previously described one is readily made with dc capability.

The Operational RMS Converter

Figure 9-18 shows the second common approach to RMS conversion. Here again, a dc voltage is delivered which is proportional to the RMS value of the input ac waveform. The conversion process is quite different from the first method, however. Actually, the scheme amounts to a dedicated calculator which solves the mathematical description of RMS value. Expressed in words, the RMS value for any waveforms is the square-root of the mean (or average) squared amplitude during a specified time interval. For repetitive waves, the time interval is one cycle. Symbolically, this relationship can be written, $RMS = \sqrt{\overline{e(t)^2}}$. Here, e(t) represents the ac waveform. That is, a wave of instantaneous amplitude, e, which is a function of time, t. The horizontal line above $e(t)^2$ denotes the mean or average value. By hand calculation, one would square the amplitudes for a number, say ten, of equi-spaced time intervals, then divide the sum of these squared amplitudes by ten. Depending upon the nature of the waveform, it may or may not be easy to perform this calculation. Mathematically, the best way to find this mean or average is by calculus. We will not concern ourselves with the detailed mathematics because, as it will be shown, our electronic RMS converter will readily do the calculation for us. Suffice it to point out that RMS is actually the acronym for "root-mean-square", which implies the square root of the average squared-amplitude of the waveform considered over one cycle of repetitive events. (Bear in mind, too, that the RMS value of an ac waveform corresponds to that dc level which produces the same heating effect in a resistance.)

Looking at Fig. 9-18, it can be seen that three identifiable circuit sections perform the operations alluded to. Thus, the input section is an analog multiplier with its two input terminal tied together, enabling this module to square the amplitude of the ac input signal. The second section is an op-amp connected as an integrator. Such a circuit has the property of averaging a varying input signal. In this respect, its behavior is like that of a low-pass filter. Finally, the third section performs the square root operation. Circuitwise, it comprises an op-amp with an analog multiplier in its feedback path. It is evident that the overall arrangement processes the ac input signal in such a way as to satisfy the mathematical definition of the RMS value.

Besides its mathematical elegance, this so-called operational method tends to have several advantages over the thermal method. It is less costly to produce. Its response time is much faster than that of the thermally-based RMS converter. It is easier to use dc coupling in the operational converter than in the thermal RMS converter and calibration tends to be a more straightforward process.

CORRECTING FOR ELECTRICAL NOISE

A common source of error in measuring RMS voltages in communications systems is the simultaneous presence of wideband noise. This noise is variously described as Johnson noise, thermal noise, semiconductor noise (except for the "flicker" variety), or Gaussian noise. There is generally not much to be gained by splitting hairs over these definitions inasmuch as none of them are encountered in pure form in practical applications. A fairly good correction for the presence of any of these noise voltages can be made with a wideband true RMS meter. The basic idea is to measure the RMS value of the desired signal together with the noise voltage, and then make a second measurement of the RMS value of the noise voltage alone. This is usually easy to do in communications equipment by simply disabling an oscillator, a driver, or a coupling provision. These two measurements enable the following relationship to be computed:

$$V_{signal} = \sqrt{V^2_{total} - V^2_{noise}}$$

Although theory requires the consideration of both infinite bandwidth and infinite crest factor in accounting for the contribution of noise voltage, acceptable results are often attained for bandwidths of several-hundred kHz to several megaHertz, and for crest factors of four or so. Such limitations on the parameters of noise voltage will invariably be imposed by the very circuitry undergoing measurement. We can live with this providing the measurement meter doesn't introduce too much further compromise. The application of this correction technique can be quite important when dealing with low-level signals in the presence of noise. It tends to be safer to use a meter with a dc-coupled RMS converter for noise evaluations. Otherwise, significant low-frequency contribution of the noise spectrum might be lost to the measurement process.

Flicker-type noise is present in both semiconductor and thermionic devices. It has the unique characteristic that its amplitude grows continuously greater at lower frequencies. Here, the use of a dc-coupled RMS converter is almost mandatory. Otherwise, the same procedure and relationship described for the "white noise" sources applies. It might prove profitable to apply filtering techniques to discriminate against the white noise background when correcting for the presence of flicker noise. This could conceivably be as simple as a capacitor shunted from the measurement point to ground. Such a filter or bypass capacitor would attenuate the higher-frequency noise energy, leaving the bulk of the energy vested in lower-frequency flicker noise relatively unaffected.

YOUR METER'S PASSBAND

The highest quality true RMS instrument will perform within its accuracy rating only for frequencies within its passband. This is a very important matter and is often overlooked. We become accustomed to viewing ac waveforms on the oscilloscope where a non-sinusoidal waveform appears to be only comprised of what we see occurring at a certain repetitive rate. However, from the Fourier Theorem of Wave Analysis, we know that in addition to the repetition frequency, non-sinusoidal waves contain higher multiple of har-

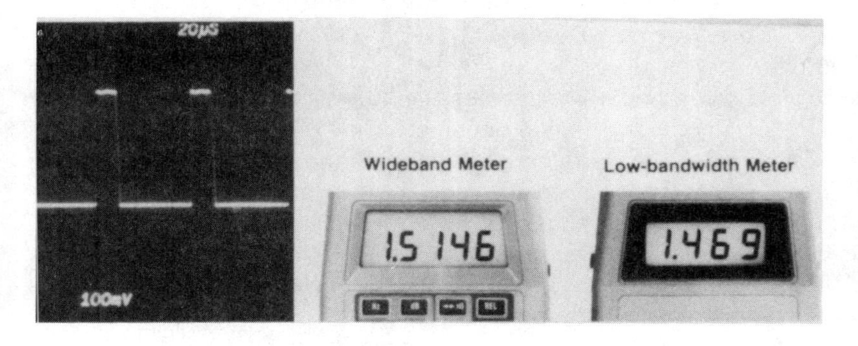

Fig. 9-19. Example of measurement error from restricted bandwidth meter. Although the repetition rate of the waveform shown above may be well within the response of the low-bandwidth meter, a significant portion of the harmonic energy lies outside this instrument's passband. Courtesy of the John Fluke Mfg. Co., Inc.

monics of the fundamental frequency or repetition rate. Indeed, a considerable portion of the energy often resides in numerous higher harmonics, thus spreading the energy over a wide frequency spectrum. This being true, a DMM with a specified bandwidth of, say 15 kHz, cannot accurately measure a non-sinusoidal wave with a repetition rate of nine or ten kHz. Even though the fundamental frequency (its repetitive rate) of the measured signal falls well within the instrument's passband, this is not true of the harmonics, which may extend into the hundreds of kHz.

Figure 9-19 illustrates how the reading provided by a narrow-band true RMS meter tends to be less "true" than the measurement made by the wideband instrument. We know by inspection that the narrow rectangular waveform is rich in harmonic content. Let's assume, in order to keep matters simple, that we are only interested in the RMS value of the ac, not the dc component of the waveform. This precludes discussions of the dc component, which is made elsewhere. It obviously behooves us to know the capabilities of our meter, and to be able to reasonably estimate the ac parameters of the signal undergoing measurement. Sometimes this means that an oscilloscope and/or a frequency counter are often valuable adjuncts to the digital multimeter. This is one reason some instrument makers feel the frequency-measuring function is very useful to the DMM user. In any event, as a rule of thumb, the usable passband of the DMM should be at least ten times the repetition rate of any non-sinusoidal waveform undergoing measurement. Of course, some non-sinusoids are richer in higher harmonic content than others, so some discretion must be used.

The above discussion pertains to true RMS meters. It applies equally well to averaging meters if such an instrument is being used to measure certain non-sinusoidal waveforms. Otherwise, the correction factors for the waves will not be valid. If, on the other hand, the averaging meter is being used to measure a sine wave, rated meter accuracy will be attainable as long as the frequency of the measured sine wave falls within the passband of the meter. This, of course, is a relatively easy condition to meet, or at least to assess.

ANOTHER PITFALL

We have covered the three main methods of RMS conversion in digital multimeters. In the average-responding type, an internal scale factor of 1.11 enables the readout to indicate the RMS value of sine waves. In the thermal method, the actual heating effect of the ac wave is monitored. In the operational method, the mathematical expression for RMS value is solved by means of analog circuits. It was also pointed out that the average-responding meter can be used for certain non-sinusoidal waveforms by applying a correction factor to the readout. For all three methods of RMS conversion, the instrument bandwidth should be on the order of ten times the frequency of the non-sinusoidal waveform being measured in order to accurately respond to the harmonic content of such waves.

It turns out that there is yet another consideration appropriate to all three types of meters. It has to do with the crest-factor of the waveform. For a given RMS value, a waveform can resemble a square wave, having a crest factor of one, or can consist of repetitive narrow spikes with a very high crest factor, say ten or even one hundred. (*Crest factor* is simply the ratio of peak to RMS value. It is unity for a square wave and the familiar 1.41 for a sine wave.) What is involved here is the instrument's ability to accommodate the waveform undergoing measurement without saturation of its input amplifier or buffer. It boils down to a matter of dynamic response and is essentially similar to the problem of amplifying certain music waveforms in audio equipment. Simply put, if the crest factor of a waveform exceeds the linear amplifying capability of the digital multimeter, the readout will be erroneous—usually grossly so. Figure 9-20 provides insight into the nature of crest-factor for different waves. In symmetrical square-waves, crest-factor, RMS, and average values are unity.

Many meters come with a crest-factor maximum in their specifications. This implies the highest crest-factor which can be handled without impairment of accuracy. This is not always simple because there may also be frequency and scale dependency. In any event, it is important to know what your particular instrument will do under actual measurement conditions. Thus far no instrument has been seen by the author which provides a warning of excessively high crest factor in an ac waveform. It would seem, however, that such an indication would be feasable and useful.

DEFINING THE DIGITAL MULTIMETER

Perhaps the most obvious difference among digital multimeters involves the number of digits in the readout. And, initially, the most confusing specification is the use of fractional digits. For example, the most popular instrument for everyday work has four visible digits, but is specified as a 3 1/2 digit meter. Why not just say it has four digits and be done with it?

The significance of the fractional-digit description has to do with the indicating capability of the most significant (the leftmost) digit. Although there are, indeed, four digits in the so-called three and one-half digit meter, the most significant digit can have only two possible values, zero and one. This is the commonplace scheme used. If the most significant digit were free to indicate from zero through nine, the instrument featuring such a readout would be termed a four digit meter. Such meters are uncommon, though technically feasable. See Fig. 9-21.

Another way to look at this situation is to view the 3 1/2 digit meter as one having 100% overange capability. Thus, on a zero-to-one "full-scale" range, it would actually be possible to make measurements up to 1.999 units, or for practical purposes, two, which corresponds to 100% overange. But this can be somewhat ambiguous because many meters have ranges taking full ad-

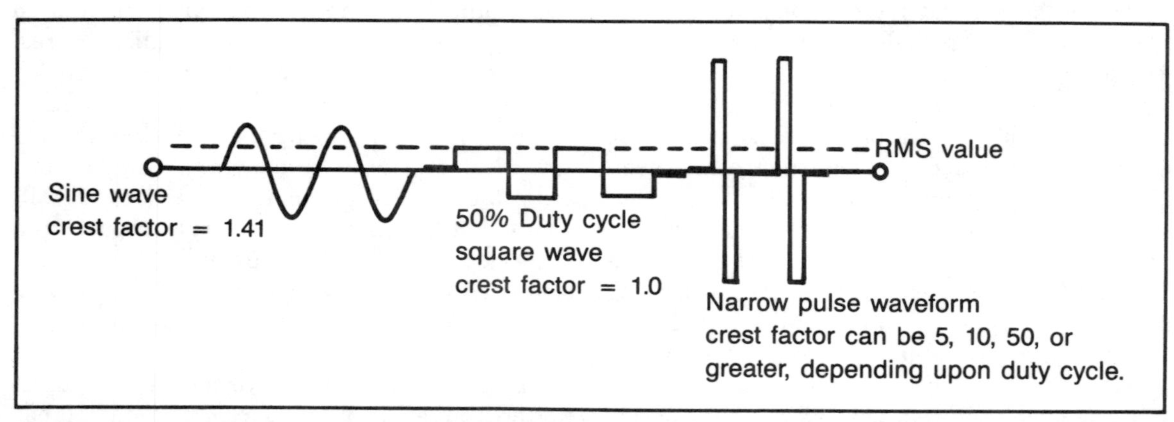

Fig. 9-20. The pitfall of dynamic response in RMS measurements. In principle, the above waveforms have identical RMS values. However, a waveform having high peak values can exceed the input capabilities of a measuring instrument even though it is set to indicate RMS values.

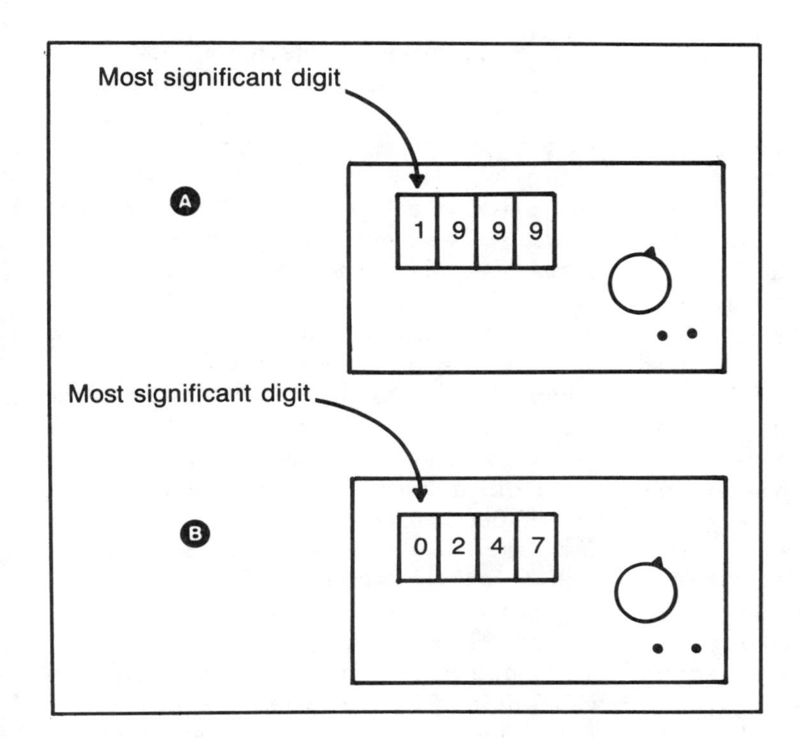

Fig. 9-21. How the fractional digit concept works in a 3 1/2 digit meter. The most significant digit can halve only two values, zero and one. A) Maximum indication. B) Intermediate indication.

vantage of the "over-range" region. Thus, we encounter 0-2, 0-20, etc. ranges.

The same reasoning applies to instruments with other numbers of digits in their readouts. Thus, when we see three digits, we have a two and one-half digit meter. Five digits would similarly be termed a four and one-half digit meter. Four and one-half and five and one-half digit meters can have potential accuracies and resolutions which put them in the class of laboratory instruments. However, to realize the accuracy and resolution capability potentially possible from a multi-digit readout, extreme care has to go into the design so that noise immunity and stability are inordinately good. Other readout schemes are occasionally encountered, but there seems to be less than universal agreement over their designation. For example, a DMM with the ability to indicate to 3999 might be designated as a 3 3/4 digit meter.

In any event, the use of fractional digits always implies some restriction in the freedom of the most-significant digit. In order to simplify our discussions, the digital multimeters dealt with in this book will involve half-digit type limitations of the most significant digit, and will generally be 3 1/2 digit meters. Such meters surpass the best analog instruments in accuracy and resolution and are more than adequate for not only service and test applications, but are suitable for much of the work in design and research as well.

Resolution

From our experience with analog instruments, we know that the smallest increment of change that can be resolved depends upon such factors as the friction of the meter's movement, the fineness and integrity of the printed scale, whether there is a mirror, and upon our powers of observation. In the digital meter, resolution tends to be a matter of how many digits are employed in the readout. Consider the popular 3 1/2 digit format. For practical purposes, the "full-scale" indication of such a readout is 2000. (For the purpose at hand, it does not matter where the decimal point may be). The indication may be read to one part in approximately two-thousand, or to 0.05 percent of the maximum value.

If you are inclined to split hairs, it might be pointed out that the inherent plus-or-minus one digit uncertainty of digital readouts actually makes the resolution less. However, the marketing departments of instrument manufacturers are not predisposed to this kind of hair-splitting!

There are other factors which adversely affect the theoretical resolution of DMMs. The vulnerability to noise causes jitter in the least significant digit(s). This is especially true in multi-digit (greater than 3 1/2) readouts. Indeed, some of the noise interference may be generated in the "front end" of the meter, itself. A notorious offender in ultra-sensitive meters in the 60-Hz power line and its harmonics. It is both, an art and a science to keep the power-line noise out of the meter. One useful technique in meters using ramp or dual-slope converters is to make integrating times equal to one period of 60 Hz. This renders the meter substantially "nblind" to not only 60 Hz, but to its harmonics, as well. All things considered, the 0.005% resolution of a 4 1/2 digit meter may be reasonably approached in a good design. And 0.0005% resolution in a laboratory-type 5 1/2 digit instrument might be a good short-term target to shoot for in a screened room, and under very favorable conditions. This would generally imply battery operation in locality far removed from industrial and manufacturing activities.

A practical conclusion that can be drawn is that the achievable resolution of digital multimeters is, at worst, a dramatic improvement over what we have been accustomed to with analog instruments.

Accuracy

There is a psychological tendency to equate the digital format with accuracy. Although the opportunity for relatively-accurate measurement capability is, undeniably, there, it does not derive exclusively from the readout. First of all, it behooves us to deal with the semantics of such words as "accurate" and "precise" (which are often used interchangeably).

Accuracy pertains to the capability of producing an indication very close to the true absolute value, as would be revealed by comparison with a primary standard. Thus, a measurement of, say, 12.54 volts is accurate to the degree that this indication is close to that value. We see at once that accuracy is, among other things, dependent upon the manufacturer's calibration techniques, the thermal stability of the instrument, the linearity of charge and discharge ramps, the integrity of internal references, the performance quality of comparators and zero-detectors, and upon the instrument's noise-rejection. If all of these factors are of high quality, then greater indicating accuracy acrues with increasing the number of digits in the readout. But, the mere ability to indicate 12.54 volts to a greater number of decimal places does not, in itself, improve the accuracy.

The word "precision" actually has a somewhat different connotation. It indicates the repeatability or reproducibility of a measurement. Thus, a DMM with drift in its internal voltage reference cannot provide precise readings to very many decimal places—making the same measurement a few minutes apart will produce varying indications even though the voltage or current source undergoing measurement has remained constant. Inasmuch as practical instruments do have temperature coefficients and drift factors, precision must be taken into account even though the word may not appear in the performance specifications. Thus, any assumed accuracy must be compatible with the conditions pertaining to the measurement. This involves ambient temperature, warm-up time for stabilization, the amount and kind of rfi and emi in the measurement environment, etc.

Because of time, temperature, and noise factors, a DMM is less than perfectly precise. And because of this lack of perfect precision, accuracy must be stated under qualifying conditions such as fifteen-minutes for warm-up, operation at room-temperature, and under noise conditions limited in the specifications. Accuracies are almost universally expressed for the basic dc measuring mode. Accuracies for other modes generally differ, tending to be less. In particular, ac accuracies for average-responding meters degrade rapidly with the harmonic "polution" of the apparent sine wave. (On a 'scope, it requires three to five percent of harmonic content to be visible.) Accuracies (dc) for 2 1/2 digit meters tend to be between 0.5 and 1.5%. For 3 1/2 digit meters it lies between 0.5 and 0.05%. Finally, a 5 1/2 digit laboratory DMM can measure with accuracy in the vicinity of 0.002%.

The most comprehensive way to express accuracy in a digital multimeter is to specify one of the dc ranges as ± % of indication, ± % of full scale, ± one digit. This would be accompanied by certain allusions to line voltage, ambient temperature, warm-up time, and possibly noise qualifications. However, for most 3 1/2 digit meters intended for the usual mundane measurement purposes, this is usually shortened to ± of full scale on a low dc range, ± one digit. This may pertain to an ambient of 25 °C with data enabling the user to apply the instrument's temperature-coefficient to correct for deviation from the 25 ° C ambient. Conversely, the accuracy specification may embrace a temperature *range*, such as 17 ° C to 35 ° C.

It is interesting to note that a general-purpose VOM may have a dc measuring accuracy of ± 3%; a laboratory-grade VOM would feature a dc measuring accuracy in the vicinity of ± 1 1/2%. Thus, an inexpensive 3 1/2 digit DMM is capable of much greater accuracy than even a costly VOM. Moreover, the readily-realizable accuracy of the DMM is not influenced by parallax, by bearing friction, by meter position, by magnetic fields, or by electrostatic phenomena. Also, a properly packaged DMM is far more tolerant to physical abuse. Although a rugged VOM may survive a drop to the floor, such an event is much more likely to impair accuracy, precision, and calibration than a similar mishap with the DMM. (Of course, such abuse should be carefully avoided with *any* measuring instrument!)

Inasmuch as other measurement modes are realized by adding converter circuits to the basic dc function of the DMM, it is only natural that the overall accuracy will be less than the basic ac accuracy. Thus, the ohmmeter accuracy can be expected to be about half as good as that of the basic dc accuracy. The degradation is brought about by a less than perfect internal constant-current source. Even so, DMM resistance measurements are far more accurate than those ordinarily attainable from VOMs. One is freed from optical parallax, cramped scales, and erratic battery action.

Ac measuring ranges are also less accurate than corresponding dc ranges. In average-responding DMMs, the ac accuracy specification often allows for a maximum harmonic distortion of 1/2% in a sine wave. (Three to five percent distortion is needed for visual detection of an oscilloscope.) A frequency range such as 20 Hz to 30 kHz, will accompany the ac accuracy specification. A true RMS measuring DMM will maintain a specified accuracy over a stated bandwidth, but will have the added stipulation that the measured non-sinusoidal waveform cannot exceed a certain crest-factor. And, as previously mentioned, some "true" RMS meters are truer than others by virtue of their ability to accommodate the dc component or dc level in an ac waveform.

Sensitivity

Sensitivity is a performance parameter that lends itself well to the ploys and game-playing tactics of the marketing department. This stems directly from our natural tendency to view the quality of any instrument in terms of how responsive it is. This attitude seems to encompass the entire gamut of detection and measuring devices. It includes telescopes, microscopes, radar devices, oscilloscopes, etc. We are, indeed, more likely to purchase the DMM with impressive sensitivity claims, then the otherwise-competitive meter unable to boast such sensitivity. This is justifiable in many instances if we are seeking the DMM having the lowest measurement range of the parameter of interest. But, if it is naively assumed that routine measurements can be readily made in the vicinity of the sensitivity rating, good luck!

The culprit is noise. Also, the inherent precision of the meter (its repeatability) may not harmonize with the number of decimal places featured in the readout. The noise interference is often at power-line frequency, or at one of its harmonics. But DMMs are vulnerable to the whole spectrum of rfi and emi. When high sensitivity is specified, the common-mode rejection ratio (CMRR) and the normal-mode rejection ration (NMRR) had better be very high. Otherwise, the "paper" sensitivity of the instrument and the *useable* sensitivity may be far apart. This tends to become more important with multi-digit instruments.

The specified sensitivity is generally derived from consideration of the least significant digit of the lowest range. For example, suppose the full-scale indication for the lowest dc voltage range is 199.9 millivolts. (Practically; this would be designated as a 200 millivolt range.) Each of the least-significant digits represents 0.1 millivolt, or 100 microvolts. Therefore, such a DMM would have a specified sensitivity of 100 microvolts.

Note the difference in the concepts of of sensitivity as applied to VOMs and DMMs. In the VOM, the higher the ohms-per-volt rating, the more sensitive is the meter. This is tantamount to saying that a sensitive VOM consumes less current from the measured source than a less sensitive VOM. But most DMMs have exceedingly-high input resistance—commonly ten megohms; all such instruments draw negligible current from most measurement points. So, the notion of sensitivity is more meaningful if viewed in a different way. Also, sensitivity and resolution are different and are not to be used interchangeably. Sensitivity is generally expressed in appropriate fractions of a volt, ampere, or ohm for the lowest range on the meter, being the, "worth" of the least significant digit on that range. *Resolution* is a percentage given by the reciprocal of the total number of counts provided by the readout. Thus, a 3 1/2 digit readout with a total possible count of 1999 (or 2000 for practical purposes) has a resolution of 1/2000, or 0.05%. Note that a high resolution DMM need not necessarily also have high sensitivity.

The converse is also true—high sensitivity in a DMM does not necessarily imply high resolution. Although the two performance parameters are independent, in practical instruments there is a tendency for the two to go together. With the careful design and high quality components needed to achieve high sensitivity, it makes good marketing sense to also design in high resolution. Sensitivity is governed primarily by the analog processing circuitry, that is, the "front-end" of the DMM. Resolution, on the other hand, is a function of the digital format of the readout. In general, the more digits, the higher the resolution.

NMRR AND CMRR-ACCURACY AND RESOLUTION

It is not wise to muster too much enthusiasm over an instrument maker's boast of a DMM's sensitivity and resolution before noting two specifications relating to measuring performance in the face of noise interference. These are NMRR and CMRR; the first stands for normal-mode rejection ratio, the second abbreviates common-mode rejection ratio. Both are ratios expressed in decibels and denote how much electrical noise can be accommodated alone with the measured quantity without degrading the accuracy of the measurement. Although these ratios usually pertain to the power-line frequency, the general inference is that the greater these ratios, the better is the immunity of DMM across the noise spectrum. This assumption is generally valid even though NMRR and CMRR both tend to decline with increasing noise frequency. These ratios are, in a loose way, similar to signal-to-noise ratio in communications work. There, too, an impressive sensitivity may not be usable if the signal-to-noise ratio of the receiver is low.

Normal-mode rejection ratio can be demonstrated by the simplified setup shown in Fig. 9-22. A reading of 1000 millivolts is indicated in a measurement of the dc source. This measurement usually is full-scale for the lowest dc voltage range,although sometimes the coverage of 1999 might be used. (It doesn't make too much difference if consistent evaluation procedures are employed. For purposes of illustration, this particular DMM can be assumed to have a lowest dc voltage measuring range of 1000 millivolts and no specified over-range. Note that there is also an ac source

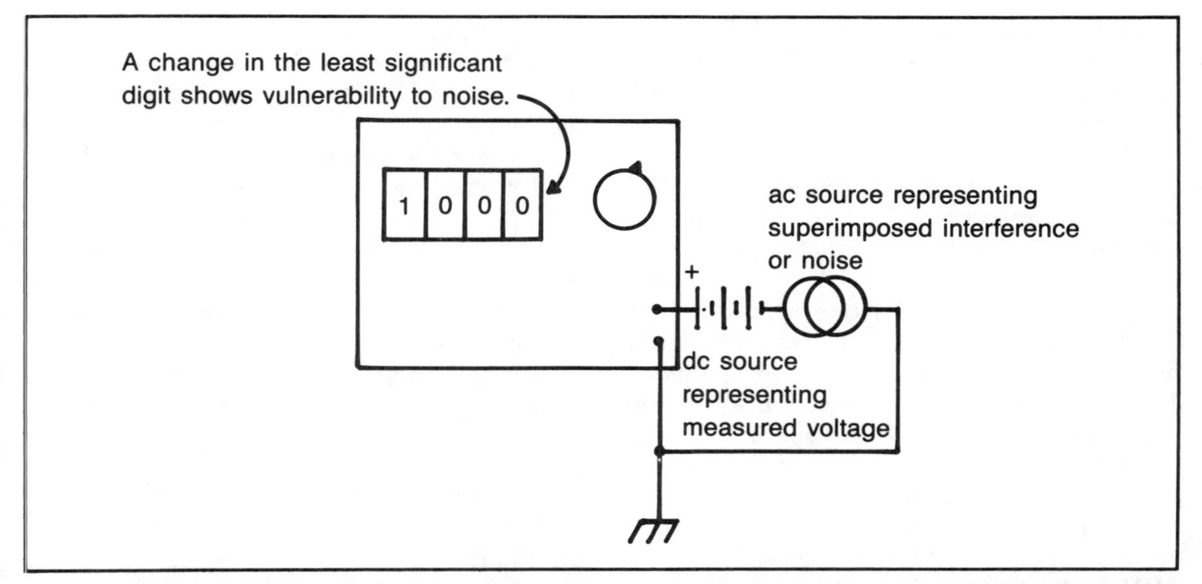

Fig. 9-22. Setup for demonstrating normal mode rejection ratio (NMRR) of a DMM. The basic idea is to increase the amplitude of the ac source until the least significant digit changes. The NMRR can then be calculated from the amplitude of the ac source and from the value of the least significant digit.

connected in series with the dc source. By increasing the amplitude of this ac source up from zero, an interference level will be found that just begins to effect the least-significantdigit. If this interference level is recorded, we have sufficient data to compute and interpret the normal-mode rejection ratio.

The normal-mode rejection ratio is the RMS value of the ac interference divided by the resolution of the meter. As pointed out, this result is usually expressed in dB. In the setup of Fig. 9-22 for example, suppose it requires 2000 millivolts of ac to just disturb the least significant (rightmost) digit. The resolution of the meter on this dc range is 1000 microvolts or one millivolt. That is, each digit displayed by the rightmost number is worth one-millivolt. The numerical ratio for the NMRR is

$$\frac{2000 \times 10^{-3}}{1 \times 10^{-3}} = 2000.$$

Then NMRR is

$$20 \times \log_{10} (2000) = 20 \times 3.3010 = 66 \text{ dB}.$$

The interpretation is that 2000 millivolts of power-line frequency superimposed on the "hot" dc input terminal is the very maximum interference that can be tolerated without degrading accuracy of the 1000 millivolt dc measurement.

Common-mode rejection ratio applies to DMMs in which the two measurement input-terminals are "floating," i.e., neither of them are connected to the chassis or to earth-ground.The basic idea here is to apply the ac interfering signal simultaneously to both input terminals with respect to earth-ground. Figure 9-23 shows how this is done in a test setup. The 1000 ohm resistance represents the internal resistance of a typical dc source. Its inclusion is necessary because the part of the common-mode signal that develops a voltage drop across this resistance actually represents a normal-mode signal. Some instrument maker's specifications take this correction into account, but others fail to do so. One argument is that for high common-mode rejection ratios, say over 100 dB, the correction is not of great importance. In any event, common-mode rejection ratio, or CMRR, is generally made at power-line frequency, as in the case with NMRR. As with NMRR, CMRR tends to degrade with increasing frequency. However, the power-line frequency specification provides a basis for comparison.

The simulated internal resistance of the dc source wasn't needed in the test setup for NMRR because its presence would not materially affect the test results. Otherwise, the basic idea involved in the two tests is identical. In Fig. 9-23, the ac source is advanced from zero until its amplitude is just sufficient to disturb the least-significant digit during full-scale readout of the lowest dc voltage range. Then the CMRR calculation is made exactly in the same manner as for the NMRR specification. Although CMRR will ordinarily be much higher than NMRR, it can still be the limiting factor to usable measurements in certain situations, such as in medical applications. When both input terminals of a DMM are ungrounded, the chassis of the instrument is often, but not always, connected to earth-ground. A little experimentation may be in order here, especially when dealing with common-mode interference of higher than power-line frequency.

When conducting critical measurements in severe noise environments with a DMM with a marginally satisfactory CMRR, it is sometimes profitable to experiment with the connections to the input terminals. The balance to ground of these terminals is not symmetrical and a transposition of the connections may provide the optimum CMRR for the particular measurement situation.

The concepts of NMRR and CMRR are not necessarily restricted to the dc voltage measurement mode. Similar evaluations can be, and sometimes are, made for the ac and ohms modes. The basic idea is retained—to indicate the degree of noise immunity. Here, too, the specification is generally given for the power-line frequency.

EXTENDING MEASUREMENT CAPABILITIES

In the evolution of the digital multimeter, the first available measurement mode was dc voltage. Then followed the other modes, enabling competition with the VOM, i.e., dc current, ac volts and current, and resistance. A continuing development involves the widespread dissemination of RMS capability for the ac measurement modes. Another observable development trend has been the expansion of the meter to include functions in addition

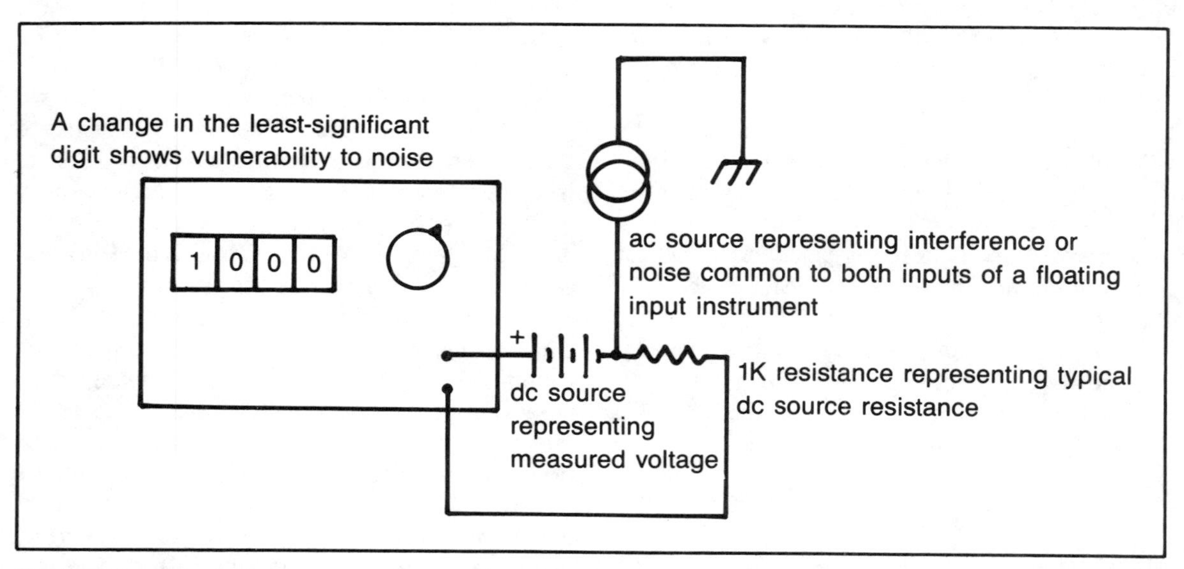

Fig. 9-23. Setup for demonstrating common mode rejection ratio (CMRR) of a DMM. The basic idea is to increase the amplitude of the ac source until the least-significant digit changes. The CMRR can then be calculated from the amplitude of the ac source and from the value of the least-significant digit.

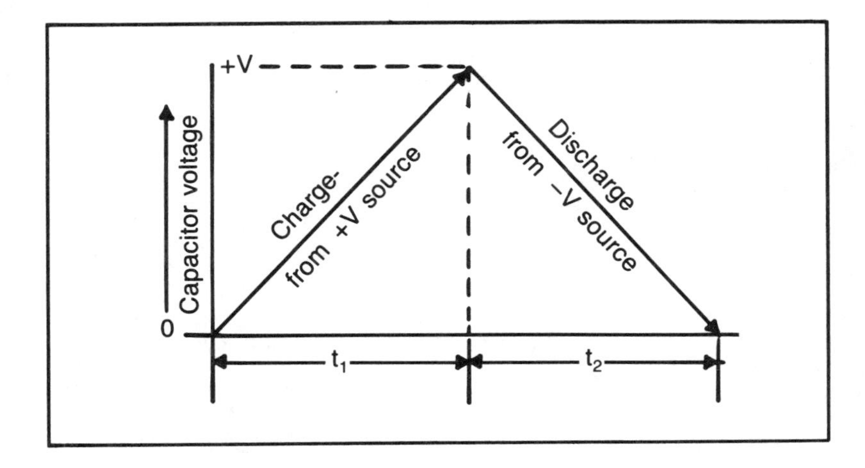

Fig. 9-24. Frequency readout in a DMM makes use of the dual-slope technique. t_1, the charge time, allows the counter to fill up from the internal clock, thereby establishing the full-scale frequency. During time t_2, which is made identical with t_1, the counter is driven by the unknown frequency source and produces the frequency readout at the end of t_2.

to the traditional voltage, current, and resistance measurements. Thus, there are instruments which also include signal and pulse generators, frequency counters, capacity measuring capability, and the ability to function as an electronic thermometer. These additional capabilities may stem from the original design, from probes or add-ons, or from clever ways to utilize the traditional built-in modes. Not only are the rf, peak voltage, high voltage, and current clamp-on units that have been used with VOMs and FET meters applicable to DMMs, but certain extended measurements derive from the unique nature of digital meters.

Consider, for example, digital meters with frequency measuring capability. This should not be too surprising, for most digital meters incorporate clocks and counters in their design architecture. In some, the incoming dc voltage is converted into frequency and the readout is scaled to indicate volts, amps, or resistance. In others, the charge and discharge of capacitors is translated into a number of counts stored in a counter, which again is read out in the desired measurement parameter. With a little ingenuity, a clever designer can make use of the counters already present to provide the instrument with frequency measuring capability.

One way in which this can be accomplished is illustrated in Fig. 9-24. Here, we are looking at the charge-discharge cycle of a dual-slope digital meter. The charge time, t_1, is made equal to the discharge time, t_2 by using equal but opposite polarity reference-voltages. During t_1, the counter fills up from the internal clock. This establishes the full-scale frequency of the measurement process. During t_2, this or another counter, is driven by the frequency undergoing measurement. At the end of t_2, the accumulated count becomes the readout. If it required 10,000 counts to define times t_1 and t_2, then the inflow of, say one fourth of that number of counts during t_2 will produce the readout of 2500 Hz.

Of course, it is not always feasable or wise to attempt modification of the circuitry of a DMM. Another approach to make the DMM function as a frequency counter is by means of an add-on frequency to voltage converter. Such an arrangement using a precision IC is shown in Fig. 9-25. The basic arrangement is intended for operation from pulse signal sources. In order to use it for sine-wave frequencies, a Schmitt trigger, or other squaring circuit should be used ahead of this arrangement. Calibration consists of adjusting the 5K variable resistance to cause a known 10 kHz source to indicate 10 kHz on the DMM.

Inasmuch as frequency measurements are such a valuable adjunct to voltage, current and resistance measurements, and because of the natural adaptation of the DMMs readout to frequency indications, another frequency-to-voltage converter circuit is shown in Fig. 9-26. The scale provisions of this circuit are particularly convenient for use with many DMMs.

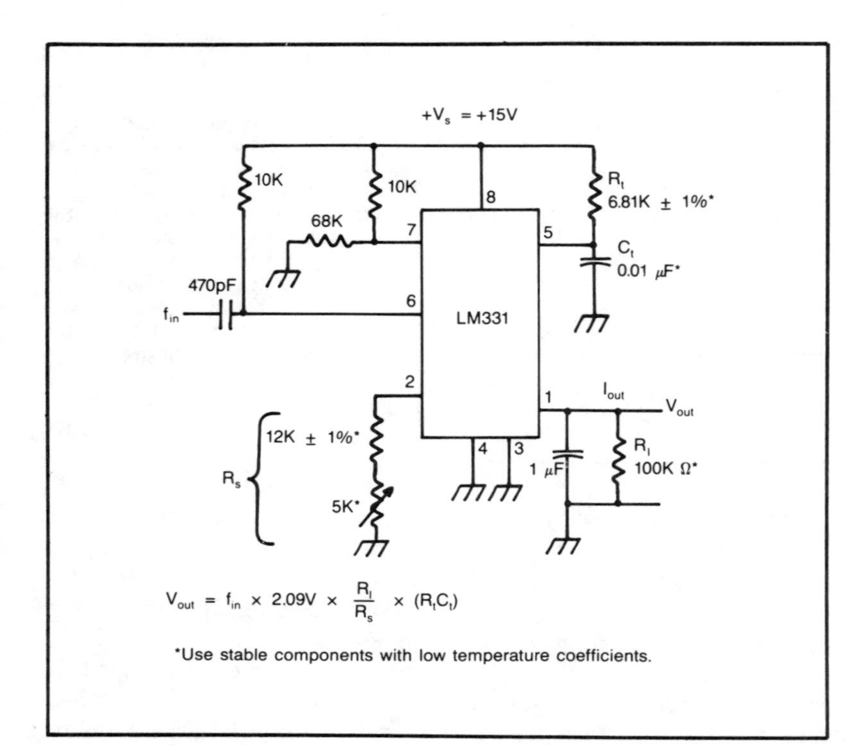

Fig. 9-25. Frequency-to-voltage converter for enabling a DMM to function as a counter. As an add-on arrangement, this circuit is interposed between the frequency source and the DMM. Courtesy of National Semiconductor Corp.

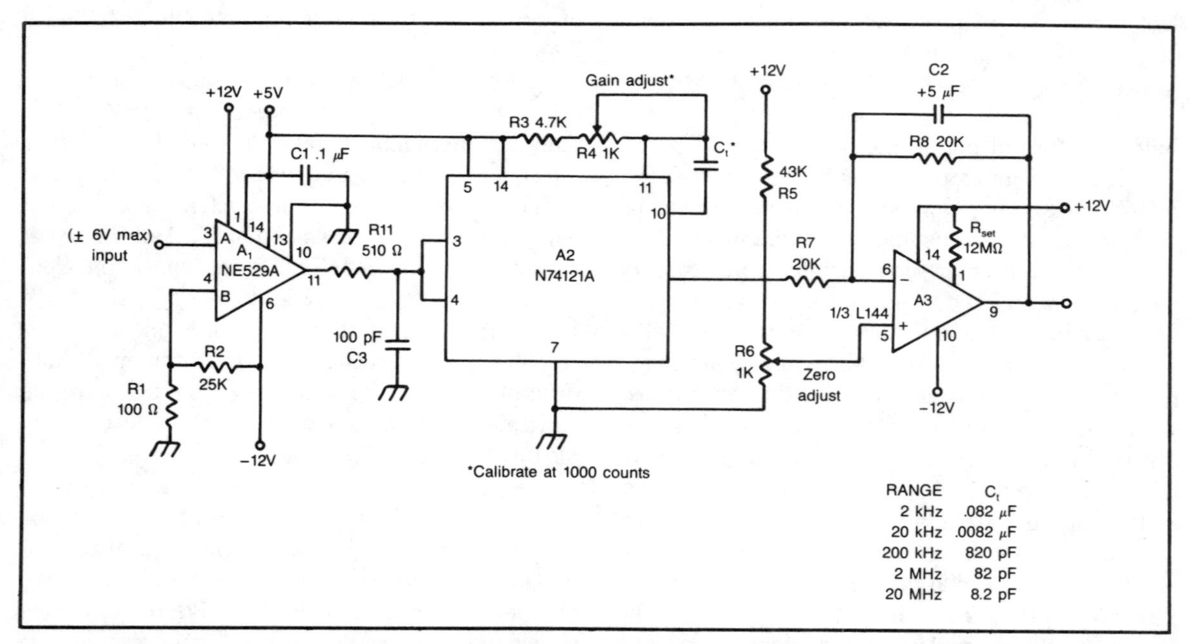

Fig. 9-26. Another adapter circuit for converting a DMM into a frequency counter. Essentially a frequency to voltage converter, this add-on unit enables the dc voltage readout of the DMM to be interpreted as repetition rate or frequency. The DMM should be set at its 2 volt dc measuring mode. Courtesy of Siliconix.

Despite the use of different ICs, both circuits average a train of constant duration pulses derived from the frequency being measured. The averaging is accomplished in both circuits by an RC low-pass filter. In the circuit of Fig. 9-26, C_t is the capacitor forming the shunt—arm of the averaging filter. By connecting C_t to a rotary switch, the DMM can be readily converted into a wide-range frequency counter. The nice thing about these add-on units is that the DMM does not become involved in any high-frequency problems, but continues to function as a digital voltmeter even though its readout is now scaled in frequency units.

MEASURING CAPACITORS

In their applications literature, the makers of digital multimeters sometimes show how such an instrument can be used for measuring capacitance. The instructions are usually intended for a particular model. However, if one understands the principle involved, it is easy to use any DMM to measure capacitance with sufficient accuracy for many electronic purposes. Departures from accuracy are not primarily due to the adaptation of the DMM for this purpose, but stem from the difficulty of timing an occurence with a stop-watch accurately. Where time intervals of at least ten seconds are involved, the accuracy can be quite acceptable, especially if the average of several measurements is used. for greater time intervals, the accuracy is easily better than the usual tolerance of electronic capacitors. Translated into practical terms, many capitance determinations can be made to an accuracy within plus or minus several percent.

It is fortuitously happens that the charging rate of a capacitor is a linear function of time when the capacitor being charged from a constant-current source, see Fig. (9-27). This greatly simplifies measurement procedures, calibration, and computation of the parameters in a simple circuit comprising a constant-current source and a capacitor. For, if we monitor the voltage developed in the capacitor, the situation is quite different from that prevailing in constant-voltage charging circuits. In the former case, the capacitor voltage is directly proportional to elapsed time. In the latter case, capacitor voltage is an exponential function of time. Simple arithmetic relates time and voltage when constant-current charging is used; the same calculations for constant-voltage charging are considerable more formidable and do not lend themselves to convenient scaling of the readout.

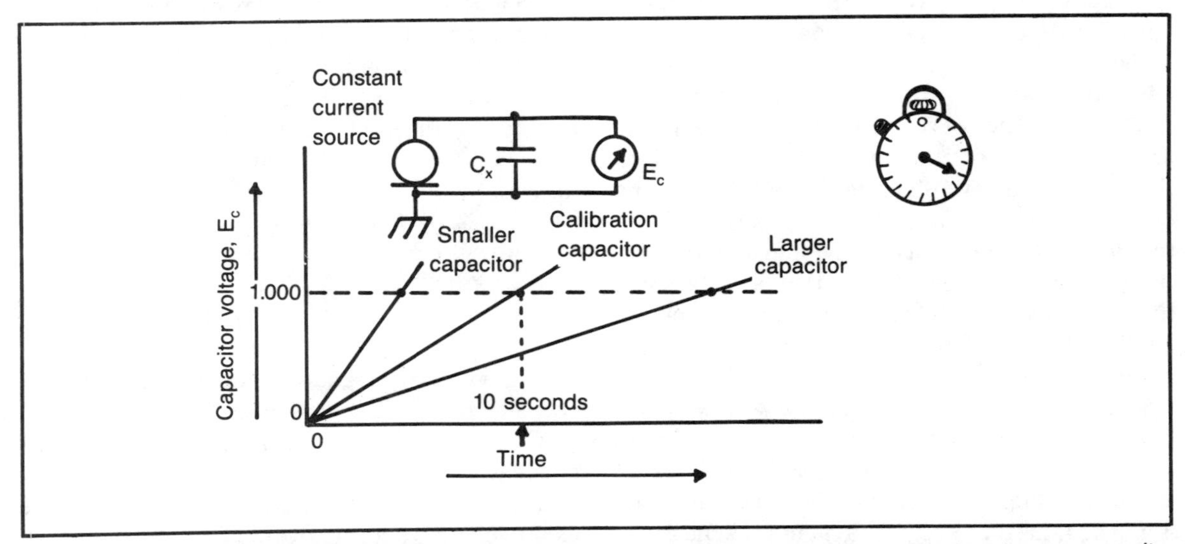

Fig. 9-27. Principle of capacitance-measuring technique in digital multimeters. The time required to charge a capacitor to a pre-determined voltage level (1.000 volt in above example.) is a simple linear function of time. This is because the charging source supplies constant current. The elapsed time is measured with a stop-watch.

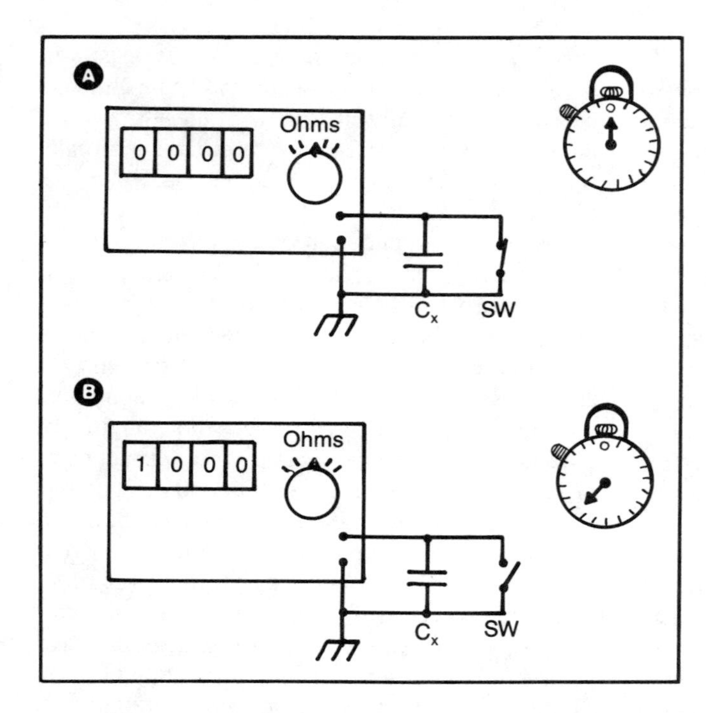

Fig. 9-28. Setup for measuring capacitance with the digital multimeter. A) Initial conditions: DMM is set at selected ohms range, Switch SW is closed, Stop watch is at zero. B) Measurement procedure: Switch SW is opened simultaneously with starting of stop-watch. Elapsed time for producing count of 1000 on DMM is recorded. C_x can then be evaluated in terms of time required for same count with a known calibration capacitor.

Further enhancing convenience, the requisite constant-current source for such measurements is found within the DMM, itself, being associated with the ohmmeter function of the instrument. Consider the relationship, $C = (I \times t)/E$. Here C is capacitance in Farads, I is the constant-current in amperes, and t is the time in seconds needed to develop voltage E in the capacitor. Current, I, is fixed in the design of the DMM and voltage, E, can arbitrarily be fixed at a convenient value, such as that represented by 1000 with the DMM set at one of its ohms ranges. (Recall that the DMM continues to operate as a *voltmeter* in the ohms mode.) Although it is not necessary that this be so, assume that the readout of 1000 happens to correspond to 1.000 volt. Let us now examine the situation more closely.

In so many words, capacitance will be determined by measuring the *time*, t, needed to charge an unknown capacitor to 1.000 volt. The test setup is shown in Fig. 9-28. The switch across the capacitor is kept in its closed position until the measurement procedure commences. (Because of dielectric hysteresis, "memory" from previous use, and charge-acquisition from electrostatic sources, capacitors often have residual charges). The time measurement, as already pointed out, is made with a stop watch.

Suppose that, with the DMM, set at its highest ohms range, the constant-current available to charge the capacitor is one-tenth of a micro-ampere, or 1×10^{-7} ampere. If we find that it requires ten seconds for the digital meter to indicate 1000, the unknown capacitor can then be evaluated from

$$C = \frac{I \times t}{E_c} = \frac{1 \times 10^{-7} \times 10}{1.000}$$

$$= 10^{6} \text{ Farad, or } 1.0\ \mu F$$

Even if one does not know the value of the constant-current, or the voltage corresponding to the readout of 1000, the measurement technique can be successfully implemented. This is accomplished by calibration with a known capacitor. In the above example, a capacitor known to be 1 μF could have been used to establish the corresponding time interval of ten seconds. Thereafter, a linear relationship would enable determination of other capacitance values. Thus, a 4 μF capacitor would require forty seconds to charge to the voltage corresponding to 1000. Conversely, a .5 μF capacitor would attain this readout in five seconds.

For lower ohms ranges, the capacitance corresponding to a ten-second interval becomes greater. Other things being equal, for each tenfold reduction in ohms range, the capacitance corresponding to a 1000 readout in ten seconds becomes ten times greater. This is because, in the simplest situation, the charging current is ten times greater. But other things are *not* always equal in DMMs with many ohms ranges. Not only does the constant-current change from range to range, but often the voltage corresponding to 1000 will also change. More specifically, some ranges may use the meter as a 2000 millivolt instrument while other ranges may utilize the 200 millivolt range. If you know what your DMM does, the equation relating current, time, and voltage may be put to good use. If the current and/or meter voltage are not known, calibration may be made from known capacitors as described above.

As may be inferred, this measurement technique is applicable to large capacitors. generally, capacitors from about a quarter-microfarad to many thousands of microfarads can be measured or checked out by this procedure. The exact range depends upon the internal currents and voltages used in the DMM. Some DMMs will permit measurements down to the vicinity of .01 μF. In any event, the polarity of electrolytic capacitors must be observed. Accordingly, it will be necessary to determine the polarity of the ohms test leads.

ELECTRONIC THERMOMETER

Temperature probes and temperature add-ons have been used for many years with VOMs, VTMs, and FET meters. Usually, these temperature sensors have taken the form of a bridge circuit with the temperature-sensitive element in one arm. And the most popular element for sensing temperature has been the thermistor. There is no reason why such temperature-sensing circuitry can't also be used with a digital multimeter. However, this type of electronic thermometer is not the best choice of available methods for taking advantage of the potential accuracy and resolution of digital meters. Temperature sensors based upon thermistors suffer from nonlinear response, hysteresis, errors from self-heating, and may be vulnerable to aging effects. It would be much nicer to interface the DMM with a temperature sensor capable of delivering a precisely proportional dc voltage with respect to sensed temperature.

An approach to this objective is attainable from the temperature-dependent voltage-drop across a pn junction operating from a constant-current source. The basic implementation of this technique is shown in Fig. 9-29. The base-emitter voltage of the silicon npn transistor will change with temperature at the rate of minus 2.3 millivolts per degree Celsius. This relationship can be used to calibrate the readout obtained on the DMM set at its one-volt or two-volt dc range. Although the actual numbers displayed on the DMM may not offer the most convenient way to read temperature, the measurements nonetheless can have considerably greater accuracy and resolution than was attainable from the thermistor-VOM combinations.

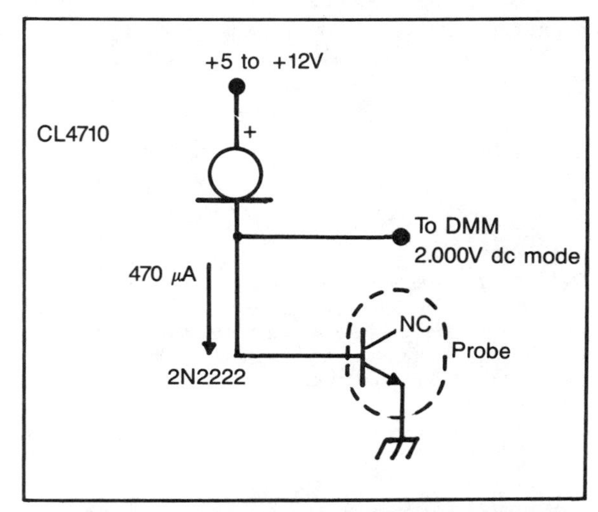

Fig. 9-29. A simple add-on temperature probe for DMMs. The reading at room temperature (21 °C to 25 °C) should be noted. Thereafter, the change in the readout will occur at the rate of −2.3 millivolts per degree Celsius.

Sophisticated extensions of the basic energy-gap principle demonstrated by the diode-connected transistor of Fig. 9-29 can produce a temperature sensor of even greater precision, and with characteristics ideally suited to interfacing with a DMM. A commercial version is shown in Fig. 9-30. There is no way we can implement such circuitry with discrete elements because precisely matched transistors with exceedingly tight thermal coupling must be used. This calls for monolithic fabrication. Fortunately, the semiconductor firms have already designed and are marketing such dedicated ICs. These ICs easily make a precision electronic thermometer out of a DMM. Although the internal circuitry of the temperature sensor is quite complex, the actual application is surprisingly simple. This can be seen in Fig. 9-31 with regard to the LM335 temperature-sensing IC. Moreover, the "built-in" characteristic of ten-millivolts per degree Kelvin enables the use of a convenient temperature scale for the DMM.

OTHER TEMPERATURE SENSORS

There are other temperature sensors which, because they can convert temperature difference into predictable dc voltage, are suitable for use with digital multimeters. The Sensor, made by Texas Instruments, for example,is made from heavily-doped silicon and behaves as a metal in the sense that it exhibits a positive temperature coefficient (as opposed to the negative temperature coefficient of thermistors). But, unlike metals, the coefficient is quite large—on the order of + 0.7% per degree Celsius. These resistor-like devices are usable over a temperature range of − 60 ° C to + 150 ° C. They are oridnarily connected in bridge circuits similar to those employed with thermistors. The bridge must be supplied from a stable voltage source.

Another device, the thermocouple, develops an electromotive force (voltage) via the contact of two dissimilar metals. Such junctions are often welded together and form a temperature sensor with exceptional mechanical, electrical, and chemical ruggedness. Moreover, some are suitable for direct immersion into various molten metals, their upper temperature limit being in the vicinity of 1800 ° C, or even higher. In laboratories, this temperature-sensing technique often takes the form of two series-connected thermocouple junctions, one being exposed to an ice bath, or to a reference temperature, and the other being used to probe the unknown temperature. In more mundane uses, this need not be done, but it is necessary to pay heed to the kind of metals used in wires and terminals of the thermocouple circuit. Many DMM manufac-

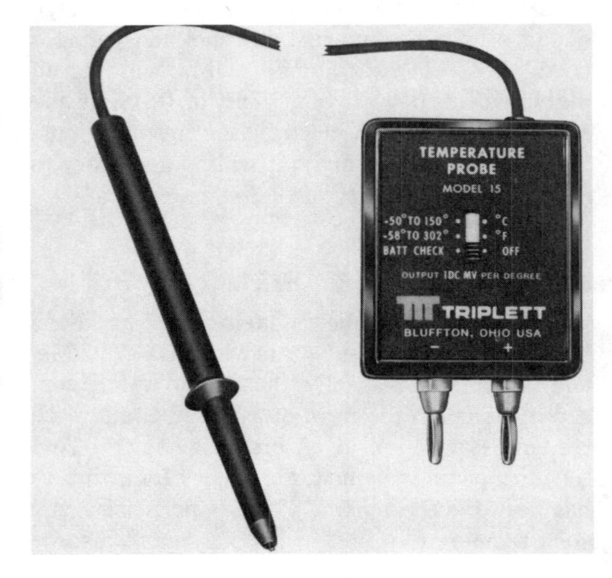

Fig. 9-30. A silicon junction temperature probe for DMMs. The unit has its self-contained battery, with a provision for checking its condition. Also, a switch-selectable option scales the DMM for either Celsius or Fahrenheit readings. Courtesy of the Triplett Corporation.

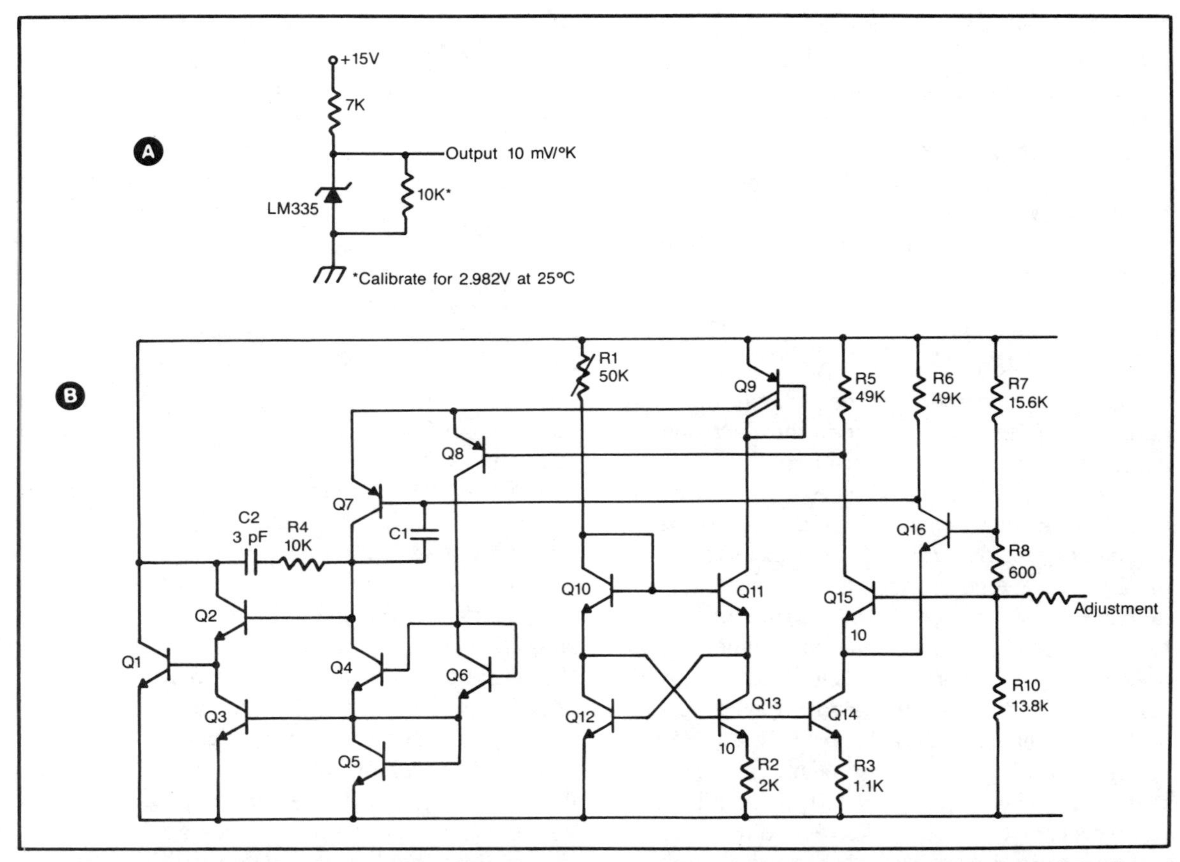

Fig. 9-31. Using a dedicated IC as a temperature sensor for a digital multimeter. A) setup and calibration data. B) Internal circuitry of LM335 precision temperature-sensor. Courtesy of National Semiconductor Corp.

turers specify the type K thermocouple temperature sensor for use with their instruments. This sensor has a temperature range of −20° C to 260° C and is particularly suitable for the thermal investigations ordinarily done in electronics. The AN595 IC is a thermocouple amplifier with on-chip compensation for the reference junction of a thermocouple temperature sensor. Made by Analog Devices, this module is suitable for DMMs; it delivers 10 mV/C.

The following is useful information when working with temperature measurements:

- ☐ Kelvin and Rankine temperature scales are absolute scales in the sense that their zero indications designate "rock-bottom"—the coldest anything can get. Calibration is based upon thermodynamic theory.
- ☐ Celsius (formerly Centigrade) and Fahrenheit scales are "practical" temperature scales; they derive their calibration from the melting point of ice and boiling point of water.
- ☐ Conversion from Celsius to Fahrenheit scales: $F = 9/5\ C + 32$
- ☐ Conversion from Fahrenheit to Celsius scales: $C = 5/9\ (F - 32)$
- ☐ Conversion from Kelvin to Celsius scales: $C = K - 273.16$
- ☐ Conversion from Rankine to Fahrenheit scales: $F = R - 459.72$
- ☐ From these relationships, the melting point of ice is approximately: 0° C, 32° F, 273° K, and

491° R. Correspondingly, the boiling point of water is: 100° C, 212° F, 373° K, and 671° R.

□ Ambient or "room-temperature" is often close to 20° C, or 68° F. This also corresponds to approximately 293° K.

□ Celsius and Kelvin degrees are the same "size" (they denote the same temperature difference). Similarly, Fahrenheit and Rankine degrees measure identical temperature differences.

CIRCUIT ARCHITECTURE

Allusions have already been made to the accuracy and precision required in the electronic circuits of the DMM, particularly in analog processing. The approximate techniques and simplistic approaches used in other meters are not satisfactory for the DMM. Inasmuch as the digital display stands ready to accurately indicate the data delivered to it, it behooves the instrument designer to make certain that the measured parameter is processed with inordinately high precision in the "front-end" of the meter. Fortuitously, op-amps and allied linear circuits can provide a wide variety of functions with close compliance to the desired mathematical operation. Indeed, the semiconductor firms have targeted on the very conversions and transformations needed in the circuit building blocks of digital meters.

Although this book emphasizes the use, rather than the design of these instruments, some insights into their circuit architecture is bound to enhance one's ability to intelligently apply these instruments in actual measurement situations. Additionally,the prospect of building digital multimeters for special purposes is not so overwhelming when it is realized that the basic function blockscan be selected from a catalog of ICs.

Several important circuits are shown in Fig. 9-32. These are precision circuits in the sense that they perform their respective functions with mathematical rigor. Thus, the integrator of Fig. 9-32A produces a highly-linear voltage ramp. The zero-detector of Fig. 9-32B very accurately "reports" the transistion of the applied signal through its zero level. The ac-to-dc converter of Fig. 9-32C is a far cry from ordinary rectifier circuits which are faulted by the energy-gap and nonlinearity of junction rectifier-diodes. This converter develops an output voltage that is linear virtually down to zero-volts. Thus, it is tailor-made for the ac measurement functions of average-responding DMMs. In Fig. 9-32D, we see how the semiconductor firm can pre-empt our needs, for here is a dedicated subsystem for the resistance-measuring mode of the DMM.

Finally, the circuit of Fig. 9-32E is that of an extremely low-driftpeak detector. As with other circuits, the output is a dc voltage. This makes sense inasmuch as the DMM is basically a dc voltmeter; other measurement modes are accomplished by converting the measured parameter, be it resistance, ac voltage or current, or dc current, to a proportionate dc voltage.

To be sure, the use of op-amps and similar circuits leads to excellent performance and greatly simplifies the manufacture of these instruments. But future progress, already in evidence, will make use of a relatively few dedicated chips. "Smart" instruments will go through their paces with the aid of a microprocessor, and there will be easy interface with the IEEE 408 general purpose instrument bus. Basic circuit operations,such as those shown in Fig. 9-32 should still be identifiable, however. That is, near-term progress appears more likely to involve the hardware implementation techniques, rather that the basic concepts used in digital multimeters.

Although it is possible to integrate the entire circuitry of a digital meter on a single dedicated chip, this would not necessarily be the most cost-effective or practical approach. An effective compromise architecture is shown in Fig. 9-33 where a digital voltmeter is configured around two dedicated ICs. Certain peripheral functions are handled by simple external circuits. The parts count of such an arrangement is dramatically reduced from what it would be in an electrically equivalent circuit using op-amps and discrete elements. Indeed, there is much more than meets the eye in such as arrangement because IC integration easily permits the incorporation of functions which would

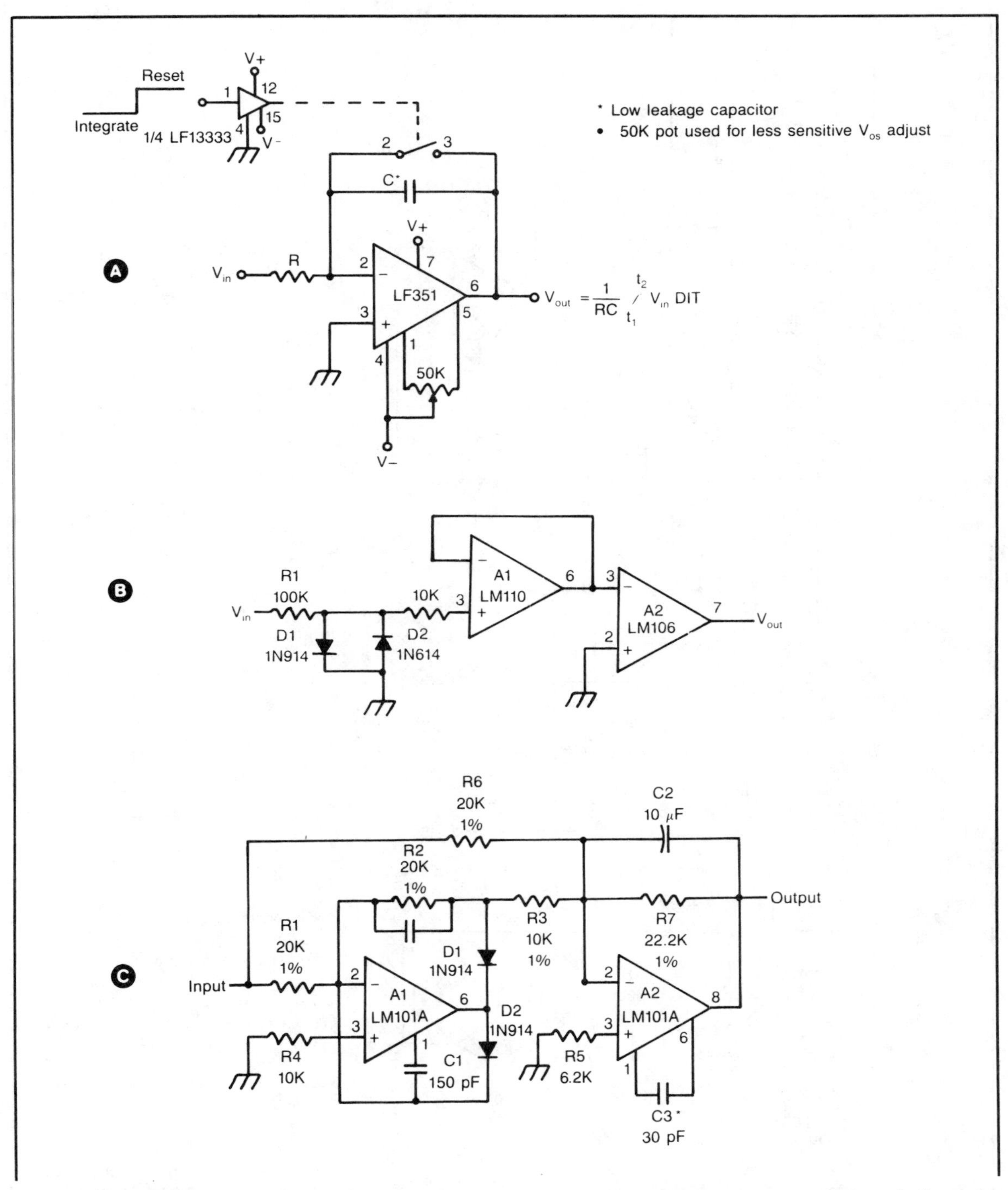

Fig. 9-32. Typical IC functions used in digital multimeters. A) Integrator. B) Zero Crossing Detector. C) Ac to dc Converter. D) Ohms to Voltage Converter. E) Peak Detector. Courtesy of National Semiconductor Corp.

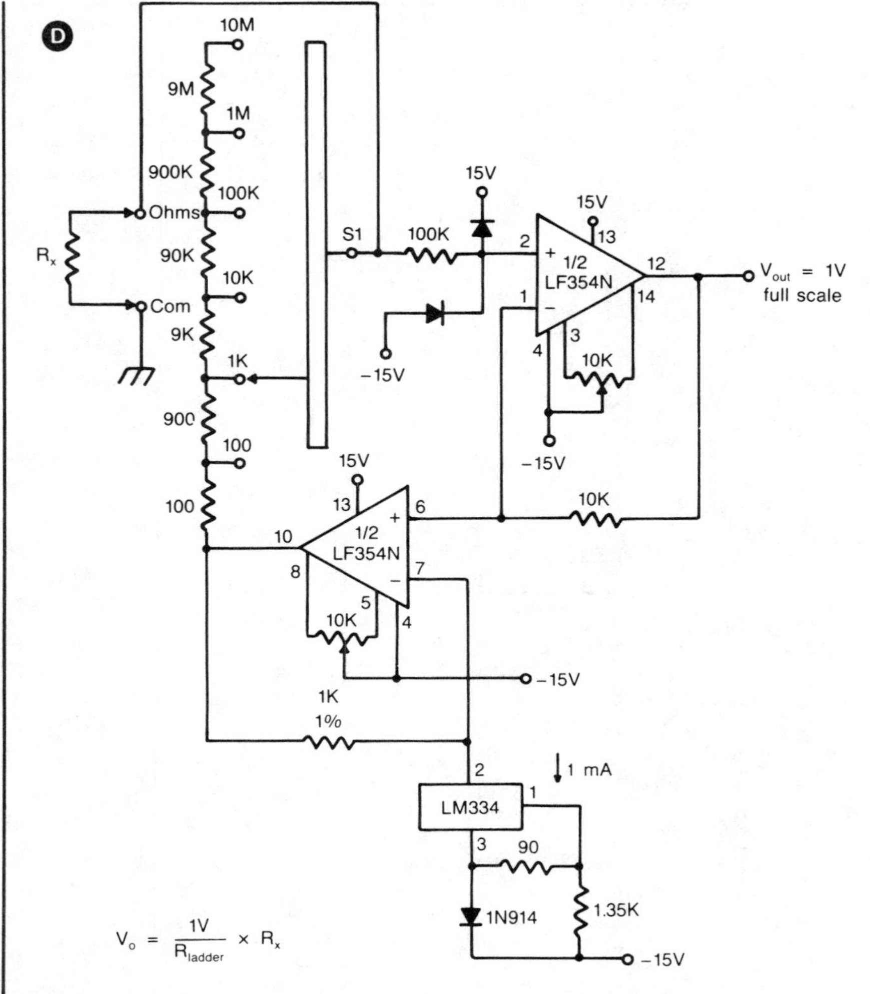

Where R_{ladder} is the resistance from switch S1 pole to pin 10 of the LF354.

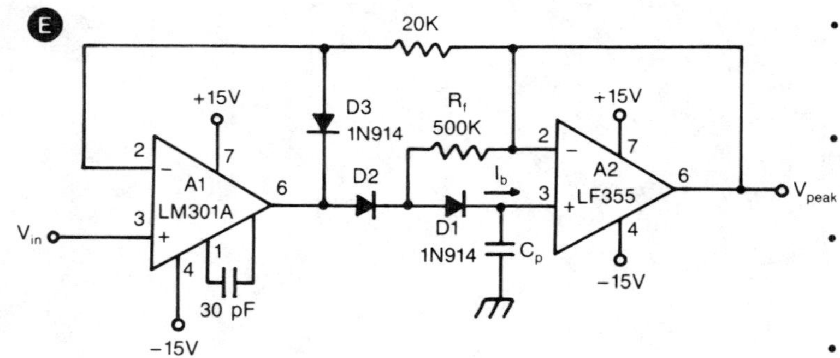

- By adding D1 and R_f, $V_{D1} = 0$ during hold mode. Leakage of D2 provided by feedback path through R_f.
- Leakage of circuit is essentially I_b (LF 155, LF156) plus capacitor leakage of C_p.
- Diode D3 clamps V_{out} (A1) to V_{in}—V_{D3} to improve speed and to limit reverse bias of D2.
- Maximum input frequency should be $\ll 1/2\pi R_f C_{D2}$ where C_{D2} is the shunt capacitance of D2.

prove too expensive or electrically awkward to implement with architecture employing many op-amps and other basic building blocks.

The LD111, for example, contains a bipolar comparator,a bipolar integrating amplifier, two MOS-FET unity-gain amplifiers, several p-channel analog switches, and level-shifting drivers to enable interfacing with the LD110.

The LD110 is even more complex. It contains control logic, a BCD counter, static latches, a multiplexer, data buffers, digit buffers, a decoder, a time-base counter, and a two-phase clock generator. It supplies information for automatic zeroing, automatic polarity, and automatic ranging. Such architecture frees the instrument designer from the expense of a long drawn out project, inasmuch as the basic meter-functions are available from such dedicated ICs. The instrument user gets greater reliability and, generally speaking better precision than is readily forthcoming from architecture using less sophisticated building blocks. Best of all, a very "brainy" digital multimeter can be marketed for a modest price. The rationale behind this is similar to that pertaining to hand-held calculators; no matter how many active devices are contained in single, or in several master-chips, once such ICs are debugged and proven out, they can be produced in extremely large quantities. Inasmuch as initial research and development costs are greatly diluted by such mass production, the manufacturing cost quickly decreases. Even more significant is the fact that it becomes virtually as easy to turn out tens of thousands as tens of dozens. Finally, the manufacturing technique tends to produce uniformly characterized units no matter how great the quantity.

Possibly the salient feature of an arrangement such as shown in Fig. 9-33 is its flexibility and adaptibility. It is basically a 3 1/2 digit, 2,000-volt panel meter, but it is easy for the instrument designer to modify its performance with various modes, functions, and ranges. Either common-anode or common-cathode LED displays can be accommodated and a straightforward peripheral circuit can be used to introduce autoranging. It is likewise simple to utilize the basic scheme as a digital thermometer, as the readout of an electronic scale, and for other applications. Although the use of a single monolithic IC for the entire digital-voltmeter or multimeter would be the most elegant design approach, it would then be difficult to retain the expandability and flexibility of the architecture used in Fig. 9-33.

DISPLAY DEVICES

The first digital display or readout devices were mechanical, consisting of numbered wheels and driven by servo-controlled motors. Next on the scene were various formats comprising miniature incandescent lamps. These evolved to a surprisingly satisfactory state and gave rise to various control techniques that now enable the popular light-emitting diodes (LEDs) and liquid-crystal displays (LCDs) to dominate instrumentation readouts. For example, the seven-bar, or seven-segment format was first devised with filamentary lamps. Such lamps yield a nice bright indication. They have certain inherent shortcomings, however, and it is intriguing to speculate what readouts would be like if LEDs and LCDs had not become technological and economic successes. Of course the weak link in the incandescent lamp is the more-or-less fragile filament. Even when operated at low currents in the interest of longevity, the thermal response time tends to pose problems in instrument readouts. Although lamps are now little used, the seven-segment display format remains in vogue. This is illustrated in Fig. 9-34.

Gas discharge plasma readouts emit light via the ionization of neon, argon, and other noble gasses. A small amount of mercury is often used to manipulate the ionization potential and to modify the spectrum. These displays have been made into many forms, some of which have external electrodes. Light output is very good and the definite on-off characteristic of plasma devices allows for convenient multiplexing although slow turn-off can pose problems. No longer the dominant readout, plasma displays continue to find application. One disadvantage of this type of readout is the high ionization-potential needed—on the order of 150 to 200 volts. This creates an interfacing problem with

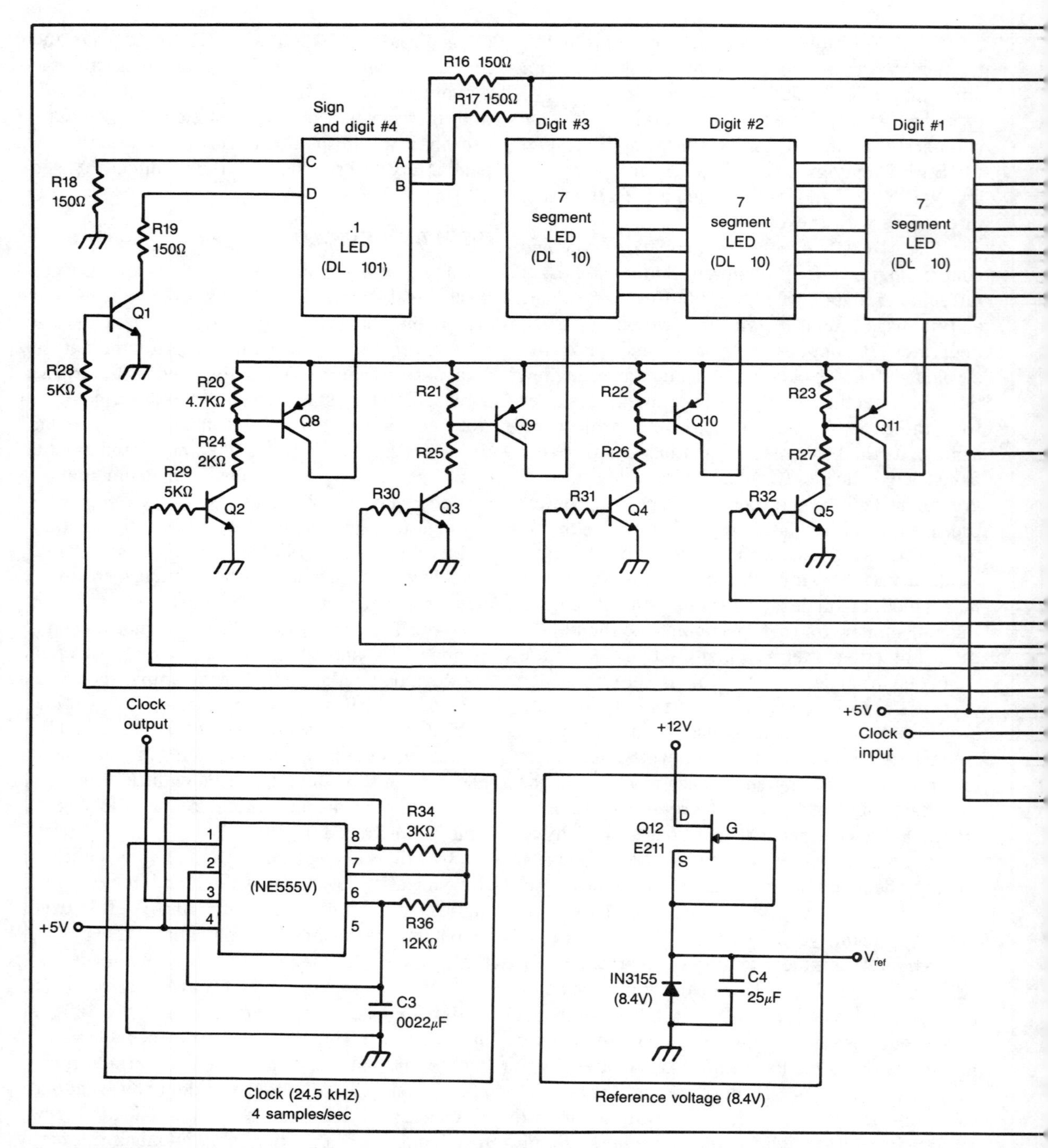

Fig. 9-33. A digital voltmeter designed around two multi-function IC modules. In this architectural scheme, most of the analog processing is accomplished within the LD111 monolithic module. Basic digital processing is provided by the LD110 monolithic module. Courtesy of Siliconix Incorporated.

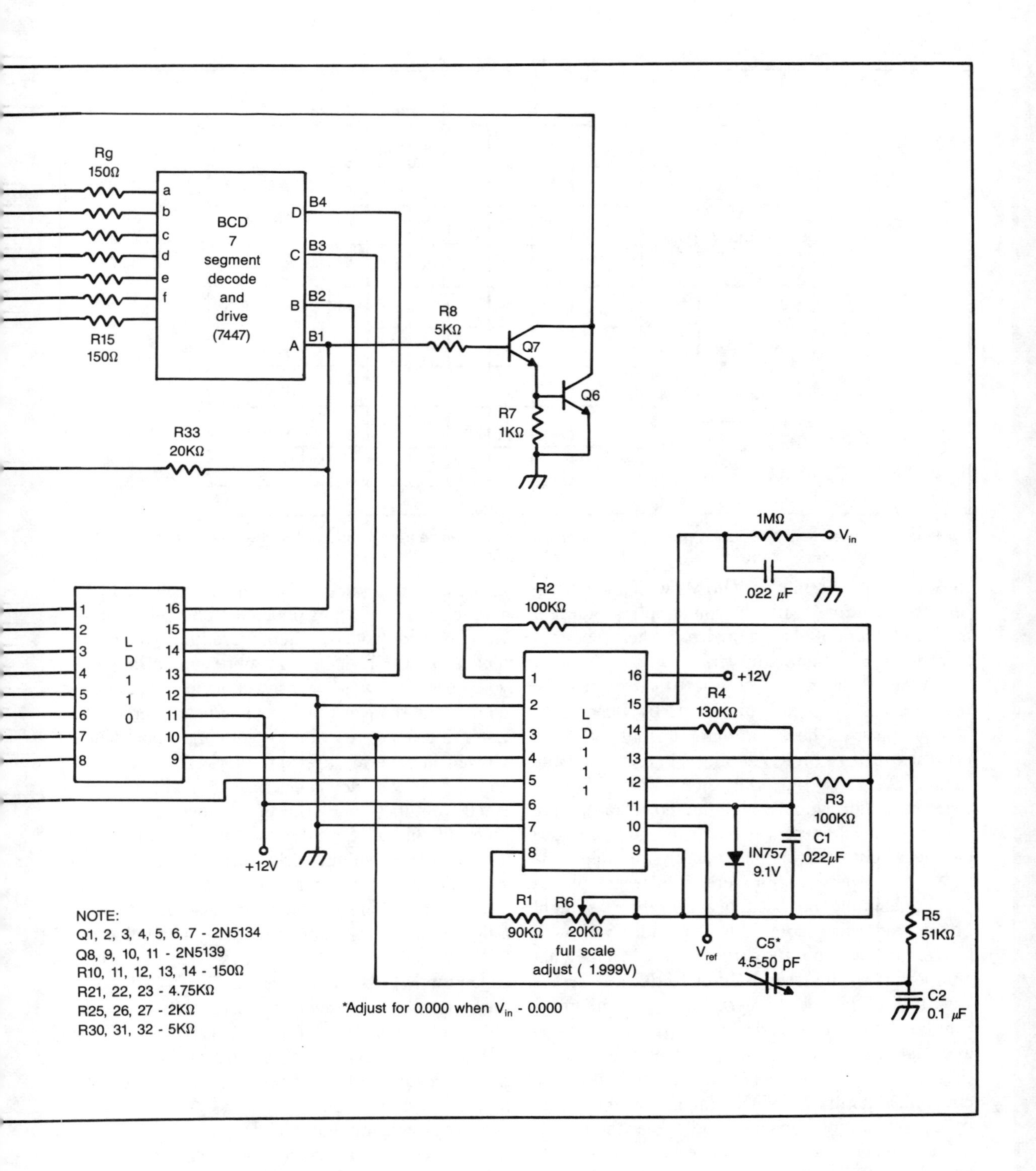
Rg
150Ω
a
b
c
d
e
f
BCD
7
segment
decode
and
drive
(7447)
D
C
B
A
B4
B3
B2
B1
R15
150Ω
R8
5KΩ
Q7
Q6
R7
1KΩ
R33
20KΩ
1MΩ
V_{in}
.022 μF
R2
100KΩ
L
D
1
1
0
1
2
3
4
5
6
7
8
16
15
14
13
12
11
10
9
L
D
1
1
1
+12V
R4
130KΩ
R3
100KΩ
C1
.022μF
IN757
9.1V
+12V
R1
90KΩ
R6
20KΩ
full scale
adjust (1.999V)
V_{ref}
C5*
4.5-50 pF
R5
51KΩ
C2
0.1 μF
NOTE:
Q1, 2, 3, 4, 5, 6, 7 - 2N5134
Q8, 9, 10, 11 - 2N5139
R10, 11, 12, 13, 14 - 150Ω
R21, 22, 23 - 4.75KΩ
R25, 26, 27 - 2KΩ
R30, 31, 32 - 5KΩ
*Adjust for 0.000 when V_{in} - 0.000

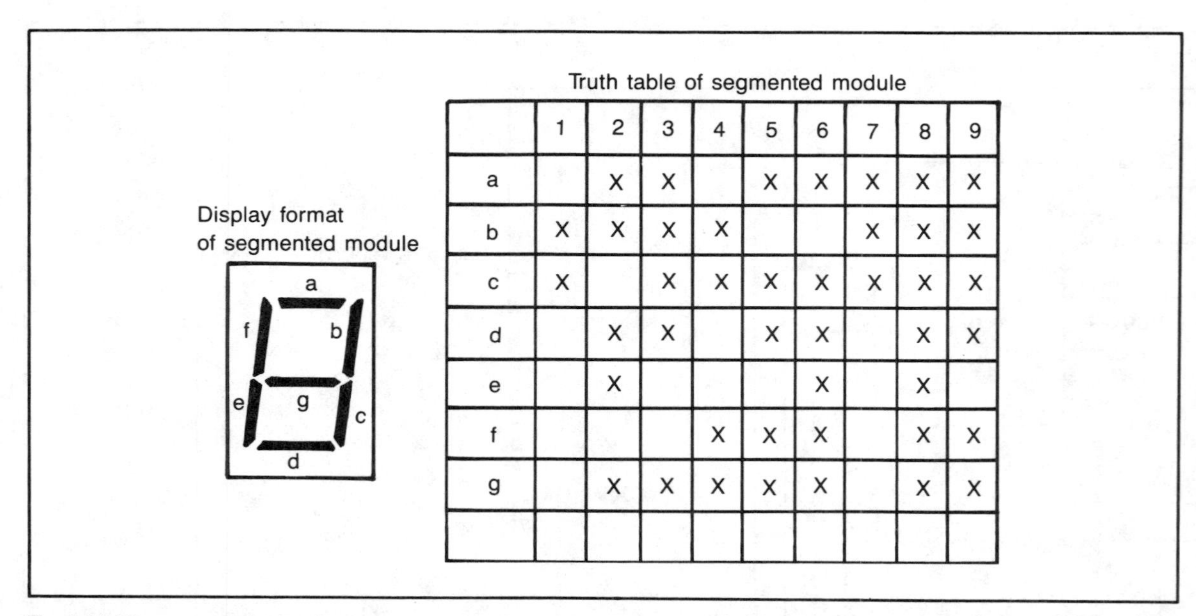

Truth table of segmented module

	1	2	3	4	5	6	7	8	9
a		X	X		X	X	X	X	X
b	X	X	X	X			X	X	X
c	X		X	X	X	X	X	X	X
d		X	X		X	X		X	X
e		X				X		X	
f				X	X	X		X	X
g		X	X	X	X	X		X	X

Fig. 9-34. The seven-segment display module. Segment illumination can be from a variety of light sources, or light modifiers.

the solid-state circuitry of DMMs which generally operate from five-to fifteen-volt supplies. Another adverse feature of plasma devices is their tendency to produce transients and rfi.

A long enduring and very satisfactory gas-tube readout was the Nixie tube made by Burroughs. The simplest of these had nine internal cathodes, each shaped as a digit. Although, one was stacked behind the other, only the selected cathode (and therefore, digit) would emit light. To be sure, there was visual interference from the nonluminous cathodes, but this was surprisingly minimal. The fact that the nine digits were in different planes wasn't desirable, but became acceptable. As might be inferred, this readout device blended well with the extant vacuum-tube technology. Somewhat compensating the large bulk of Nixie tube arrays was the large and high-visible digital display. Similar tubes were made by Amperex Raytheon and National Electronics. The continued use of these readout devices would not have permitted the compact and miniature instruments made possible by the oncoming transistor and IC technology.

Fluorescent readouts enjoyed a brief burst of popularity. These were electron tubes in which an anode or a multiplicity of anodes were coated with a green-light emitting phosphor phosphor. Some were grid-controlled, others were in the format of diodes. These devices were more compatible with voltages and power levels available in vacuum-tube equipment than in the later solid-state products. Later devices were available with phosphors which glowed in several different colors. A basic shortcoming of these devices is that they are basically electron tubes and therefore incorporate the physical and electrical penalties of filamentary devices. The readout using these tubes had considerable esthetic appeal, however.

Electroluminescent displays also make use of a light-emitting phosphor. Instead of the electron tube structure, these displays are essentially capacitors. A phosphor-bearing film is excited by the capacitance displacement current when ac is applied to transparent, but conductive plates. Light intensity tends to be less than that attainable from the fluorescent tubes, but the panel structure format is very convenient.

Light emitting diodes (LEDs) have become the workhorse of digital displays. They are bright, cost-effective, amenable to multiplexing, compatible with transistor and IC voltage and current levels, long-lasting, fast acting, relatively insensitive to temperature, and are available in at least four colors—red, orange, yellow, and green. Being junction diodes, they have well defined electrical characteristics.

Liquid-crystal displays (LCDs) do not emit light, but alter the optical characteristics of light impinging upon them. As a consequence, their power requirement is by far the lowest of any electrically-actuated display. This accounts for their popularity in battery-operated, hand-held, and portable instrumentation. Being in the liquid phase they are the least tolerant display with respect to hot and cold temperatures. although disabled at subfreezing levels, modern LCDs tend to recover their performance capability when the temperature returns to normal. The same tends to be true regarding exposure to excessively high temperatures. In the interest of long life, LCD instruments probably should be operated within the temperature range of 0 ° C to 60 ° C. Storage temperature should preferably be not too far outside of this range. Degradation is known to occur from high humidity, but this is a variable depending upon the marker's packaging skill.

Optically, LCDs shine where LEDs falter. In bright sunshine, LCDs are at their best with regard to visibility. Conversely, LEDs "wash out" and can scarcely be seen. On the other hand the LCD display becomes shadowy in twilight environments, whereas the LEDs offer maximum readabililty in total darkness.

The bottom line on digital readouts is that it would be rash to dismiss any display technology as an historical artifact; even those which have apparently been eclipsed from popularity may surprisingly make a comeback. Because of evolving techniques, new ICs, better materials, and changing requirements, we are likely to see new versions of filamentary, plasma, and fluorescent displays. Nor has the last word been written regarding liquid crystal displays. Until quite recently, these were regarded as immune to successful multiplexing schemes. Now, multiplexed LCD formats are common.

In multiplexing, the digits are sequentially, but repetitively illuminated or displayed. Because of persistence of vision, the viewer percieves the overall display as a constant readout. The virtue of this technique is that it greatly reduces the number and complexity of interconnections and it effects a worthwhile saving in power consumption. Figure 9-35 shows the general circuit format for multiplexing 7-segment display modules. Note the parallel connections of like-segments. As a consequence, the logic controlling the segment lines decides what number will be displayed, whereas the logic controlling the digit lines decides where that number will appear in the readout. In this regard, note that the backplane leads to the display modules are not connected in parallel, but may be selectively addressed. The more digits in the readout, the more dramatic is the saving in power and the reduction in the interconnections. Most display technologies, including LCDs (but, with some difficulties) can be multiplexed in this fashion.

Electroluminescent displays, however, are not readily multiplexed by the simple scheme of Fig. 9-35. Additional circuitry must be introduced to prevent capacitive coupling between segments and it then becomes questionable whether overall advantages accrue from multiplexing. Whereas the segments in a plasma display cannot turn on unless supplied with ionization voltage, the segments in an electroluminescent display can glow at reduced intensity solely from capacitive leakage currents.

This should suffice to provide the user with practical insight into the nature of the digital readouts on DMMs. If this text were design-oriented, there would obviously be much more to be concerned with, such as the drive and decoding circuitry, the management of decimals and designation symbols, the design of switching circuits, etc.

Much of the diagnostic and servicing activity connected with digital logic circuits involves the detection of high and low logic states. From such basic information, functional integrity of circuits and systems can be evaluated; most malfunctions are caused by faulty logic. Faults can derive from opens and shorts with ICs or from solder bridges or breaks in the PC wiring. Other types of faults occur too. For example, glitches and transients often produce inappropriate changes of state in logic circuitry. And out-of-spec voltages from power supplies effect the operation of gates, flip-flops, and other digital functions. Whatever the cause of defective operation, and whatever the effect, static voltage measurements can often pinpoint the source of the trouble.

Consequently, it would seem that the traditional test instruments—scopes, VOMs and DMMs would be useful in finding the source and nature of trouble in digital systems. This, indeed, is true. It is also true, however, that these instruments often take a back seat to test instruments of less precision, and which simply reveal the presence of logic highs and lows, the existence of pulse activity, and similar information of a qualitative nature. It would seem that the traditional measuring instruments, like over-age engineers, are overqualified for the task! A reason closer to the reality of the situation is that it is more convenient to explore the numerous and dense pin-connections of a circuit board with a tiny hand-held logic probe. This isparticularly true because one need not remove visual contact with the pins being checked out—the visual indication of logic state is right on the probe.

But this does not mean we should junk our precision measuring-instruments when graduating from analog to digital systems. The very best performance checks and troubleshooting procedures usually involve a combination of simple state-testers and precision measuring-instruments. And many trouble-shooting techniques can be accomplished just with scopes, VOMs, or DMMs. Having just completed our discussion of DMMs, let's investigate a few situations where a DMM can be profitably deployed to troubleshoot logic circuitry.

FINDING FAULTS

In the early days of electrical technology, there

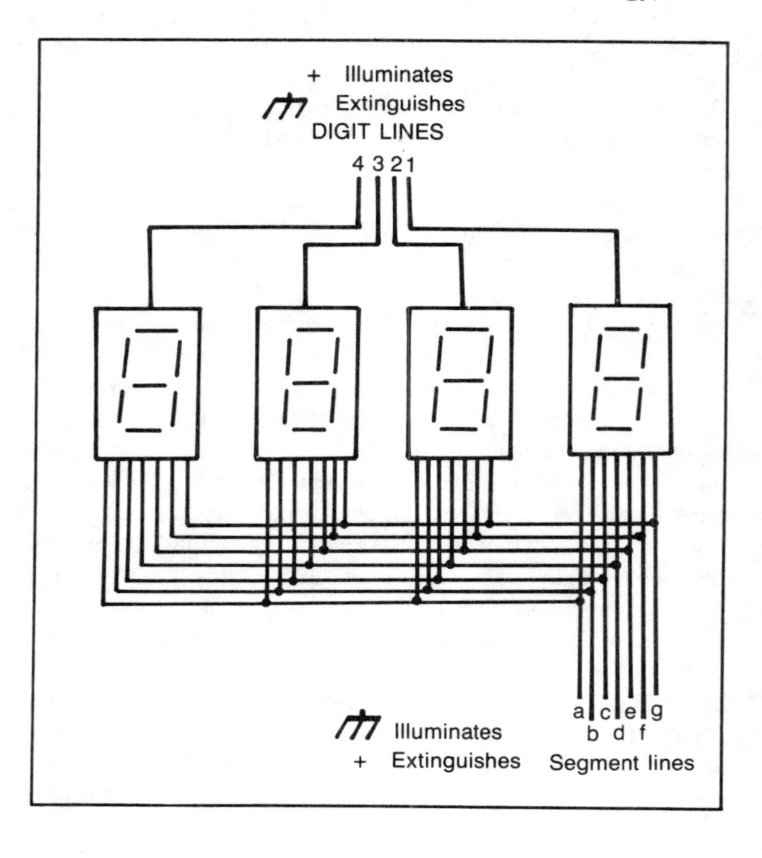

Fig. 9-35. Multiplexing format for seven-segment readout devices. Circuit interconnections and power drain are reduced by this technique. Only one digit at a time is displayed, but at a sequence faster than the persistence of the eye which perceives a steady readout at normal brightness.

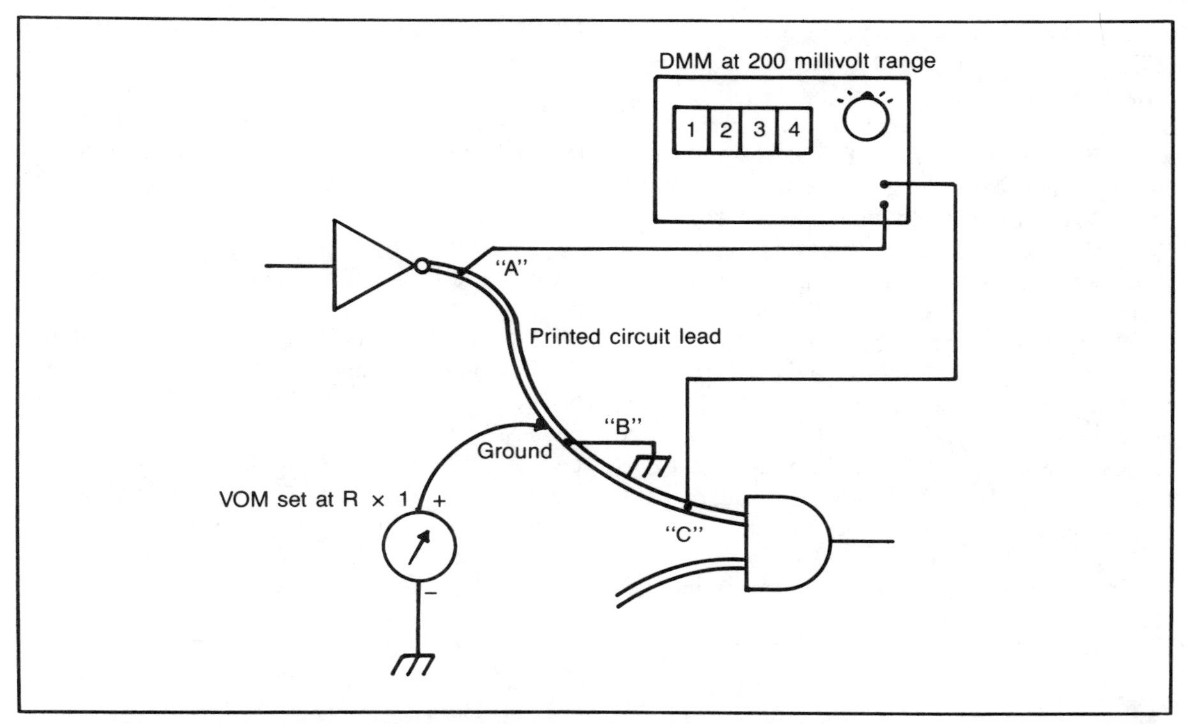

Fig. 9-36. Method of locating ground fault on a PC board. The VOM is used as a current source. The exploring probe of the VOM will cause increasing readings of one polarity on the DMM as it is moved towards "A" and of the opposite polarity as it is moved towards "C". With the probe at "B", the location of the round, the DMM will read zero. The logic circuit power-supply must be off for this test.

was considerable focus on techniques for detecting and locating open and short-circuits and grounds in telegraph, telephone, and power lines. These were often buried and it required much testing and mathematical ingenuity to pinppoint a fault from limited measurement data. Test devices generally were limited to galvanometers, voltmeters, and current meters which were used in conjunction with bridge circuits derived from the Wheatstone bridge. The most common of these were the Varley loop, the Murray loop, the Kelvin bridge, and the slide-wire bridge. Descriptions of these, together with the balancing equations can be found in the older engineering handbooks. The same basic concepts can be applied to digital circuit boards when prospecting for grounds, shorts, and opens in the wiring pattern. In most instances, we need only gather qualitative data and use common sense to locate such faults; this is because the PC wiring is not buried in the earth, and conductor runs can be readily observed and accessed.

One of the basic ideas in digital testing is to try to avoid the need to remove any ICs unless certain tests definitely point to a defective one. Shorts, opens, and grounds often are the result of solder bridges or microscopic cracks in the foil. Additionally these faults can occur *within* IC modules. Whenever they occur, they tend to manifest themselves as stuck highs, stuck lows, or as logic bad-levels. These malfunctions can be detected with logic probes, or with the aid of a scope, VOM, or a DMM. Assuming that such a malfunction has been found, a testing technique such as shown in Fig. 9-36 can often prove useful. Here, the VOM is set at its R × 1 scale and is actually used as a source of current, rather than as a measuring instrument. A VOM capable of supplying at least 200 milliamperes is best for this use. The DMM is set at its most sensitive dc voltage range, generally 200 millivolts. One DMM lead connects as closely as possible to the pin of the stuck gate or logic circuit. The other DMM lead connects as far away on the same PC lead as possible. If the free lead from the VOM is now moved along this lead, the polarity of the voltage drop thereby being measured will go through zero and reverse sign while passing the grounded point of the pc lead. The zero indication corresponds to the location of the ground.

The polarity shown for the VOM current-source avoids low impedance semiconductor conduction paths in the logic ICs. (In most VOMs, the actual polarity is opposite to that indicated alongside the pin jacks. This, of course, can be confirmed with the DDM.) Before conducting this test, it must be ascertained that the V_{CC} supply to the digital circuit board is off.

It should be noted that this test procedure can infer to the presence of the ground-short within the ICs themselves. In such instances, the DMM reading will not go through zero and reverse polarity as the exploratory probe is moved along the conductor. But a minimum reading at locations "A" or "C" probably indicates that the closest IC has an internal ground-fault.

The test set-up shown in Fig. 9-37 can be used when it has been determined that a logic state is in a stuck-high situation. This digital condition can be the result of a short between a node-lead and the V_{CC} line. The short may be caused by a solder bridge or by a metallic "whisker" between physically close conductors. Such shorts can also be internal to the ICs. As in the previous test, this test is made in the absence of V_{CC} voltage from the power supply. If the exploratory probe is moved from "A" to "C" in Fig. 9-37, the DMM reading will decrease, go through zero at "B," the location

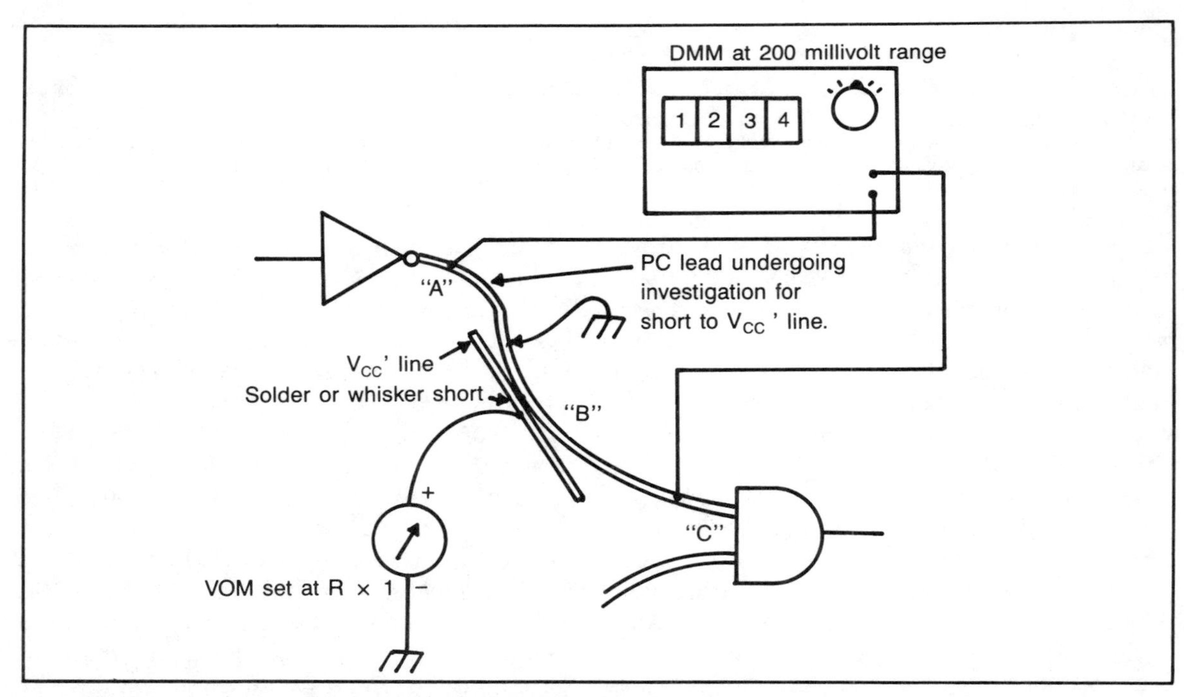

Fig. 9-37. Method of locating short between V_{CC} line and logic node line. The logic circuit power-supply must be off for this test. As the ground probe is moved away from "A", the potential-drop indicated by the DMM will decrease, and will become zero at "B", the physical location of the short. As the ground probe is moved toward "C", the potential-drop will again increase, but of opposite polarity.

of the short, then increase in the opposite polarity as "C" is approached.

If the short is within one of the ICs rather than in the external PC wiring, it will not be possible to go through zero and reverse polarity as with external faults. However, the IC which is closest to the exploratory probe, when a minimum reading is obtained on the DMM, is probably internally-faulted.

For the test procedures of Figs. 9-36 and 9-37, best results are obtained when:

- ☐ The PC conductor between "A" and "C" is long, narrow, and of thin-gauge material.
- ☐ The injected current is high.
- ☐ The sensitivity of the DMM is high. A 200-millivolt range is often satisfactory, but when the PC conductor resistance is inordinately low and/or the current available from the VOM is relatively low (there is considerable variation in VOMs), a more sensitive DMM range will be required in order to get meaningful readings.
- ☐ The internal battery of the VOM is fresh; otherwise current variation during the test can cause confusion.

A ten-centimeter run of popularly-used PC wiring can have a resistance on the order of 0.05 ohm. With 200 milliamps of injected current, the potential drop available for these test procedures would therefore be ten millivolts. Thus, if one is dealing with a short conductor and is using a low-current VOM, the DMM voltage-sensitivity may be the decisive factor in obtaining meaningful measurements from these tests.

Obtaining Precise Circuit Information

There is nothing like a DMM for resolving some of the troublesome malfunctions that crop up in digital circuits. Although qualitative determinations of logic states and various "go, no-go" tests are quick, and are conveniently made, they sometimes fail to tell the whole story. Consider, for example the matter of power supply voltage, V_{CC} in TTL logic. This is often assumed to be a nominal 5 volts and otherwise of relative unimportance. The fact of the matter is that some logic circuits frequently fail to perform properly for V_{CC} deviations of + 5% from 5 volts. Indeed some logic circuits are susceptable to damage at voltages greater than 5.25 volts. A DMM should be used to monitor V_{CC} during actual operation. If the PC board is inoperative, this may be easier said than done,but it should be ascertained that the power supply is capable of maintaining its voltage regulation under conditions simulating the load of the circuitry. Sometimes an adjustment of a quarter of a volt in power supply voltage makes the difference between a malfunctioning and a properly operating digital system. Often this will prompt some kind of a decision to be made regarding marginal IC's—those with inordinate voltage-sensitivity.

Positive identification of bad regions, i.e., stuck voltage levels between the high and low logic thresholds, are readily made with the DMM. By the same token, high and low logic levels which have survived logic probe tests may be found to be only marginally acceptable when investigated with the DMM. Such marginal levels tend to be inimical to reliable operation and should be investigated. The cause will often be found to be a logic IC which has for some reason been damaged, and which is likely to become completely inooperative after an unpredictable time.

Because of its constant-current characteristics, the ohms function of a DMM provides an excellent means of comparing the "goodness" of an IC with one known to be operative, or surmised to be good because it is brand new. (This, of course, is not always a valid assumption.) Comparative measurements can be quickly made with respect to the V_{CC} pin and also with respect to the ground pin. The suspected bad IC must be freed from the PC board for this kind of test. Such comparison tests are particularly attractive for PC boards with sockets. Although a mere substitution of another IC will identify one that is defective, the ohmmeter comparison procedure will tell us something of the nature of the internal defect. This can be useful in implementing remedial measures for preventing a recurrance of the problem when a new IC is installed.

Example of the Resistance-Comparison Test Procedure

Figure 9-38 depicts the basic idea of this test procedure. Fictitious ICs are dealt with inasmuch as the actual numerical values recorded from each pin measurement are not, in themselves, significant. Indeed, different resistance values will be measured with different DMMs because of different currents available from their test prods. Even a wilder range of resistance measurement would result from the use of a VOM because these simple instruments do not provide constant current. What is important is the pin-to-pin comparison with a known good IC when both are tested under like conditions. Some discretion must be exercised; the resistance values need only be in the "same ballpark" to qualify a test as OK. Most of such measurements will be in agreement within + 15%, particularly if the compared ICs are of the same brand. It is generally not important to be concerned with divergent values of high resistances, say over 200,000 ohms. Lower resistances, those under 20,000 ohms should be in approximate agreement, however.

In the diagram of Fig. 9-38 the V_{CC} terminals (pin 1) of the ICs are connected in order to facilitate the test procedure. After measurements are recorded for both polarities of the ohmmeter leads, a similar test procedures should be conducted with the "stationary" ohmmeter lead clipped to the ground (8) pins of the ICs. Do not connect V_{CC} pins and ground pins together at the same time. This, instead of making things easier, can lead to confused results because the defective IC can alter the measurements on some of the pins of the good IC.

Naturally, if this test procedure quickly identifies defective pins, it may be construed unnecessary to carry out all four sets of measurements. On the other hand, the additional measurements may prove useful in analyzing the detailed nature of the internal fault.

Initially, a little experimentation will pay dividends when working with actual ICs. There is nothing sacred about the 20K resistance range; some ICs will certainly provide better tests with lower ranges. In any event, on some pins, and for some defects, it will be found expedient to take readings for more than one resistance range. Note that this has been done for the pin 6 measurements of our fictitious ICs. In general, the "low-power" ohms mode will not be used inasmuch as the basic idea is to forward-bias the pn junctions whenever the polarity is appropriate. Very complex multipinned ICs can be tested by this procedure. Damaged junctions, internal shorts, and opened bonding wires will generally make themselves evident before completion of the four sets of comparison checks. These tests can also be useful when a good IC is not available. But, then, one must have at hand a schematic diagram of the internal circuitry and the tester's analytical ability assumes prime importance in the interpretation of the measurements.

Bad Levels—A Symptom of a Circuit Defect

Servicing and debugging digital circuitry is often involved with the detection of "bad" levels. These are voltages between those that properly define logic high and logic low states. In normal operation the bad region which encompasses these "forbidden" voltage levels is traversed very rapidly as logic states switch from high to low and vice versa. But faulty operation will often have a bad voltage level as one of its symptoms. Bad levels are readily detectable with a DMM as is shown in Fig. 9-39. Numerical data defining the bad regions of various logic families is depicted in Table 9-1.

Note that both the input and output circuits of logic devices have bad regions. These regions are, in essence, "windows"—they should be transparent rather than visible. The input threshold range defines the noise immunity of the logic device. Thus, the HTL (high threshold logic) devices have the highest input threshold range, 6 volts, and, therefore the greatest noise immunity of the bipolar transistor devices. In practice, it turns out that the noise immunity of a high-speed logic device will not be as good as that of a low-speed device even

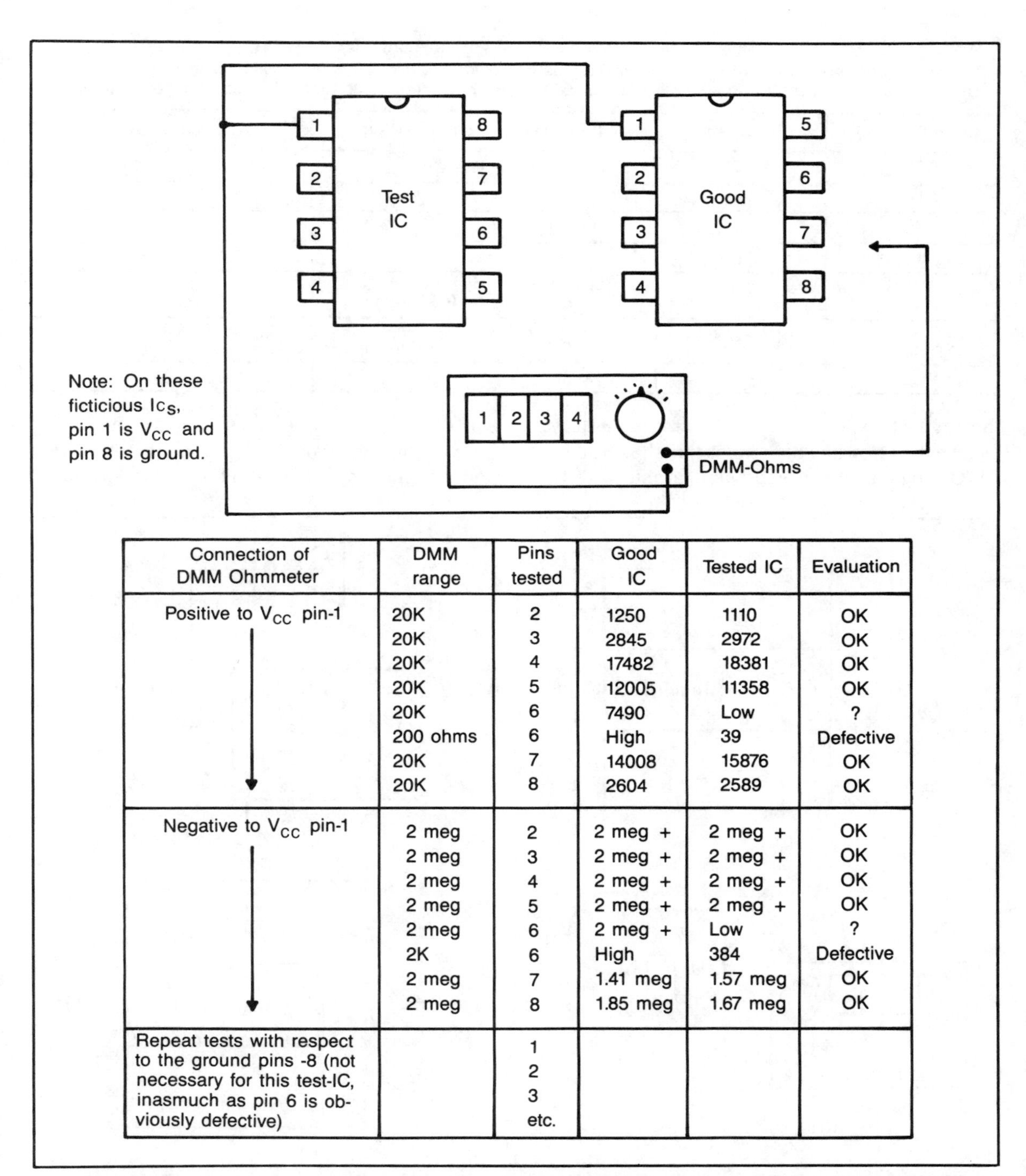

Connection of DMM Ohmmeter	DMM range	Pins tested	Good IC	Tested IC	Evaluation
Positive to V_{CC} pin-1 ↓	20K	2	1250	1110	OK
	20K	3	2845	2972	OK
	20K	4	17482	18381	OK
	20K	5	12005	11358	OK
	20K	6	7490	Low	?
	200 ohms	6	High	39	Defective
	20K	7	14008	15876	OK
	20K	8	2604	2589	OK
Negative to V_{CC} pin-1 ↓	2 meg	2	2 meg +	2 meg +	OK
	2 meg	3	2 meg +	2 meg +	OK
	2 meg	4	2 meg +	2 meg +	OK
	2 meg	5	2 meg +	2 meg +	OK
	2 meg	6	2 meg +	Low	?
	2K	6	High	384	Defective
	2 meg	7	1.41 meg	1.57 meg	OK
	2 meg	8	1.85 meg	1.67 meg	OK
Repeat tests with respect to the ground pins -8 (not necessary for this test-IC, inasmuch as pin 6 is obviously defective)		1 2 3 etc.			

Fig. 9-38. Example of comparative resistance tests of digital IC. Constant-current characteristic of DMM ohms-function helps make such comparisons meaningful. The 2 meg + readings are, for most practical purposes, infinite resistance paths. The polarity of the DMM leads can be determined with another meter.

Table 9-2. DC Voltage Levels of Various Logic Families: TTL, Transistor-Transistor Logic; DTL, Diode-Transistor Logic; RTL, Resistance-Transistor Logic; HTL, High Threshold Logic; CMOS, Complementary Metal-Oxide Semiconductor Logic; and ECL, Emitter-Coupled Logic.

Logic Family	V_{CC}-Volts	Input Threshold Range-Volts	Output Logic Swing-Volts
TTL	5.0	1.2 (from 0.8 to 2.0)	3.5 (from 0.4 to 3.9)
DTL	5.0	1.2	4.5
RTL	3.6	0.5	1.0
HTL	15.0	6.0	13.0
CMOS	3.0 to 10.0	30% of V_{CC} ± 0.5 V (low) to 70% of V_{CC} ± 0.5 V (high)	0 to V_{CC}
CMOS	10.0 to 18.0	30% of V_{CC} ± 1.0 V (low) to 70% of V_{CC} ± 1.0 V (high)	0 to V_{CC}
ECL	8.0	1.3	2.0

though the two devices have the same input threshold range. Thus, the high input threshold range of CMOS logic is not directly comparable with that of HTL devices on the basis of noise immunity.

The logic symbols in Fig. 9-39 do not include power supply connections. This is standard prac-

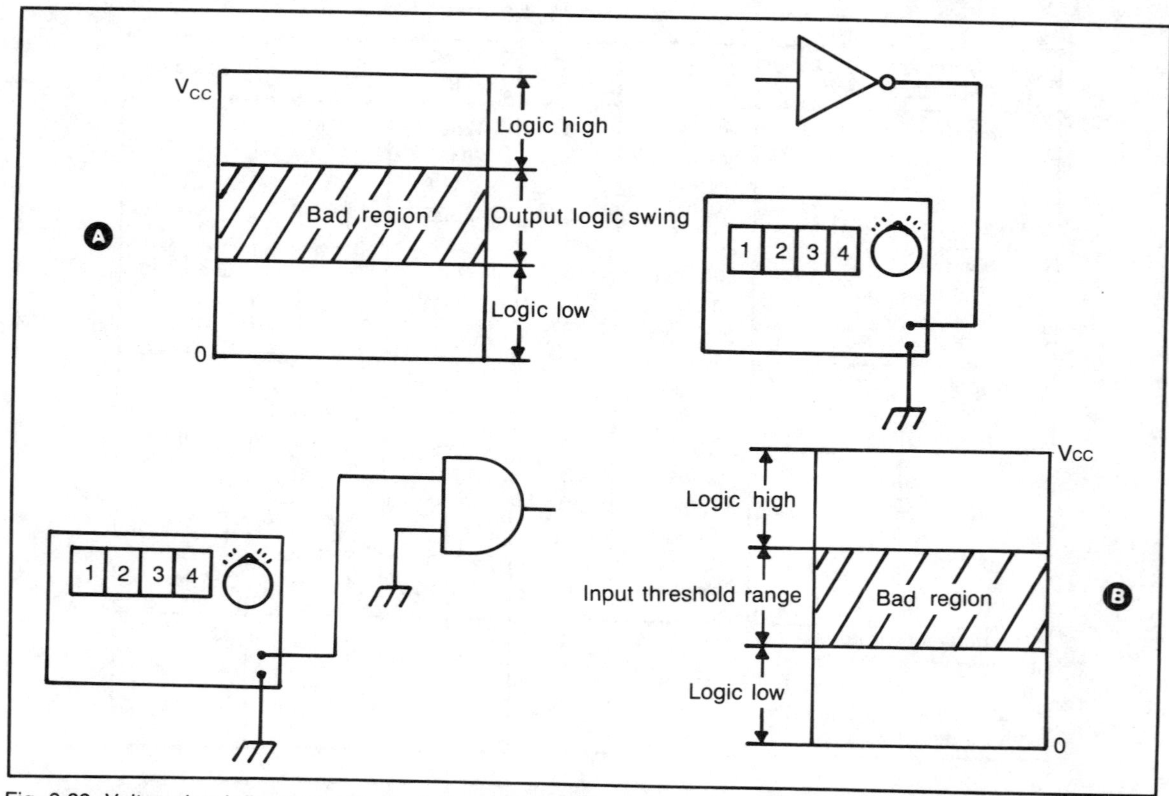

Fig. 9-39. Voltage-level diagrams of logic devices. A) The relative voltage levels at the output of logic devices. B) The relative voltage levels at the input of logic devices. In normal operation, the bad region is rapidly traversed. In some modes of faulty operation, a fixed voltage level within the bad regions may be measured.

tice and enables the symbols to present any logic family. However, the voltage level diagram of Fig. 9-39B is somewhat idealized for bipolar transistor devices in that logic low will never be zero, but generally several tenths of a volt, corresponding to a saturated pn junction of the output transistor. Zero voltage, or some value very close to zero, would be indicative of a defect either within the IC or external to it.

The interpretation of zero volts for logic low at the input of a logic device such as is shown in Fig. 9-39A is different. In this case, the device *would* be properly responsive to a zero voltage level at its input. In bipolar transistor families such a zero-level low will not be available from the preceding logic device. Therefore, the detection of zero voltage at the input of a logic device is usually indicative of a circuit defect where bipolar transistor devices are used.

The detection of a bad voltage level is significant only if it is ascertained that no pulse activity is going on. Only a "stuck" value of dc voltage qualifies as a bad level.

TESTING DIGITAL LOGIC SYSTEMS AND CIRCUITS

Sophisticated and costly test equipment certainly have their place in the evaluation of digital logic systems and circuits. Nonetheless, much can be deduced from simple, and often, primitive testers. In lieu of complex instrumentation, the powerful ingredient is human intuition, experience, and interpretive ability. Given these essentials, it is often possible to dispense with high-frequency oscilloscopes, multi-channel analyzers, and specialized digital format generators.

To begin with, most of the elements in a digital system are two-state devices; in essence, they are electronic switches which, under predetermined conditions, are either on or off. If we can confirm the truth table of a given logic element, it is highly probable that the cause of faulty operation in the system lies elsewhere. Such a state test is not altogether infallible because it does not duplicate the dynamics of actual operation. But even here, as will be shown, it is easily possible to simulate rapid transistions and to detect transients and glitches, as well as to determine the logic state produced by stimulation of a circuit node. The procedure is reminiscent of signal tracing in radios, TV sets, stereo amplifiers, and other analog circuitries where the path of an injected signal is followed until it suffers an unintended modification. At that point, it becomes easy to home-in on the defective device or component. In similar fashion, the various gates, flip-flops, inverters, and other logic circuits can be investigated for their proper responses. Even though hundreds, or thousands of logic elements may be involved in a system or subsystem, the failure of a single element can fault the overall operation of the entire system. Thus, the ability to ferret out and isolate a stuck gate, shorted IC, or other fault is a very worthwhile accomplishment.

At first thought, it would seem that the testing of high and low-logic states could be readily carried out with a voltmeter or with an oscilloscope. In a pinch, such ordinary instruments can be used, but such testing would be found to be inconvenient if not impractical. It is highly desirable that one's vision should not be distracted, even momentarily, from the array of tiny and closely spaced IC pins being tested. A slip of the test prod while glancing at a voltmeter or oscilloscope could produce erroneous results, or even worse, could wipe out the IC. For the sake of psychological continuity, one needs to quickly and effortlessly see the result of touching IC pins with test prods. For this purpose, logic probes have been developed which yield useful digital information via the actuation of LEDs on the logic probe itself. The mere touching of the probe to the IC pin reveals the logic state of the circuit node corresponding to that pin. As mentioned, other vital information may also be derived from such a simple test procedure.

Simple Logic Probe and Logic Pulser

Two very simple test devices for logic circuits are shown in Figs. 9-40 and 9-41. The basic structural element for both is the barrel of an old ball-point pen. Otherwise, construction methods are left

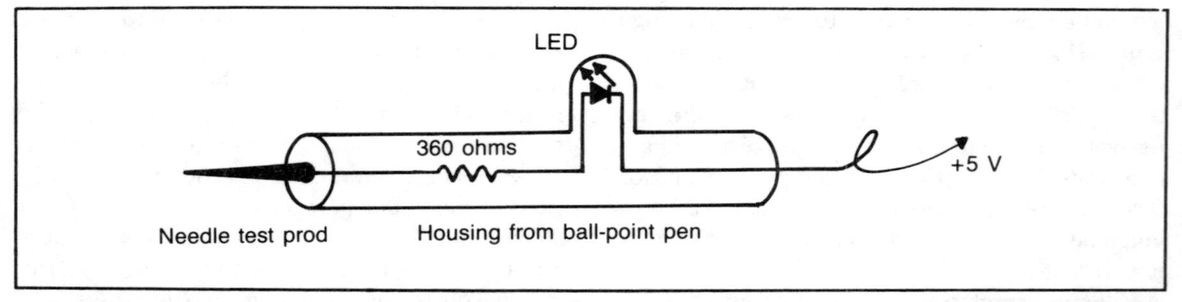

Fig. 9-40. The stone-age logic probe. Despite its simplicity, meaningful test results can be derived from observation of the LED.

to the ingenuity of the experimenter. The device illustrated in Fig. 9-40 is a logic probe. For convenience, the end of the test lead which connects to the +5 volt side of the logic circuit's power supply, can be terminated with an alligator clip. This probe will detect the low-logic state of any digital circuit node being probed. The use of a needle for the test prod facilitates probing the small and closely packed pins of IC. It may be desirable to cover all but the protruding point of the needle with plastic tubing.

This logic probe positively identifies the low state of a logic element which is in working order. It can be fooled, however by a short circuit. Additionally, the fact the LED remains off for logic high is not an unambiguous indication, for an open circuit also allows the LED to remain off. These uncertainties can be somewhat overcome in certain logic circuits if the simple logic pulser of Fig. 9-41 is used in conjunction with the logic probe. The capacitor in the logic pulser is charged to +5 volts, but releases a pulse of current as it discharges into the input circuit of the logic gate being stimulated. (After such an event, the capacitor regains its fully-charged condition in about one-quarter of a second after the pulser is removed from the pin of the logic gate.) Having thus touched one of the logic gate pins with the logic pulser, the logic probe is next deployed to determine the logic state of the gate's output. It is obvious that some types of defects can be detected with these "stone-age" testers. Because these testers are powerful tools within their limitations, it is not surprising that many commercial versions of these testers incorporate improvements for eliminating uncertainties and for extending the tests into the more meaningful dynamic domain where logic ICs can be "exercised" with repetitive pulsing, and where the presence and effect of glitches and transients can be investigated.

Also, in real-world logic systems, there are other evaluations to be made. There are three-state logic elements in which the third "state" is a very high impedance. And there is a third or bad "state" in certain defective logic elements. The best of the commercial logic probes greatly reduce the

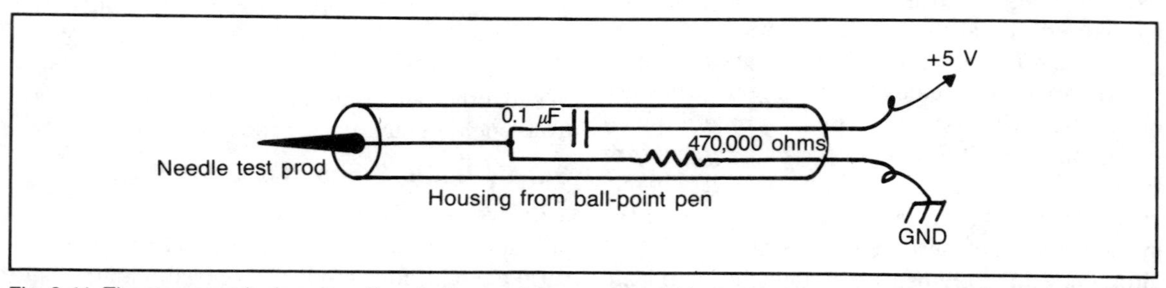

Fig. 9-41. The stone-age logic pulser. Despite its simplicity, meaningful logic transitions can be provoked in digital circuits by this pulse source.

likelihood of being fooled by these situations. Similarly, commercial logic pulsers use active circuits to generate a logic element in its actual circuit—it is not necessary to unsolder the pins. Commercial logic probes and commercial logic pulsers are generally made to accommodate more than a single logic family. Various other features are embodied in these hand-held logic test devices but most share one thing in common with the "stone-age" testers—operating voltage is obtained from the logic system or circuit undergoing test.

More Sophisticated Logic Probes

Commercial logic probes are more sophisticated than the ultra-simple "stone-age" probe previously described. Yet, they remain relatively simple instruments when compared to such measuring instruments as the oscilloscope, the frequency counter, or the digital multimeter. Figures 9-42, 9-43, 9-45, and 9-46 show some basic circuitries which are similar to those used in commercial logic-probes.

In Fig. 9-42, the input transistor is normally nonconductive, but is turned on when probing a logic level of approximately 0.65 to 0.70 volt. Thus, any voltage level lower than the turn-on value is interpreted as a logic low. Conversely, the entire voltage range between the turn-on value and V_{CC} is detected as logic high. The hex-inverters following the input transistor operate in the following way: When the probe is not contacting a logic device, the LOW (Green) LED is on. This LED is also on when the probe is contacting a logic device pin which is at a logic low. If the pin of the logic device is at logic high, the green LED turns off, while the logic high LED (RED) turns on. Note that there can be ambiguous indications involving a circuit node at logic low, shorted to ground, or open—all three situations result in the green LED being on. Nearly all logic probes produce some kinds of am-

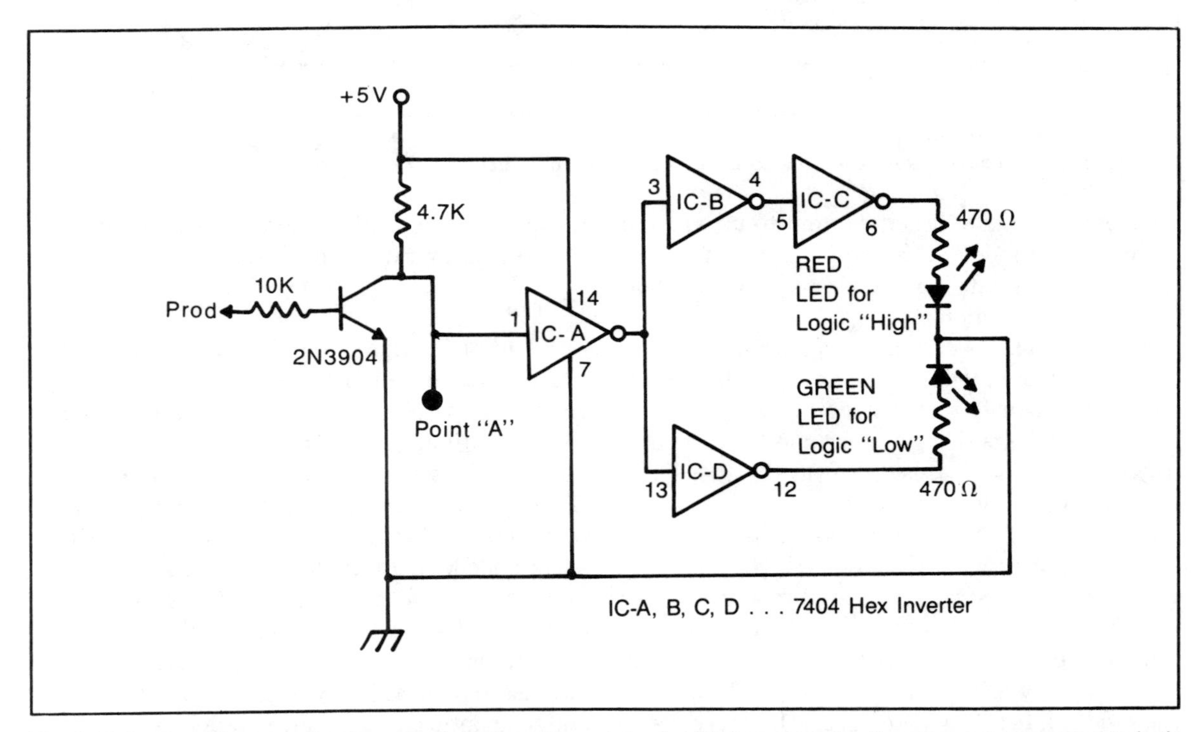

Fig. 9-42. Logic probe for indication of both high and low states in TTL logic circuits. If prod is not connected to logic circuit, the green LED glows. If prods monitors a logic "high", the red LED glows, the green LED turns off. If prod monitors a logic "low", the green LED glows. Point "A" connects to optional glitch-catcher circuit—see Fig. 9-41.

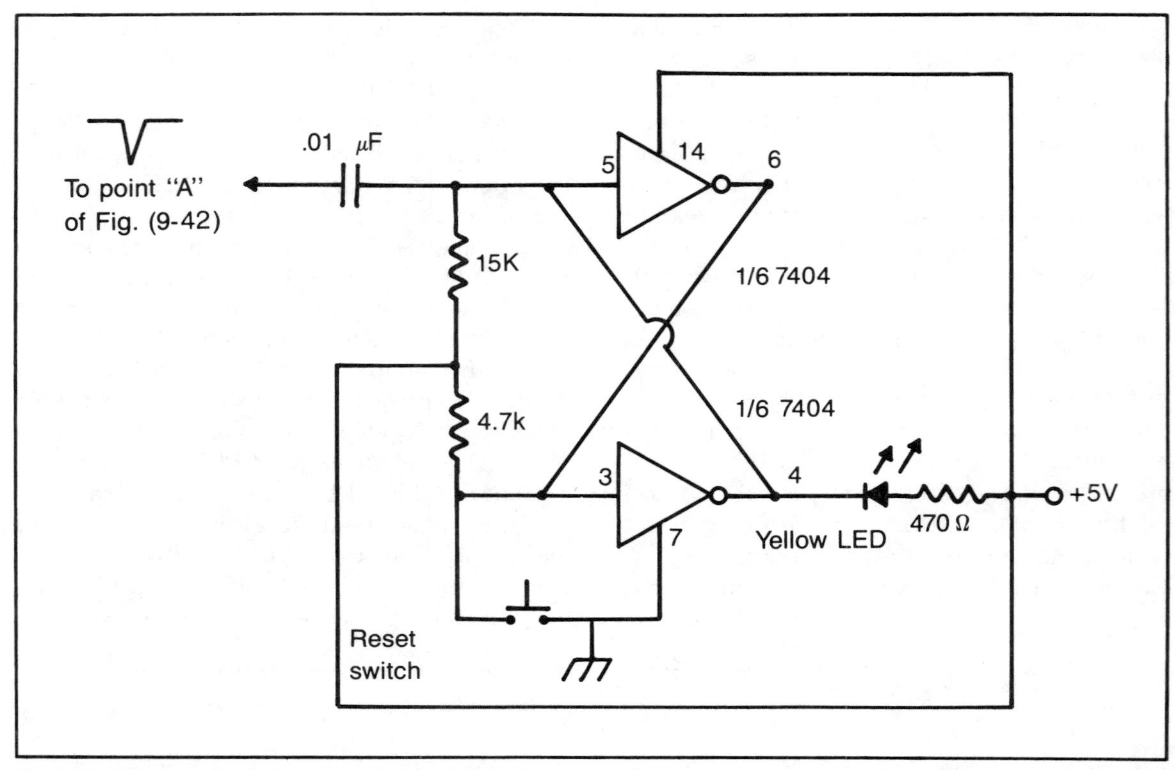

Fig. 9-43. Glitch or transient catching add-on circuit for logic probe of Fig. 9-42. A very narrow pulse at the input of this latch causes it to change state and provide a current path for the LED.

biguous indications. It is important to understand the operating mode of the probe being used, and to devise test strategems to circumvent or defeat ambiguous indications.

The usefulness of a basic logic probe, such as the circuit of Fig. 9-42 can be extended by the addition of a *memory* responsive to glitches or very narrow transients. This can assume the form of a triggered latch, such as is shown in Fig. 9-43. The input of the latch connects to point "A" of the schematic of Fig. 9-42. In use, the normal logic transitions or pulses will, of course, trigger the latch. However, if these transitions are brought to a halt by an appropriate disruption of the data stream somewhere in the system, the latch then is free to "capture" any narrow transient or glitch. Indeed, it provides a better way to reveal the presence of such disturbances than can ordinarily be provided by an oscilloscope. Oscilloscopic investigation is generally more fruitful *after* the transient or glitch is detected by this probe feature.

Glitches

A brief excursion into the nature and generation of glitches is in order. Glitches often occur in logic systems because of non-ideal timing. When we draw timing diagrams on paper, the usual assumption is that the pulses are square or rectangular with instantaneous rise and fall times, and that the arrival and departure of the logic states is as intended. With such assumptions, the truth tables seem to be on our side, ensuring the planned operation of the digital circuitry. In actual practice, these timing waveforms are trapezoidal, having finite rise and fall times. Moreover, pulse trains may suffer unanticipated or unequal delays due to device propagation-times and stray circuit parameters. We therefore are obliged to consult truth tables to see what might happen when "forbidden" logic transitions occur, even though these states may endure for very brief intervals.

Figure 9-44 shows how glitches may be generated by non-ideal waveforms impressed at AND-circuit gates. In A the output of the AND gate remains at logic low despite the logic transitions occurring at the gates. This is because the logic states at the gates always comply with the truth table conditions of logic-low output.

In B of Fig. 9-44 there are brief times when *both* gates "see" a logic high. These times can be discerned within the circled segments of the gate waveforms. During these momentary intervals, the AND gate obediently (to its truth table) delivers a logic high, which identifies as the glitch. Note that the probe memory circuit of Fig. 9-43 will "capture" the first of such glitches it detects at output C of the AND gate.

Other digital circuit glitches may have occasional rather than repetitive occurrences. They are, nevertheless, a serious threat to the proper performance of logic circuitry, especially where there are "triggered" logic elements, such as flip-flops, registers, counters, monostable multivibrators, etc. Many commercial logic probes have memory or glitch-catching circuits capable of detecting glitches in the 50 to 300 nanosecond range, while the better-performing ones will respond to 5 and 10 nanosecond glitches. Frequently, it is the very narrow glitches—those easy to overlook with oscilloscopes—that upset logic systems.

There are other sources of glitches, but non-ideal timing usually remains the root cause of these transients. For example, because of semiconductor storage time, the internal transistors of a logic IC may overlap switching functions and thereby output a glitch under certain operating conditions. In a somewhat different category, but with similar results are the transients which may enter via the power line, or which may be produced by switching power supplies.

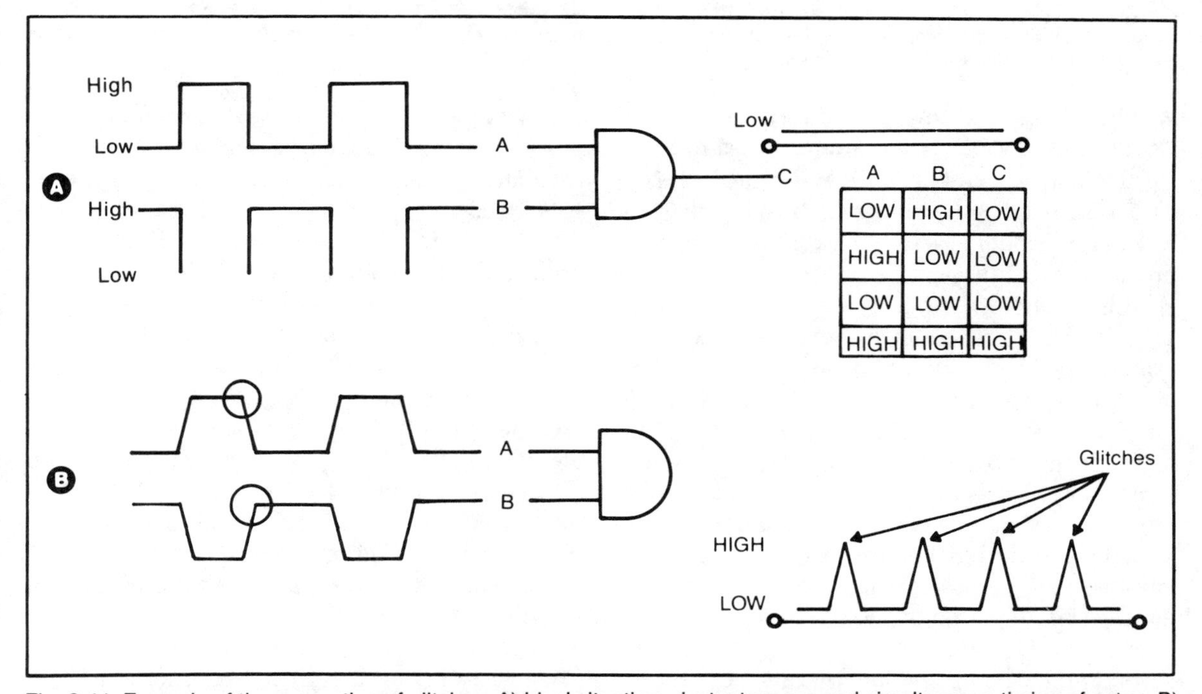

Fig. 9-44. Example of the generation of glitches. A) Ideal situation—instantaneous and simultaneous timing of gates. B) Because of finite rise and fall times and/or timing inequalities, the AND gate experiences brief intervals when unintended logic-highs are produced at output C.

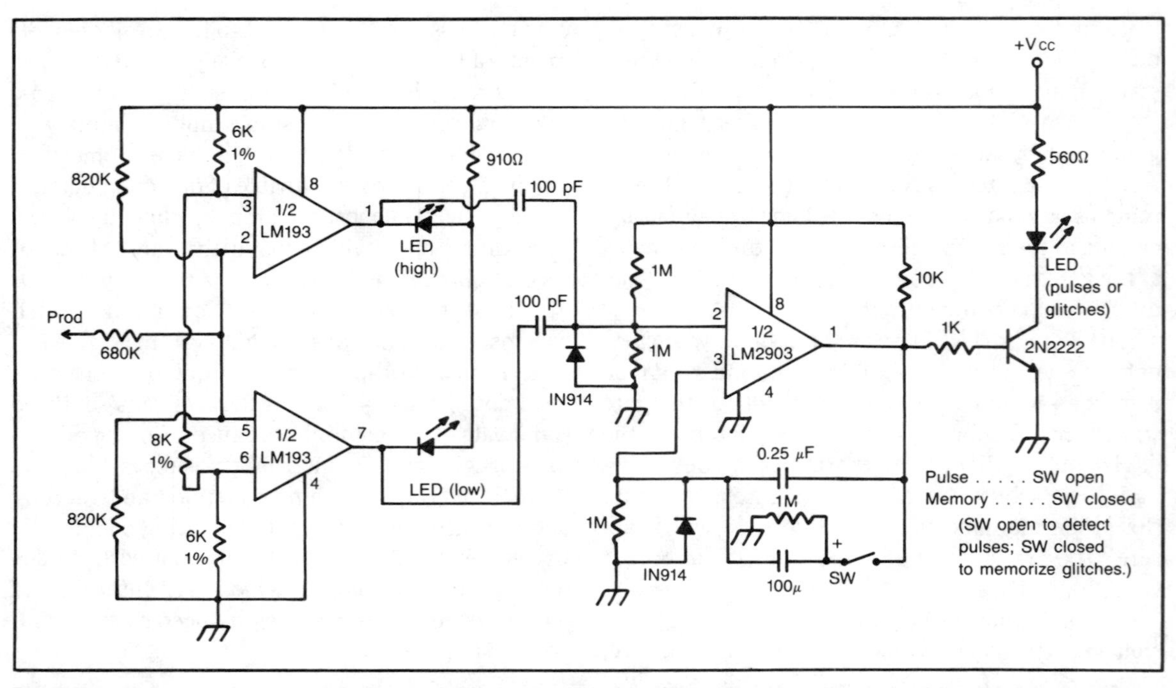

Fig. 9-45. A logic probe with versatile features. This probe can be used for RTL, DTL, TTL, or CMOS Logic Circuits. Pulses, as well as glitches and transients can be detected. However, because of the 30 percent and 70 percent thresholds, the best use of this probe is for CMOS logic.

A Probe with Pulse-Detection Mode and CMOS Logic Accommodation

The probe circuit depicted in Fig. 9-45 uses op-amp comparators to detect high and low logic levels. The input resistance network is arranged to provide for HIGH response at 70% V_{CC} and for response to logic LOW at 30% V_{CC}. Because of this, as well as the probe's high input resistance, this probe's best use is with CMOS logic. The memory, or glitch-catching circuit, is a monostable multivibrator configured about the LM2903 op-amp. When Sw is set at MEMORY, the circuit performs somewhat as the latch of Fig. 9-43. However, the memory is not permanent; rather, it is something on the order of one hundred seconds. For most practical purposes, this serves the same needs as the infinite memory of a true latch; the "stretching" of a 50 nanosecond glitch to 100 seconds of LED illumination, reveals the presence of the glitch just about as satisfactory as if the LED remained on continuously following a response.

With Sw open (at the PULSE mode), this probe provides an important test technique not within the capability of the previously described probe, even when used in conjunction with the latch-type glitch-catcher of Fig. 9-43. In the PULSE mode, the probe is used to monitor or search for normal pulse activity. When this test strategy is pursued, the time-constant of the monostable multivibrator is on the order of several tenths of a second—just sufficient to enable comfortable perception of the turn-on time of the PULSE LED. This constant blinking rate means that the probe is sampling the much-faster transitions of a logic pulsetrain. In this test mode, the probe can be "fooled" in some instances by a pulsetrain of glitches or transients. One way to determine that the blinking PULSE LED is responding to "healthy" logic transitions is to observe the behavior of the HIGH and LOW LEDs. For a 50% duty-cycle logic pulsetrain, these LEDs will be at less than full intensity and will provide equal light output. For duty-cycles favoring logic-high, the HIGH LED will be brighter than the LOW LED. The converse is true if the duty cycle favors logic low. However, repetitive glitches will *not* produce responses in either the HIGH or the LOW LED.

A Simple Three-State Logic Probe

The simple logic probe shown in Fig. 9-46 has the desirable feature that it indicates three voltage states—those corresponding to logic high and logic low, and the approximate voltage range corresponding to the "bad" region between input threshold levels. High logic is indicated by the appropriate LED. Similarly, low logic is indicated by the lighting of the other LED. The bad region is indicated by both LEDs being OFF. Note also, that when the probe is not contacting a logic device, *both* LEDs remain OFF.

This mode of operation tends to be more useful than that of the logic probe of Fig. 9-42. This is because of the ease with which low logic and the bad region can be identified. The bad level response of this probe is approximately from 0.65 volt to 2.25 volts. The exact edges of the "window" can be manipulated by "tweaking" some of the components, or by experimenting with different transistors or diodes. Almost any general-purpose npn transistor will be found suitable, but those with betas at the high end of their tolerance range are likely to prove most satisfactory. As with the other logic probes, this one is intended to operate from the V_{CC} supply of the digital circuit being tested.

The circuit operation is such that the two input transistors (those closest to the probe) alternate their conductive states for high and low logic levels. Before the probe contacts the pins of a digital de-

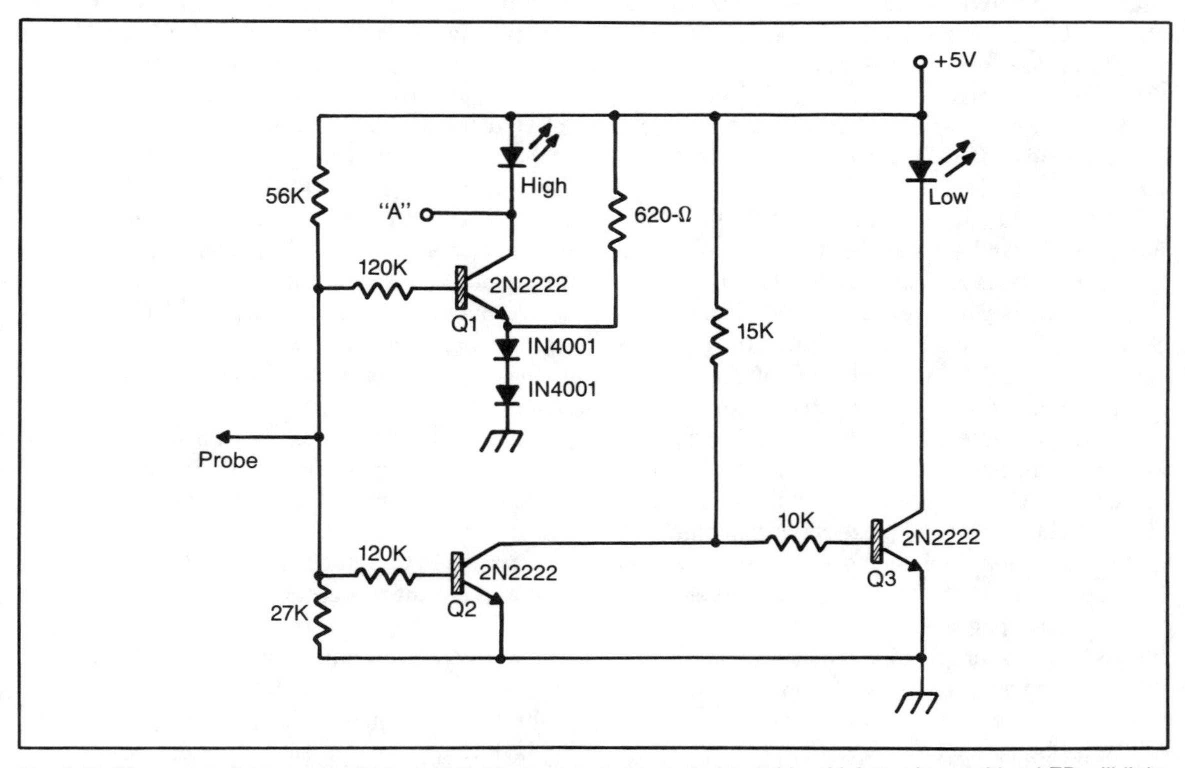

Fig. 9-46. Three-state logic probe. State 1: If the logic state being tested is neither high nor low, neither LED will light. Also, if the probe is not contacting a high or low logic voltage, neither LED will light. State 2: However, the detection of a logic high will be indicated by the lighting of the HIGH LED. State 3: Similarly, the detection of a logic low will be indicated by the lighting of the LOW LED.

Table 9-3. Comparative Response of Three Logic-Probes.

LOGIC PROBE	PROBE IS NOT TOUCHING PINS OF LOGIC IC	PROBE IS SENSING LOGIC-LOW	PROBE IS SENSING LOGIC-HIGH	PROBE IS SENSING BAD-REGION
Fig. (9-42) 5 V. TTL or DTL systems	High LED is OFF Low LED is ON	High LED is OFF Low LED is ON	High LED is ON Low LED is OFF	High LED is ON Low LED is OFF
Fig. (9-45) CMOS, but usable for TTL and DTL	High LED is OFF Low LED is ON	High LED is OFF Low LED is ON	High LED is ON Low LED is OFF	Both LEDs are OFF for voltage levels between 30% and 70% of V_{CC}
Fig. (9-46) 5 V. TTL or DTL	Both LEDs are OFF	High LED is OFF Low LED is ON	High LED is ON Low LED is OFF	Both LEDs are OFF for voltage levels greater than 0.65 volt, but less than 2.25 volt.

vice, the upper input transistor, Q1, is biased off and the lower input transistor, Q2, is biased on. A high logic level (above 2.25 volts) will bias Q1 on, causing the HIGH LED to glow. This will not change the condition of Q2, which is already conducting. A low logic state (below 0.65 volt) will deprive Q2 of its forward bias, causing it to turn off, thereby turning Q3 and the LOW LED on. But, Q1, being already off, will not change its conductive state.

A glitch-catching circuit, such as the one shown in Fig. 9-43, can be connected to point "A" of this logic probe. In any event, logic probes require getting used to; they often evidence quite distinct "personalities," especially when monitoring pulse activity. If feasible, the behavior of the LED indicators should be investigated with a pulse generator capable of supplying 50% duty-cycle pulses and also pulsetrains with high and low duty-cycles. The observed response will then facilitate interpretations of actual pulse activity in digital circuits. Table 9-2 depicts the response to various logic situations of the three logic probes shown in the previous pages of this chapter. Unless one knows the responses of a particular probe, it is possible to make erroneous interpretations for such situations as grounded nodes, bad regions and faulty probe contact.

Simulation of Logic Circuits with a Pulser

The logic pulser is a valuable adjunct to the logic probes just described. While an elemental logic pulser has been described in the form of a charged capacitor, commercial pulsers are considerably more sophisticated and allow for more penetrating testing techniques. The basic idea of most of these logic pulsers is to provide for injection of powerful, but short-duration pulses at selected circuit nodes in the digital system. Because such a pulse source has very low impedance, it can override the logic state of a device. It has the capability of pulsing the logic device into its opposite state, notwithstanding the logic information it may otherwise be subjected to. Thus, the logic device may be "exercised" and monitored with a logic probe. If the device behaves according to its truth table, in most instances all is well. But an improper reaction to the stimulation from the logic pulser indicates that the troubleshooter is very close to the fault is a misbehaving digital system.

Commercial models of these digital-diagnostic instruments are shown in the opening photos of Chapter 8 and Chapter 9. Note that both are hand-held and circuit-powered.

The best diagnostic use of the pulser probe pair results when the probe contains pulse-detection circuitry, such as the LM2903 monostable multivibrator depicted with the logic probe of Fig. 9-45. In this particular circuit, when switch Sw is open, the circuit will "stretch" very short pulses to a duration sufficient to be visible on the pulse LED. A single narrow pulse resulting from stimulation by the logic pulser will be detected as a single flash of light from the pulse LED on the probe because the pulse duration will be stretched out to a quarter, or a third of a second, or so. The response of the pulse LED on the logic probe to a pulse train of narrow pulses originated by the logic pulser will be a flashing light at a rate of three Hz, or so. Figure 9-47 illustrates the stimulus response testing of a NAND gate. Note that the tested device need not be unsoldered from its PC board.

Most logic pulsers provide the option of injecting a single narrow pulse or a pulsetrain of narrow pulses. Observation of the pulse detector, or pulse stretcher on the logic probe then reveals the state of health of the logic device. One need not be concerned with the polarity or amplitude of the injected pulses—the design of the logic pulser is such that logic devices will be driven into their *opposite* logic states. An exception to this statement occurs when the node receiving the injected pulses is shorted to ground. But the very absence of pulse activity under these circumstances still provides diagnostic information.

THE LOGIC MONITOR/CLIP

The logic monitor or logic clip is, in essence, a multiple logic probe featuring the ability of indicating the logic state of a number of IC pins simultaneously. In its most common format, it simply slips over the IC module undergoing test; dual rows of LEDs then reveal the logic state of each pin of the IC. In many models it is not even necessary to make any connections to any dc power source because the device contains steering-diodes that automatically seek out the V_{CC} and ground pins of the digital IC. Logic monitors may accommodate eight, sixteen, or even 40-pin ICs. Some use liquid crystal displays in order to conserve power. Figure 9-48 shows a logic monitor as part of a logic analysis kit, which also includes a logic probe and a logic pulser.

The logic monitor has a single threshold. It indicates logic low by the absence of a glowing LED. But it cannot distinguish between logic high and a bad level. Inasmuch as it lacks pulse-stretching circuitry, it cannot respond to the narrow pulses from

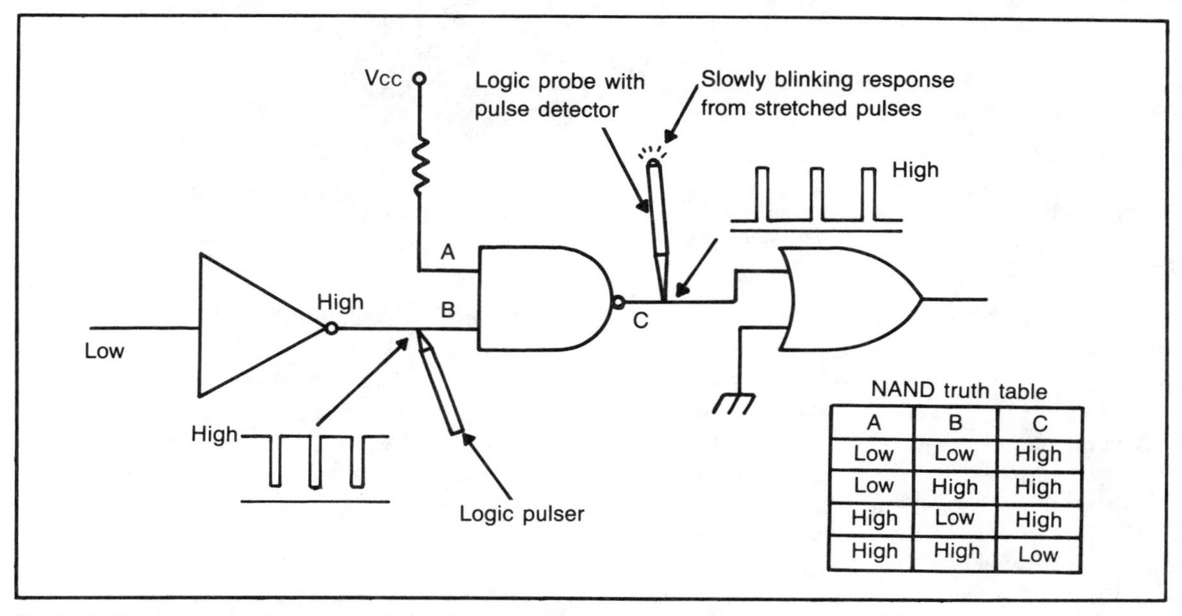

Fig. 9-47. Stimulus-response testing with logic pulser and logic probe. The NAND gate undergoing test is "exercised" by the pulser. Output pulse activity can be inferred by slowly blinking LED in pulse detector (stretcher) of the logic probe. The narrow but powerful pulses over-ride the prevailing logic state at B. Single pulses, as well as pulsetrains are available from the pulser.

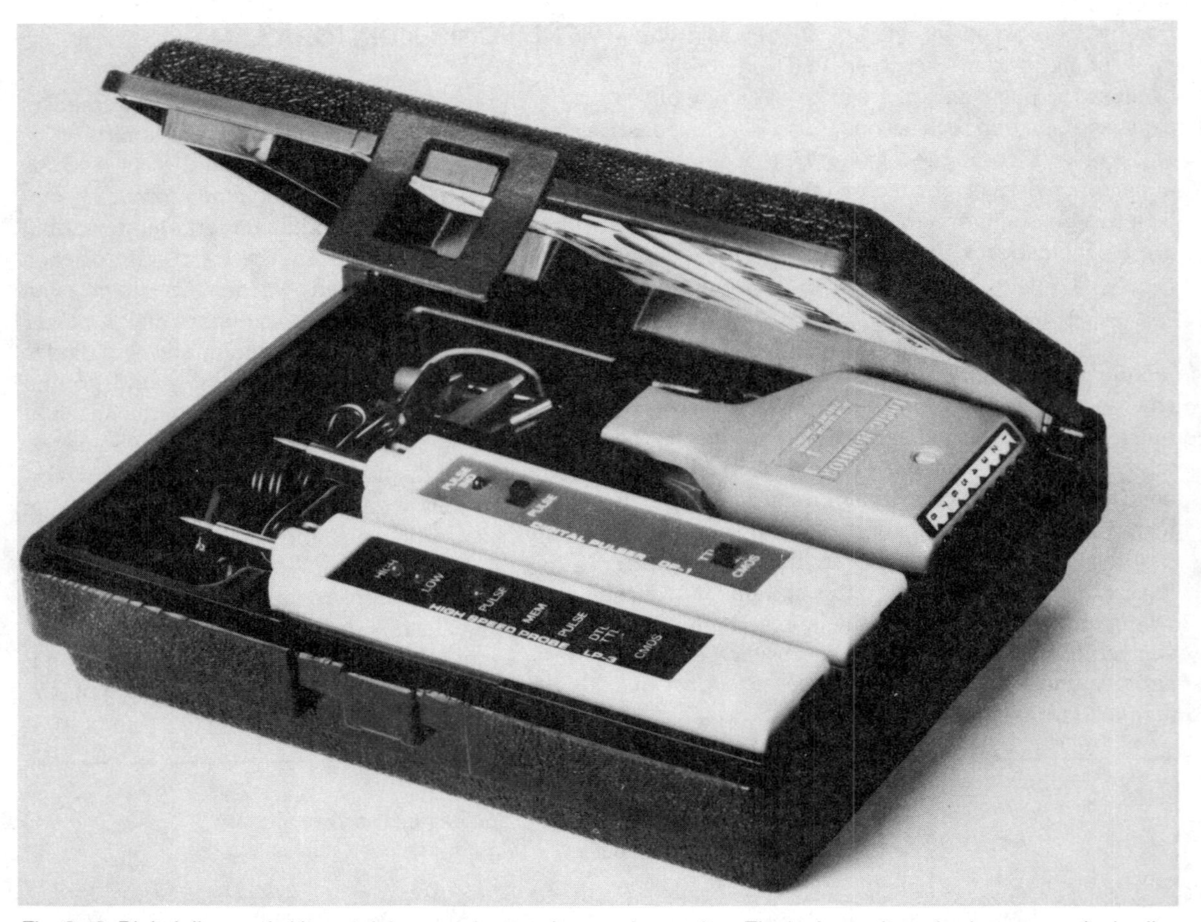

Fig. 9-48. Digital diagnostic kit containing a probe, a pulser, and a monitor. The logic monitor, also known as a logic clip, fits over a DIP integrated-circuit module. It reveals the logic state of each pin. For such a determination, it can be considered equivalent to sixteen logic probes simultaneously deployed to indicate the logic level of each IC pin. Courtesy of Global Specialties Corporation.

a logic pulser. And, inasmuch as it also lacks memory, it cannot reveal the presence of glitches or transients. Moreover, it is not suitable for checking fast-changing logic states because of the difficulty of keeping track of so many indications. It is, however, exceedingly convenient and quick for determining the overall logic "health" of a multiple pin IC under static, or slowly changing conditions. As can be surmised by the logic analyzer kit, the logic monitor, the logic probe, and the logic pulser have complementary test functions. In addition to its testing speed and convenience, the logic monitor has the advantage over the logic probe in that the relative timing of a multiplicity of logic states can be observed.

Figure 9-49 is a partial schematic of a simple logic monitor. Besides the multi-contact connecting clip which fits over the IC undergoing test, connections are made by alligator-clip test leads to V_{CC} and ground of the digital circuit. These latter connections are not necessary in many commercial versions of the logic monitor because steering diodes are used to sense the appropriate pins of the IC being tested. In these monitors, operating voltage is automatically obtained when the multicontact connector is fitted to the IC. However, the simpler scheme depicted in Fig. 9-49 is straightforward and anyone familiar with a logic probe probably wouldn't mind making the additional two connections. With the Darlington transistors and the resistance values shown, the logic threshold is about 2.25 volts. Although intended for testing the logic states of TTL ICs, modification of the input resistance network can optimize performance for other logic families.

When using a logic monitor, it is well to keep in mind that logic transitions faster than about 15 Hz will be perceived as steady LED indications, although at reduced intensity.

QUESTIONS

9-1. The specifications for a digital multimeter include an accuracy of ± 0.25% for the 2.000 volt "full-scale" indication. Although manufacturers often use differing methods for showing the performance of their instruments, what is lacking, on a conceptual basis, in the above accuracy description?

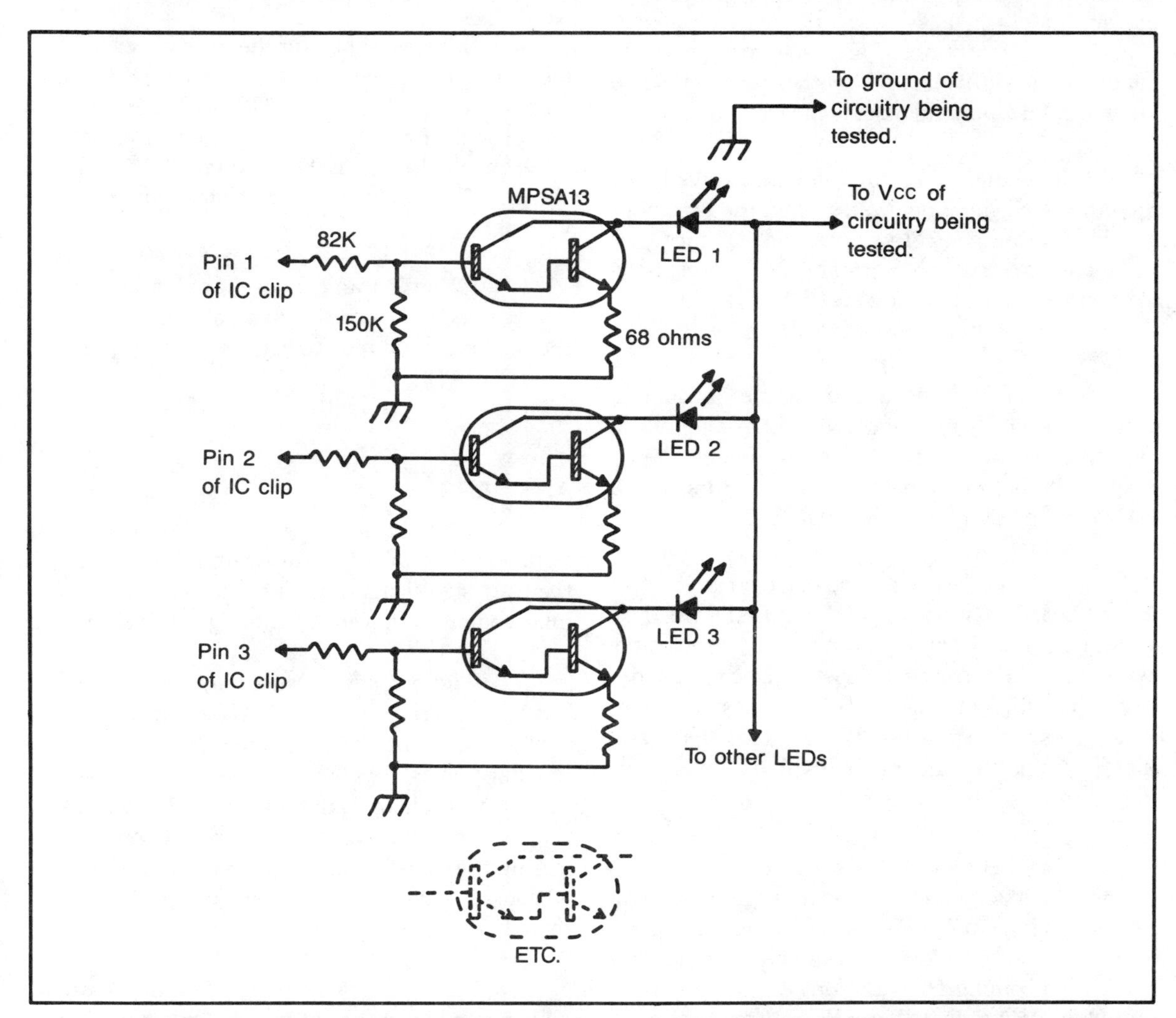

Fig. 9-49. Partial circuit of a simple logic monitor. V_{CC} and ground connections must be made to digital circuitry undergoing test. The connections between the 82 K resistors and multicontact IC clip can be made with ribbon-cable.

9-2. A true RMS digital multimeter was used to measure the ac RMS value of a 10 kHz symmetrical square wave. Later, it was found that the reading was considerably in error. The meter was not defective, nor was the measurement procedure responsible for the bad readout. What should be one of the first inquiries made regarding this situation?

9-3. A certain digital multimeter is claimed to have true RMS capability for ac measurements. Before actually using this instrument for measurements in a system with numerous nonsinusoidal waveforms, one should ascertain that its bandwidth is sufficient. What other item should be known before making circuit measurements?

9-4. Two digital multimeters are available for monitoring the ac input voltage and the load voltage in a Triac control system. One of these instruments is a true RMS type, the other responds to average values but indicates RMS values. How should these meters be deployed?

9-5. A VOM is used to measure and match resistors used in a differential amplifier and the accuracy is sufficient for the requirements. Why might it be better to match the transistors with a DMM rather than with the VOM?

9-6. Resistance checks are made on an inactive circuit board with both a VOM and a DMM. Discounting the inherent difference in accuracies between the two instruments,most of the circuit nodes test substantially the same. Some nodes, however, yield grossly different resistance values on the two meters. What is the likely explanation of this divergence of test results?

9-7. One section of a hex-inverter on a digital circuit board is found to be inoperative. Investigation with a logic pulser and a logic probe shows the input to be stuck at logic low and no pulse activity to be forthcoming from the output, although the output is stuck at logic high. Prior to actual confirmation, what is the likely cause of this condition?

9-8. A digital darkroom timer has a piezobuzzer controlled by a latch circuit. In turn, the latch circuit is triggered by the high output of an AND gate. One of the two inputs to the AND gate is maintained at logic high; the other input is at logic low until expiration of the set time interval, whereupon it, too, is made logic high. This, indeed, is the observed mode of operation, but occasionally, the alarm sounds prematurely. What is a likely cause and how should an appropriate test be performed?

9-9. In using a logic probe, it is observed that most circuit nodes of a digital system produce equal intensity in the HIGH and LOW LED indicators on the probe. However, at some nodes, either the HIGH or the LOW LED is predominantly brilliant. What is the likely interpretation of these observations?

9-10. What parameter of digital circuit operation can be much more conveniently ascertained by the use of a logic monitor/clip, rather than a logic probe?

ANSWERS

9-1. The basic accuracy of all digital instruments is subject to plus-or-minus one count of the least significant digit. This basic uncertainty is inherent in the nature of the digital readout.

9-2. The error was caused by the insufficient bandwidth of the digital multimeter. The bandwidth should usually be great enough to embrace at least ten harmonics of nonsinusoidal waveforms. Although this rule of thumb may be subject to variation in some cases, it at least serves its purpose by raising the question of bandwidth. Certain waves, of course, have more of their energy invested in their harmonics than others.

9-3. One should determine whether the digital multimeter is direct or ac coupled when used in its ac measuring modes. From a purist's standpoint, "true" RMS measurements must include dc levels and dc components. However, many true RMS labeled instruments are ac coupled, thereby excluding any dc from the measurement. With such instruments, a separate measurement of the dc portion of the wave must be made. Then the following computation is necessary: True RMS = $\sqrt{(V_{ac})^2 + (V_{dc})^2}$.

9-4. Use the true RMS meter to monitor the non-sinusoidal load voltage. Use the average-responding meter to monitor the sinusoidal input voltage. If the converse deployment of these instruments were implemented, the average-responding type would provide erroneous readings because its RMS indications are valid only for sine waves.

9-5. The measurement of the effective resistance of the pn junctions of the transistors with the VOM is not likely to be as accurate and reproducible as was the measurement of resistance with that instrument. This is because the forward resistance of the pn junction is a nonlinear function of current. The advantage of the DMM is that its ohms function provides a precisely regulated constant-current.

9-6. Evidently, the relatively high voltage at the prods of the VOM turned on certain pn junctions in semiconductor devices on the circuit board. Sometimes, the reversing of the polarity of the test prods circumvents this problem, but this technique is not always applicable. Many DMMs have a so-called "low-power" ohms function in which the test prod voltage is well below the forward-conduction threshold of semiconductor junctions, especially the now predominant silicon devices.

9-7. The evidence points to a ground fault at the input. This could be within the inverter, itself, in the PC wiring, or at the output of a preceding logic device. An ohms check usually provides confirmation of such defects.

9-8. It would appear that an occasional glitch or transient is delivered to the normally low input of the AND gate. The momentary logic high that consequently outputs from the AND gate then triggers the latch circuit and sets off the alarm. A logic probe with a glitch trapping circuit should be set at its MEMORY mode and applied to the normally low input of the AND gate for perhaps several minutes. The advent of the suspected glitch or transient will be evidenced if the pulse LED on the probe is turned on. The basic idea is that very narrow pulses that elude ordinary observation can upset a digital system.

9-9. When the duty-cycle of the logic pulsetrain is 50%, the LEDs on the logic probe will glow at the same intensity. At other duty-cycles, unequal intensities will be observed. Thus, if the duty-cycle of the sampled logic pulsetrain is, say 70% with the long durations in favor of logic high, the HIGH LED will glow with greater intensity and this will be at the expense of light output from the LOW LED. These observations are generally made for pulse rates greater than 15 or 16 Hz, which defines the persistance-of-vision boundary of the eye.

9-10. With the logic monitor/clip, the relative timing of the logic states at the pins of a multipin IC can be quickly observed. This can be done by "exercising" the IC at a slow rate. But, even a static evaluation is useful in detecting commonly occurring faults in the logic state of certain pins relative to the logic state of other pins.

Index

W

Z